Advances in Delays and Dynamics

Volume 1

Series Editor

Silviu-Iulian Niculescu, Laboratory of Signals and Systems, Gif-Sur-Yvette, France

For further volumes:
http://www.springer.com/series/11914

Tomáš Vyhlídal · Jean-François Lafay
Rifat Sipahi
Editors

Delay Systems

From Theory to Numerics and Applications

Springer

Editors
Tomáš Vyhlídal
Centre for Applied Cybernetics
Dept. of Instrumentation and Control Eng
Faculty of Mechanical Engineering
Czech Technical University in Prague
Prague
Czech Republic

Rifat Sipahi
Department of Mechanical and Industrial Engineering
Northeastern University
Boston
Massachusetts
USA

Jean-François Lafay
Ecole Centrale de Nantes
Institut de Recherche en Communications
et Cybernétique de Nantes
Nantes Cedex 3
France

ISSN 2197-117X ISSN 2197-1161 (electronic)
ISBN 978-3-319-34685-4 ISBN 978-3-319-01695-5 (eBook)
DOI 10.1007/978-3-319-01695-5
Springer Cham Heidelberg New York Dordrecht London

Printed on acid-free paper

Springer is part of Springer Science+Business Media (www.springer.com)

Preface

This volume is the first of a new series by Springer entitled: *Advances in Delays and Dynamics*. We are honored that Silviu-Iulian Niculescu as the Founding Editor of the Series, has entrusted us with the responsibility to compose this first volume which, like with any initialization of series, is of particular importance.

The purpose of this new series is to bring together recent results on the study of dynamics associated with the explicit consideration of delays in the models of systems, where delays arise in general due to the time needed to transmit energy and/or information among subsystems of systems. This research field is very rich and has been heavily explored over the past twenty years as progresses in automatic control have led engineers to be confronted with more and more complex and large-scale systems with delays, for which explicit consideration of delays is crucial in order to thoroughly understand and improve such systems. Networked systems are good illustrations of this, and their actual development significantly and critically overlaps with time-delay systems research area that then directly affects the modern society across applications such as telecommunications, health care operations, transportation systems, supply chains and logistics, population dynamics, etc. Many theoretical advances are now converging towards these applications, for which solutions are sought and from which new problems as well as novel methodologies arrive. It is also equally important not to lose sight of the applicability of such methodologies. That is why, even if many contributions collected in this volume concern the analysis of continuous time-delay systems, we have also included some new trends on sampled data systems as well as recent results on computational and software tools.

This volume presents the most recent trends as well as new directions in the field of *modelling, analysis and control synthesis of time delay systems*. Most of the chapters are based on preliminary contributions presented at the IFAC (International Federation of Automatic Control) workshops with the focus on *Time Delay Systems* (TDS), particularly at *2012 IFAC TDS Workshop* held in Boston, USA, and *2013 IFAC TDS Workshop* held in Grenoble, France, organised within the IFAC Joint conference with the *5th Symposium on System Structure and Control* and *6th Workshop on Fractional Differentiation and Its Applications*. Moreover, some chapters are incorporated based on the results from *time delay sessions* organized within the *2011 IFAC World Congress* held in Milan, Italy, and some other are invited. It is worthy to note that the contents of this volume do not have significant overlap with those presented at IFAC events since the authors contributing in this volume extended their contributions beyond what they presented at the past IFAC events.

The book is collected under the following five parts:

Part I - Stability Analysis and Control Design

Stability analysis and control design for systems with delays have long been pursued, and this topic has many rich examples in the literature, yet it never lost its critical role in the field. This first part of the book covers recent results in various aspects of stability theory and control design methods:

In the first chapter by CARLOS CUVAS, ADRIAN RAMÍREZ, ALEXEY EGOROV and SABINE MONDIÉ, the authors exclusively utilize the Lyapunov matrix of the delay system in order to prove necessary conditions for the exponential stability of a one-delay linear system. The results are based on the substitution of a special initial function into a Lyapunov-Krasovskii functional. An illustration example is included to show the effectiveness of the proposed conditions.

The second chapter by NICOLE GEHRING, JOACHIM RUDOLPH, and FRANK WOITTENNEK addresses the control of linear time-invariant systems with incommensurate lumped and distributed delays, for which the authors design a prediction-free tracking controller that assigns an arbitrary finite spectrum to the closed loop system. An illustrative example of a heat accumulator is provided to demonstrate the application of results.

In the third chapter by ALEXANDRE SEURET and FRÉDÉRIC GOUAISBAUT, new useful inequalities in combination with a simple Lyapunov-Krasovskii functional lead to new stability criteria for linear time delay systems expressed in terms of LMIs. On numerical examples, improvements are demonstrated.

Fourth chapter is authored by ALI FUAT ERGENC, where a matrix method is introduced for determining robust-stability zones of the general linear time invariant discrete-time dynamics with large delays against parametric uncertainties. The technique employs Kronecker Product and unique properties of palindrome polynomials by which a sufficient condition for robust stability and dominant pole assignment is possible.

Chapter five, which is contributed by PAVEL ZÍTEK, JAROMÍR FIŠER and TOMÁŠ VYHLÍDAL presents a novel method for dominant pole placement for a time delay PID control loop, while minimizing the absolute error integral. The quality of the disturbance rejection response is taken as the decisive criterion in the presented design of the time delayed plant control.

In the sixth chapter, stability of linear delayed systems subjected to digital control is analyzed by the authors DAVID LEHOTZKY and TAMAS INSPERGER, using the semi-discretization method. Different approaches are presented and discussed based on the number of discretization points and on the order of the approximation of the delayed term. In case studies, stability charts using the method are obtained.

Part II - Networks and Graphs

In the past decade, many studies focused on understanding the effects of networks, graphs, and dynamical systems. This research can reveal interesting findings about how the interactions between these parameters/concepts are interrelated, and these findings can be used to better understand coupled systems, their functionality, and to better engineer such systems. To capture this important trend in the field, this part presents six chapters:

The first chapter is by ALEXANDRE KRUSZEWSKI, BO ZHANG and JEAN-PIERRE RICHARD and considers the problem of teleoperation over an unreliable communication network. The authors study both the stability of the two interconnected systems in the teleoperation application, and also ensure some performances and robustness based on a robust control design via delay-dependent Lyapunov-Krasovskii framework.

In the second chapter, authored by WEI QIAO and RIFAT SIPAHI, the authors explore how various graph operations can be used to create large-scale graphs for interconnected dynamical systems while still keeping the delay margin of the arising graphs as large as possible.

In section three, a leaderless consensus control protocol for double integrators with multiple time delays is studied by RUDY CEPEDA-GOMEZ and NEJAT OLGAC, where stability analysis is performed using a recent technique called the Cluster Treatment of Characteristic Roots (CTCR), and a much different stability display is created using the Spectral Delay Space (SDS) as an overture to the CTCR. Examples are provided to display the application of CTCR and SDS.

In chapter four, authored by MEHMET EREN AHSEN, HITAY ÖZBAY and SILVIU-IULIAN NICULESCU, a dynamical model of a gene regulatory network (GRNs) with a time delay in the feedback path including static nonlinearities with negative Schwarzian derivatives is analyzed under positive feedback. A set of conditions is derived for the global stability of the class of GRNs considered, extending the results in cyclic biological processes involving time delayed feedback.

In chapter five, contributed by TOSHIKI OGUCHI and EIICHI UCHIDA, formation of oscillatory patterns in networks of identical nonlinear systems with time-delays is considered. Both linear and nonlinear analysis are presented, and the authors propose a design method of networks to achieve prescribed oscillation profiles from these networks.

In chapter six authored by WENLIAN LU, FATIHCAN M. ATAY and JÜRGEN JOST, the authors present an analysis of consensus problem in networks of multi-agents systems. The study involves both the linear linear consensus dynamics as a special example, as well as the case when the multi-agent systems at hand become random dynamic systems. Particularly, sufficient conditions for both consensus at uniform value and synchronization at periodic trajectories are developed.

Part III - Time Delay and Sampled-Data Systems

Application of control algorithms in real time systems is one of the main goals, which would inevitably invite discrete time systems and sampling. In such cases, while the dynamical system can be continuous, the peripherals around it can be digital, creating a mix type closed loop system, in which various questions about stability and control need to be answered. This part of the book is dedicated for this type of problems, and is inspired from the invited session *Time Delay and Sampled-Data Systems* held at the *IFAC World Congress 2011*, which was organized by EMILIA FRIDMAN. The only chapter outside this session is chapter three.

The first chapter is contributed by KUN LIU, EMILIA FRIDMAN, LAURENTIU HETEL and JEAN-PIERRE RICHARD, where the authors analyze the exponential stability of Networked Control Systems (NCSs), in particular using a static output feedback controller for linear systems with sensor nodes distributed over a network. The approach here presents the closed-loop system at hand as a switched system with multiple delayed samples, which ultimately enables exponential stability conditions derived using Lyapunov framework. The efficiency of the method is illustrated on the classical cart-pendulum benchmark problem.

The second chapter contributed by ERIK I. VERRIEST proposes two approaches, namely, lossless and forgetful causalization, for modeling discrete delay systems with time varying delay. While tackling the problem of the potential ill-posedness due to rapidly increasing delay in the system, the approach utilized here considers the time delay system using an extension of the state space. Structural problems, such as stability and reachability, are investigated using time-invariant theory, and a special reflecto-difference equation is analyzed in more detail as an example of a system with unbounded nonlocal behavior.

The third chapter by HIROSHI ITO, PIERDOMENICO PEPE and ZHONG-PING JIANG deals with interconnected systems described by retarded nonlinear equations with discontinuous right-hand side. The problem of feedback control redesign to achieve input-to-state and integral input-to-state stability is solved. It is shown that it is possible to design a decentralized controller accomplishing the robustification whenever a small-gain condition is satisfied.

In the fourth chapter, MATTHEW M. PEET and ALEXANDRE SEURET consider the problem of global stability of nonlinear sampled-data systems. A recently introduced Lyapunov approach is used to derive stability conditions for both the synchronous and asynchronous cases. This approach requires the existence of a Lyapunov function which decreases over each sampling interval. To enforce this constraint a form of slack variable over the sampling period is used allowing the Lyapunov function to be temporarily increasing. Several numerical examples are included to illustrate this approach.

Chapter five is authored by PASQUALE PALUMBO, PIERDOMENICO PEPE, SIMONA PANUNZI and ANDREA DE GAETANO, where a closed-loop control law for the glucose-insulin system is proposed based on a delay differential equation model of the system. Asymptotic tracking of a desired time evolution for the blood glucose concentration is achieved by means of the derived nonlinear control law, and the results are also supported by simulations.

Part IV - Computational and Software Tools

Advancements in the field of time delay systems are also heavily dependent on our ability to reliably produce numerical results, simulations, design approaches, etc.

Therefore, progress in computational and software tools is extremely important in order to support the field and potentially move it towards real-world implementation. For this purpose, this part brings together five chapters:

In the first chapter contributed by WIM MICHIELS and SUAT GUMUSSOY, an eigenvalue based framework is developed for the stability analysis and stabilization of coupled systems with time-delays, investigating the spectral and stability properties, while taking into account the effect of small delay perturbations. Authors also briefly address numerical methods for stability assessment and for designing stabilizing controllers for such systems.

In the second chapter by SUAT GUMUSSOY and PASCAL GAHINET, Computer Aided Control System Design (CACSD) of complex interconnected time delay systems is addressed together with the illustration of the functionality of Control System Toolbox in MATLAB. The chapter serves well as an tutorial on CACSD functionalities and opens effective directions in computer algorithm design for control of time delay systems.

The third chapter by SUN YI, SHIMING DUAN, PATRICK W. NELSON and A. GALIP ULSOY provides an overview of the Lambert W function approach for analysis and control of time delay system with a constant delay. The use of the MATLAB-based open source software in the LambertWDDE Toolbox is also introduced using numerical examples.

In chapter four, a Matlab toolbox YALTA for the H_∞-stability analysis of classical and fractional systems with commensurate delays is presented by the authors DAVID AVANESSOFF, ANDRÉ R. FIORAVANTI, CATHERINE BONNET and LE HA VY NGUYEN, covering both the neutral and retarded systems. Four detailed examples are included to demonstrate how to use the toolbox.

In chapter five, an updated QPmR algorithm implementation for computation and analysis of the spectrum of quasi-polynomials is presented by the authors TOMÁŠ VYHLÍDAL and PAVEL ZÍTEK. The authors demonstrate how QPmR is able to compute all the zeros of a quasi-polynomial located in a given region of the complex plane, and also analyse the spectrum distribution feature of both retarded and neutral type quasi-polynomials, along with case studies.

Part V - Applications

Many problems studied in the field of time delay systems are actually derived and inspired from real-world problems. This part aims to present a glimpse of the many applications studied in the community, bringing together six chapters encompassing applications from biology, human-in-the-loop control, neural networks, control in the context of automotive engines, haptic control, and synthetic biology:

This part starts with the chapter presenting results on modeling of cell dynamics in Acute Myeloid Leukemia contributed by JOSE LOUIS AVILA, CATHERINE BONNET, JEAN CLAIRAMBAULT, HITAY ÖZBAY, SILVIU-IULIAN NICULESCU, FATEN MERHI, ANNABELLE BALLESTA, RUOPING TANG and JEAN-PIERRE MARIE. In the chapter, the dynamical model is reduced to two coupled nonlinear equations with four internal sub-systems involving distributed delays. Equilibrium and local stability analysis of this model are performed and several simulations illustrate the results.

In the second chapter, JOSHUA VAUGHAN and WILLIAM SINGHOSE focus on the influence of time delays on crane operator performance. The work is motivated by remote control of cranes where the operator controls the oscillatory payload while suffering from decreased perception of the environment and the potential time delays caused by remote operation. Input shaping control is shown to improve operator completion times over a large range of operating conditions and communication time delays.

The third chapter is on the effects of time delays on nonlinear dynamics of neural networks, authored by GÁBOR OROSZ. A decomposition method is utilized to derive modal equations that allow one to analyze the dynamics around synchronous states. It is shown that for sufficiently strong coupling there exist delay ranges where stable equilibria coexist with stable oscillations which allow neural systems to respond to different environmental stimuli with different spatiotemporal patterns.

In the fourth chapter, DELPHINE BRESCH-PIETRI, THOMAS LEROY, JONATHAN CHAUVIN and NICOLAS PETIT provide an overview and study of the low-pressure burned gas recirculation in spark-ignited engines for automotive powertrain. It is shown, that a linear delay system permits to capture the dominant effects of the system dynamics. The modeled transport delay is defined by implicit equations stemming from first principles and can be calculated online. This model is shown to be sufficiently accurate to replace a sensor that would be difficult and costly to implement on commercial engines.

The fifth chapter contributed by QUOC VIET DANG, ANTOINE DEQUIDT, LAURENT VERMEIREN and MICHEL DAMBRINE deals with the issues of design and control for a force feedback haptic device in the case of interaction with a virtual wall. The results, derived from Linear Matrix Inequality conditions, are utilized in optimal design method for an electromechanical haptic device with high performances, followed by observer-based force feedback architecture used to improve the stability of haptic system taking into account variations of the communication delay.

In the last chapter EDWARD LAMBERT, EDWARD J. HANCOCK and ANTONIS PAPACHRISTODOULOU discuss how oscillators of the first type can be designed to meet frequency and amplitude requirements. The authors also discuss how coupling heterogeneous populations of delayed oscillators can produce oscillations with robust amplitude and frequency. The analysis and design is rooted in techniques from control theory and dynamical systems. The motivation for the work comes from the emerging field of synthetic biology, where oscillators are one of the best studied synthetic genetic circuits.

Last but not least, we would like to thank the editors of the ADD@S book series at Springer for handling this volume. Besides, we would like to express full acknowledgment of IFAC - International Federation of Automatic Control, and thank all the organizers of the past IFAC events of

- 18th IFAC World Congress, Milan, Italy, 2011
- 10th IFAC Workshop on Time-Delay Systems, Boston, US, 2012
- 11th IFAC Workshop on Time-Delay Systems, Grenoble, France, 2013

at which some preliminary results of some of the chapters in this volume were presented.

Prague,	Tomáš Vyhlídal
Nantes,	Jean-François Lafay
Boston,	Rifat Sipahi
	June 2013

Contents

Part I Stability Analysis and Control Design

Part II Networks and Graphs

Part III Time-Delay and Sampled-Data Systems

Acronyms

A/D	Analog Digital
ADD@S	Advances in Delays and Dynamics at Springer
AE	Adjunct Equation
AML	Acute Myeloid Leukemia
BB	Building Block
CACSD	Computer Aided Control System Design
CE	Characteristic Equation
CPG	Central Pattern Generator
CTCR	Cluster Treatment of Characteristic Roots
DDAE	Delay Differential Algebraic Equations
DDE	Delay-Differential Equation
DDE-BIFTOOL	Matlab Package for Bifurcation Analysis of DDE
DMCM	Delay Margin Contour Map
DNA	Deoxyribonucleic Acid
DTDS	Discrete-Time Delayed System
EGR	Exhaust Gas Recirculation
FAR	Fuel-to-Air Ratio
GAS	Globally Asymptotically Stable
GLTI	Generalized Linear Time Invariant
GPS	Global Positioning System
GRN	Gene Regulatory Network
HANSO	Matlab package: Hybrid Algorithm for Non-Smooth Optimization
HGO	Hepatic Glucose Output
HSC	Hematopoiteic Stem Cell
IAE	Integral Absolute Error
IDR	Insulin Delivery Rate
IFAC	International Federation of Automatic Control
IMEP	Indicated Mean Effective Pressure
IQC	Integral Quadratic Constraint
iISS	Integral Input-to-State Stability

ISS	Input-to-State Stability
LambertWDDE Toolbox	Matlab package for the analysis of LTI DDEs
LFT	Linear Fractional Transformation
LKF	Lyapunov-Krasovskii Functionals
LMI	Linear Matrix Inequality
LP	Low Pressure
LTI	Linear Time Invariant
MAD	Maximum Allowable Delay
MATI	Maximum Allowable Transmission Interval
MIMO	Multi Input Multi Output
mRNA	Messenger Ribonucleic Acid
NCS	Networked Control Systems
NP-hard	Non-Polynomial Hard
NTP	Network Time Protocol
ODE	Ordinary Differential Equation
PAC	Programmable Automation Controllers
PD	Proportional Derivative
PDE	Partial Differential Equations
PI	Proportional Integral
PID	Proportional Integral Derivative
QPmR	Matlab Package: Quasi-Polynomial Mapping Based Rootfinder
RDS	Random Dynamical System
RE	Responsible Eigenvalue
RWP	Random Waypoint Model
SC	Sub-Cutaneous
SDS	Spectral Delay Space
SeDuMi	Software Package for Solving Optimization Problems over Symmetric Cones
SI	Spark Ignited
SOS	Sum-of-Squares
SOSTOOLS	Matlab toolbox for solving Sum-of-Squares Optimization Programs
STDS	Single Time Delayed System
T1DM	Type 1 Diabetes Mellitus
T2DM	Type 2 Diabetes Mellitus
TDS	Time Delay Systems
VVT	Variable Valve Timing
YALTA	Matlab Package: Yet Another LTI TDS Algorithm
ZOH	Zero Order Hold

Part I
Stability Analysis and Control Design

Necessary Stability Conditions for One Delay Systems: A Lyapunov Matrix Approach

Carlos Cuvas, Adrian Ramírez, Alexey Egorov, and Sabine Mondié

Abstract. Necessary conditions for the exponential stability of one delay linear systems are proved. These conditions depend exclusively on the Lyapunov matrix of the delay system, thus improving previous results which were expressed not only in terms of the Lyapunov matrix, but also on system matrices. They are obtained via the substitution of a special initial function into a Lyapunov-Krasovskii functional whose existence, when the system is exponentially stable, is established. An illustrative example shows the effectiveness of the proposed conditions in determining candidate stability regions in the space of parameters. Finally, a procedure for improving these necessary conditions is outlined.

1 Introduction

The form of the functional having a prescribed derivative associated to exponentially stable linear delay systems have been studied, in the Lyapunov Krasovskii framework [11], in the work of Repin [19] and Datko [2], followed by the contributions of Infante and Castelan [6], Huang [5], where a cubic lower bound was found, and Louisell [12], where substantial advances were achieved.

During the past decade, Kharitonov and Zhabko [7], Kharitonov [8], [10] clarified fundamental concepts and results of this approach: the Lyapunov matrix delay function is obtained as the solution of the dynamic, symmetric and algebraic properties which are the analogue of the Lyapunov equation; the existence and uniqueness of solutions, provided a spectrum Lyapunov like condition is satisfied, is established in

Carlos Cuvas · Adrian Ramírez · Sabine Mondié
Departamento de Control Automático, Cinvestav, IPN, México D.F.
e-mail: {ccuvas,aramirez,smondie}@ctrl.cinvestav.mx

Alexey Egorov
Saint-Petersburg State University, Saint-Petersburg, Russia
e-mail: alexey3.1416@gmail.com

T. Vyhlídal, J.-F. Lafay, and R. Sipahi (eds.), *Delay Systems*,
Advances in Delays and Dynamics 1,
DOI: 10.1007/978-3-319-01695-5_1, © Springer International Publishing Switzerland 2014

Kharitonov and Plischke [9]. This lead to the presentation of a functional, named of complete type, that satisfies a quadratic lower bound if the system is exponentially stable.

These results represent indeed an extension of the well known Lyapunov theory for delay free case, where the Sylvester criteria allows the characterization of the stability of the system in terms of the positivity of the Lyapunov matrix. The idea of extending such a powerful result to linear delay systems has produced a number of stability results in recent contributions. Conditions establishing, for the one delay scalar equation, the coincidence of the well known stability region with Lyapunov function conditions were recently obtained in [17]. Instability conditions were determined in [15] for retarded, neutral type and distributed delay systems. Finally, necessary conditions in terms of the Lyapunov matrix are reported in [14] for the case of single delay systems. These conditions are significantly conservative when matrix A_1 is singular. The fact that this situation often arise in control, as in the problem of delayed output feedback, or in the proportional retarded control of systems studied respectively in [13] and [20], is a strong motivation for looking for better suited conditions.

In this contribution, we analyze the stability of linear time delay systems of the form

$$\dot{x}(t) = A_0 x(t) + A_1 x(t-h), \tag{1}$$

where $A_0, A_1 \in \mathbb{R}^{n\times n}$, $h \geq 0$ is the delay and the initial condition is

$$x(\theta) = \varphi(\theta), \quad -h \leq \theta \leq 0, \ \varphi \in \mathscr{PC}[-h,0].$$

The organization of this contribution is as follows: in section 2, preliminaries on the theoretical framework of Lyapunov Krasovskii functionals with prescribed derivative are recalled, and some technical results are introduced. The main contribution of this work, necessary conditions that depend on the Lyapunov matrix, is proved in section 3. In section 4, a nontrivial illustrative example suggests a procedure for improving the necessary conditions. The contribution ends with some concluding remarks.

Notation: the Euclidian norm for vectors is denoted $\|\cdot\|$. *For a given initial condition* $\varphi(\theta)$ *in the set of piecewise functions defined on the interval* $[-h,0]$, $\mathscr{PC}([-h,0],\mathbb{R}^n)$, $x_t(\varphi) = \{x(t+\theta,\varphi), \theta \in [-h,0]\}$ *denotes the restriction of the solution* $x(t,\varphi)$ *of system (1) to the interval* $[t-h,t]$. *When the initial condition is not crucial, the argument* φ *is omitted. The set of piecewise functions is equipped with the norm* $\|\varphi\|_h = \sup_{\theta\in[-h,0]} \|\varphi(\theta)\|$.

For a symmetric matrix $Q \in \mathbb{R}^{n\times n}$, *the notation* $Q > R$ $(Q \geq R)$ *means that* $Q - R$ *is positive definite (positive semidefinite).*

2 Preliminaries

In this section stability results on Lyapunov Krasovskii functionals with prescribed derivative for one delay systems are recalled and some auxiliary results are introduced.

2.1 Theoretical Framework

System (1) is said to be exponentially stable if there exist constants $\gamma \geq 1$ and $\beta > 0$ such that for every initial function $\varphi \in \mathscr{PC}([-h,0],\mathbb{R}^n)$, the solution $x(t,\varphi)$ satisfies $\|x(t,\varphi)\| \leq \gamma e^{-\beta t} \|\varphi\|_h$.

The *basic* functional $v_0(x_t)$ with prescribed time derivative $-x^T(t)Wx(t)$, which is crucial in our developments, was introduced in [5].

Lemma 1. *Given a positive definite matrix W, the functional $v_0(x_t)$ whose time derivative along the trajectories of system (1) is equal to $-x^T(t)Wx(t)$, is of the form*

$$
\begin{aligned}
v_0(\varphi) &= \varphi^T(0)U(0)\varphi(0) + 2\varphi^T(0)\int_{-h}^{0} U^T(h+\theta)A_1\varphi(\theta)d\theta \\
&\quad + \int_{-h}^{0} \varphi^T(\theta_2)\int_{-h}^{0} A_1^T U(\theta_2-\theta_1)A_1\varphi(\theta_1)d\theta_1 d\theta_2.
\end{aligned} \tag{2}
$$

Then, if system (1) is exponentially stable, the matrix function $U(\theta)$, $\theta \in [-h,h]$, is the unique solution of the dynamic equation

$$
U'(\theta) = U(\theta)A_0 + U(\theta-h)A_1, \quad \theta \geq 0, \tag{3}
$$

with boundary conditions, called symmetric and algebraic properties,

$$
U(-\theta) = U^T(\theta), \quad \theta \geq 0, \tag{4}
$$

$$
A_0^T U(0) + U(0)A_0 + A_1^T U(h) + U^T(h)A_1 = -W. \tag{5}
$$

It was shown in [5], [4], that this functional admits a local cubic lower bound:

Theorem 1. *If the system (1) is exponentially stable, then for any $\alpha > 0$ there is a constant $c > 0$ such that*

$$
v(\varphi) \geqslant c\|\varphi(0)\|^3, \quad \|\varphi\|_h \leqslant \alpha.
$$

Moreover, examples showing that the functional (2) does not admit a quadratic lower bound leading to the presentation of a functional named *complete* that satisfies one in [7].

Here, we consider a particular case of the additional term introduced in [7]. In this case the functional is not of complete type, yet it satisfies a quadratic lower bound.

Theorem 2. *Let the delay system (1) be exponentially stable and let a positive definite matrix W be given. Then, for any $0 < Q < W$, the functional*

$$v(\varphi) = v_0(\varphi) + \tilde{v}(\varphi),$$

where v_0 is the basic functional introduced in Lemma 1 and $\tilde{v}$ is defined as

$$\tilde{v}(\varphi) = \int_{-h}^{0} \varphi^T(\theta) Q \varphi(\theta) d\theta,$$

is such that there exist positive scalars $\beta(Q)$, $\alpha_1(Q)$ and $\alpha_2(Q)$ such that

$$\frac{d}{dt} v(x_t) \leq -\beta(Q) \|x(t)\|^2,$$

$$\alpha_1(Q) \|\varphi(0)\|^2 \leq v(\varphi) \leq \alpha_2(Q) \|\varphi\|_h^2. \tag{6}$$

In addition, $\alpha_1(0) = 0$.

Proof. See [16] for a detailed proof.

Next, we remind some useful properties of the Lyapunov delay matrix.

Lemma 2. *[9] The matrix $U(\theta)$, $\theta \in [-h, 0]$ is infinitely many times differentiable on $(-h, 0)$. Its first derivative has a jump discontinuity at $\theta = 0$.*

Lemma 3. *[7] The first derivative of the Lyapunov matrix satisfies*

$$U'(\theta) = U(\theta) A_0 + U^T(h - \theta) A_1, \quad \theta \geq 0, \tag{7}$$

$$U'(\theta) = -A_0^T U(\theta) - A_1^T U(h + \theta), \quad \theta < 0. \tag{8}$$

Proof. The expression (7) follows straighforwardly from (3) and (4). The symmetric property implies that $-U'(-\tau) = [U'(\tau)]^T$, $\tau > 0$. The result follows from the dynamic property and a change of variable.

We also prove the following equality.

Lemma 4. *For $\theta \in [-h, h]$ the Lyapunov matrix satisfies*

$$\begin{aligned} U''(\theta) = {} & U'(\theta) A_0 - A_0^T U'(\theta) + A_0^T U(\theta) A_0 - A_1^T U(\theta) A_1 \\ & + \delta(\theta) \left[U(\theta) A_0 + U^T(h - \theta) A_1 + A_0^T U(\theta) + A_1^T U(h + \theta) \right], \end{aligned} \tag{9}$$

where $\delta(\theta)$ is the Dirac delta function.

Proof. The expressions (7) and (8) can be summarized as

$$\begin{aligned} U'(\theta) = {} & \chi(\theta \geq 0) \left[U(\theta) A_0 + U^T(h - \theta) A_1 \right] \\ & + (1 - \chi(\theta \geq 0)) \left[-A_0^T U(\theta) - A_1^T U(h + \theta) \right], \end{aligned}$$

where $\chi(\theta \geq 0)$ denotes the Heaviside step function whose value is zero for negative arguments and one for arguments greater or equal to zero. The second derivative is

$$\begin{aligned} U''(\theta) &= \chi(\theta \geq 0)\left[U'(\theta)A_0 - \left[U^T(h-\theta)\right]' A_1\right] \\ &+ (1-\chi(\theta \geq 0))\left[-A_0^T U'(\theta) - A_1^T U'(h+\theta)\right] \\ &+ \delta(\theta)\left[U(\theta)A_0 + U^T(h-\theta)A_1 + A_0^T U(\theta) + A_1^T U(h+\theta)\right]. \end{aligned} \tag{10}$$

Notice that for $\theta \in [0,h]$

$$\begin{aligned} &U'(\theta)A_0 - \left[U^T(h-\theta)\right]' A_1 \\ &= U'(\theta)A_0 - \left[U(h-\theta)A_0 + U^T(\theta)A_1\right]^T A_1 \\ &= U'(\theta)A_0 - A_0^T U'(\theta) + A_0^T U(\theta)A_0 - A_1^T U(\theta)A_1 \end{aligned} \tag{11}$$

and that for $\theta \in [0,h]$

$$\begin{aligned} &-A_0^T U'(\theta) - A_1^T U'(h+\theta) \\ &= -A_0^T U'(\theta) - A_1^T\left[U(h+\theta)A_0 + U^T(-\theta)A_1\right] \\ &= -A_0^T U'(\theta) + U'(\theta)A_0 + A_0^T U(\theta)A_0 - A_1^T U(\theta)A_1. \end{aligned} \tag{12}$$

Finally, (9) follows by substituting (11) and (12) into (10).

2.2 Auxiliary Results

Next, we present some useful technical results concerning matrix $U(\theta)$, $\theta \in [-h,h]$.

Proposition 1. *The following equality holds for nonnegative scalars a,b*

$$II_b^a = 2\int_{-b}^{-a} e^{A_0^T\theta} A_1^T U(h+\theta)d\theta = -2e^{-A_0^T a}U(-a) + 2e^{-A_0^T b}U(-b). \tag{13}$$

Proof. Using (8) yields

$$II_b^a = -2\int_{-b}^{-a} e^{A_0^T\theta}\left[U'(\theta) + A_0^T U(\theta)\right]d\theta = -2\int_{-b}^{-a}\frac{d}{d\theta}\left(e^{A_0^T\theta}U(\theta)\right)d\theta.$$

Integration by parts leads to

$$II_b^a = -2\, e^{A_0^T\theta}U(\theta)\Big|_{-b}^{-a}$$

and (13) follows.

Proposition 2. *The following equality holds for nonnegative scalars a,b,c,d such that $(a,b)\cap(c,d)$ is a set of measure zero:*

$$
\begin{aligned}
III_{b,d}^{a,c} &= \int_{-b}^{-a} e^{A_0^T\theta_1} \int_{-d}^{-c} A_1^T U(\theta_1-\theta_2)A_1 e^{A_0\theta_2} d\theta_2 d\theta_1 \\
&= e^{-A_0^T a}U(-a+c)e^{-A_0 c} - e^{-A_0^T a}U(-a+d)e^{-A_0 d} \\
&\quad -e^{-A_0^T b}U(-b+c)e^{-A_0 c} + e^{-A_0^T b}U(-b+d)e^{-A_0 d}.
\end{aligned}
\tag{14}
$$

Proof. In this case, $\theta_2-\theta_1$ does not vanish hence (9) yields

$$
\begin{aligned}
III_{b,d}^{a,c} = \int_{-b}^{-a} e^{A_0^T\theta_1} \int_{-d}^{-c} &\left[U'(\theta_1-\theta_2)A_0 - A_0^T U'(\theta_1-\theta_2)\right. \\
&\left. + A_0^T U(\theta_1-\theta_2)A_0 - U''(\theta_1-\theta_2)\right] e^{A_0\theta_2} d\theta_2 d\theta_1,
\end{aligned}
$$

equivalently,

$$
III_{b,d}^{a,c} = \int_{-b}^{-a} \frac{\partial}{\partial\theta_1}\left(e^{A_0^T\theta_1}\int_{-d}^{-c}\frac{\partial}{\partial\theta_2}\left(U(\theta_1-\theta_2)e^{A_0\theta_2}\right)d\theta_2\right)d\theta_1.
$$

Integration by parts with respect to θ_2 yields

$$
\begin{aligned}
III_{b,d}^{a,c} &= \int_{-b}^{-a} \frac{\partial}{\partial\theta_1}\left(e^{A_0^T\theta_1}\left\{U(\theta_1-\theta_2)e^{A_0\theta_2}\Big|_{-d}^{-c}\right\}\right)d\theta_1 \\
&= \int_{-b}^{-a} \frac{\partial}{\partial\theta_1}\left(e^{A_0^T\theta_1}\left\{U(\theta_1+c)e^{-A_0 c} - U(\theta_1+d)e^{-A_0 d}\right\}\right)d\theta_1.
\end{aligned}
$$

Integration by parts with respect to θ_1 yields

$$
III_{b,d}^{a,c} = e^{A_0^T\theta_1}\left\{U(\theta_1+c)e^{-A_0 c} - U(\theta_1+d)e^{-A_0 d}\right\}\Big|_{-b}^{-a}
$$

and (14) follows.

Proposition 3. *The following equality holds for nonnegative scalars a,b*

$$
\int_{-b}^{-a} e^{A_0^T\theta_1}\int_{-b}^{-a} A_1^T U(\theta_1-\theta_2)A_1 e^{A_0\theta_2} d\theta_2 d\theta_1 = III_b^{a(1)} + III_b^{a(2)},
$$

where

$$
\begin{aligned}
III_b^{a(1)} &= e^{-A_0^T a}U(-a+c)e^{-A_0 c} - e^{-A_0^T a}U(-a+d)e^{-A_0 d} \\
&\quad -e^{-A_0^T b}U(-b+c)e^{-A_0 c} + e^{-A_0^T b}U(-b+d)e^{-A_0 d}
\end{aligned}
$$

and

$$
III_b^{a(2)} = -\int_{-b}^{-a} e^{A_0^T\theta} W e^{A_0\theta} d\theta.
$$

Proof. The equality (9) yields

$$III_b^{a(1)} = \int_{-b}^{-a} e^{A_0^T\theta_1} \int_{-b}^{-a} \left[U'(\theta_1-\theta_2)A_0 - A_0^T U'(\theta_1-\theta_2) \right.$$
$$\left. + A_0^T U(\theta_1-\theta_2)A_0 - U''(\theta_1-\theta_2)\right] e^{A_0\theta_2} d\theta_2 d\theta_1$$

and

$$III_b^{a(2)} = \int_{-b}^{-a} e^{A_0^T\theta_1} \int_{-b}^{-a} \delta(\theta_1-\theta_2) \left[U(\theta_1-\theta_2)A_0 + U^T(h-\theta_1+\theta_2)A_1 \right.$$
$$\left. + A_0^T U(\theta_1-\theta_2) + A_1^T U(h+\theta_1-\theta_2)\right] e^{A_0\theta_2} d\theta_2 d\theta_1.$$

The term $III_b^{a(1)}$ is obtained by following the steps in Proposition 2.
Next, $III_b^{a(2)}$ reduces to

$$III_b^{a(2)} = \int_{-b}^{-a} e^{A_0^T\theta_1} \left[U(0)A_0 + U^T(h)A_1 + A_0^T U(0) + A_1^T U(h)\right] e^{A_0\theta_1} d\theta_1.$$

Substituting (5) yields

$$III_b^{a(2)} = -\int_{-b}^{-a} e^{A_0^T\theta} W e^{A_0\theta} d\theta.$$

3 Necessary Conditions

Necessary conditions for the exponential stability of system (1), that also hold when A_1 is singular are presented. They are obtained by substituting into the functional a set of initial functions of exponential form, and using the results of the previous sections allowing the elimination of matrix A_1. Integration by parts, and the quadratic lower bound of Theorem 2 are also substantial elements of the proof.

Lemma 5. *If the delay system (1) is stable then the matrix*

$$\mathscr{K}(\tau) = \begin{pmatrix} U(0) & U(\tau) & U(h) \\ U^T(\tau) & U(0) & U(h-\tau) \\ U^T(h) & U^T(h-\tau) & U(0) \end{pmatrix} \tag{15}$$

is such that

$$\mathscr{K}(\tau) \geq 0, \quad \tau \in [0,h], \tag{16}$$

and

$$U(0) > 0, \tag{17}$$

with $U(\tau)$ the solution of (3), (4) and (5).

Proof. We start with the functional (2) and we consider the special initial function

$$\bar{\varphi}(\theta) = \begin{cases} e^{A_0\theta}\mu, \ \theta \in [-h, -\tau), \\ e^{A_0\theta}\eta, \ \theta \in [-\tau, 0), \\ \gamma, \qquad \theta = 0, \end{cases} \qquad \tau \in (0, h), \quad \mu, \eta, \gamma \in \mathbb{R}^n. \tag{18}$$

The first term of $v_0(\varphi)$ reduces to

$$v_I(\bar{\varphi}) = \bar{\varphi}^T(0)U(0)\bar{\varphi}(0) = \gamma^T U(0)\gamma. \tag{19}$$

The simple integral term in $v_0(\varphi)$ can be written as

$$\begin{aligned} v_{II}(\varphi) &= 2\varphi^T(0)\int_{-h}^{0} U^T(h+\theta)A_1\varphi(\theta)d\theta \\ &= 2\int_{-h}^{-\tau} \varphi^T(\theta)A_1^T U(h+\theta)d\theta\varphi(0) + 2\int_{-\tau}^{0} \varphi^T(\theta)A_1^T U(h+\theta)d\theta\varphi(0). \end{aligned}$$

Substituting the initial function (18) yields

$$v_{II}(\bar{\varphi}) = 2\int_{-h}^{-\tau} \mu^T e^{A_0^T\theta}A_1^T U(h+\theta)d\theta\gamma + 2\int_{-\tau}^{0} \eta^T e^{A_0^T\theta}A_1^T U(h+\theta)d\theta\gamma$$

and Proposition 1 implies that

$$\begin{aligned} v_{II}(\bar{\varphi}) = &-2\mu^T e^{-A_0^T\tau}U(-\tau)\gamma + 2\mu^T e^{-A_0^T h}U(-h)\gamma \\ &-2\eta^T U(0)\gamma + 2\eta^T e^{-A_0^T\tau}U(-\tau)\gamma. \end{aligned} \tag{20}$$

The double integral, is decomposed as

$$\begin{aligned} v_{III}(\varphi) = &\int_{-\tau}^{0} \varphi^T(\theta_1)\int_{-h}^{-\tau} A_1^T U(\theta_1-\theta_2)A_1\varphi^T(\theta_2)d\theta_2 d\theta_1 \\ &+\int_{-h}^{-\tau} \varphi^T(\theta_1)\int_{-\tau}^{0} A_1^T U(\theta_1-\theta_2)A_1\varphi^T(\theta_2)d\theta_2 d\theta_1 \\ &+\int_{-\tau}^{0} \varphi^T(\theta_1)\int_{-\tau}^{0} A_1^T U(\theta_1-\theta_2)A_1\varphi^T(\theta_2)d\theta_2 d\theta_1 \\ &+\int_{-h}^{-\tau} \varphi^T(\theta_1)\int_{-h}^{-\tau} A_1^T U(\theta_1-\theta_2)A_1\varphi^T(\theta_2)d\theta_2 d\theta_1. \end{aligned}$$

Substituting the initial function (18) yields

$$v_{III}(\bar{\varphi}) = \eta^T \int_{-\tau}^{0} e^{A_0^T\theta_1} \int_{-h}^{-\tau} A_1^T U(\theta_1-\theta_2) A_1 e^{A_0\theta_2} d\theta_2 d\theta_1 \mu$$
$$+\mu^T \int_{-h}^{-\tau} e^{A_0^T\theta_1} \int_{-\tau}^{0} A_1^T U(\theta_1-\theta_2) A_1 e^{A_0\theta_2} d\theta_2 d\theta_1 \eta$$
$$+\eta^T \int_{-\tau}^{0} e^{A_0^T\theta_1} \int_{-\tau}^{0} A_1^T U(\theta_1-\theta_2) A_1 e^{A_0\theta_2} d\theta_2 d\theta_1 \eta$$
$$+\mu^T \int_{-h}^{-\tau} e^{A_0^T\theta_1} \int_{-h}^{-\tau} A_1^T U(\theta_1-\theta_2) A_1 e^{A_0\theta_2} d\theta_2 d\theta_1 \mu.$$

Using Proposition 2 in the two first terms and Proposition 3 in the two last ones gives $v_{III}(\bar{\varphi}) = v_{III_1}(\bar{\varphi}) + v_{III_2}(\bar{\varphi})$, with

$$\begin{aligned} v_{III_1}(\bar{\varphi}) = \eta^T & \Big\{ U(\tau)e^{-A_0\tau} - U(h)e^{-A_0 h} \\ & - e^{-A_0^T\tau}U(0)e^{-A_0\tau} + e^{-A_0^T\tau}U(h-\tau)e^{-A_0 h} \Big\} \mu \\ & + \mu^T \Big\{ e^{-A_0^T\tau}U(-\tau) - e^{-A_0^T\tau}U(0)e^{-A_0\tau} \\ & - e^{-A_0^T h}U(-h) + e^{-A_0^T h}U(-h+\tau)e^{-A_0\tau} \Big\} \eta \\ & + \eta^T \big\{ U(0) - U(\tau)e^{-A_0\tau} \\ & - e^{-A_0^T\tau}U(-\tau) + e^{-A_0^T\tau}U(0)e^{-A_0\tau} \big\} \eta \\ & + \mu^T \Big\{ e^{-A_0^T\tau}U(0)e^{-A_0\tau} - e^{-A_0^T\tau}U(h-\tau)e^{-A_0 h} \\ & - e^{-A_0^T h}U(\tau-h)e^{-A_0\tau} + e^{-A_0^T h}U(0)e^{-A_0 h} \Big\} \mu \end{aligned} \tag{21}$$

and

$$v_{III_2}(\bar{\varphi}) = -\mu^T \int_{-h}^{-\tau} e^{A_0^T\theta} W e^{A_0\theta} d\theta \mu - \eta^T \int_{-\tau}^{0} e^{A_0^T\theta} W e^{A_0\theta} d\theta \eta. \tag{22}$$

Next, observe that substituting the initial function (18) into $\tilde{v}$ gives

$$\tilde{v}(\bar{\varphi}) = \mu^T \int_{-h}^{-\tau} e^{A_0^T\theta} Q e^{A_0\theta} d\theta \mu + \eta^T \int_{-\tau}^{0} e^{A_0^T\theta} Q e^{A_0\theta} d\theta \eta. \tag{23}$$

Now, adding the terms (19), (20), (21) and (22) corresponding to v_0 with (23) corresponding to $\tilde{v}$, and rearranging yields

$$v(\bar{\varphi}) = v_0(\bar{\varphi}) + \tilde{v}(\bar{\varphi}) = \Phi^T \mathscr{M}(\tau) \Phi + v_{III_2}(\bar{\varphi}) + \tilde{v}(\bar{\varphi}),$$

where

$$\Phi = \begin{pmatrix} \gamma^T & \eta^T e^{-A_0^T\tau} & \mu^T e^{-A_0^T h} & -\eta^T & -\mu^T e^{-A_0^T\tau} \end{pmatrix}^T$$

and

$$\mathscr{M}(\tau)=\begin{pmatrix} U(0) & U(\tau) & U(h) & U(0) & U(\tau) \\ U^T(\tau) & U(0) & U(h-\tau) & U^T(\tau) & U(0) \\ U^T(h) & U^T(h-\tau) & U(0) & U^T(h) & U^T(h-\tau) \\ U(0) & U(\tau) & U(h) & U(0) & U(\tau) \\ U^T(\tau) & U(0) & U(h-\tau) & U^T(\tau) & U(0) \end{pmatrix}.$$

The above can be written as

$$v(\bar{\varphi})=\Psi^T\left\{\mathscr{H}^T(\tau)\mathscr{K}(\tau)\mathscr{H}(\tau)\right\}\Psi+v_{III_2}(\bar{\varphi})+\tilde{v}(\bar{\varphi})$$

with

$$\Psi=\begin{pmatrix}\gamma \\ e^{-A_0\tau}\eta \\ e^{-A_0h}\mu\end{pmatrix},\ \mathscr{H}(\tau)=\begin{pmatrix} I & -e^{A_0\tau} & 0 \\ 0 & I & -e^{A_0(h-\tau)} \\ 0 & 0 & I\end{pmatrix}$$

and $\mathscr{K}(\tau)$ is defined in (15). As $0<Q<W$, the quadratic term

$$v_{III_2}(\bar{\varphi})+\tilde{v}(\bar{\varphi})=\mu^T\int_{-h}^{-\tau}e^{A_0^T\theta}(Q-W)e^{A_0\theta}d\theta\mu+\eta^T\int_{-\tau}^{0}e^{A_0^T\theta}(Q-W)e^{A_0\theta}d\theta\eta$$

is negative, therefore, in view of the quadratic lower bound (6), if the system (1) is stable it necessary holds that

$$\Psi^T\left\{\mathscr{H}^T(\tau)\mathscr{K}(\tau)\mathscr{H}(\tau)\right\}\Psi\geq 0 \text{ for } \tau\in[0,h].$$

As the vectors γ,η,μ are arbitrary and the exponential terms in Ψ are non singular

$$\mathscr{H}^T(\tau)\mathscr{K}(\tau)\mathscr{H}(\tau)\geq 0 \text{ for } \tau\in[0,h].$$

Finally, as the orthogonal transformation $\mathscr{H}(\tau)$ is nonsingular for all τ, the condition (16) follows. In addition, for the special case $\eta=0,\mu=0$, the condition (17) is satisfied.

Remark 1. It is worthy of mention that it was shown in [1]that for the scalar case ($n=1$) the conditions (16-17) provide a stability criterion, i.e. they are necessary and sufficient conditions.

4 Illustrative Example and Additional Considerations

In this section, an example illustrates the fact that the new conditions improve previously obtained results. The imaginary axis crossing loci of the characteristic quasipolynomial of the system are determined using the well known D-subdivision techniques, see [18]. These boundaries are depicted, along with points in the parametric space for which the necessary conditions reported in [14] or the new conditions (16-17) hold. Notice that the matrix A_1 is singular. Additional examples are available in [16].

Example 1. Let us consider the σ-stabilization of a second order system via proportional retarded output feedback analyzed in [20]. The closed loop system is in the form (1) with

$$A_0 = \begin{pmatrix} 0 & 1 \\ -(\sigma^2+\nu^2-2\delta\nu\sigma+bk_p) & -2(\delta\nu-\sigma) \end{pmatrix}, \; A_1 = \begin{pmatrix} 0 & 0 \\ bk_r e^{h\sigma} & 0 \end{pmatrix}.$$

Here ν and δ are the frequency and the damping coefficient of the second order system, σ is the desired exponential decay and k_r, k_p, h are the controller parameters. The closed loop quasipolynomial of this system is

$$p(s) = s^2 + 2(\delta\nu-\sigma)s + (\sigma^2+\nu^2-2\delta\nu\sigma+bk_p) - bk_r e^{h\sigma}e^{-hs}.$$

The exact stability domain is delimited by the boundaries described by the parametric equations

$$k_r(\omega) = -2\omega(\delta\nu-\sigma)/be^{h\sigma}\sin(h(\omega)\omega),$$

$$h(\omega) = (1/\omega)\cot^{-1}\frac{(-\omega^2+\sigma^2+\nu^2-2\delta\nu\sigma+bk_p)}{(-2\omega(\delta\nu-\sigma))} + n\frac{\pi}{\omega}, \quad n = 0,1,3,4,$$

and by $k_r = (\sigma^2+\nu^2-2\delta\nu\sigma+bk_p)/be^{h\sigma}$. On Figures 1 and 2, the fixed parameter values are $\sigma = 2, b = 31, \nu = 17.6, \delta = 0.0128$ and $k_p = 22.57$.

A few comments concerning the above example are in order: clearly, the conditions (16-17) improve the upper estimate of the stability region obtained with the conditions in [14], yet the shaded region does not match the known stability region, hence we conclude that these conditions are not sufficient.

On the one hand, as the number of unstable roots in a given region resulting from the partition of the space of parameters delimited by imaginary axis crossing hypersurfaces in the space of parameters is constant, see [18], the necessary stability conditions must hold at all points of a given region. As a consequence, one can conclude that if at least one point does not satisfy the necessary conditions, such a region in the partitioned space of parameters is actually an instability region.

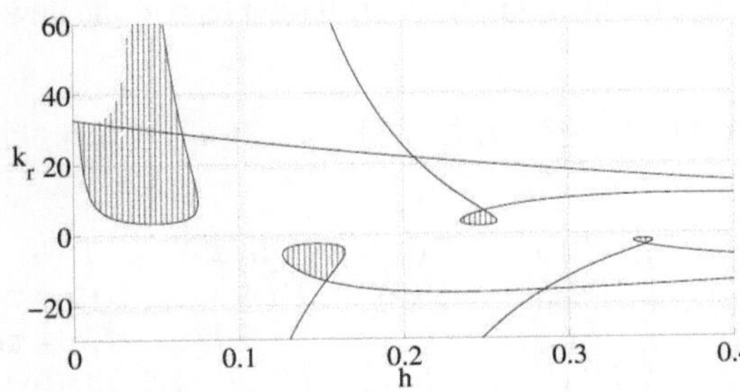

Fig. 1 Example 1, conditions in [14]

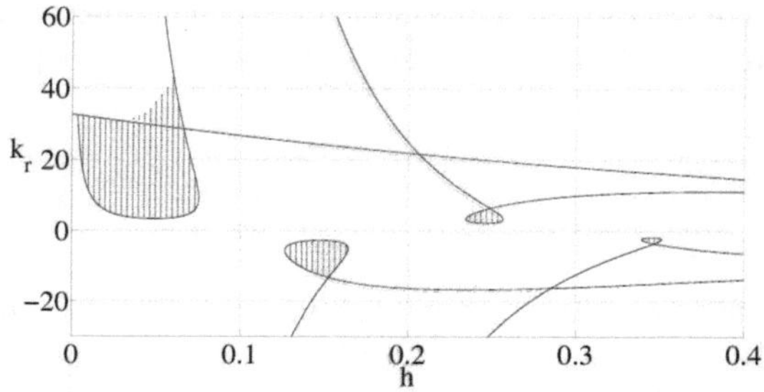

Fig. 2 Example 1, conditions (16, 17)

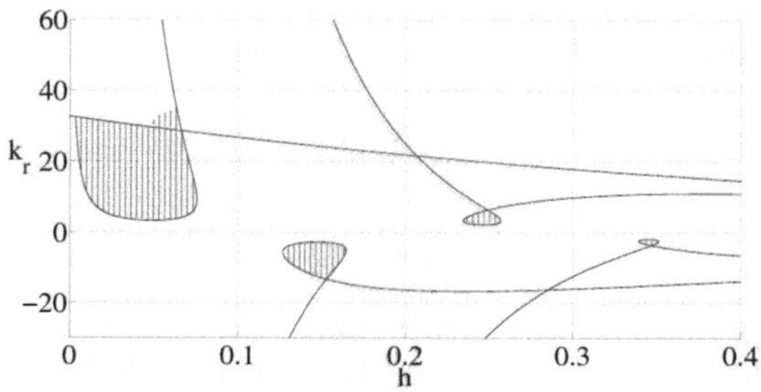

Fig. 3 Example 1, conditions (17, 24)

On the other hand, a natural query is wether or not it is possible to refine these necessary conditions. One observes that further improvements may be achieved by employing the initial condition

$$\tilde{\varphi}(\theta) = \begin{cases} e^{A_0\theta}\gamma_4, \ \theta \in [-h, -\tau_2), \\ e^{A_0\theta}\gamma_3, \ \theta \in [-\tau_2, -\tau_1), \\ e^{A_0\theta}\gamma_2, \ \theta \in [-\tau_1, 0), \\ \gamma_1, \qquad \theta = 0, \end{cases} \qquad \begin{array}{c} \tau_1, \tau_2 \in (0, h), \\ \tau_1 < \tau_2, \\ \gamma_1, \gamma_2, \gamma_3, \gamma_4 \in \mathbb{R}^n. \end{array}$$

The steps of the proof presented in this contribution lead to the new necessary conditions (16) and

$$\mathcal{K}_4(\tau_1, \tau_2) \geq 0, \ \tau_1 \in [0, h], \ \tau_2 \in [0, h], \ \tau_2 > \tau_1, \tag{24}$$

with

$$\mathcal{K}_4(\tau_1, \tau_2) = \begin{pmatrix} U(0) & U(\tau_1) & U(\tau_2) & U(h) \\ U^T(\tau_1) & U(0) & U(\tau_2 - \tau_1) & U(h - \tau_1) \\ U^T(\tau_2) & U^T(\tau_2 - \tau_1) & U(0) & U(h - \tau_2) \\ U^T(h) & U^T(h - \tau_1) & U^T(h - \tau_2) & U(0) \end{pmatrix}.$$

For these conditions, as shown on Figure 3, the estimate of the stability regions for Example 1 is improved compared to conditions (16,17).

Moreover, one can obtain in a similar manner the necessary conditions (17) and

$$\mathscr{K}_6(\tau) \geq 0, \; \tau \in \left(0, \frac{h}{3}\right), \tag{25}$$

where

$$\mathscr{K}_6(\tau) = \begin{pmatrix} U(0) & U(\tau) & U(\frac{h}{3}) & U(\tau+\frac{h}{3}) & U(\frac{2h}{3}) & U(\tau+\frac{2h}{3}) \\ U^T(\tau) & U(0) & U(\frac{h}{3}-\tau) & U(\frac{h}{3}) & U(\frac{2h}{3}-\tau) & U(\frac{2h}{3}) \\ U^T(\frac{h}{3}) & U^T(\frac{h}{3}-\tau) & U(0) & U(\tau) & U(\frac{h}{3}) & U(\tau+\frac{h}{3}) \\ U^T(\tau+\frac{h}{3}) & U^T(\frac{h}{3}) & U^T(\tau) & U(0) & U(\frac{h}{3}-\tau) & U(\frac{h}{3}) \\ U^T(\frac{2h}{3}) & U^T(\frac{2h}{3}-\tau) & U^T(\frac{h}{3}) & U^T(\frac{h}{3}-\tau) & U(0) & U(\tau) \\ U^T(\tau+\frac{2h}{3}) & U^T(\frac{2h}{3}) & U^T(\tau+\frac{h}{3}) & U^T(\frac{h}{3}) & U^T(\tau) & U(0) \end{pmatrix},$$

as in [3], which improves the above results (see Figure 4).

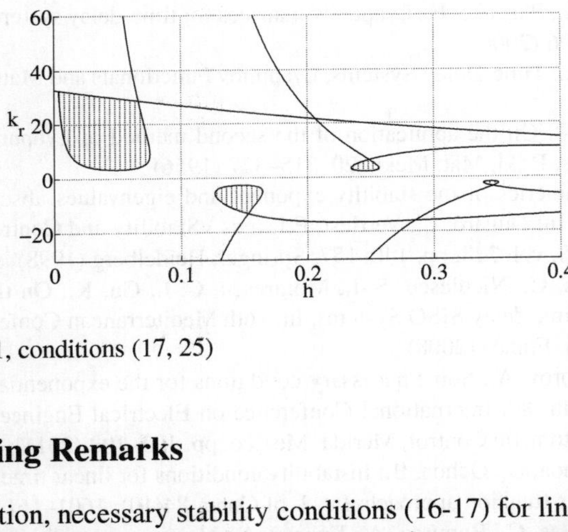

Fig. 4 Example 1, conditions (17, 25)

5 Concluding Remarks

In this contribution, necessary stability conditions (16-17) for linear one delay systems that are also valid when matrix A_1 is singular are presented. They are expressed in terms of the Lyapunov matrix of the system, thus extending fundamental results on delay free systems. Although the obtained conditions are shown not to be sufficient, a procedure that provides a better estimate of the stability region is outlined.

Acknowledgements. The presented research has been supported by Conacyt, grant 180725.

References

1. Cuvas, C., Mondié, S.: Stability Criteria for the Single Delay Equation. In: 11th IFAC Workshop on Time Delay Systems, Grenoble, France, pp. 7–11 (2013)
2. Datko, R.: An algorithm for computing Liapunov functionals for some differential difference equations. Ordinary Differential Equations, pp. 387–398. Academic Press, New York (1972)
3. Egorov, A., Mondié, S.: Necessary Conditions for the Stability of Multiple Time-Delay Systems Via the Delay Lyapunov Matrix. In: 11th IFAC Workshop on Time Delay Systems, Grenoble, France, pp. 12–17 (2013)
4. Hale, J.K.: Theory of functional differential equations. Springer, New-York (1977)
5. Huang, W.: Generalization of Liapunov's theorem in a linear delay system. J. of Math. Anal. and Appl. 142, 83–94 (1989)
6. Infante, E.F., Castelan, W.B.: A Liapunov functional for a matrix difference-differential equation. J. of Dif. Equ. 29, 439–451 (1978)
7. Kharitonov, V.L., Zhabko, A.P.: Lyapunov-Krasovskii approach for robust stability of time delay systems. Automatica 39, 15–20 (2003)
8. Kharitonov, V.L.: Lyapunov matrices for a class of time delay systems. Syst. & Contr. Lett. 55, 610–617 (2006)
9. Kharitonov, V.L., Plischke, E.: Lyapunov matrices for time-delay systems. Syst. & Contr. Lett. 55, 697–706 (2006)
10. Kharitonov, V.L.: Time-Delay Systems, Lyapunov Functionals and Matrices. Birkhauser, Basel (2013)
11. Krasovskii, N.N.: On the application of the second method of Lyapunov for equations with time delays. Prikl. Mat. Mekh. 20, 315–327 (1956)
12. Louisell, J.: Numerics of the stability exponent and eigenvalues abscissas of a matrix delay equation. In: Dugard, L., Verriest, E.I. (eds.) Stability and Control of Time Delay Systems. LNCIS, vol. 228, pp. 140–157. Springer, Heidelberg (1998)
13. Mendez-Barrios, C., Niculescu, S.-I., Morarescu, C.-I., Gu, K.: On the fragility of PI controllers for time delay SISO Systems. In: 16th Mediterranean Conference on Control and Automation, France (2008)
14. Mondié, S., Egorov, A.: Some necessary conditions for the exponential stability of one delay systems. In: 8th International Conference on Electrical Engineering, Computing Science and Automatic Control, Merida, Mexico, pp. 103–108 (2011)
15. Mondié, S., Ochoa, G., Ochoa, B.: Instability conditions for linear time delay systems: a Lyapunov matrix function approach. Int. J. of Contr. 84(10), 1601–1611 (2011)
16. Mondié, S., Cuvas, C., Ramírez, A., Egorov, A.: Necessary conditions for the stability of one delay systems: a Lyapunov matrix approach. In: 10th IFAC Workshop on Time Delay Systems, Boston, USA, pp. 13–18 (2012)
17. Mondié, S.: Assessing the exact stability region of the single delay scalar equation via its Lyapunov function. IMA J. of Math. Control Info. 29(4), 459–470 (2012)
18. Neimark, J.: D-subdivisions and spaces of quasi-polynomials. Prikl. Mat. Meh. 13, 349–380 (1949)
19. Repin, M.Y.: Quadratic Lyapunov functionals for systems with delay. Prik. Mat. Meh. 29, 564–566 (1965)
20. Villafuerte, R., Mondié, S., Garrido, R.: Tuning of Proportional Retarded Controllers: Theory and Experiments. IEEE Transactions on Control System Technology 21(3), 983–990 (2013)

Control of Linear Delay Systems: An Approach without Explicit Predictions

Nicole Gehring, Joachim Rudolph, and Frank Woittennek

Abstract. The control of linear time-invariant systems with incommensurate lumped and distributed delays is addressed. Using a module-theoretic point of view where these systems are modules over the ring of entire functions in $\mathbb{R}(s)[\mathrm{e}^{-\boldsymbol{\tau}s}]$ necessary and sufficient conditions for the freeness of these modules are presented. If these conditions hold a module basis can be used to design a tracking controller that assigns an arbitrary finite spectrum to the closed loop. Though the controller is infinite dimensional, in general, it does not involve any explicit predictions. This generalizes the so-called reduction approach, by which for certain state representations predictions can be calculated exactly and thus finite spectrum assignment can be achieved. Examples illustrate the main results.

1 Introduction

Most commonly, linear time-invariant systems with lumped (or pointwise) delays are expressed in terms of a representation (e.g. [13])

$$\dot{\boldsymbol{x}}(t) - \sum_{i=1}^{r} C_i \dot{\boldsymbol{x}}(t-\tau_i) = A_0 \boldsymbol{x}(t) + \sum_{i=1}^{r} A_i \boldsymbol{x}(t-\tau_i) + B_0 \boldsymbol{u}(t) + \sum_{i=1}^{r} B_i \boldsymbol{u}(t-\tau_i) \qquad (1)$$

with $\boldsymbol{x}(t) \in \mathbb{R}^n$ and input $\boldsymbol{u}(t) \in \mathbb{R}^m$, where $A_i \in \mathbb{R}^{n\times n}$, $B_i \in \mathbb{R}^{n\times m}$, $i = 0,\ldots,r$, and $C_i \in \mathbb{R}^{n\times n}$, $i = 1,\ldots,r$ are matrices of constants. The delay amplitudes $0 < \tau_i \in \mathbb{R}$, $i = 1,\ldots,r$ are *commensurate* if there exists a common divisor τ such that $\tau_i = \alpha_i \tau$,

Nicole Gehring · Joachim Rudolph
Lehrstuhl für Systemtheorie und Regelungstechnik, Universität des Saarlandes, Campus A5 1, 66123 Saarbrücken, Germany
e-mail: {n.gehring,j.rudolph}@lsr.uni-saarland.de

Frank Woittennek
Institut für Regelungs- und Steuerungstheorie, Technische Universität Dresden, 01062 Dresden, Germany
e-mail: frank.woittennek@tu-dresden.de

T. Vyhlídal, J.-F. Lafay, and R. Sipahi (eds.), *Delay Systems*,
Advances in Delays and Dynamics 1,
DOI: 10.1007/978-3-319-01695-5_2,

$\alpha_i \in \mathbb{Z}$, and *incommensurate* otherwise. The latter more general case is addressed here. For linear systems with both lumped and distributed delays, weighted integrals of $\boldsymbol{x}$ and $\boldsymbol{u}$ over finite intervals, such as

$$\int_0^{\tau_i} F(\sigma)\boldsymbol{x}(\sigma)\,d\sigma, \tag{2}$$

appear in addition to the terms on the right-hand side of (1).

Equation (1) is a *retarded* delay system if all matrices C_i, $i = 1, \ldots, r$ are identical to zero. Otherwise it is called a *neutral* delay system[1].

More generally, a lumped delay system is simply a set of difference-differential (and difference-algebraic) equations. As in [21], in the present setting, such a system is a finitely generated module over the polynomial ring[2] $\mathbb{R}[\frac{d}{dt}, \boldsymbol{\delta}] = \mathbb{R}[\frac{d}{dt}, \delta_1, \ldots, \delta_r]$ with r delay operators δ_i, $i = 1, \ldots, r$, where for every $f : \mathbb{R} \to \mathbb{R}$ one has $(\delta_i f)(t) = f(t - \tau_i)$. A ring $\mathbb{R}[s, \mathrm{e}^{-\boldsymbol{\tau} s}] = \mathbb{R}[s, \mathrm{e}^{-\tau_1 s}, \ldots, \mathrm{e}^{-\tau_r s}]$ is defined by the mappings $\frac{d}{dt} \mapsto s$, $\delta_i \mapsto \mathrm{e}^{-\tau_i s}$, $i = 1, \ldots, r$, where s is the parameter of the Laplace transform. This yields a ring isomorphism between $\mathbb{R}[\frac{d}{dt}, \boldsymbol{\delta}]$ and $\mathbb{R}[s, \mathrm{e}^{-\boldsymbol{\tau} s}]$. By the Paley-Wiener Theorem (e.g. [25]), the Laplace transform of a distributed delay, i.e., a linear convolution operator with kernel of bounded support (as (2) for example), is an entire function[3]. A typical example has the form

$$\frac{1 - \mathrm{e}^{-\tau_i(s+\beta)}}{s + \beta}.$$

It is not an element of $\mathbb{R}[s, \mathrm{e}^{-\boldsymbol{\tau} s}]$ but it belongs to an extended ring comprising all entire functions in $\mathbb{R}(s)[\mathrm{e}^{-\boldsymbol{\tau} s}]$, the ring of polynomials in $\mathrm{e}^{-\tau_i s}$, $i = 1, \ldots, r$ with coefficients that are rational in s. Denoting by $\mathscr{O}$ the ring of all entire functions, linear systems with distributed (and lumped) delays are modules over the ring

$$\mathscr{H}_r = \mathbb{R}(s)[\mathrm{e}^{-\boldsymbol{\tau} s}] \cap \mathscr{O},$$

the subring of entire functions in $\mathbb{R}(s)[\mathrm{e}^{-\boldsymbol{\tau} s}]$. In the following, the notion of a module is used as a synonym for a system [7].

In order to represent a module a matrix notation is used, as in [21]. For that, let $\mathscr{R}$ be a ring (e.g. $\mathscr{R} = \mathscr{H}_r$) and $\mathscr{M}$ a finitely presented[4] $\mathscr{R}$-module. Then $\mathscr{M}$ can be expressed by a set of $\bar{q}$ equations in terms of

$$P_{\mathscr{M}} \boldsymbol{w} = \boldsymbol{0} \tag{3}$$

[1] A retarded system may always be transformed into neutral form. Since the inverse is not true, a more prudent way of characterizing a delay system (1) is by considering its characteristic equation (w.r.t. $\boldsymbol{x}$) which is a quasi-polynomial in s and $\mathrm{e}^{-\tau_i s}$, $i = 1, \ldots, r$. Then, the system is neutral if the coefficient of s^n is not a real number but instead depends on $\mathrm{e}^{-\tau_i s}$.

[2] All discussions are restricted to the field $\mathbb{R}$ of real numbers, though they still hold for $\mathbb{C}$.

[3] More precisely, it is an entire function of exponential type satisfying certain growth conditions on the imaginary axis.

[4] All modules introduced in the following are assumed to be finitely presented.

with system variables $\boldsymbol{w} = (w_1, \dots, w_l)^T$, where $P_{\mathscr{M}}$ is a $(\bar{q} \times l)$-matrix with entries from $\mathscr{R}$. Note that input and output variables may be included in $\boldsymbol{w}$. In general, the $\bar{q}$ equations cannot be assumed to be $\mathscr{R}$-linear independent [26]. Hence $q = \mathrm{rk}_{\mathscr{R}} P_{\mathscr{M}} \leq \min(\bar{q}, l)$.

When it comes to the control of delay systems (1) the so-called reduction approach presents a valuable tool. Based on a retarded state representation with lumped and distributed delays of both inputs and states, the well-known works [1, 17, 19] showed that exact predictions can be obtained by invoking the systems solution[5]. The basic idea is to transform the state of a delay system such that a state representation without delays is obtained. By applying methods known for linear systems without delays, a stabilizing or tracking controller assigning a potentially finite spectrum to the closed loop can then be designed in a straightforward manner. The control law can be characterized as being prediction-free in that no explicit predictions occur.

Spectral controllability has been shown to be necessary in order for the reduction approach to work (e.g. [6]). Over the years, the thought of explicitly calculating a predicted value has been extended to other classes of systems such as those involving multiple incommensurate delays [24]. A rather recent detailed overview can be found in [18].

All the previously mentioned contributions base their results on retarded state representations. However, several technological processes, including networked control systems and population dynamics, are modeled by neutral delay systems, a more general class of delay systems [14, 15]. In the module-theoretic framework, both classes of delay systems are simply modules over $\mathscr{H}_r$. This more general perspective on delay systems is chosen here. The main observation is that if a module over $\mathscr{H}_r$ satisfies certain controllability conditions corresponding to freeness of associated modules, a basis (or basic or flat output) can be derived. This allows one to design a prediction-free tracking control, i.e., one without (explicit) predictions. The results are independent of a specific system representation (e.g. state representation or transfer function) and the choice of a control input.

In particular, first, essential controllability properties are given in Section 2. Before presenting the necessary and sufficient conditions for a prediction-free control in Section 4, some seminal remarks on modules over the ring $K[\boldsymbol{\delta}]$ concerning the choice of appropriate delay operators $\boldsymbol{\delta}$ are outlined in Section 3. Different configurations of a heat accumulator serve as illustrative examples in Section 5.

2 Controllability Properties

As in the case of delay-free systems, certain controllability properties are of special significance in the design of stabilizing (or tracking) controllers. In the context of linear delay systems, spectral controllability is well known and rather popular (see e.g. [21] and references therein). It can be considered as one possible extension of

---

[5] This idea possibly originated in [20], where linear systems with lumped delays were considered.

the controllability concept for linear systems without delays. A generalized Hautus test is usually used for its characterization [2,3]. The definition given below is based on the one in [21] and generalizes this characterization to $\mathscr{H}_r$-modules with a matrix representation as in (3).

Definition 1. A $\mathscr{H}_r$-module Σ is spectrally controllable iff

$$\mathrm{rk}_{\mathbb{C}} P_\Sigma(s, \mathrm{e}^{-\tau_1 s}, \ldots, \mathrm{e}^{-\tau_r s}) = q, \quad \forall s \in \mathbb{C}. \tag{4}$$

In the module-theoretic framework controllability properties are closely related to concepts like torsion-freeness and freeness of certain modules, and by that to the existence of a basis. While for linear finite-dimensional systems many of these module properties coincide, the situation changes for delay systems. An extensive study of the subject was performed in [9,21].

In general, an $\mathscr{R}$-module $\mathscr{M}$ is called *torsion-free* if the set $t\mathscr{M}$ of all so-called torsion-elements $0 \neq z \in \mathscr{M}$ with $\alpha z = 0$, $\alpha \in \mathscr{R}$ is void [5]. Simply speaking, an $\mathscr{R}$-module $\mathscr{M}$ is torsion-free if for a representation (3) there does not exist an $\mathscr{R}$-linear combination of system variables $\boldsymbol{w}$ satisfying an autonomous equation.

A *basis* of an $\mathscr{R}$-module $\mathscr{M}$ is an $\mathscr{R}$-linear independent family of elements which spans $\mathscr{M}$ [5]. Hence, for a representation (3) a basis $\boldsymbol{v} = M\boldsymbol{w}$, $M \in \mathscr{R}^{(l-q)\times l}$ is an $\mathscr{R}$-linear combination of the system variables $\boldsymbol{w}$ such that one has $\boldsymbol{w} = T\boldsymbol{v}$, $T \in \mathscr{R}^{l\times(l-q)}$. A module $\mathscr{M}$ which has a basis is called *free*.

Modules over the ring $\mathbb{R}(s)[\mathrm{e}^{-\boldsymbol{\tau} s}]$ are particularly important in the following. Below, generalized Hautus tests are given that allow one to check a representation (3) of an $\mathbb{R}(s)[\mathrm{e}^{-\boldsymbol{\tau} s}]$-module for freeness and a special case of the so-called π-freeness[6]. The definitions are adapted from [9,21].

Proposition 1. *An $\mathbb{R}(s)[\mathrm{e}^{-\boldsymbol{\tau} s}]$-module Σ is free iff*

$$\mathrm{rk}_{\mathbb{C}} P_\Sigma(z_1, \ldots, z_r) = q, \quad \forall (z_1, \ldots, z_r) \in \mathbb{C}^r.$$

Proposition 2. *Let $\pi = \mathrm{e}^{-\mu_1\tau_1 s - \cdots - \mu_r\tau_r s}$, $\mu_1, \ldots, \mu_r \in \mathbb{N}$. An $\mathbb{R}(s)[\mathrm{e}^{-\boldsymbol{\tau} s}]$-module is π-free iff*

$$\mathrm{rk}_{\mathbb{C}} P_\Sigma(z_1, \ldots, z_r) = q,\ \forall (z_1, \ldots, z_r) \in \mathbb{C}^r, z_k \neq 0, k \in \{j = 1, \ldots, r | \mu_j \neq 0\}.$$

In contrast to the concept of spectral controllability these characterizations do not account for the algebraic dependence between the delay operators δ_i and the differentiation operator $\frac{d}{dt}$ that are linked by the mappings $\delta_i \mapsto \mathrm{e}^{-\tau_i s}$ and $\frac{d}{dt} \mapsto s$ (see [21]). Instead, the operators are replaced by complex numbers z_i and s. Note that due to the isomorphism between $K[\mathrm{e}^{-\boldsymbol{\tau} s}]$ and $K[\boldsymbol{\delta}]$ (with a field K) Propositions 1 and 2 also characterize $K[\boldsymbol{\delta}]$-modules that are basically discrete-time systems.

6 In [9, 21, 22] the liberation polynomial π is an element of $\mathbb{R}[\mathrm{e}^{-\boldsymbol{\tau} s}]$ (or $\mathbb{R}[\boldsymbol{\delta}]$). Here, π denotes a monomial from that ring. This specific class of π-free systems is often called δ-free (e.g. [22]).

3 Remarks on Modules over the Ring $K[\boldsymbol{\delta}]$

Modules over the ring $K[\boldsymbol{\delta}]$, where K is an arbitrary field, can be represented by difference equations. However, when faced with such equations the question arises of how to define the incommensurate delay amplitudes τ_i in order to obtain a $K[\boldsymbol{\delta}]$-module. While in some physical systems the choice of the delay amplitudes might seem natural, the freeness of a $K[\boldsymbol{\delta}]$-module strongly depends on their definition.

Example 1. An $\mathbb{R}[\delta_1,\delta_2]$-module represented by

$$\delta_1 w_1 + \delta_2 w_2 = 0$$

is δ_1-free (or δ_2-free) with a δ_1-basis w_2 (or a δ_2-basis w_1). Assuming that $\tau_1 < \tau_2$ and introducing new delay operators $\bar{\delta}_1 = \delta_1$, $\bar{\delta}_2 = \delta_1^{-1}\delta_2$ with positive delay amplitudes $\bar{\tau}_1$, $\bar{\tau}_2$ a relation

$$\bar{\delta}_1 w_1 + \bar{\delta}_1\bar{\delta}_2 w_2 = \bar{\delta}_1(w_1 + \bar{\delta}_2 w_2) = 0$$

is derived. The corresponding $\mathbb{R}[\bar{\delta}_1,\bar{\delta}_2]$-module has a torsion-element $z = w_1 + \bar{\delta}_2 w_2$ with $\bar{\delta}_1 z = 0$. However, allowing for a simple time shift, i.e., a multiplication with $\bar{\delta}_1^{-1}$, results in an $\mathbb{R}[\bar{\delta}_1,\bar{\delta}_2]$-module that is torsion-free. The case where $\tau_1 > \tau_2$ can be treated in a similar way.

The previous example motivates the concept of the so-called δ-equivalence (see [22]). Basically, two modules are called δ-equivalent if they can be represented by the same set of equations up to a multiplication with an operator in the monoid $S = \{\delta_1^{\mu_1}\cdots\delta_r^{\mu_r} \,|\, \mu_1,\ldots,\mu_r \in \mathbb{N}\}$. Then by localization $S^{-1}K[\boldsymbol{\delta}] = K[\boldsymbol{\delta},\boldsymbol{\delta}^{-1}]$.

Definition 2. Two $K[\boldsymbol{\delta}]$-modules Λ_1 and Λ_2 are called δ-equivalent if

$$K[\boldsymbol{\delta},\boldsymbol{\delta}^{-1}] \otimes_{K[\boldsymbol{\delta}]} \Lambda_1 = K[\boldsymbol{\delta},\boldsymbol{\delta}^{-1}] \otimes_{K[\boldsymbol{\delta}]} \Lambda_2.$$

Hence, $\Lambda_1/t_S\Lambda_1 = \Lambda_2/t_S\Lambda_2$ where $t_S\Lambda_i$ is the kernel of the mapping $\Lambda_i \to S^{-1}K[\boldsymbol{\delta}] \otimes_{K[\boldsymbol{\delta}]} \Lambda_i$, $i = 1,2$, i.e., the torsion-submodule w.r.t. the monoid S.

Example 2. Consider the system

$$\delta_1 w_1 + (1+\delta_2) w_2 = 0.$$

The corresponding $\mathbb{R}[\delta_1,\delta_2]$-module is δ_1-free with a δ_1-basis w_2. However, assuming $\tau_1 < \tau_2$ new operators $\bar{\delta}_1 = \delta_1$, $\bar{\delta}_2 = \delta_1^{-1}\delta_2$ with positive delay amplitudes $\bar{\tau}_1$, $\bar{\tau}_2$ can be introduced, with which the system equation can be rewritten as

$$\bar{\delta}_1 w_1 + (1+\bar{\delta}_1\bar{\delta}_2) w_2 = 0.$$

The resulting $\mathbb{R}[\bar{\delta}_1,\bar{\delta}_2]$-module is free with basis $w_1 + \bar{\delta}_2 w_2$.

A similar approach can be taken in the case where $\tau_1 > \tau_2$. First, introduce operators $\tilde{\delta}_1 = \delta_1\delta_2^{-\gamma}$, $\tilde{\delta}_2 = \delta_2$ with γ maximal in the sense that $0 < \tilde{\tau}_1 < \tilde{\tau}_2$:

$$\tilde{\delta}_1\tilde{\delta}_2^{\gamma} w_1 + (1+\tilde{\delta}_2)w_2 = 0.$$

Then, defining $\bar{\delta}_1 = \tilde{\delta}_1$, $\bar{\delta}_2 = \tilde{\delta}_1^{-1}\tilde{\delta}_2$ yields a free $\mathbb{R}[\bar{\delta}_1, \bar{\delta}_2]$-module represented by

$$\bar{\delta}_1(\bar{\delta}_1\bar{\delta}_2)^{\gamma}\bar{w}_1 + (1+\bar{\delta}_1\bar{\delta}_2)\bar{w}_2 = 0.$$

(The determination of a basis depends on the value of γ.)

The basic idea in introducing new delay operators $\bar{\delta}_i$, $i = 1,\dots,r$ is to make use of the knowledge of the delay amplitudes τ_i of the operators δ_i and to allow for multiplications with δ_i^{-1}, i.e., the inverse of δ_i, as long as the resulting product of delay operators has a strictly positive (delay) amplitude $\bar{\tau}_i$. This idea was previously proposed in [11] and is stated here without proof. Implicitly, it has also been used in other contributions (e.g., [16]).

As above, let $\bar{\boldsymbol{\delta}} = (\bar{\delta}_1,\dots,\bar{\delta}_r)$ be delay operators with strictly positive amplitudes $\bar{\tau}_i$ where for every $f:\mathbb{R}\to\mathbb{R}$ one has $(\bar{\delta}_i f)(t) = f(t-\bar{\tau}_i)$, $i = 1,\dots,r$. Then for a ring $K[\bar{\boldsymbol{\delta}}]$ it follows $K[\boldsymbol{\delta}] \subseteq K[\bar{\boldsymbol{\delta}}] \subset K[\boldsymbol{\delta},\boldsymbol{\delta}^{-1}]$. Also, define the $K[\bar{\boldsymbol{\delta}}]$-module $\bar{\Lambda} = K[\bar{\boldsymbol{\delta}}] \otimes_{K[\boldsymbol{\delta}]} \Lambda$ and the monoid $\bar{S} = \{\bar{\delta}_1^{\mu_1}\cdots\bar{\delta}_r^{\mu_r} \,|\, \mu_1,\dots,\mu_r \in \mathbb{N}\}$ (in the same manner as S above).

Lemma 1. *If a $K[\boldsymbol{\delta}]$-module Λ is π-free with $\pi \in S$, then there exists a ring $K[\bar{\boldsymbol{\delta}}]$ such that the $K[\bar{\boldsymbol{\delta}}]$-module $\bar{\Lambda}/t_{\bar{S}}\bar{\Lambda}$ is free.*

Lemma 1 makes use of the concept of δ-equivalence. In a very natural way, for a π-free $K[\boldsymbol{\delta}]$-module, $\pi \in S$, it allows one to use an extension of scalars to produce a free $K[\bar{\boldsymbol{\delta}}]$-module. By the mapping $\delta_i \mapsto \mathrm{e}^{-\tau_i s}$ the results directly translate to modules over $K[\mathrm{e}^{-\boldsymbol{\tau}s}]$ and thus to modules over $\mathbb{R}(s)[\mathrm{e}^{-\boldsymbol{\tau}s}]$.

4 Prediction-Free Control

In this section necessary and sufficient conditions for the freeness of an $\mathscr{H}_r$-module are stated, both for the commensurate and for the incommensurate case. If these conditions hold, the design of a prediction-free tracking controller is rather straightforward. Due to page limitations, only a sketch of proof is given for all theorems presented. In particular, constructive proof elements are skipped. These details will be given elsewhere.

4.1 Conditions in the Commensurate Case

Systems with commensurate delays are modules over $\mathscr{H}_1 =: \mathscr{H}$, the subset of entire functions in $\mathbb{R}(s)[\mathrm{e}^{-\tau s}]$. In [4] it was proven that in the commensurate case $\mathscr{H}$ is a Bézout domain, i.e., any finitely generated ideal in this domain is principal. The same result was obtained independently in the behavioral context [12].

Theorem 1. *Let Σ be an $\mathscr{H}$-module. The following statements are equivalent:*

1. *Σ is torsion-free.*
2. *Σ is free.*
3. *Σ is spectrally controllable and $\mathbb{R}(s)[\mathrm{e}^{-\tau s}] \otimes_{\mathscr{H}} \Sigma$ is free.*

While the equivalence between the first two statements is due to $\mathscr{H}$ being a Bézout domain, the equivalence to the third one can be shown in a constructive manner. In the context of the reduction approach, especially the property of spectral controllability has been proven to be necessary in order to assign an arbitrary finite spectrum to a closed loop (see e.g. [6]). While this property is also sufficient for a broad class of state representations this is no longer the case for general representations (3). As a matter of fact, spectral controllability of Σ is necessary and sufficient for a δ-equivalent module of Σ to be free, i.e., the condition of freeness of $\mathbb{R}(s)[\mathrm{e}^{-\tau s}] \otimes_{\mathscr{H}} \Sigma$ can be dropped if δ-equivalence is employed.

4.2 Conditions in the Incommensurate Case

As presented in [11], in contrast to the subset of entire functions in $\mathbb{R}(s)[\mathrm{e}^{-\tau s}]$, the set $\mathscr{H}_r = \mathbb{R}(s)[\mathrm{e}^{-\boldsymbol{\tau} s}] \cap \mathscr{O}$ does no longer have the Bézout property, in general. Hence, torsion-freeness of an $\mathscr{H}_r$-module does not imply freeness of that module.

Theorem 2. *An $\mathscr{H}_r$-module Σ is free iff $\mathbb{R}(s)[\mathrm{e}^{-\boldsymbol{\tau} s}] \otimes_{\mathscr{H}_r} \Sigma$ is free and Σ is spectrally controllable.*

The necessity of the two conditions is rather obvious. Since $\mathscr{H}_r \subset \mathbb{R}(s)[\mathrm{e}^{-\boldsymbol{\tau} s}]$, freeness of an $\mathscr{H}_r$-module Σ implies freeness of $\mathbb{R}(s)[\mathrm{e}^{-\boldsymbol{\tau} s}] \otimes_{\mathscr{H}_r} \Sigma$. Also, by contradiction, if Σ were not spectrally controllable then $\mathscr{H}_r$-torsion-elements would exist and Σ could not be free. As in the commensurate case, the sufficiency can be shown constructively.

Taking into account Lemma 1 for modules over the ring $K[\boldsymbol{\delta}]$, Theorem 2 can be extended to π-free modules $\mathbb{R}(s)[\mathrm{e}^{-\boldsymbol{\tau} s}] \otimes_{\mathscr{H}_r} \Sigma$ with $\pi = \mathrm{e}^{-\mu_1 \tau_1 s - \cdots - \mu_r \tau_r s}$, $\mu_1, \ldots, \mu_r \in \mathbb{N}$. For that, denote by $\bar{\mathscr{H}}_r = \mathbb{R}(s)[\mathrm{e}^{-\bar{\boldsymbol{\tau}} s}] \cap \mathscr{O}$ the subring of entire functions belonging to $\mathbb{R}(s)[\mathrm{e}^{-\bar{\boldsymbol{\tau}} s}]$, $\mathrm{e}^{-\bar{\boldsymbol{\tau}} s} = (\mathrm{e}^{-\bar{\tau}_1 s}, \ldots, \mathrm{e}^{-\bar{\tau}_r s})$ where delay amplitudes $\bar{\boldsymbol{\tau}}$ are defined based on Lemma 1. Furthermore, let $\bar{\Sigma}$ be the $\bar{\mathscr{H}}_r$-module $\bar{\mathscr{H}}_r \otimes_{\mathscr{H}_r} \Sigma$ and $\bar{S}_e$ the monoid following from $\bar{S}$ (see Section 3) by the mapping $\delta_i \mapsto \mathrm{e}^{-\bar{\tau}_i s}$.

Theorem 3. *The $\bar{\mathscr{H}}_r$-module $\bar{\Sigma}/t_{\bar{S}_e}\bar{\Sigma}$ is free iff $\mathbb{R}(s)[\mathrm{e}^{-\boldsymbol{\tau} s}] \otimes_{\mathscr{H}_r} \Sigma$ is π-free with $\pi = \mathrm{e}^{-\mu_1 \tau_1 s - \cdots - \mu_r \tau_r s}$, $\mu_1, \ldots, \mu_r \in \mathbb{N}$ and Σ is spectrally controllable.*

4.3 Control Design

If the conditions in Theorems 1 for the commensurate case and 2 resp. 3 for the incommensurate case hold, an $\mathscr{H}_r$-basis of Σ can be found (respectively an $\bar{\mathscr{H}}_r$-basis of $\bar{\Sigma}$). In the commensurate case [4, 12, 23] solved Bézout equations in order to determine a basis. Alternatively, new system variables can be introduced in (3) simplifying the structure of the system equations. This approach corresponds to the transformation of a matrix into an appropriately defined row Hermite form where a basis is obtained as a byproduct. Note that the idea of the reduction approach also is to transform the state of a delay system such that a state representation without delays is obtained.

Since a basis parametrizes all system variables, including the input variables, both, an open-loop and a closed loop tracking controller can be designed in a straightforward manner[7]. A resulting controller will not involve any (explicit) predictions but instead it will incorporate distributed delays, basically representing the exact predicted values. Thus, the controller is of infinite dimension, in general.

5 Example: A Heat Accumulator

The heat accumulator[8] sketched in Figure 1 is basically a fluid-filled tank with some piping. The tank is constantly fed through an inlet pipe of length $L_1 > 0$. An outlet pipe is controlled such that the volume in the tank remains constant. The fluid in the tank is assumed to be ideally mixed, with temperature T. By a loop of length $L_2 > 0$, the fluid is recycled. While the tank and the pipes are isolated, the temperature T_0 of the inlet pipe's jacket can be freely assigned.

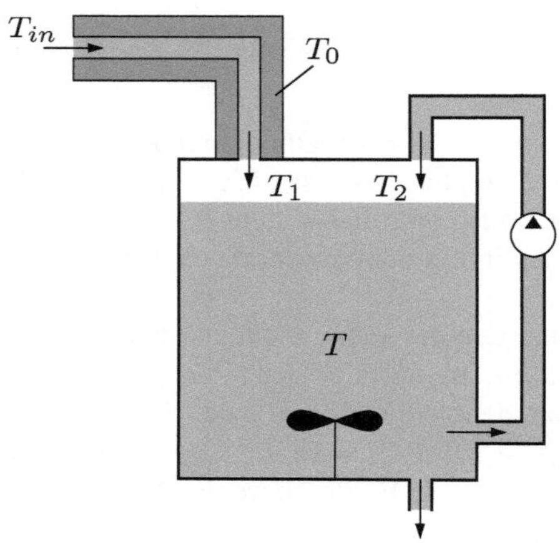

Fig. 1 Schematics of a heat accumulator with a recycle loop

Assuming constant specific heat capacity for the fluid in the heat accumulator and a constant cross-section of the pipes, from a heat balance transport equations for the temperature θ_1 in the inlet pipe and θ_2 in the recycle loop are obtained:

[7] For nonlinear systems this concept is known as flatness-based control [8].

[8] This example was previously studied in [11].

$$\frac{\partial\theta_1}{\partial t}(z,t)+v_1\frac{\partial\theta_1}{\partial z}(z,t)=\beta_1 T_0(t), \qquad z\in[0,L_1] \tag{5a}$$

$$\frac{\partial\theta_2}{\partial t}(z,t)+v_2\frac{\partial\theta_2}{\partial z}(z,t)=0, \qquad z\in[0,L_2], \tag{5b}$$

with the constant velocities $v_1, v_2 > 0$. Since all temperatures are defined relative to the ambient temperature, the boundary conditions read

$$\theta_1(0,t)=\beta_2 T_{in}(t), \qquad \theta_2(0,t)=T(t).$$

The quantities

$$T_1(t)=\theta_1(L_1,t), \qquad T_2(t)=\theta_2(L_2,t)$$

denote the temperatures at the end of the inlet pipe and at the end of the recycle loop, respectively. Finally, the heat balance for the tank together with the solution of (5) (under homogeneous initial conditions) gives the system equations

$$T_1(t)=\beta_2 T_{in}(t-\tau_1)+\beta_1\int_0^{\tau_1} T_0(t-\sigma)\,d\sigma \tag{6a}$$

$$T_2(t)=T(t-\tau_2) \tag{6b}$$

$$\dot{T}(t)=\alpha_1\big(T_1(t)-T(t)\big)+\alpha_2\big(T_2(t)-T(t)\big), \tag{6c}$$

with $\alpha_i\in\mathbb{R}$, $\tau_i=L_i/v_i$, $i=1,2$. Parameters $\beta_1,\beta_2\in\{0,1\}$ are introduced in order to activate or deactivate the control inputs T_0 and T_{in}.

While a controller for (6) can be designed using the reduction approach, the aim here is to illustrate the general results from Section 4. Therefore different configurations of inputs are addressed. For more complex examples of neutral delay systems, the reader is referred to [10, 11].

5.1 Control via the Jacket Temperature T_0

In order to consider the case where the jacket temperature T_0 is the (only) control input, set $\beta_1=1$ and $\beta_2=0$. By that the inlet temperature T_{in} equals the ambient temperature (and is thus zero). Furthermore, assume that no recycle loop is present, i.e., $\alpha_2=0$. Then, from (6) a relation between the jacket temperature T_0 and the temperature T in the reactor is obtained:

$$\dot{T}(t)=-\alpha_1 T(t)+\alpha_1\int_0^{\tau_1} T_0(t-\sigma)\,d\sigma. \tag{7}$$

This equation is equivalent to (6) (for $\beta_1=1$, $\beta_2=0$) and can be viewed as a delayed state representation w.r.t. the input T_0.

The Laplace transform of (7) (with zero initial condition) reads[9]

$$(s+\alpha_1)\hat{T}-\alpha_1\frac{1-\mathrm{e}^{-\tau_1 s}}{s}\hat{T}_0=0$$

[9] All transformed system variables are written with a hat.

and yields an $\mathscr{H}_1$-module Σ. Both conditions in Theorem 1 are fulfilled: Σ is spectrally controllable and $\mathbb{R}(s)[\mathrm{e}^{-\tau_1 s}] \otimes_{\mathscr{H}_1} \Sigma$ is free with basis $\hat{T}$. Thus Σ is free.

As to determine an $\mathscr{H}_1$-basis it is helpful to eliminate the exponential function $\mathrm{e}^{-\tau_1 s}$, corresponding to the delay operator δ_1, from the representation. Introducing

$$\hat{w}_1 = \frac{1}{\alpha_1}\hat{T}_0 \tag{8a}$$

$$\hat{w}_2 = \hat{T} - \frac{\alpha_1(1-\mathrm{e}^{-\tau_1 s}) - s(\mathrm{e}^{\tau_1\alpha_1}-1)}{s(s+\alpha_1)}\hat{T}_0 \tag{8b}$$

by a unimodular transformation over $\mathscr{H}_1$ between $(\hat{T},\hat{T}_0)$ and $(\hat{w}_1,\hat{w}_2)$ gives

$$(s+\alpha_1)\hat{w}_2 - (\mathrm{e}^{\tau_1\alpha_1}-1)\hat{w}_1 = 0. \tag{9}$$

Note that the factor of $\hat{T}_0$ in (8b) is an entire function and thus belongs to $\mathscr{H}_1$. Then, from (9) it is obvious that $\hat{w}_2 =: \hat{y}$ is an $\mathscr{H}_1$-basis, defined by (8b) in terms of $\hat{T}$ and $\hat{T}_0$. Taking the inverse Laplace transform of (8b), in the time domain one has

$$y(t) = T(t) + \frac{\mathrm{e}^{\alpha_1\tau_1}}{\alpha_1}\int_0^{\tau_1}(\mathrm{e}^{-\alpha_1\sigma} - \mathrm{e}^{-\alpha_1\tau_1})T_0(t-\sigma)\,d\sigma. \tag{10}$$

The parameterizations of T and T_0 in terms of the basis y follow from (8b) and (9):

$$T(t) = \frac{\alpha_1}{\mathrm{e}^{\tau_1\alpha_1}-1}\int_0^{\tau_1} y(t-\sigma)\,d\sigma \tag{11a}$$

$$T_0(t) = \frac{1}{\mathrm{e}^{\tau_1\alpha_1}-1}\big(\dot{y}(t) + \alpha_1 y(t)\big). \tag{11b}$$

Equation (11b) can be viewed as a state representation w.r.t. the input T_0. It is a linear system without delays that is equivalent to the original one (7). Note that the basis (10), which is the state in this case, is the same variable that would be introduced in the reduction approach (e.g., see example 5.3 in [1] with $x(t) \in \mathbb{R}$).

In order to design a controller that tracks a reference trajectory $t \mapsto y_r(t)$ a first-order dynamics (corresponding to the order of the system (11b))

$$\dot{e} + ke = 0, \qquad k > 0 \tag{12}$$

for the error $e = y - y_r$ is prescribed. Replacing $\dot{y}$ using (11b) yields a control law that does not require any predictions and that assigns a finite spectrum to the closed loop:

$$T_0(t) = \frac{1}{\mathrm{e}^{\tau_1\alpha_1}-1}\big(\dot{y}_r(t) - ke(t) + \alpha_1 y(t)\big).$$

Though this controller makes use of the knowledge of a basis, which is usually not directly measurable, y can be calculated from (10) if T is measured and T_0 is known on the interval $[-\tau_1, 0)$. In that case, the controller involves distributed delays and is thus infinite dimensional.

5.2 Control via the Inlet Temperature T_{in}

Let the inlet pipe be isolated, i.e. $\beta_1 = 0$, and the temperature T_{in} serve as the control input. For convenience, the parameter $\beta_2 = 1$ is normalized. In this case (6) can be written as

$$\dot{T}(t) = -(\alpha_1 + \alpha_2)T(t) + \alpha_2 T(t-\tau_2) + \alpha_1 T_{in}(t-\tau_1) \tag{13}$$

and is a system with lumped delays only. Note that for $\tau_2 > \tau_1$ a control law defining T_{in} requires a predicted value of T. The Laplace transform of (13) (with zero initial condition) reads

$$(s + \alpha_1 + \alpha_2 - \alpha_2 \mathrm{e}^{-\tau_2 s})\hat{T} - \alpha_1 \mathrm{e}^{-\tau_1 s}\hat{T}_{in} = 0 \tag{14}$$

and yields an $\mathbb{R}[s, \mathrm{e}^{-\tau_1 s}, \mathrm{e}^{-\tau_2 s}]$-module. However, in the present context the system is considered as an $\mathscr{H}_2$-module Σ, a system with both lumped and distributed delays.

An inspection of (14) reveals that the $\mathscr{H}_2$-module Σ is spectrally controllable since the exponential function $\mathrm{e}^{-\tau_1 s}$ has no zeros in $\mathbb{C}$. Also, the $\mathbb{R}(s)[\mathrm{e}^{-\tau_1 s}, \mathrm{e}^{-\tau_2 s}]$-module $\mathbb{R}(s)[\mathrm{e}^{-\tau_1 s}, \mathrm{e}^{-\tau_2 s}] \otimes_{\mathscr{H}_2} \Sigma$ is $\mathrm{e}^{-\tau_1 s}$-free with $\mathrm{e}^{-\tau_1 s}$-basis $\hat{T}$. Hence, according to Theorem 3 there exists a $\mathscr{H}_2$-basis. For the purpose of determining such a basis, first, new system variables

$$\hat{w}_1 = \hat{T} \tag{15a}$$

$$\hat{w}_2 = \alpha_1 \hat{T}_{in} + \alpha_2 \mathrm{e}^{-(\tau_2 - \tau_1)s}\hat{T} = \alpha_1 \hat{T}_{in} + \alpha_2 \mathrm{e}^{-\bar{\tau}_2 s}\hat{T} \tag{15b}$$

are introduced such that (14) can be written as

$$(s + \alpha_1 + \alpha_2)\hat{w}_1 - \mathrm{e}^{-\tau_1 s}\hat{w}_2 = (s + \alpha_1 + \alpha_2)\hat{w}_1 - \mathrm{e}^{-\bar{\tau}_1 s}\hat{w}_2 = 0. \tag{16}$$

In order to avoid negative delay amplitude in (15b), i.e. predictions, assumptions concerning the delay amplitudes had to be made. Here, the previously mentioned case with $\tau_2 > \tau_1$ was chosen. New delay operators $\mathrm{e}^{-\bar{\tau}_1 s}$ and $\mathrm{e}^{-\bar{\tau}_2 s}$ with delay amplitudes $\bar{\tau}_1 = \tau_1$ and $\bar{\tau}_2 = \tau_2 - \tau_1$ are specified in (15b) and (16) resulting in a free $\mathscr{H}_2$-module (cf. Lemma 1).

In a second step the exponential function $\mathrm{e}^{-\bar{\tau}_1 s}$ can be eliminated from (16) by defining

$$\hat{\hat{w}}_1 = \hat{w}_1 + \frac{\mathrm{e}^{\bar{\tau}_1(\alpha_1+\alpha_2)} - \mathrm{e}^{-\bar{\tau}_1 s}}{s + \alpha_1 + \alpha_2}\hat{w}_2 \tag{17a}$$

$$\hat{\hat{w}}_2 = \hat{w}_2. \tag{17b}$$

From the resulting system equation

$$(s + \alpha_1 + \alpha_2)\hat{\hat{w}}_1 - \mathrm{e}^{\bar{\tau}_1(\alpha_1+\alpha_2)}\hat{\hat{w}}_2 = 0,$$

it can easily be verified that $\hat{\bar{w}}_1$ is a basis, or $\hat{y} := \mathrm{e}^{-\bar{\tau}_1(\alpha_1+\alpha_2)}\hat{\bar{w}}_1$. The basis $\hat{y}$ can be expressed in terms of the original system variables $\hat{T}$ and $\hat{T}_{in}$ by inverting the transformations (15) and (17). Then, taking the inverse Laplace transform gives

$$y(t) = \alpha_1 \int_0^{\tau_1} \mathrm{e}^{-(\alpha_1+\alpha_2)\sigma} T_{in}(t-\sigma)\,d\sigma + \mathrm{e}^{-(\alpha_1+\alpha_2)\tau_1} T(t) + \alpha_2 \int_{\tau_2-\tau_1}^{\tau_2} \mathrm{e}^{-(\alpha_1+\alpha_2)(\sigma+\tau_1-\tau_2)} T(t-\sigma)\,d\sigma. \quad (18)$$

The temperatures in the tank and at the inlet are defined as

$$T(t) = y(t-\tau_1)$$
$$T_{in}(t) = \frac{1}{\alpha_1}\big(\dot{y}(t) + (\alpha_1+\alpha_2)y(t) - \alpha_2 y(t-\tau_2)\big).$$

Obviously, the basis y is simply a predicted value of T in which the prediction is calculated from past values of T and T_{in} (see (18)). As in the previous section where T_0 was the input, the expression of the input T_{in} in terms of the basis yields a linear system that can easily be controlled. The close relation to the reduction approach is evident. Based on a first-order error dynamics (12) a control law without predictions can be obtained.

5.3 *Two Control Inputs*

Even though the single input case is more challenging (at least for this example), a configuration with two control inputs is interesting in order to emphasize the generality of the results in Section 4. For simplicity assume $\beta_1 = \beta_2 = 1$. Then, the Laplace transform of (6) (with zero initial condition) reads

$$\big(s + \alpha_1 + \alpha_2 - \alpha_2 \mathrm{e}^{-\tau_2 s}\big)\hat{T} - \alpha_1 \mathrm{e}^{-\tau_1 s}\hat{T}_{in} - \alpha_1 \frac{1-\mathrm{e}^{-\tau_1 s}}{s}\hat{T}_0 = 0.$$

Some inspection of this equation reveals that the $\mathscr{H}_2$-module Σ is spectrally controllable and that $\mathbb{R}(s)[\mathrm{e}^{-\tau_1 s}, \mathrm{e}^{-\tau_2 s}] \otimes_{\mathscr{H}_2} \Sigma$ is free. Without longer calculations a (two-component) basis

$$\hat{y}_1 := \hat{T}_0 - s\hat{T}_{in}, \qquad \hat{y}_2 := \frac{1}{\alpha_1}\hat{T}$$

can be found giving the parameterization

$$T_{in}(t) = -\int_0^{\tau_1} y_1(t-\sigma)\,d\sigma - \alpha_2 y_2(t-\tau_2) + \dot{y}_2(t) + (\alpha_1+\alpha_2)y_2(t) \quad (19a)$$
$$T_0(t) = y_1(t-\tau_1) - \alpha_2\dot{y}_2(t-\tau_2) + \ddot{y}_2(t) + (\alpha_1+\alpha_2)\dot{y}_2(t) \quad (19b)$$
$$T(t) = \alpha_1 y_2(t) \quad (19c)$$

in the time domain.

Therefore, using T_{in} and T_0 as the input variables, desired trajectories $t \mapsto y_{1,r}(t)$ and $t \mapsto y_{2,r}(t)$ can be tracked by choosing a second order error dynamics

$$\ddot{e}_2 + 2\delta \dot{e}_2 + (\delta^2 + \omega^2) e_2 = 0 \tag{20}$$

for $e_2 = y_2 - y_{2,r}$ with real parameters $\delta > 0$ and ω as well as $e_1 = y_1 - y_{1,r} = 0$. The control laws directly follow from (19a) and (19b) by replacing y_1 with $y_{1,r}$ and $\ddot{y}_2$ using (20).

References

1. Artstein, Z.: Linear systems with delayed controls: A reduction. IEEE Trans. Autom. Control 27, 869–879 (1982)
2. Bartosiewicz, Z.: Approximate controllability of neutral systems with delays in control. J. Differ. Equ. 51, 295–325 (1984)
3. Bhat, K.P.M., Koivo, H.N.: Modal characterizations of controllability and observability in time delay systems. IEEE Trans. Autom. Control, 292–293 (1976)
4. Brethé, D., Loiseau, J.-J.: A result that could bear fruit for the control of delay-differential systems. In: Proc. 4th IEEE Mediterranean Symp. Control Automation, pp. 115–128 (1996)
5. Cohn, P.M.: Algebra, vol. 1. Wiley (1982)
6. Fiagbedzi, Y.A., Pearson, A.E.: A multistage reduction technique for feedback stabilizing distributed time-lag systems. Automatica 23, 311–326 (1987)
7. Fliess, M.: Some basic structural properties of generalized linear systems. Syst. Contr. Lett. 15, 391–396 (1990)
8. Fliess, M., Lévine, J., Martin, P., Rouchon, P.: Flatness and defect of non-linear systems: introductory theory and examples. Int. J. Contr. 61, 1327–1361 (1995)
9. Fliess, M., Mounier, H.: Controllability and observability of linear delay systems: an algebraic approach. ESAIM: Control, Optimization and Calculus of Variations 3, 301–314 (1998)
10. Gehring, N., Rudolph, J., Woittennek, F.: Controllability properties and prediction-free control of linear systems with incommensurate delays. In: PAMM - Proc. in Appl. Mathem. and Mechanics, vol. 11, pp. 809–810 (2011)
11. Gehring, N., Rudolph, J., Woittennek, F.: Prediction-free tracking control for systems with incommensurate lumped and distributed delays: Two examples. In: Proc. 10th IFAC Workshop on Time Delay Systems, pp. 37–42 (June 2012)
12. Glüsing-Lüerßen, H.: A behavioral approach to delay-differential systems. SIAM J. Control Optim. 35, 480–499 (1997)
13. Gu, K., Kharitonov, V.L., Chen, J.: Stability of Time-Delay Systems. Birkhäuser Boston (2003)
14. Hale, J.K., Verduyn Lunel, S.M.: Introduction To Functional Differential Equations. Springer-Verlag (1993)
15. Kolmanovskii, V., Myshkis, A.: Introduction to the Theory and Applications of Functional Differential Equations. Kluwer Academic Publishers (1999)
16. Kosugi, N., Suyama, K.: Finite spectrum assignment of systems with general delays. Int. J. Contr. 84, 1983–1995 (2011)
17. Kwon, W., Pearson, A.: Feedback stabilization of linear systems with delayed control. IEEE Trans. Autom. Control 25, 266–269 (1980)

18. Loiseau, J.-J.: Algebraic tools for the control and stabilization of time-delay systems. Annu. Rev. Contr. 24, 135–149 (2000)
19. Manitius, A.Z., Olbrot, A.W.: Finite spectrum assignment problem for systems with delays. IEEE Trans. Autom. Control 24, 541–552 (1979)
20. Mayne, D.Q.: Control of linear systems with time delay. Electron. Lett. 40, 439–440 (1968)
21. Mounier, H.: Algebraic interpretations of the spectral controllability of a linear delay system. Forum Math. 10, 39–58 (1998)
22. Rudolph, J.: Beiträge zur flachheitsbasierten Folgeregelung linearer und nichtlinearer Systeme endlicher und unendlicher Dimension. Shaker Verlag (2003)
23. Rudolph, J., Woittennek, F.: Flatness-based control without prediction: example of a vibrating string. In: PAMM – Proc. in Appl. Mathem. and Mechanics, vol. 10, pp. 629–630 (2010)
24. Watanabe, K.: Finite spectrum assignment of linear systems with a class of noncommensurate delays. Int. J. Contr. 47, 1277–1289 (1988)
25. Yosida, K.: Functional Analysis. Springer-Verlag (1980)
26. Zerz, E.: Topics in Multidimensional Linear Systems Theory. Springer-Verlag (2000)

New Integral Inequality and Its Application to Time-Delay Systems

Alexandre Seuret and Frédéric Gouaisbaut

Abstract. In the last decade, the Jensen's inequality has been intensively used in the context of time-delay or sampled-data systems since it is an appropriate tool to obtain tractable stability conditions expressed in terms linear matrix inequalities (LMI). However, it is also well-known that this inequality unavoidably introduces an undesirable conservatism in the stability conditions and looking at the literature, reducing this gap is still an open problem. In this paper, we propose an alternative inequality based on the Fourier Theory, more precisely on the Wirtinger's inequalities. It is showed in this chapter that they allow deriving a new integral inequality which is proved to encompass the Jensen's inequality. In order to illustrate the potential gain of employing this new inequality with respect to the Jensen's one, an application to time-delay analysis is provided.

1 Introduction

The last decade has shown an increasing research activity on time-delay systems analysis and control due to both emerging adapted theoretical tools and also practical issues in the engineering field and information technology (see [20], [14] and references therein). In the case of linear system, many techniques allow to derive efficient criteria proving the stability of such systems. Among them, two frameworks, different in their spirits have been recognized as a powerful methodologies. The first one relies on Robust Analysis. In this framework, the time delay/sampled data system is transformed into a closed loop between a stable nominal system and an uncertainty depending either on the delay or the sampling process (which

Alexandre Seuret · Frédéric Gouaisbaut
CNRS, LAAS, 7 avenue du Colonel Roche, 31077 Toulouse, France
e-mail: {aseuret,fgouaisb}@laas.fr
Alexandre Seuret
Univ de Toulouse, LAAS, F-31400 Toulouse, France

Frédéric Gouaisbaut
Univ de Toulouse, UPS, LAAS, F-31400, Toulouse, France

T. Vyhlídal, J.-F. Lafay, and R. Sipahi (eds.), *Delay Systems*,
Advances in Delays and Dynamics 1,
DOI: 10.1007/978-3-319-01695-5_3, © Springer International Publishing Switzerland 2014

is also modelled by a time varying delay). The perturbation element is then embedded into some norm-bounded uncertainties and the use of scaled small gain theorem [5, 15], IQCs [13], or separation approach [2] allows then to derive efficient stability criteria. The challenge is then to reduce the conservatism either by constructing elaborated interconnections which generally include state augmentation [1] or by using finer L_2 induced norm upperbound evaluation [15], often based on Cauchy-Schwartz inequality [2]. Another popular approach is the use of a Lyapunov function to prove stability. For sampled-data systems and time delay systems as well, the last decade has seen a tremendous emergence of research devoted to the construction of Lyapunov-Krasovskii functionals which aim at reducing the inherent conservatism of this approach. Several attempts have been done concerning the choice of the structure of V by choosing extended state based Lyapunov-Krasovskii functional [1, 11], discretized Lyapunov functionals [8] or discontinuous Lyapunov functions [18]. Apart the choice of V, an important source of conservatism relies also on the way to bound some cross terms arisen when manipulating the derivative of V. According to the literature on this subject (see [9, 17, 21] for some recent papers), a common feature of all these techniques is the use of slack variables [10] and more or less refined Jensen inequality [14, 21, 22]. At this point, it is clear that for both the two frameworks - Robust Analysis and Lyapunov functionals, a part of the conservatism comes from the use of Jensen inequality or Cauchy Schwartz inequalities, usually used to get tractable criteria. The objective of the present chapter is then to show how to use another class of inequalities called Wirtinger inequalities, which are well known in Fourier Analysis. Notice that this class of inequalities has been recently used to exhibit a new Lyapunov function to prove stability of sampled-data system [14]. In this paper, its use combined with some special properties of sampled data systems has led to some interesting criteria expressed in terms of LMIs, which are less conservative at least on examples. In the present paper, contrary to the work of [14], we do not construct some new Lyapunov functional. We aim rather at developing new inequalities to be used to reduce the conservatism when computing the derivative of V. Wirtinger inequalities allow to consider a more accurate integral inequalities which can include the Jensen's one as a special case. The resulting inequality depends not only on the state $x(t)$ and the delayed or sampled state but also on the integral of the state over a delay or sampling interval. This new signal is then directly integrated into a suitable classical Lyapunov function, highlighting so the features of Wirtinger inequality. Hence, it results some new stability criteria for time delay systems directly expressed in terms of LMIs.

Notations: Throughout the paper $\mathbb{R}^n$ denotes the n-dimensional Euclidean space with vector norm $|\cdot|$, $\mathbb{R}^{n\times m}$ is the set of all $n \times m$ real matrices, and the notation $P \succ 0$, for $P \in \mathbb{R}^{n\times n}$, means that P is symmetric and positive definite. The symmetric matrix $\begin{bmatrix} A & B \\ * & C \end{bmatrix}$ stands for $\begin{bmatrix} A & B \\ B^T & C \end{bmatrix}$. Moreover, for any square matrix $A \in \mathbb{R}^{n\times n}$, we define $\mathrm{He}(A) = A + A^T$.

2 Preliminaries

In the sequel, the following notations will be considered. For any real numbers $a < b$, we consider a differentiable function $\omega : [a,\ b] \to \mathbb{R}^n$ and the vector $\Omega(a,b)$ given by

$$\Omega(a,b) = \left[\omega^T(b),\ \omega^T(a),\ \tfrac{1}{b-a}\int_a^b \omega^T(u)\mathrm{d}u \right]^T.$$

2.1 Jensen's Inequality

Let recall the well-known Jensen's inequality.

Lemma 1. *For given symmetric positive definite matrices $R > 0$ and for any differentiable signal ω in $[a,\ b] \to \mathbb{R}^n$, the following inequality holds:*

$$\begin{aligned} \int_a^b \dot{\omega}(u)R\dot{\omega}(u)\mathrm{d}u &\geq \tfrac{1}{b-a}\left(\int_a^b \dot{\omega}(u)\mathrm{d}u\right)^T R \left(\int_a^b \dot{\omega}(u)\mathrm{d}u\right) \\ &\geq \tfrac{1}{b-a}\Omega^T(a,b)W_1^T RW_1\Omega(a,b), \end{aligned} \tag{1}$$

where $W_1 = \left[I\ {-I}\ 0 \right]$.

The proof is omitted and can be found in several reference books (see [8, 16]). In the context of time-delay systems and sampled-data systems, this inequality has been the core of several important contributions (see [8, 9] for time delay systems or [14] and references therein for sampled data systems) : it is usually used to bound some integral terms of the form $\int_a^b \dot{\omega}(u)R\dot{\omega}(u)\mathrm{d}u$ which arise when calculating the derivative of Lyapunov function. Naturally, it is likely to entail some inherent conservatism and several works have been devoted to the reduction of such a gap [3, 7]. In the present paper, we propose to use a different class of inequalities called Wirtinger inequalities in order to obtain new bounds for this integral and therefore to improve the results for the stability analysis of time-delay and/or sampled-data systems. In [19], a first result on the use of Wirtinger inequality was presented. The present paper proposes a more accurate analysis of this class of inequalities and its application to a larger class of problems.

2.2 Different Wirtinger Inequalities

In the literature [12], Wirtinger inequalities are referred as inequalities which estimate the integral of the derivative function with the help of the integral of the function. Often proved using Fourier analysis, it exists several versions which depend on the characteristics or constraints we impose on the function. Let firstly recall the Wirtinger inequalities adapted to our purpose.

Lemma 2. *Let z be a continuous functions and which admits a piecewise continuous derivative and for all matrix $R = R^T > 0$ in $\mathbb{R}^{n\times n}$, the following inequalities holds*

- *If* $z(a) = z(b)$ *and* $\int_a^b z(u)du = 0$, *then,*

$$\int_a^b \dot{z}^T(u)R\dot{z}(u)\mathrm{d}u \geq \frac{4\pi^2}{(b-a)^2}\int_a^b z^T(u)Rz(u)\mathrm{d}u, \tag{2}$$

and the equality holds when $z(u) = a_1 \sin\left(\frac{2\pi(u-a)}{b-a}\right) + b_1 \cos\left(\frac{2\pi(u-a)}{b-a}\right)$, *for any* a_1 *and* b_1 *in* $\mathbb{R}^n$.
- *If* $z(a) = z(b) = 0$, *then*

$$\int_a^b \dot{z}^T(u)R\dot{z}(u)\mathrm{d}u \geq \frac{\pi^2}{(b-a)^2}\int_a^b z^T(u)Rz(u)\mathrm{d}u, \tag{3}$$

and equality holds when $z(u) = a_1 \sin\left(\frac{\pi(u-a)}{b-a}\right)$ *for any* a_1 *in* $\mathbb{R}^n$.
- *If* $z(a) = 0$, *then*

$$\int_a^b \dot{z}^T(u)R\dot{z}(u)\mathrm{d}u \geq \frac{\pi^2}{4(b-a)^2}\int_a^b z^T(u)Rz(u)\mathrm{d}u, \tag{4}$$

and equality holds when $z(u) = a_1 \sin\left(\frac{\pi(u-a)}{2(b-a)}\right)$ *for any* a_1 *in* $\mathbb{R}^n$.

Proof. The proofs are based on the one-dimensional Wirtinger inequality provided in [12] and adapted to the case of vector function using the same method as in [14].

Remark 1. The inequality (4) has been already employed in [14], leading to a new type of Lyapunov-Krasovskii functionals for sampled-data systems. Our approach differs significantly from [14] since we only use this inequality for estimating an upper-bound of the derivative of the Lyapunov functional. An interesting future feature should the extension of our work by considering the techniques proposed by Fridman et al [14].

Finally, we have proposed three different inequalities which are very closed to Jensen's inequality in its essence. Nevertheless, the function has to meet several constraints which are not generally satisfied if, for instance, the function z is related to the states of a dynamical system. The next section shows how to overcome such a problem and how to construct relevant new inequalities.

3 Application of the Wirtinger's Inequalities

The objective of this section is twofold. On the first hand, based on Lemma 2 and the inequalities (2), (3) and (4), we aim at providing new tractable inequalities, which can be easily implemented into a convex optimization scheme. On the other hand, we propose some inequalities which are proved to be less conservative than Jensen's one. Indeed, recall that the objectives of the present paper are to obtain new lower bounds of the integral $\int_a^b \dot{\omega}(u)R\dot{\omega}(u)\mathrm{d}u$, in order to be consistent with the Jensen's inequality. Thus a first step consists in defining appropriate function z such that this integral appears naturally in the developments. Thus a necessary condition is that the function z as the following form

$$z(u) = \gamma\omega(u) - y(u), \tag{5}$$

where $\omega \in W[a,b]$ is the vector function which was employed in the original Jensen's inequality in Lemma 1, γ is a constant and $y(u)$ is a function of u and are chosen so that the function z satisfies the different constraints imposed in the first inequality of Lemma 2.

Based on the first Wirtinger inequality and choosing an appropriate function z, we propose a first corollary:

Corollary 1. *For a given symmetric positive definite matrix $R > 0$, any differentiable function ω in $[a,b] \to \mathbb{R}^n$, then the following inequality holds:*

$$\int_a^b \dot{\omega}(u)R\dot{\omega}(u)\mathrm{d}u \geq \frac{1}{(b-a)}\Omega^T(a,b)W_1^T RW_1\Omega(a,b), \tag{6}$$

where $W_1 = \begin{bmatrix} I & -I & 0 \end{bmatrix}$.

Proof. The main contribution here is to propose an appropriate interesting signal z of the form given in (5) which satisfies the conditions of of the first inequality of Lemma 2. Consider the following signal

$$z(u) = \omega(u) - \frac{1}{b-a}\int_a^b \omega(u)\mathrm{d}u - \left[\frac{u-a}{b-a} - \frac{1}{2}\right](\omega(b) - \omega(a)),$$

which has been built in order to satisfy the condition of Lemma 2. Then, computing inequality (2) leads to

$$\int_a^b \dot{z}^T(u)R\dot{z}(u)\mathrm{d}u = \int_a^b \dot{\omega}^T(u)R\dot{\omega}(u)\mathrm{d}u - \frac{1}{b-a}(\omega(b)-\omega(a))^T R(\omega(b)-\omega(a)).$$

Furthermore, applying the Jensen's inequality to the righthand side of (2) leads to

$$\int_a^b z^T(u)Rz(u)\mathrm{d}u \geq \frac{1}{b-a}\int_a^b z^T(u)\mathrm{d}uR\int_a^b z(u)\mathrm{d}u.$$

Noting that $\int_a^b z(u)\mathrm{d}u = 0$ allows to conclude the proof.

Remark 2. It is important to stress that the previous corollary is equivalent to the classical Jensen's inequality. Indeed the Jensen's inequality is included in the left-hand-side of the original Wirtinger's inequality when using the proposed signal z. Consequently, the use of this lemma seems not relevant as it is presented now. However it can be noticed that the previous corollary provides another proof of the Jensen's inequality.

The main problem comes from the constraint $\int_a^b z(u)\mathrm{d}u = 0$ which does not allow to give a lower bound of the left-hand side of (2). In the following corollary, we propose to use inequality (3) in which this constraint has been removed.

Corollary 2. *For a given symmetric positive definite matrix $R > 0$, any differentiable function ω in $[a,b] \to \mathbb{R}^n$, then the following inequality holds:*

$$\int_a^b \dot{\omega}(u) R \dot{\omega}(u) \mathrm{d}u \geq \frac{1}{b-a} \Omega^T(a,b) \left[W_1^T R W_1 + \pi^2 W_2^T R W_2 \right] \Omega(a,b), \tag{7}$$

where W_1 is given in Lemma 1, and $W_2 = \left[I/2 \; I/2 \; -I \right]$.

Proof. For any function $\omega \in W[a,b]$, consider a signal z given by

$$z(u) = \omega(u) - \frac{u-a}{b-a}\omega(b) - \frac{b-u}{b-a}\omega(a), \qquad \forall u \in [a,\ b].$$

By construction, the function $z(u)$ satisfies the conditions required to satisfy the inequality (3), i.e. $z(a) = z(b) = 0$. The computation of the left-hand-side of the inequality stated in the second inequality of Lemma 2 leads to:

$$\begin{array}{rl} \int_a^b \dot{z}^T(u) R \dot{z}(u) \mathrm{d}u & = \int_a^b \dot{\omega}^T(u) R \dot{\omega}(u) \mathrm{d}u - \frac{1}{b-a}(\omega(b) - \omega(a))^T R (\omega(b) - \omega(a)) \\ & = \int_a^b \dot{\omega}^T(u) R \dot{\omega}(u) \mathrm{d}u - \frac{1}{b-a} \Omega^T(a,b) W_1^T R W_1 \Omega(a,b). \end{array} \tag{8}$$

Consider now the right-hand side of the inequality (3). Applying the Jensen's inequality, it yields

$$\frac{\pi^2}{(b-a)^2} \int_a^b z^T(u) R z(u) \mathrm{d}u \geq \frac{\pi^2}{(b-a)^3} \left(\int_a^b z(u) \mathrm{d}u \right)^T R \left(\int_a^b z(u) \mathrm{d}u \right). \tag{9}$$

The last step of the proof consists in the computation of the integral $\int_a^b z(u)\mathrm{d}u$, which is obtained as follows

$$\begin{array}{rl} \int_a^b z(u) \mathrm{d}u & = -\left(\int_a^b \frac{(u-a)}{(b-a)} \mathrm{d}u \right) \omega(b) - \left(\int_a^b \frac{(b-u)}{(b-a)} \mathrm{d}u \right) \omega(a) + \int_a^b \omega(u) \mathrm{d}u \\ & = -(b-a) \left[\frac{1}{2}(\omega(b) + \omega(a)) - \frac{1}{b-a} \int_a^b \omega(u) \mathrm{d}u \right] \\ & = -(b-a) W_2 \Omega(a,b). \end{array} \tag{10}$$

Then applying (3), we obtain

$$\begin{array}{rl} \int_a^b \dot{\omega}^T(u) R \dot{\omega}(u) \mathrm{d}u \geq & \frac{1}{b-a} \Omega^T(a,b) W_1^T R W_1 \Omega(a,b) \\ & + \frac{\pi^2}{b-a} \Omega^T(a,b) W_2^T R W_2 \Omega(a,b), \end{array}$$

which concludes the proof of Corollary 2.

Remark 3. The previous corollary has been already presented in [19]. However its proof has been remarkably shorten.

Remark 4. The inequality (7) encompasses the Jensen's inequality since the matrix $\pi^2 W_2^T R W_2$ is positive definite and the term $\frac{1}{b-a}\Omega^T(a,b) W_1^T R \; W_1 \Omega(a,b)$ is exactly the right-hand of the Jensen's inequality. It is also worth noting that this improvement is allowed by using an extra signal $\int_a^b \omega(u)du$ and not only the signals $\omega(b)$ and $\omega(a)$. Therefore, it suggests that in order to be useful, this inequality should be combined with a Lyapunov functional where the signal $\int_a^b \omega(u)du$ appears explicitly.

The following corollary is based on the third inequality showed in Lemma 2, where only one constraint is imposed.

Corollary 3. *For a given symmetric positive definite matrix $R > 0$, any differentiable function ω in $[a,b] \to \mathbb{R}^n$, then the following inequality holds:*

$$\int_a^b \dot{\omega}^T(u)R\dot{\omega}(u)\mathrm{d}u \geq \frac{\pi^2}{4(b-a)}\Omega^T(a,b)W_3^T RW_3,\Omega(a,b) \tag{11}$$

where $W_3 = \begin{bmatrix} 0 & -I & I \end{bmatrix}$.

Proof. As proposed in [14], consider the signal $z(u) = \omega(u) - \omega(a)$. This function satisfies the condition of Lemma 2c. Then, we have

$$\int_a^b \dot{\omega}^T(u)R\dot{\omega}(u)\mathrm{d}u \geq \frac{\pi^2}{4(b-a)^2}\int_a^b (\omega(u)-\omega(a))^T R(\omega(u)-\omega(a))\mathrm{d}u.$$

Applying the Jensen's inequality yields the results.

Remark 5. It is relevant to try a comparison between Jensen's inequality and the previous one. But, since the matrix $\frac{\pi^2}{4}W_3^T RW_3 - W_1^T RW_1$ is not definite positive, the quantity

$$\Omega^T(a,b)\left(\frac{\pi^2}{4}W_3^T RW_3 - W_1^T RW_1\right)\Omega^T(a,b),$$

may be positive or negative, depending on the components of $\Omega(a,b)$. In that case, we cannot state that we improve Jensen's inequality. Consequently, this inequality will not be consider in the sequel.

4 Appropriate Inequalities for Robust Stability Analysis

In all the three inequalities (6), (7) and (11), the resulting lower bound is rational with respect to $(b-a)$, which is ill-posed when this quantity tends to zero. At a price of an increasing computational burden, an equivalent formulation depending linearly on $b-a$ can be drawn as follows. Noting that, for all matrices Y_i, $i \in \{1,2,3\}$ in $\mathbb{R}^{n\times 3n}$, the matrix $\frac{1}{b-a}(RW_i - (b-a)Y_i)^T R^{-1}(RW_i - (b-a)Y_i)$ is positive for all $i \in \{1,2,3\}$, it yields

$$\tfrac{1}{(b-a)}W_i^T RW_i \geq Y_i^T W_i + W_i^T Y_i - (b-a)Y_i^T R^{-1} Y_i, \quad i \in \{1,2,3\}.$$

This inequality turns out to be relevant in the context of time-delay systems or sampled-data systems as it will be explained in Section 5. Applying the same inequality to the second term of the inequalities (6), (7) and (11) , one modifies the previous corollaries as follows.

Corollary 4. *For any matrix $R>0$, any differentiable signal ω in $[a,b]\to\mathbb{R}^n$ and for any matrices Y_1 in $\mathbb{R}^{n\times 3n}$, the following inequality holds:*

- *for any matrices Y_1 in $\mathbb{R}^{n\times 3n}$, the following inequality holds:*

$$\int_a^b \dot{\omega}(u)R\dot{\omega}(u)\mathrm{d}u \geq \Omega^T(a,b)\left[\mathrm{He}\{Y_1^T W_1\} - (b-a)\left(Y_1^T R^{-1} Y_1\right)\right]\Omega(a,b). \quad (12)$$

- *for any matrices Y_1 and Y_2 in $\mathbb{R}^{n\times 3n}$, the following inequality holds:*

$$\begin{aligned}\int_a^b \dot{\omega}(u)R\dot{\omega}(u)\mathrm{d}u \geq \Omega^T(a,b)&\left[\mathrm{He}\{Y_1^T W_1 + \pi^2 Y_2^T W_{20}\}\right.\\ &\left.-(b-a)\left(Y_1^T R^{-1} Y_1 + \pi^2 Y_2^T R^{-1} Y_2\right)\right]\Omega(a,b).\end{aligned} \quad (13)$$

- *for any matrices Y_3 in $\mathbb{R}^{n\times 3n}$, the following inequality holds:*

$$\int_a^b \dot{\omega}(u)R\dot{\omega}(u)\mathrm{d}u \geq \frac{\pi^2}{4}\Omega^T(a,b)\left[\mathrm{He}\{Y_3^T W_3\} - (b-a)\left(Y_3^T R^{-1} Y_3\right)\right]\Omega(a,b). \quad (14)$$

Remark 6. In the literature, several by-products of the Jensen's inequality have been proposed and employed (see for example [11,17] and references therein). Obviously, the same by-products could be derived from the Corollaries proposed in this section and therefore will not be presented in the present chapter.

In the following, we will show how these inequalities can be applied to the stability analysis of time-delay systems. As expected, we will show that the use of these new inequalities reduces the conservatism of the stability conditions. It has to be noticed that, in theses new inequalities, the functions to be considered are $\omega(b)$, $\omega(a)$ and $\frac{1}{b-a}\int_a^b \omega(u)du$. The two first signals appear naturally in the context of discrete time-delay or sampled-data systems but not the last one. It only appears in the context of distributed time-delay systems. At first sight, an expected consequence is that these new inequalities only help in the context of distributed time-delay systems. However we will present two solutions dealing with the context of discrete-time delay. The objective will be to show how this third signal is in relation with these classes of systems.

5 Application to the Stability Analysis of Time-Delay Systems

Before entering into the details of this section, it is important to stress that the present paper does not focus on the development of new Lyapunov-Krasovskii functionals. The present section on the stability analysis of time-delay systems highlights the potential gains of applying the inequalities showed in the previous section. In particular the following basic problems are under consideration:

- Stability analysis of systems with discrete and distributed constant delays,
- Robust stability analysis of systems with unknown delays,

5.1 Systems with Constant and Known Delay

We present in this sub-section a first stability result for time delay systems, which is based on the use of the Wirtinger inequalities developed in Sections 3 and 4. This approach is based on a slightly modified Lyapunov-Krasovskii functional and allows us to establish the main theorem for the robust delay range stability analysis. Consider a linear time-delay system of the form:

$$\begin{cases} \dot{x}(t) = Ax(t) + A_d x(t-h) + A_D \int_{t-h}^{t} x(s)ds, & \forall t \geq 0, \\ x(t) = \phi(t), & \forall t \in [-h, 0], \end{cases} \tag{15}$$

where $x(t) \in \mathbb{R}^n$ is the state vector, ϕ is the initial condition and A, A_d, $A_D \in \mathbb{R}^{n\times n}$ are constant matrices. The delay is assumed to be known and constant.

Remark 7. In the literature, systems with discrete and distributed delays are subject to different delays, i.e., the value of the discrete delay and the distributed delay are not considered equal. Despite, in the present article we will only consider the case of a single value of the two delays. The motivation to this restrictive choice comes form the following statement. Our goal is to show that a generic structure of Lyapunov-Krasovskii functionals can deal with these two problems with the same efficiency.

Based on the previous inequality and classical results on Lyapunov-Krasovskii functionals, two stability theorems are provided.

Theorem 1. *For a given constant delay h, assume that there exist $n \times n$ matrices $P = P^T > 0$, $S = S^T > 0$, $R = R^T > 0$, Q and $Z = Z^T$ such that the following LMIs are satisfied*

$$\Pi_1(h) = \begin{bmatrix} P & Q \\ * & Z + S/h \end{bmatrix} > 0, \tag{16}$$

$$\Pi_2^1(h) = \Pi_2^0(h) - \tfrac{1}{h}\left[W_1^T R W_1 + \pi^2 W_2^T R W_2\right] < 0, \tag{17}$$

where

$$\Pi_2^0(h) = \begin{bmatrix} \Delta_2^0 & PA_d - Q & h(PA_D + A^T Q + Z) \\ * & -S & h(A_d^T Q - Z) \\ * & * & h^2(A_D Q + Q^T A_D^T) \end{bmatrix} + h \begin{bmatrix} A^T \\ A_d^T \\ hA_D^T \end{bmatrix} R \begin{bmatrix} A^T \\ A_d^T \\ hA_D^T \end{bmatrix}^T,$$

and $\Delta_2^0 = PA + A^T P + S + Q + Q^T$. Then the system (21) is asymptotically stable for the constant delay h.

Proof. Consider a Lyapunov-Krasovskii functional of the form

$$V(x_t, \dot{x}_t) = \begin{bmatrix} x(t) \\ \int_{t-h}^{t} x(s)\mathrm{d}s \end{bmatrix}^T \begin{bmatrix} P & Q \\ * & Z \end{bmatrix} \begin{bmatrix} x(t) \\ \int_{t-h}^{t} x(s)\mathrm{d}s \end{bmatrix} + \int_{t-h}^{t} x^T(s)Sx(s)\mathrm{d}s \\ + \int_{t-h}^{t} (h - t + s)\dot{x}^T(s)R\dot{x}(s)\mathrm{d}s \tag{18}$$

The previous functional refers to a classical type of functionals to derive delay-dependent stability conditions (see for instance [8]). It is interesting to note that this class of functionals employed a signal of the form $\int_{t-h}^{t} x(s)ds$ which is related to the third signal introduced in the previous section. If the matrices Q and Z are set to zero, one recovers the usual functional employed in the literature. First of all, following [8] and using Jensen inequality, a lower-bound for V can be easily found:

$$V(x_t,\dot{x}_t) \geq \begin{bmatrix} x(t) \\ \int_{t-h}^{t} x(s)\mathrm{d}s \end{bmatrix}^T \Pi_1(h) \begin{bmatrix} x(t) \\ \int_{t-h}^{t} x(s)\mathrm{d}s \end{bmatrix} \\ + \int_{t-h}^{t}(h-t+s)\dot{x}^T(s)R\dot{x}(s)\mathrm{d}s, \tag{19}$$

and it is clear that the positive definiteness of the matrices P, S, R and $\Pi_1(h)$ implies the positive definiteness of the functional V. Classical computations show that the derivative of the functional along the trajectories of the system (15) satisfies

$$\dot{V}(x_t,\dot{x}_t) = \xi^T(t)\Pi_2^0(h)\xi(t) - \int_{-h}^{0}\dot{x}^T(t+s)R\dot{x}(t+s)\mathrm{d}s.$$

where Π_2^0 is defined in Theorem 1 and

$$\xi(t) = \begin{bmatrix} x(t) \\ x(t-h) \\ \frac{1}{h}\int_{t-h}^{t} x(s)ds \end{bmatrix}.$$

Applying Corollary 2, respectively, the following upper-bounds of the derivative of the functional is then obtained:

$$\dot{V}(x_t,\dot{x}_t) \leq \xi^T(t)\Pi_2^1(h)\xi(t), \tag{20}$$

where $\Pi_2^1(h)$ is defined in (17). Then if the stability condition from Theorem 1 is satisfied, the system (15) is asymptotically stable.

5.2 Systems with Constant and Unknown Delay: Delay Range Stability

Consider the case of a system with a single discrete delay (i.e. $A_D = 0$). Then we have

$$\begin{cases} \dot{x}(t) = Ax(t) + A_d x(t-h) & \forall t \geq 0, \\ x(t) = \phi(t) & \forall t \in [-h_{\max}, 0], \end{cases} \tag{21}$$

The delay h is a positive constant scalar which satisfies, from now on, the constraint $h \in [h_{\min}, h_{\max}]$ where $h_{\min}$, $h_{\max}$ are given positive constants. In the following, we aim at assessing stability of system (21) with the delay constraints described above via the an appropriate Lyapunov-Krasovskii functional. The following result holds.

Theorem 2. *For an uncertain constant delay $h \in [h_{min}, h_{max}]$, assume that there exist $n \times n$-matrices $P = P^T > 0$, $S = S^T > 0$, $R = R^T > 0$ Q and $Z = Z^T$ and two $3n \times n$-matrices Y_1 and Y_2, such that $\Pi_1(h_{max}) > 0$ and*

$$\Pi_3(h) = \begin{bmatrix} \Pi_3^0(h) - He\{Y_1W_1 + \pi^2 Y_2 W_2\} & hY_1 & \pi^2 hY_2 \\ * & -hR & 0 \\ * & * & -\pi^2 hR \end{bmatrix} < 0, \tag{22}$$

for all $h \in \{h_{min}, h_{max}\}$ where Π_1 is given in (16) where $\Pi_3^0 = \Pi_2^0$ with $A_D = 0$. Then, the system (21) is asymptotically stable for all constant delay $h \in [h_{min},\ h_{max}]$.

Proof. The proof uses the same Lyapunov-Krasovskii functional as in Theorem 1. Similar calculations lead to

$$\dot{V}(x_t, \dot{x}_t) = \xi^T(t)\Pi_3^0(h)\xi(t) - \textstyle\int_{-h}^{0} \dot{x}^T(t+s)R\dot{x}(t+s)\mathrm{d}s.$$

Applying Corollary 4 and the Schur complement ensure that the derivative of the Lyapunov-Krasovskii functional along the trajectories of (21) is negative definite if $\Pi_3(h)$ is negative definite for this h. Since the matrix Π_3 is affine with respect to h. The conditions $\Pi_3(h_{min}) < 0$ and $\Pi_3(h_{max}) < 0$ ensures that $\Pi_3(h) < 0$ for all $h \in [h_{min},\ h_{max}]$.

5.3 Examples

The purpose of the following section is to show how the inequalities given in Section 3 leads to a relevant reduction of conservatism in the stability condition. In it is important to stress that, our goal is not to find the best result on several examples. Our goal is to show the gap between existing results based on the Jensen's inequality and the ones proposed in the article.

5.3.1 Constant Discrete Delay Case

In this section, we will consider the two following examples. On the first hand, the linear time-delay systems (21) with the matrices with the matrices

$$A = \begin{bmatrix} -2 & 0 \\ 0 & -0.9 \end{bmatrix}, \ A_d = \begin{bmatrix} -1 & 0 \\ -1 & -1 \end{bmatrix}, \ A_D = \begin{bmatrix} 0 & 0 \\ 0 & 0 \end{bmatrix}. \tag{23}$$

is under consideration. This system is a well-known delay dependent stable system, that is the delay free system is stable and the maximum allowable delay $h_{\max} = 6.1721$ can be easily computed by delay sweeping techniques. To demonstrate the effectiveness of our approach, results are compared to the literature and are reported in Table 1. All papers except [13] use Lyapunov theory in order to derive stability criteria. Many recent papers give the same result since they are intrinsically based on the same Lyapunov functional and use the same bounding cross terms technique i.e. Jensen inequality. Some papers [2, 22], which use an augmented Lyapunov can go further but with a numerically increasing burden, compared to our proposal. The robust approach [13] gives a very good upper-bound with a similar computational complexity than our result. The discretized Lyapunov functional proposed by [8] gives a delay upperbound very closed to the maximum allowable delay with an increasing numerical complexity.

Table 1 Results for Example (23) for constant delay h

Theorems	h_{max}	number of variables
[6]	4.472	$1.5n^2+1.5n$
[9]	4.472	$3n^2+3n$
[21]	4.472	$2.5n^2+1.5n$
[22]	4.472	$3n^2+3n$
[13]	6.1107	$1.5n^2+9n+9$
[2]	5.120	$7n^2+4n$
[22]	5.02	$18n^2+18n$
[11]	4.97	$69n^2+5n$
[8] (N=1)	6.059	$5.5n^2+2.5n$
[1]	5.120	$6.5n^2+3.5n$
Th.1	5.901	$3n^2+2n$
Th.1 with $Q=Z=0$	4.472	$1.5n^2+1.5n$

Theorem 2 addresses also the stability of systems with *interval delays*, which may be unstable for small delays (or without delays) as it is illustrated with the second example.

$$A=\begin{bmatrix}0 & 1\\ -2 & 0.1\end{bmatrix},\quad A_d=\begin{bmatrix}0\ 0\\ 1\ 0\end{bmatrix},\quad A_D=\begin{bmatrix}0\ 0\\ 0\ 0\end{bmatrix}. \tag{24}$$

As $Re(eig(A+A_d))=0.05>0$, the delay free system is unstable and in this case, the results to assess stability of this system are much more scarce. They are often related to robust analysis [2] or discretized Lyapunov-Krasovskii functionals [8]. The results are reported in Table 2. In this example, our result gives better result than [8] and [2] with a fewer numbers of variables to be optimized. Notice that with the discretization technique from [8], increasing N yields to a better result approaching the analytical bound.

Table 2 Results for Example (25) for constant delay h

Theorems	h_{min}	h_{max}	number of variables
[9]	$\emptyset$	$\emptyset$	$3n^2+3n$
[2]	0.102	1.424	$7n^2+4n$
[8] (N=1)	0.1006	1.4272	$5.5n^2+2.5n$
Th.1	0.1006	1.473	$3n^2+2n$
Th.1 with $Q=Z=0$	$\emptyset$	$\emptyset$	$1.5n^2+1.5n$

5.3.2 Constant Distributed Delay Case

Consider the linear time-delay systems (15) with the matrices taken from [23]

$$A = \begin{bmatrix} 0.2 & 0 \\ 0.2 & 0.1 \end{bmatrix}, A_d = \begin{bmatrix} 0 & 0 \\ 0 & 0 \end{bmatrix}, A_D = \begin{bmatrix} -1 & 0 \\ -1 & -1 \end{bmatrix} \tag{25}$$

In [23], stability is guaranteed for delays over the interval [0.209, 1.194]. The stability conditions proposed in [4] ensures stability for any constant delay $h \in$ [0.2001, 1.633]. Using our new inequality, Theorem 1 ensures stability for all constant delay which belongs to the interval [0.200, 1.877] which is much larger than the interval found in the literature. Note that an eigenvalue analysis provides that the system remains stable for all constant delays in the interval [0.200, 2.04]. This shows the potential of the new inequality.

6 Conclusions

In this paper, we have provided new useful inequalities which encompass the Jensen's inequality. In combination with a simple Lyapunov-Krasovskii functional, this inequality leads to new stability criteria for linear time delay system systems. This new result has been expressed in terms of LMIs and has shown on numerical examples a large improvement of existing results using only a limited number of matrix variables. More generally, this new inequality could be coupled to more elaborated Lyapunov functionals.

References

1. Ariba, Y., Gouaisbaut, F.: An augmented model for robust stability analysis of time-varying delay systems. International Journal of Control 82(11), 1616–1626 (2009)
2. Ariba, Y., Gouaisbaut, F., Johansson, K.H.: Stability interval for time-varying delay systems. In: 49th IEEE Conference on Decision and Control (CDC), pp. 1017–1022 (2010)
3. Briat, C.: Convergence and Equivalence results for the Jensen's inequality - Application to time-delay and sampled-data systems. IEEE Trans. on Automatic Control 56(7), 1660–1665 (2011)
4. Chen, W.-H., Zheng, W.X.: Delay-dependent robust stabilization for uncertain neutral systems with distributed delays. Automatica 43(1), 95–104 (2007)
5. Fujioka, H.: Stability analysis of systems with aperiodic sample-and-hold devices. Automatica 45(3), 771–775 (2009)
6. Gouaisbaut, F., Peaucelle, D.: A note on stability of time delay systems. In: 5th IFAC Symposium on Robust Control Design, ROCOND 2006 (2006)
7. Gouaisbaut, F., Peaucelle, D.: Robust stability of time-delay systems with interval delays. In: 46th IEEE Conference on Decision and Control (2007)
8. Gu, K., Kharitonov, V.-L., Chen, J.: Stability of time-delay systems. Birkhauser (2003)
9. He, Y., Wang, Q.G., Xie, L., Lin, C.: Further improvement of free-weighting matrices technique for systems with time-varying delay. IEEE Trans. on Automatic Control 52(2), 293–299 (2007)

10. Jiang, X., Han, Q.L.: Delay-dependent robust stability for uncertain linear systems with interval time-varying delay. Automatica 42(6), 1059–1065 (2006)
11. Kim, J.H.: Note on stability of linear systems with time-varying delay. Automatica 47(9), 2118–2121 (2011)
12. Kammler, W.D.: A first Course in Fourier Analysis. Cambridge University Press, Cambridge (2007)
13. Kao, C.Y., Rantzer, A.: Stability analysis of systems with uncertain time-varying delays. Automatica 43(6), 959–970 (2007)
14. Liu, K., Fridman, E.: Wirtinger's Inequality and Lyapunov-Based Sampled-Data Stabilization. Automatica 48(1), 102–108 (2012)
15. Mirkin, L.: Exponential stability of impulsive systems with application to uncertain sampled-data systems. IEEE Trans. on Automatic Control 52(6), 1109–1112 (2007)
16. Niculescu, S.I.: Delay Effects on Stability. A Robust Control Approach. LNCIS, vol. 269. Springer, Heidelberg (2001)
17. Park, P., Ko, J.W., Jeong, C.: Reciprocally convex approach to stability of systems with time-varying delays. Automatica 47(1), 235–238 (2011)
18. Seuret, A.: A novel stability analysis of linear systems under asynchronous samplings. Automatica 48(1), 177–182 (2012)
19. Seuret, A., Gouaisbaut, F.: On the use of the Wirtinger's inequalities for time-delay systems. In: Proc. of the 10th IFAC Workshop on Time Delay Systems, IFAC TDS 2012 (2012)
20. Sipahi, R., Niculescu, S., Abdallah, C.T., Michiels, W., Gu, K.: Stability and Stabilization of Systems with Time Delay. IEEE Control Systems 31(1), 38–65 (2011)
21. Shao, H.: New delay-dependent stability criteria for systems with interval delay. Automatica 45(3), 744–749 (2009)
22. Sun, J., Liu, G.P., Chen, J., Rees, D.: Improved delay-range-dependent stability criteria for linear systems with time-varying delays. Automatica 46(2), 466–470 (2010)
23. Zheng, F., Frank, P.M.: Robust control of uncertain distributed delay systems with application to the stabilization of combustion in rocket motor chambers. Automatica 38(3), 487–497 (2002)

A Matrix Technique for Robust Controller Design for Discrete-Time Time-Delayed Systems

Ali Fuat Ergenc

Abstract. A matrix method is introduced for determination of robust-stability zones of the general linear time invariant discrete-time dynamics with large delays against parametric uncertainties. The technique employs Kronecker Product and unique properties of palindrome polynomials, which are subset of self-inversive polynomials. These polynomials possess interesting features on the distribution of its zeros. The main motivation in this chapter is to develop a practical tool for determination of robust stability zones against parametric uncertainties and dominant pole assignment of systems in discrete-time domain. A sufficient condition for robust stability and dominant pole assignment is presented. The procedure for the solution is demonstrated via some example case studies.

1 Introduction

In industry, most of the processes and systems incorporate unavoidable time delays originated from the nature of the system or the feedback mechanisms. Several decades, many researchers have investigated over the design of controllers to tackle the stability and performance of the closed-loop systems with time-delays. In many studies, the controllers are designed in continuous time domain and then discretized to be programmed on digital controllers such as industrial computers and Programmable Automation Controllers(PAC). Another way is to discretize the mathematical model of the process with a sampling period of realization and design the controllers in z-domain [1, 17]. Wide use of computers lead to design of digital controllers for various processes, which is a very well studied topic in the literature [11–14, 22, 28, 29]. In many studies, Lyapunov based approaches are used to guarantee the stability of the time delayed systems [7, 27]. Linear Matrix Inquality(LMI)'s solutions are also utilized to solve robustness problems in discrete-time

Ali Fuat Ergenc
Istanbul Technical University, Control Engineering Department, 34469 Istanbul
e-mail: ali.ergenc@itu.edu.tr

T. Vyhlídal, J.-F. Lafay, and R. Sipahi (eds.), *Delay Systems*,
Advances in Delays and Dynamics 1,
DOI: 10.1007/978-3-319-01695-5_4, © Springer International Publishing Switzerland 2014

time delayed systems [8, 9, 15]. Digital Smith predictors may provide nondelayed-feedback to overcome the time delay effects in the processes [1, 13], but they are very sensitive to the structural variations of the system. There are also some studies with pre-chosen controller types such as PID controllers for time delayed systems in the literature [6, 17, 25, 26, 28, 29]. In industry PID controllers are widely used due to its vast literature and already built in libraries in commercial controllers. Most of the time, sampling period is chosen with respect to dynamics of the systems, where time delays are relatively small. Since the degree of the discrete-time system is a function of the delay and the sampling period, it drastically increases when the delays are larger with respect to the sampling period. Some of the methods, using state space approach for the stability analysis, offer a technique that assigns additional states that corresponding delayed states [17]. In these methods, the size of the state-space matrices increases excessively when the delay is large. Even though there are relatively simple and effective methods in the literature, their approaches are conservative [19–21]. Furthermore, the main concern behind these studies is to keep the stability of the system intact. The location of the dominant poles of such systems is not very well studied. Linear time-invariant single time delayed system (LTI-STDS) with parametric uncertainties is presented in state space as follows,

$$\dot{\mathbf{x}}_c(t) = \mathbf{A}_c(\mathbf{q})\mathbf{x}_c(t) + \mathbf{B}_c(\mathbf{q})\mathbf{x}_c(t-\tau) \tag{1}$$

where $\mathbf{x}_c \in \mathbb{R}^n$, $\mathbf{A}_c(\mathbf{q})$, $\mathbf{B}_c(\mathbf{q})$ are matrices in $\mathbb{R}^{n\times n}$, $\mathbf{q} \in \mathbb{R}^p$ and the time delay $\tau \in \mathbb{R}^+$.

In this study, we address the dominant pole placement problem of linear time invariant, discrete-time time delayed system (LTI-DTDS), which is derived by discretization of (1), in the general form of following,

$$\mathbf{x}(k+1) = \mathbf{A}(\mathbf{q})\mathbf{x}(\mathbf{k}) + \mathbf{B}(\mathbf{q})\mathbf{x}(k-m) \tag{2}$$

where $\mathbf{x} \in \mathbb{R}^n$, $\mathbf{A}(\mathbf{q})$, $\mathbf{B}(\mathbf{q})$, are matrices in $\mathbb{R}^{n\times n}$, $\mathbf{q} \in \mathbb{R}^p$ and the time delay is $m \in \mathbb{Z}^+$. The time delay m, which denotes the value of the delay in terms of the number of sampling period (T_s, $(m)T_s \leq \tau < (m+1)T_s$ is subset of integer numbers. We indicate vector and matrix quantities in boldface capital notation. In the text, open unit disc, unit circle and outside of unit circle are referred as $\mathbb{D}, \mathbb{T}, \mathbb{S}$, respectively. Therefore, $\mathbb{D} \bigcup \mathbb{T} \bigcup \mathbb{S} = \mathbb{C}$ represents the entire complex plane.

The problem of dictating stability and dominant pole assignment is transformed into assignment of a certain number of zeros in $\mathbb{D}$ of a polynomial derived from the system equations. The novelty in this method is that the dominant pole assignment problem of high degree discrete-time system is reduced to a problem of root distribution of a polynomial with a smaller degree. The vector $\mathbf{q}$ includes both uncertainties of the system and the feedback controller parameters. The method is based on unique properties of palindrome polynomials which are natural outcomes of the Kronecker Product operations on the state-space matrices. In the procedure described in following sections, the polynomial, which is derived from state matrices, presents interspersed zeros on the unit-circle under certain conditions.

The paper is divided in five sections. In section 2 problem statement of the study is given. In section 3 robust-stability criteria for LTI-DTDS under parametric uncertainties is presented. Section 4 is on dominant-pole assignment. In the final section, section 5, illustrative example case studies are given.

2 The Problem Statement

The characteristic equation of the discrete-time system in (2) is,

$$CE(z,\mathbf{q}) = \det\left[z\mathbf{I} - \mathbf{A}(\mathbf{q}) - \mathbf{B}(\mathbf{q})z^{-m}\right] = \Sigma_{j=1}^{m+n} a_j z^j$$
$$= a_{m+n}z^{m+n} + a_{m+n-1}z^{m+n-1} \ldots a_1 z + a_0 = 0 \tag{3}$$

where $\mathbf{A}(\mathbf{q})$ and $\mathbf{B}(\mathbf{q})$ are the system matrices with the parametric uncertainties and tunable feedback controller parameters, m is the order of the delay in terms of the sampling period T_s and n is the dimension of the dynamics.

Definition 1. The stability posture of the system in (2) is determined by the number of characteristic roots of discrete-time system (3) in $\mathbb{D}$. This number is naturally a function of the delay, the parametric uncertainties and controller parameters, which are the parameters in (2). The system is stable when all of the roots of (3) is on $\mathbb{D}$. For any stability switching, a characteristic root z must exist on $\mathbb{T}$. This is a known fact of the root continuity argument in the parametric space. In our study, we assume that the delay is constant, thus the order of the polynomial is fixed

In order to asses the stability of the system we have to determine all the control parameter set which present robustness against parametric uncertainties. Furthermore, we aim to assign a boundary for the dominant poles of the system to obtain desired performance of closed-loop system. The desired performance of the system here is described as the upper bound of the settling time of the output to the reference. In continuous-time domain ($s-domain$), the measure of settling time is the distance of the dominant pole from the imaginary axis. In other words, the real part of the dominant pole determines the settling time. In discrete-time domain ($z-domain$), it is the distance from the unit circle $\mathbb{T}$. Strictly saying, the dominant poles of the discrete-time system should lie on predetermined circle with a radius of $r < 1$. If the radius is chosen as unit length ($r = 1$) this problem is only a stability problem. In order to achieve this, first, we convert the problem of examining crossing of characteristic roots of the system of the unit circle into crossing a predetermined circle with radius r. Naturally, if the delay is too large compared to sampling period(T_s) the number of the roots to be determined is excessively large. At this point, we utilize Kronecker Multiplication to reduce the number of the roots to ease the procedure for dominant pole assignment.

Kronecker Product of Two Matrices:

Kronecker Product of two square matrices $\mathbf{A}(n_1 \times n_1)$ and $\mathbf{B}(n_2 \times n_2)$ is defined as [2, 3],

$$\mathbf{A} \otimes \mathbf{B} = \begin{bmatrix} a_{1,1}B & \cdots & a_{1,n_1}B \\ a_{2,1}B & \cdots & a_{2,n_1}B \\ \vdots & \ddots & \vdots \\ a_{n_1,1}B & \cdots & a_{n_1,n_1}B \end{bmatrix}$$

where $\mathbf{A} \in \mathbb{R}^{n_1 \times n_1}, \mathbf{B} \in \mathbb{R}^{n_2 \times n_2}$. Here $\otimes$ denotes *Kronecker product* operation. The important property of the Kronecker Product of $\mathbf{A}$ and $\mathbf{B}$ is that this new square matrix

$$\mathbf{A} \otimes \mathbf{B} \in \mathbb{R}^{(n_1 \cdot n_2) \times (n_1 \cdot n_2)}$$

has $n_1 \cdot n_2$ eigenvalues which are indeed pair-wise combinatoric multiplications of the n_1 eigenvalues of $\mathbf{A}$ and n_2 eigenvalues of $\mathbf{B}$. That is, the Kronecker product operation, in fact, induces the "eigenvalue multiplication" character to the matrices. This property is used in the core of our technique.

Definition 2. Adjunct Equation (AE) of the system in (2), is defined as follows :

$$AE(z, \mathbf{q}) = \det\left[(z\mathbf{I} - \mathbf{A}(\mathbf{q})) \otimes (z^{-1}\mathbf{I} - \mathbf{A}(\mathbf{q}) - \mathbf{B}(\mathbf{q}) \otimes \mathbf{B}(\mathbf{q}))\right] = 0 \tag{4}$$

Theorem 1. *For the system given in (2) if the system is switching stability the following findings are equivalent:*

1. *At least a pair of unitary complex numbers* $\mathbf{z} = \{z_j\} \in \mathbb{T}$, $|z| = 1$, *satisfies AE.*
2. *There exists at least one pair of unitary characteristic roots, z, of (3).*
3. *There exists a corresponding parameter vector* $\mathbf{q} \in \mathbb{R}^p$, *where* $\langle \mathbf{q}, z \rangle$ *holds.*

Proof. Proof of the theorem can be found in [5]

Equation (4) is a necessary condition for a unimodular eigenvalue ($z \in \mathbb{T}$) of (2) for a certain $\mathbf{q} \in \mathbb{Q}^p$ generating a stability switching. The equation has a degree of $2n^2 << (n+m)$ for large delays and it is smaller sized than original characteristic equation ($e.g., \tau > 10T_s$. The procedure is now considerably simplified to find $\mathbf{z} \in \mathbb{T}$ solutions of (4) with respect to certain $\mathbf{q}$.

After stating the basis, we propose a new approach for the determination of the parameter space $\mathbb{Q}^p$ where the system (2) has dominant poles on a predetermined disk with the radius r.

3 Robust Stability Analysis

A linear time delayed discrete-time system is "robust stable" when all the characteristic roots of (3) lie on the $\mathbb{D}$ for $\forall \mathbf{q} \in \mathbb{Q}^p$. Determination of the characteristic roots of (2) for all $\mathbf{q} \in \mathbb{R}^p$ is very hard and almost impossible when the system has high degree due to large time-delay in the system. In this study, we present a relatively simple method which states the sufficient conditions for robust stability of (2).

Theorem 2. *(Robust Stability) A system given in (2) is stable for* $\forall \mathbf{q} \in \mathbb{Q}^p$ *if*

1. *There exists a parameter vector* $\mathbf{q} \in \mathbb{Q}^p$ *that renders all of the roots of the following equation lie on* $\mathbb{D}$ *(* $\forall z \in \mathbb{D}$*)*
$$\det\left[z\mathbf{I} - \mathbf{A}(\mathbf{q}) - z^{-m}\mathbf{B}(\mathbf{q})\right] = 0$$
2. $AE(z, \mathbf{q})$ *has no roots* $\mathbf{z} = \{z_j\} \in \mathbb{T}$ *for any* $\mathbf{q} \in \mathbb{Q}^p$*.*

Proof. Proof of the theorem can be derived using unitary root crossing concept. In the first item, we state initial condition for the stability of the system for existence of such set. In the second item, stating that AE has no roots on $\mathbb{T}$, we guarantee that characteristic equation (3) has no $|z| = 1$ roots on the unit-circle.

This is a conventional theorem for robust-stability. The main idea is that if AE does not have any unimodular roots then CE has no stability switching roots. In rest of the paper, we emphasize on some distinctive properties of AE which will ease the robustness analysis of the time delayed discrete-time systems against parametric variations. AE belongs to exceptional class of polynomials called palindromes (i.e. reciprocal polynomials). Furthermore, it is generated by Kronecker Multiplication, thus it has the even degree $n = 2k$ $(k = 1, 2, ...l)$. In order to explain the structural properties, we formally rewrite the AE in terms of z :

$$AE(z, \mathbf{q}) = \sum_{j=0}^{2n^2} b_j(\mathbf{q}) z_k^j \tag{5}$$

It is necessary to state some definitions and lemmas for analysis of this polynomial.

Definition 3. A polynomial P is called palindrome if

$$P^r(z) = z^n P(1/z) = P(z) \; for \; \forall z \neq 0. \tag{6}$$

Equivalently, writing $P(z) = \sum_j^p b_j z^j$ we have for a palindrome polynomial

$$\sum_{j=0}^{p} b_{p-j} z^j = \sum_{j=0}^{p} b_j z^j \; for \; \forall z \neq 0. \tag{7}$$

and therefore $b_{p-j} = b_j$ for $0 \leq j \leq p$.

This type of the polynomials are special kind of self-inversive polynomials where the coefficients (b_j) are real numbers [4]. Inherently, (5) is a palindrome polynomial in terms of z^j's as mentioned above. The zeros of this class of polynomials lie either on the unit circle $\mathbb{T}$ or occur in pairs conjugate to $\mathbb{T}$ (symmetric pair of roots wrt unit circle). This is a distinctive feature of *AE* that we take advantage when we determine crossing roots of (3). Inversely, for robust stability we demand that none of the roots of (5) lie on the unit circle. Our objective is to determine control and uncertainty parameter space which satisfies this condition. Although it seems as a straightforward analysis, determination of $\mathbf{q}$ space for which (5) have no unimodular zeros is not trivial. In the literature, there are several methods derived from Schur-Cohn criteria to find roots whereabouts wrt to $\mathbb{T}$, however they fail in this problem due to the nature of palindrome polynomials [18].

At the current standpoint, it is very useful to mention some of the friendly properties of palindrome polynomials. Critical points of the polynomial P (i.e. zeros of P', which is the derivative of P wrt z) and the zeros of P have a relationship which is stated as in the theorem below [24].

Theorem 3. *Let P be a palindrome polynomial of degree n. Suppose that P has exactly β zeros on the unit circle $\mathbb{T}$ (multiplicity included) and exactly μ critical points in the closed unit disc $\mathbb{U}$ (counted according to multiplicity). Then*

$$\beta = (2\mu + 1) - n. \tag{8}$$

Proof. The proof of this theorem can be found in [24] in details.

Combining this theorem with the theorem taken from [23] and [19], which is given below can provide us a sufficient condition for robust stability.

Theorem 4. *Let P(z) a polynomial equation,*

$$P(z) = \sum_{j=0}^{p} b_j z^j \tag{9}$$

where $b_j \in \mathbb{C}$ and $b_p \neq 0$ If

$$|b_k| > \sum_{j \neq k}^{p} |b_j| \tag{10}$$

then P(z) has exactly k zeros in the unit circle and noting that P(z), under the above condition has no unimodular zeros (i.e. $z \in \mathbb{T}$).

Proof. Proof of this theorem is easily achieved by substituting $r = R = 1$ in Pellet's Theorem which can be found in [18]. The proof is also given in the studies of Rajan&Reddy [23] and Mori [19].

Following the theorem, the derivation of the following corollary for the conditions of robust stability is simple.

Corollary 1. *A linear time invariant time delayed discrete time system described in (2) is robust stable if*

1. *There exists a parameter vector* $\mathbf{q} \in \mathbb{Q}^p$ *that renders all of the roots of the following equation lie on* $\mathbb{D}$ *(* $\forall z \in \mathbb{D}$*)*

$$\det\left[z\mathbf{I} - \mathbf{A}(\mathbf{q}) - z^{-m}\mathbf{B}(\mathbf{q})\right] = 0$$

2. *Critical equation* $D(z,\mathbf{q})$ *of* $AE(z,\mathbf{q})$ *satisfies*

$$|b_\mu \mathbf{q})| > \sum_{j\neq\mu}^{p} |b_j \mathbf{q}| \tag{11}$$

where $\mu \leq (p/2) - 1$ *and* μ *is an integer number.*

In the theorem, it is stated that if the inequality condition is satisfied the AE has no unimodular roots. It is a handy condition for our goal considering that we desire to find μ number of critical roots of AE in the unit circle. Although the theorem can be used to obtain sufficient conditions for robust stability, it is very conservative. In addition, finding the set $\mathbb{Q}^p$ that satisfies the condition is unlikely. This method is used in previous work of the author in [4] for delay-independent stability of LTI-TDS but it is less effective for discrete-time systems, especially for the dominant pole assignment problems.

Since the method above may not derive a valuable set $\mathbb{Q}^p$ it is necessary to state less conservative approach using other features of *palindrome polynomials*. The condition for an arbitrary palindrome polynomial having unimodular roots is stated in the theorem below.

Theorem 5. *Let P(z) be a palindrome having even degree n. The polynomial P(z) has a unimodular root if and only if the following cosine polynomial has real roots.*

$$f(x) = b_{n/2} + \sum_{j=0}^{n/2-1} 2b_j cos((n/2 - j)x); \tag{12}$$

Proof. Proof of this theorem is given in [10].

A necessary and sufficient condition for robust stability can be attained by using the theorem above. Followed by the theorems we shall collect the steps of the design as a procedure to clarify the method.

Procedure:

1. Compute AE of the system using Kronecker Matrix Method as described in (4)
2. Find an initial point for $\mathbf{q}$ that satisfies $\forall z \in \mathbb{D}$ roots of (3)
3. Evaluate (10) for $\mathbf{q} \in \mathbb{Q}^p$ and check if the inequality is satisfied
4. If previous step fails, evaluate (12) for $\mathbf{q} \in \mathbb{Q}^p$ and check if (12) has roots between $0 \leq x < 2\pi$

The procedure above is a tool to investigate the parameter space $\mathbf{q} \in \mathbb{Q}^p$ that discrete time system is stable.

4 Dominant Pole Assignment

In previous chapter, we stated how to attain the parameter space for robust stability. Employing a simple transformation as following

$$z = r.t \tag{13}$$

where $r \in \mathbb{R}$ and $t \in \mathbb{T}$ turns this robust stability problem into dominant pole assignment problem. This transformation reduces the diameter of the circle of interest by selecting a diameter $r < 1$. Because, the dominant poles of discrete time systems are the characteristic roots which are the closest to the unit circle $\mathbb{T}$. The problem of the right most root assignment in continuous systems is the dual problem with the largest modulus root assignment in discrete time systems. Thus, if we can achieve that the largest modulus eigenvalues of (2) lie in a circle with radius $r < 1$ then restricted settling time of the response of the system is achieved. We will elaborate on that in the example section for the better explanation.

5 Example Case Studies

Example 1: A second order ($n = 2$) discrete time system with sampling period of $T = 0.01s$ is considered. The state space equation of the system is

$$\mathbf{x}(k+1) = \mathbf{A}\mathbf{x}(k) + \mathbf{G}(\alpha,\eta)\mathbf{x}(k-m) \tag{14}$$

where,

$$\mathbf{A} = \begin{bmatrix} 0.997 & 0.0099 \\ -0.1584 & 0.9826 \end{bmatrix} \quad \mathbf{B} = \begin{bmatrix} 0 & 0 \\ -0.0134\eta & -0.0122\alpha \end{bmatrix} \tag{15}$$

Let's set the time delay to $\tau = 200ms$ which corresponds $m = 20$. The characteristic equation of the system is constructed as

$$CE(z,\alpha,\eta) = z^{22} - 1.9796z^{21} + 0.98129z^{20} + 0.01224\alpha z - 0.0122\alpha + 0.000133379\eta.$$

It is a 22nd order polynomial and root finding is cumbersome. Notice that, if the sampling period $T_s = 0.001s$ for the same delay the order of the CE would be equal to 202. The conservative conditions can be applied for the robust stability derived by [19, 21] but it is unlikely that a satisfactory outcome would be achieved. On the other hand, converting every delayed state into a new state as in [17] would enlarge the size of the system matrices drastically. If we apply Kronecker Product method described in previous sections we have the corresponding adjunct characteristic equation

$$
\begin{aligned}
AE(z,\alpha,\eta) = {} & 0.962938z^8 + (-7.69788 + 0.00014658\alpha^2 - 0.0000016\eta\alpha)z^7 + \\
& (26.9286 - 0.000879\alpha^2 + 0.00000959\eta\alpha - 0.000000017\eta^2)z^6 + \\
& (-53.8402 + 0.002197\alpha^2 - 0.0000239\eta\alpha + 0.000000069\eta^2)z^5 + \\
& (67.2933 + 0.0000319\eta\alpha - 0.000000104\eta^2 - 0.00292912\alpha^2)z^4 + \\
& (-53.8402 + 0.002197\alpha^2 - 0.0000239\eta\alpha + 0.000000069\eta^2)z^3 + \\
& (26.9286 - 0.000879\alpha^2 + 0.00000959\eta\alpha - 0.000000017\eta^2)z^2 + \\
& (-7.6979 + 0.00014658\alpha^2 - 0.0000016\eta\alpha)z + 0.962938
\end{aligned}
$$

Keeping in mind that AE is 8th order, the number of roots of $D(z,\alpha,\eta)$ that should lie in unit circle is equal to 3 (using (8)). The inequality (1) can be constructed for k=3. But for this system there is no parameter set $\mathbb{Q}^p$ that satisfies this inequality. Due to this reason, it is necessary to utilize less conservative technique. In our procedure, third step fails so we pursue to fourth step. We form the cosine function for AE as

$$
\begin{aligned}
f(x,\alpha,\eta) = {} & 67.29331 - 0.0000001046\eta^2 - 0.0029291\alpha^2 + 0.000031956\eta\alpha + \\
& 21.9258769\cos(4x) + (-0.000001602\eta\alpha - 7.6978848 + 0.000146584\alpha^2)\cos(3x) + \\
& 2(-0.00000001745\eta^2 + 0.000009598\eta\alpha - 0.0008790778\alpha^2 + 26.92855)\cos(2x) + \\
& 2(-53.84 + 0.00219\alpha^2 - 0.000024\eta\alpha + 0.0000000698\eta^2\cos(x)
\end{aligned}
$$

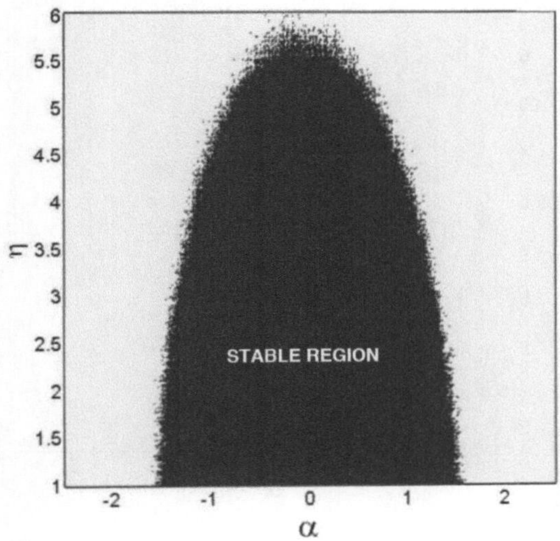

Fig. 1 Region of parameters (α,η) where roots of AE $z \notin \mathbb{T}$

Here, the system given is stable for the $<\alpha,\eta>$ values those don't create real roots ($x \notin \mathbb{R}$) for the $f(x,\alpha,\eta)$. Fig.1 exhibits the region for non-real roots for function f which also corresponds to the region no unimodular roots ($z \notin \mathbb{T}$) of AE. The system is guaranteed to be stable for $<\alpha,\eta>$ parameters valued from shaded region. The algebraic manipulations are done using MAPLE 14™. Matlab™R14 is utilized for numerical computations on a PC equipped with Intel Core™I3 2.4MHz processor and Windows 7™32 bit edition.

Example 2: The system in the first example is considered for dominant pole placement. Our desired settling time for this system is $T_{set} < 2.5s$ within 2% band. The corresponding dominant root of CE should lie in circle with a radius $r = 0.984$. If we apply the transformation (13) to the AE of our system we have

$$\begin{aligned}AE(z,\alpha,\eta) = {} & 0.9028z^8 + (-7.2173 - 0.00000290\eta\alpha + 0.000266\alpha^2)z^7 + \\ & (-0.0000000322\eta^2 + 0.0000175\eta\alpha + 25.247 - 0.00159\alpha^2)z^6 + \\ & (-50.4777 + 0.000000128\eta^2 - 0.00004388\eta\alpha + 0.00398\alpha^2)z^5 + \\ & (-0.00000019\eta^2 + 0.0000585\eta\alpha - 0.005316421\alpha^2 + 63.09)z^4 + \\ & (-0.00004388\eta\alpha - 50.4777 + 0.00398\alpha^2 + 0.000000128\eta^2)z^3 + \\ & (-0.0000000322\eta^2 - 0.00159\alpha^2 + 25.247 + 0.0000175\eta\alpha)z^2 + \\ & (-7.2173 - 0.00000290\eta\,\alpha + 0.000266\alpha^2)z + 0.9028\end{aligned}$$

The corresponding $f(x,\alpha,\eta)$ again evaluated for $r = 0.984$ and stability map of the system wrt parameters is obtained as in Fig. 2. Noticed that the parameter space has shrunk. That was an expected outcome as we contract the region for the root locations.

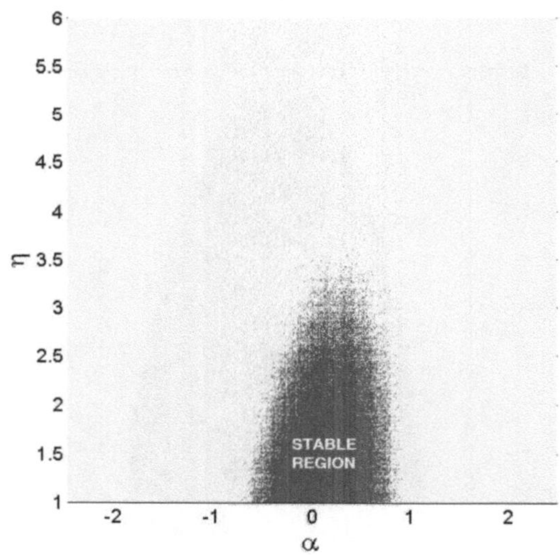

Fig. 2 The parameter space (α,η) for roots of CE $|z| < r = 0.984$

6 Conclusion and Discussion

In this study, a new matrix solution of robust stability problem against parametric uncertainties for linear time invariant discrete-time time delayed systems is presented. The method reduces the size of polynomial of interest which decreased computation time drastically compared to other methods. Furthermore, although the method presents a sufficient condition, the conservatism is much less than the other techniques mentioned in the introduction section. The unique properties of palindrome polynomials are used to construct the main idea. Characteristic equation of the LTI discrete-time systems with large time delay is associated with a palindrome polynomial and their coinciding roots distribution relative to the unit-circle is investigated. It is shown that the relationship between the palindromic polynomial and its "cosine function" is a very useful tool to analyze root locations of the adjunct equation of the system. A practical condition is presented which, in turn, resolves the stability robustness of LTI discrete-time systems with large delays against parametric uncertainties. An easy transformation is also incorporated to locate the dominant poles of the system.

References

1. Bahill, A.T.: A simple adaptive smith-predictor for controlling time-delay systems. Control Systems Magazine, 16–22 (May1983)
2. Bernstein, D.S.: Matrix Mathematics. Princeton University Press (2005)
3. Brewer, J.W.: Kronecker products and matrix calculus in system theory. IEEE Transactions on Circuits and Systems CAS-25, 772–781 (1978)
4. Ergenc, A.F.: A new method for delay-independent stability of time-delayed systems. IFAC TDS Workshop, Prague, Czech Republic (2010)
5. Ergenc, A.F.: A Matrix Technique for Dominant Pole Placement of Discrete-Time Time-Delayed Systems. IFAC TDS Workshop, Boston, USA (2012)
6. Eris, O., Kurtulan, S.: A new PI tuning rule for first order plus dead-time systems. In: IEEE AFRICON Conference (2011)
7. Fridman, E., Shaked, U.: Stability and guaranteed cost control of uncertain discrete delay systems. International Journal of Control 78(4), 235–246 (2005)
8. Gonzalez, A., Garcia, P., Albertos, P., Castillo, P., Lozano, R.: Robustness of a discrete-time predictor-based controller for time-varying measurement delay. Control Engineering Practice 20(2), 102–110 (2012)
9. He, Y., Wu, M., Liu, G.-P., She, J.: Output feedback stabilization for a discrete-time system with a time-varying delay. IEEE Transactions on Automatic Control 53(10), 2372–2377 (2008)
10. Konvalina, J., Matache, V.: Palindrome-polynomials with roots on the unit circle. C. R. Math. Acad. Sci. Soc. R. Can. 26(2), 39–44 (2004)
11. Lampe, B.P., Rosenwasser, E.N.: H2-optimisation of mimo sampled-data systems with pure time-delays. International Journal of Control 82(10), 1899–1916 (2009)
12. Landau, I.D.: Robust digital control of systems with time delay. In: Proceedings of the IEEE Conference on Decision and Control, vol. 1, pp. 865–869 (1994)
13. Landau, I.D.: Robust digital control of systems with time delay (the Smith predictor revisited). International Journal of Control 62(2), 325–347 (1995)
14. Landau, I.D., Karimi, A.: Robust digital control using pole placement with sensitivity function shaping method. International Journal of Robust and Nonlinear Control 8(2), 191–210 (1998)
15. Leite, V.J.S., Tarbouriech, S., Peres, P.L.D.: Robust h state feedback control of discrete-time systems with state delay: An lmi approach. IMA Journal of Mathematical Control and Information 26(3), 357–373 (2009)
16. Louisell, J.: A matrix method for determining the imaginary axis eigenvalues of a delay system. IEEE Transactions on Automatic Control 46(12), 2008–2012 (2001)

17. Madsen, J.M., Shieh, L.-S., Guo, S.-M.: State-space digital pid controller design for multivariable analog systems with multiple time delays. Asian Journal of Control 8(2), 161–173 (2006)
18. Marden, M.: The Geometry of the Zeros of a Polynomial in a Complex Variable. American Mathematical Society (1949)
19. Mori, T.: Note on the absolute value of the roots of a polynomial. IEEE Transactions of Automatic Control AC-29(1), 54–55 (1984)
20. Mori, T.: Bound for radius of stability-preserving hypersphere in parameter space for Schur polynomials. International Journal of Systems Science 20(9), 1697–1702 (1989)
21. Mori, T., Fukuma, N., Kuwahara, M.: Delay-independent stability criteria for discrete-delay systems. IEEE Transactions on Automatic Control AC-27(4), 964–966 (1982)
22. Park, J.-J., Choi, G.-S., Shieh, L.-S.: Digital modeling and control of multiple time-delayed systems via SVD. In: Kim, T.-H., Gelogo, Y. (eds.) MulGraB 2011, Part I. CCIS, vol. 262, pp. 243–252. Springer, Heidelberg (2011)
23. Rajan, P.K., Reddy, H.C.: Comments on note on the absolute value of the roots of a polynomial. IEEE Transactions of Automatic Control AC-30(1), 80 (1985)
24. Sheil-Small, T.: Complex Polynomials. Cambridge University Press, Cambridge (2002)
25. Viteckova, M., Vitecek, A., Babiuch, M.: Unified approach to analog and digital two-degree-of-freedom pi controller tuning for integrating plants with time delay. Acta Montanistica Slovaca 16(1), 89–94 (2011)
26. Yucelen, T., Kaymakci, O., Kurtulan, S.: Self-tuning pid controller using Ziegler-Nichols method for programmable logic controllers. In: IFAC Proceedings Volumes (IFAC-PapersOnline), vol. 1, pp. 11–16 (2006)
27. Zhang, B., Xu, S., Zou, Y.: Improved stability criterion and its applications in delayed controller design for discrete-time systems. Automatica 44(11), 2963–2967 (2008)
28. Zhang, Y., Shieh, L.-S., Akujuobi, C.M., Ali, W.H.: Digital PID controller design for delayed multivariable systems. Asian Journal of Control 6(4), 483–495 (2004a)
29. Zhang, Y., Shieh, L.-S., Liu, C.R., Guo, S.-M.: Digital pid controller design for multivariable analogue systems with computational input-delay. IMA Journal of Mathematical Control and Information 21(4), 433–456 (2004b)

Dominant Trio of Poles Assignment in Delayed PID Control Loop

Pavel Zítek, Jaromír Fišer, and Tomáš Vyhlídal

Abstract. Besides its original use in the state space based design the pole assignment is also applied to tuning the PID controllers. However, infinite spectrum of poles caused by the usual assumption of a time delay in the plant model highlights the requirement that the prescribed pole positions have to result in the *dominant* pole assignment to be effective in tuning the control loop with time delay. With respect to three parameters of PID controller just three poles can be placed by the assignment and a dominance guarantee of their prescription is crucial in this tuning method. A novel method of selecting a trio of numbers $p_{1,2,3}$ to make them the dominant poles of the control loop is dealt with in the chapter with an additional minimizing the absolute error integral. The dominance of each of the placement trials is checked by an argument increment criterion and an optimum of relative damping of the response is assessed to minimize the control error integral. The quality of the disturbance rejection response is taken as the decisive criterion in the presented design of the time delay plant control.

1 Introduction

Pole assignment is a widely-used approach to state space system design that has also been widely applied in the last two decades in methods for tuning PID controllers. Three parameters of the PID controller principle can handle the assignment of just *three poles*, and this makes assumptions of sophisticated, detailed or higher- order process models unrealistic. The time delay effect can be assumed to be a general process property, but this leads to the appearance of an *infinite spectrum* of control loop poles, and to a need to investigate their dominance as a crucial issue in pole placement.

Pavel Zítek · Jaromír Fišer · Tomáš Vyhlídal
Center for Applied Cybernetics and Dept. of Instrumentation and Control Eng., Faculty of Mechanical Eng., Czech Tech. Univ. in Prague, Technická 4, 166 07 Praha 6, Czech Republic
e-mail: {pavel.zitek,jaromir.fiser,tomas.vyhlidal}@fs.cvut.cz

T. Vyhlídal, J.-F. Lafay, and R. Sipahi (eds.), *Delay Systems*, Advances in Delays and Dynamics 1,
DOI: 10.1007/978-3-319-01695-5_5,

The dependence of the dominant pair of closed-loop poles on the controller parameters was first investigated by Hwang and Chang [1] by means of the Taylor series expansion about the critical gain. Instead of *dominant*, the term leading poles was preferred in this paper. Dominant pole placement design was introduced somewhat differently by Persson and Åström [2] and was further explained in Åström and Hägglund [3]. At about the same time, Hwang and Fang [4] published an extensive optimization study on dominant pole placement for first and second order time delay plants. Numerous methods with modified specifications of tuning conditions were presented subsequently, and a survey was presented by O'Dwyer [5].

Applying the pole assignment approach to systems with a delay (i.e. with an infinite spectrum of poles) has led to the specific problem of how to select a proper prescription of pole positions so that they will be capable candidates for becoming dominant poles that really do determine the behaviour of the system. Any pole placement in a time delay system is always connected with a risk that, although the prescribed poles are achieved in the system spectrum, they may lose any meaning because some *other poles* spontaneously assume the dominant role in the infinite system spectrum. Consequently, any result of a pole assignment of this kind can be approved as valid only after checking that the placed poles really have assumed the dominant positions. To the best of the authors knowledge, no general theorem is yet available that guarantees in advance that a chosen prescription of poles for a time delay system will reliably result in the group of system dominant poles.

The pole placement approach in a control loop with delay is to be considered only for placing a small group of *dominant poles* either a complex conjugate pair or a three-pole group, usually one pair with a real pole. The key issue for this design is to select the prescribed poles in a way that guarantees their dominant position [6]. Because of the number of three-controller parameters, only the three-pole option $p_{1,2,3}$ can reasonably be prescribed in applying the pole assignment to tuning the PID controller for a time delay plant. However, in a considerable number of papers, the dominant pole placement in PID tuning has also been considered for a *single pair* of complex conjugate poles $p_{1,2} = -\alpha \pm j\Omega$, $\alpha > 0$, assigned to take up the rightmost position in the system spectrum and satisfying an additional requirement of the frequency response specification of control synthesis [7], [8]. The case of placing three prescribed poles, a complex conjugate pair $p_{1,2}$and a real pole $p_3 = -\beta$, as dominant was solved by Hwang and Fang [4]. A guarantee of dominance in pole placement based on the root locus and Nyquist plot applications was presented by Wang et al. [6]. Dominant pole placement may also be performed in an iterative way, as a series of attempts that shift the prescribed poles to the left as in [9], [10], or a modified optimization in [11].

The contribution of our work we envisage in the following three aspects. First, unlike the usual approach the tuning is focused on the disturbance rejection as usually the primary task in industrial process control. Secondly, integral absolute error (IAE) minimization is applied simultaneously with checking the dominance of the assigned poles. Finally, the use of the same value of real parts for all the three poles is proved as well-preventing the assignment from the loss of dominance.

The rest of the paper is organized as follows. In Section 2 the selection of the prescribed candidate trio of dominant poles is discussed. The procedure of placing the poles is described in Section 3 and a novel method of checking the achievement of their dominant position is presented in Section 4. The complete procedure of solving the dominant pole assignment and two application examples are presented in Section 5 and several concluding remarks are added in Section 6.

2 Selecting the Candidate Group of Dominant Poles

In order really to achieve the dominant position for the poles to be placed the prescribed values of $p_{1,2,3}$ are to be selected with a careful respect to the specific dynamic properties of the plant controlled. As Åström and Hägglund [3] revealed particularly the ultimate frequency ω_K of the control loop – usually obtained by means of the ideal relay feedback application – has to be taken into account. The ultimate frequency at which the relay feedback control loop is oscillating determines the bounds within which the attainable frequency of the control loop can be expected. Let the following three poles be considered for the pole assignment

$$p_{1,2} = -\alpha \pm j\Omega = \Omega(-\delta \pm j), p_3 = -\beta = -\kappa\delta\Omega \tag{1}$$

where α, β, Ω are supposed positive and introducing the ratios $\delta = \alpha/\Omega$ (relative damping) and $\kappa = \beta/\alpha$ helps to characterize the $p_{1,2,3}$ group. The assignment of these poles can be accepted as valid only if besides $p_{1,2,3}$ the whole rest of the system spectrum lies to the left of the prescribed poles. Note that the following consideration serves only for a relative comparison of various versions of $p_{1,2,3}$ group selection, not for particular finding the response.

Lemma 1. Consider a *stable* time delay plant described by a transfer function

$$G(s) = \frac{k}{A(s)}\exp(-s\tau), \qquad A(s) = \sum_{i=0}^{n} a_i s^i \tag{2}$$

fitting only the dominant modes of the plant, with $n \geq 2$. Assume that in the PID control loop on this plant the pole placement of $p_{1,2,3}$ has been as successful that no of the other poles influence substantially the control loop response. With regard to the integral action in the PID controller the *disturbance rejection* response – shifted by the dead time – is then generally composed of the following three functions

$$h(t+\tau) = \exp(-\alpha t)\,[C_1 \cos\Omega t + C_2 \sin\Omega t] + C_3 \exp(-\beta t) \tag{3}$$

$t \geq 0$, where the coefficients $C_{1,2,3}$ are to satisfy the initial conditions

$$h(0) = 0, \;\; h'(0) = 0, \;\; h''(0) = -p_1 p_2 p_3 \tag{4}$$

(with respect to the disturbance derivative in the control loop equation). For this type of the response the absolute error integral $I_{AE} = \int_0^\infty |h(t)|\,dt$ for $\kappa \leq 1$ is independent of the parameters δ, Ω and κ and equal to 1 and is *larger than* 1, $I_{AE} > 1$, for $\kappa > 1$.

Proof. From the first condition $h(0) = 0$ the equality $C_3 = -C_1$ easily results. The condition $h'(0) = 0$ leads to the equation $C_2 = -C_1\delta(\kappa - 1)$ pointing out that with the choice $\kappa = 1$ the middle term in (3) will disappear. After exploiting the last condition $h''(0) = -p_1p_2p_3 = \beta(\alpha^2 + \Omega^2)$ the following relationship is obtained for C_1 and C_3

$$C_3 = -C_1 = \Omega \frac{\kappa\delta(1+\delta^2)}{1+\delta^2(\kappa-1)^2} \tag{5}$$

Let us take the integral $\tilde{I} = \int_0^\infty h(t)dt$ as an auxiliary evaluation means. This integral can be expressed analytically with constant result

$$\tilde{I} = \int_0^\infty \{\exp(-\alpha t)\,[C_1 \cos\Omega t + C_2 \sin\Omega t] + C_3 \exp(-\beta t)\}\,dt = \tag{6}$$

$$= \frac{C_1\kappa\delta^2 + C_2\kappa\delta + C_3(1+\delta^2)}{\kappa\delta\Omega\,(1+\delta^2)} = \frac{1+\delta^2(\kappa-1)^2}{1+\delta^2(\kappa-1)^2} = 1 = \int_0^\infty h(t)dt$$

independent of κ, δ, Ω values. However, the auxiliary integral $\tilde{I}$ is equal to I_{AE} in case that $h(t)$ is free of negative overshoots, i.e. if $h(t) \geq 0,\ \forall t \geq 0$. With respect to the values of $C_{1,2,3}$ this type of function (3) is obtained only for $\kappa \leq 1$. On the contrary for $\kappa > 1$ the function (3) is oscillating with alternating $sgn\,h(t)$ and therefore the absolute error criterion $I_{AE} = \int_0^\infty |h(t)|\,dt$ is larger than $\tilde{I}$, $I_{AE} > \tilde{I}$. QED

On the other hand it is apparent that the choice of $\kappa < 1$ brings about an overdamped response with dominating role of p_3 with a slower rejection of the disturbance. That is why the choice of $\kappa = 1$ is to be considered as the most preferable option for the three pole assignment from the point of view of the disturbance rejection making the function (3) as simple as $h(t+\tau) = C\exp(-\alpha t)\,(1-\cos\Omega t)$.

Concluding this section one has to realize the very approximate character of the above speculation. In spite of a successful tuning of the PID controller the control loop keeps its time delay and therefore the infinite nature of its spectrum. Hence the assumed response (3) can be applied as an auxiliary specimen pointing out the essential role of $p_{1,2,3}$ in the control loop response. Only in case the poles $p_{1,2,3}$ prove to be the dominant ones their choice is confirmed as proper, the better the dominance of the poles $p_{1,2,3}$, the more justifiable the use of pattern (3). Of course the result of the pole placement primarily depends on the plant (2) dynamics.

3 Three Pole Dominant Placement in Delayed PID Feedback Loop

Consider a PID control loop composed of the time delay plant as in (2) with the controller transfer function

$$C(s) = \frac{K_P s + K_I + K_D s^2}{s} \tag{7}$$

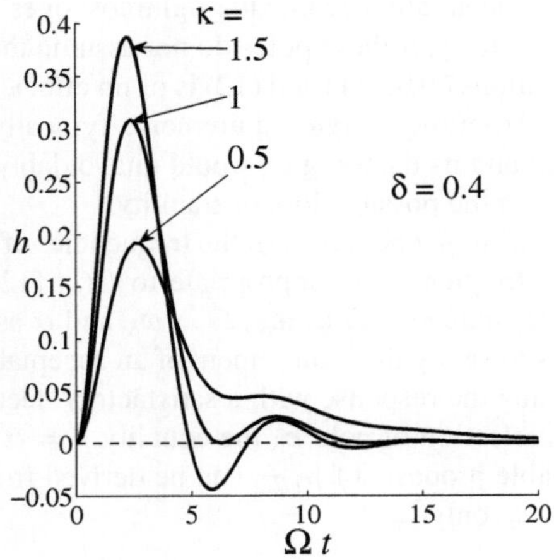

Fig. 1 Comparison of responses (3) for $\kappa = 0.5, 1.0, 1.5$

and with the closed loop characteristic equation

$$1 + C(s)G(s) = 0 \tag{8}$$

The aim is to select and position three poles $p_{1,2,3}$ in order to achieve the control loop approaching in its response the character given by (3). Due to the delay in $G(s)$ the equation (8) is transcendental and as such it admits an infinite spectrum of roots.

Suppose equation (8) to be satisfied by the roots $p_{1,2,3}$. By inserting $p_1 = -\alpha + j\Omega$ into (8) one obtains – after a slight rearrangement – the complex valued equality

$$K_P(-\alpha + j\Omega) + K_I + K_D(\alpha^2 - \Omega^2 - j2\alpha\Omega) = -\frac{p_1}{G(p_1)} \tag{9}$$

This equation can be decomposed into two particular equations for the real and imaginary parts respectively

$$-K_P\alpha + K_I + K_D(\alpha^2 - \Omega^2) = -Re\left(\frac{p_1}{G(p_1)}\right) \tag{10}$$

$$K_P\Omega - K_D 2\alpha\Omega = -Im\left(\frac{p_1}{G(p_1)}\right) \tag{11}$$

Similarly for the third prescribed pole $p_3 = -\beta$ the following equality is obtained

$$-K_P\beta + K_I + K_D\beta^2 = -\frac{p_3}{G(p_3)} \tag{12}$$

From the three conditions (10), (11) and (12) the three controller parameters can be easily determined. However, the essential aim of the controller adjustment is really met only in case these roots finally become the rightmost ones from the whole infinite spectrum. On the contrary, if these poles do not assume the rightmost position the fulfilling of the equations (10), (11) and (12) is of no effect, the behaviour of the loop becomes different from the design requirement. Typically, inadequately high prescribed frequency Ω and its damping α, would unavoidably lead to missing the design aim including even the possible loss of stability.

It is well known that the upper boundary of the frequencies of the system response is near to the ultimate frequency ω_K appropriate to $G(s)$ [12]. That is why it is possible to prescribe the frequency Ω as ω_K, $\Omega = \omega_K$ and α as the damping of the appropriate oscillations to set by the requirement of an acceptable relative damping ratio $\delta = \alpha/\Omega$, providing the response with a satisfactory decay rate. In Section 2 as the most preferable selection of pole p_3 the equality $\beta = \alpha$ was recommended. In this way the acceptable proposal of $p_{1,2,3}$ can be derived from an assessment of the ultimate frequency ω_K only.

3.1 Ultimate Frequency Assessment

In order to assess the ultimate gain and the corresponding ultimate frequency a proportional gain $K_P = K_K$ is used instead of $C(s)$. From the characteristic equation of the control loop $1 + K_K G(s) = 0$ the ultimate gain is obtained if the stability margin, i.e. non-damped oscillations at frequency ω_K arise, i.e. for

$$1 + K_K G(j\omega_K) = 0 \tag{13}$$

The decomposition into the equalities of real and imaginary parts yields

$$K_K \cos \omega_K \tau = -Re\left(\frac{1}{G_0(j\omega_K)}\right) = -R_0(\omega_K) \tag{14}$$

$$K_K \sin \omega_K \tau = Im\left(\frac{1}{G_0(j\omega_K)}\right) = I_0(\omega_K) \tag{15}$$

The ultimate gain K_K can be excluded evaluating the tangent function

$$\tan \omega_K \tau = -\frac{I_0(\omega_K)}{R_0(\omega_K)} \tag{16}$$

where the ultimate frequency can be determined from. The only unknown in this equation is ω_K and due to the periodicity of tangent function and with respect to the strict properness of $G(s)$ this equation has infinitely many real roots. But the ultimate frequency ω_K is to be the smallest of positive roots of (16).

Using the assessed ω_K the proposal of $p_{1,2,3}$ and appropriate computation of the controller parameters K_P, K_I, K_D can be made. However, a prompt checking procedure is necessary to prove whether some other undesirable poles do not take over the role of dominant poles. This can be verified by direct assessment of the rightmost part of the spectrum, using e.g. the algorithm proposed in [13]. A simpler way of verifying the dominance is presented below.

4 Argument Increment Based Check to Prove the Dominance

This section deals with a method of checking the dominance of the prescribed $p_{1,2,3}$. The necessary condition for this dominance is that the whole rest of $M(s)$ spectrum except $p_{1,2,3}$ lies *to the left* of the prescribed poles. This configuration of the poles can be proved by an application of the argument increment rule.

Lemma 2. Consider the characteristic quasi-polynomial of a PID control loop

$$M(s) = s^{n+1} + \sum_{i=0}^{n} [c_i + d_i \exp(-s\tau)]\, s^i, \quad n \geq 2 \tag{17}$$

where $d_0 = K_I k$, $d_1 = K_P k$, $d_2 = K_D k$ and let s be fixed to the straight line L parallel to Im axis $s = -\xi + j\omega$, $\xi > \beta$ is optional. If the argument increment of $M(s)$ along this straight line from the starting point $M(-\xi)$, $(\omega = 0)$, for ω growing from zero to infinity, $\omega \to \infty$, reaches the following limit

$$\lim_{\omega \to \infty} \arg M(s)\,|_{s=-\xi+j\omega} - \arg M(-\xi) = (n-5)\frac{\pi}{2} \tag{18}$$

then the whole rest of the $M(s)$ spectrum lies to the left of the prescribed $p_{1,2,3}$.

Proof. Assume that poles $p_{1,2,3}$ lie inside a region enclosed by a Jordan curve composed of

- a circle arc C of radius R, $s = R\exp(j\varphi)$, where $\varphi \in \langle -0.5\pi - \gamma,\, 0.5\pi + \gamma \rangle$, $\gamma = \arcsin(\xi/R)$,
- and a straight line L, $s = -\xi + j\omega$, where ω ranges from $-R\cos\gamma$ to $R\cos\gamma$

If just *only* $p_{1,2,3}$ zeros of $M(s)$ lie inside the considered region the total *argument increment* along C and L is 6π. To evaluate this argument increment let $M(s)$ be factorized as follows

$$M(s) = s^{n+1}\left[1 + \sum_{i=0}^{n} (c_i + d_i \exp(-s\tau)) s^{i-n-1}\right] = s^{n+1} m(s) \tag{19}$$

For the first factor s^{n+1} the argument increment along C is

$$\underset{C}{\Delta} \arg M(R\exp(j\varphi)) = (n+1)\, 2\left(\frac{\pi}{2} + \gamma\right) \tag{20}$$

where γ approaches zero for $R \to \infty$ and therefore the limit of this increment for $R \to \infty$ is $(n+1)\pi$. For the second factor $m(s)$, $s = R\exp(j\varphi)$ the values of each of its terms except 1 is vanishing for $R \to \infty$ due to negative powers of R. Therefore for $R \to \infty$ factor $m(R\exp(j\varphi)) \to 1$ and the appropriate argument increment is zero.

Hence the whole argument increment of $M(R\exp(j\varphi))$ along C is given by (20) and for $R \to \infty$ it holds

$$\lim_{R\to\infty} \Delta_C \arg M(R\exp(j\varphi)) = (n+1)\pi \tag{21}$$

If there are just only three zeros $p_{1,2,3}$ inside the Jordan curve the argument increment along L has to be given by the difference between 6π and (21), i.e. $(5-n)\pi$, if L is oriented downwards. Finally, if for practical use and with respect to the symmetry the original interval $\omega \in \langle -R\cos\gamma,\ R\cos\gamma\rangle$ is replaced by only the positive half oriented upwards $\omega \in \langle 0,\ R\cos\gamma\rangle$ the required argument increment is of half value and opposite sign, i.e. $(n-5)\pi/2$ as in (18). QED

The application of this criterion is as follows. For α and β in $p_{1,2,3}$ a value of $\xi > \beta$ is chosen and $M(-\xi + j\omega)$ is computed for ω growing from zero to some $\omega = \omega_m$, $\omega_m >> \omega_K$. In principle, the starting real value $M(-\xi)$ for $\omega = 0$ may happen to be both negative and positive for various types and orders of $M(s)$. The argument increment is evaluated in an analogous way as in Mikhaylov criterion application but its value can result both negative ($n < 5$, clock-wise direction) and positive ($n > 5$). In case of $n = 5$ the argument increment is zero. The higher ratio ξ/β for which the condition (18) is satisfied, the stronger the dominance of $p_{1,2,3}$. In the next sections two examples of application are demonstrated.

5 Relative Damping Optimization in the PID Parameter Setting

As shown above only for $\kappa \leq 1$ the auxiliary function (3) is free of negative overshoots and the equality between the integrals $\tilde{I} = I_{AE}$ holds. However, after closing the PID feedback with the plant model (2) the disturbance step response $h_M(t)$ different from (3) results. Its Laplace transform is the following meromorphic function

$$H_M(s) = \frac{k\exp(-s\tau)}{sA(s) + k\exp(-s\tau)\,(K_I + K_P s + K_D s^2)} \tag{22}$$

As in (6) the integral $\tilde{I}_M = \int_0^\infty h_M(t)dt$ (different from the absolute error integral I_{MAE}) can be investigated. Using the limit theorem the following value of the integral is obtained

$$\tilde{I}_M = \lim_{s\to 0} \frac{k\exp(-s\tau)}{sA(s) + k\exp(-s\tau)\,(K_I + K_P s + K_D s^2)} = \frac{1}{K_I} \tag{23}$$

i.e. independently of the model $G(s)$ it is given by the inverse of controller integration gain K_I. In case of no negative overshoots on $h_M(t)$ this integral is equal to the absolute error criterion I_{MAE} again and the value of $\tilde{I}_M = 1/K_I$ can be used for evaluation of the obtained disturbance rejection response: the higher K_I the smaller the error integral $\tilde{I}_M$.

5.1 Damping Optimization

The difference between the integrals $\tilde{I}_M$ and $\tilde{I}$ can be viewed as a criterion of satisfying the ideal assumption given by (3). Anyway the lower $\tilde{I}_M$ the better the PID tuning – if simultaneously the argument increment check (18) and the condition $h_M(t) \geq 0$, $\forall t > 0$, are satisfied. Since the $p_{1,2,3}$ position parameters Ω, κ are already fixed as $\Omega = \omega_K$ and $\kappa = 1$, only the damping δ is left to optimize the $p_{1,2,3}$ selection i.e. to finding a minimum of $\tilde{I}_M$ or the maximum of K_I. In this optimization the interval of δ admissible values is relatively close, $\delta \in \langle 0.25,\ 0.6 \rangle$. The following examples will demonstrate this optimization.

5.2 Example 1 - Controlling Second Order System

Consider a second-order oscillatory time delay plant described by the transfer function

$$G(s) = \frac{\exp(-0.5s)}{s^2 + s + 2}, \tag{24}$$

Using the equation (16) the following ultimate frequency is determined $\omega_K = 1.84\ s^{-1}$. On the basis of this primary knowledge let $\Omega = 1.84\ s^{-1}$and the poles $p_{1,2,3}$ let be prescribed as $p_{1,2} = 1.84(-\delta \pm j)$ and $p_3 = -1.84\ \delta$. Now repeat the pole placement for varying values of δ beginning, say from $\delta = 0.3$ and increasing it by 0.01, and comparing the obtained values of K_I. The optimization procedure is recorded in Fig. 2 where the optimum damping as $\delta = 0.38$ is assessed. The corresponding setting of the controller is the following $K_P = 1.175$, $K_I = 1.227\mathrm{s}^{-1}$, $K_D = 0.976\mathrm{s}$.

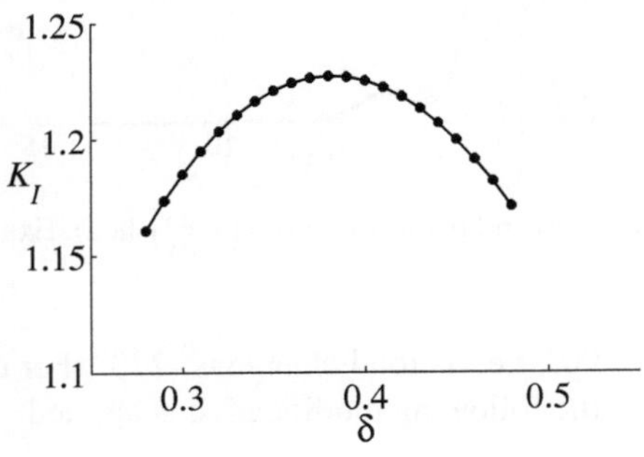

Fig. 2 Damping optimization in Example 1

The obtained controller setting is immediately tested on satisfying the dominance requirement by argument increment procedure with $\xi = 1.5\beta$. For $\delta = 0.38$ the dominance check is recorded in Fig. 3 where the argument increment decreases

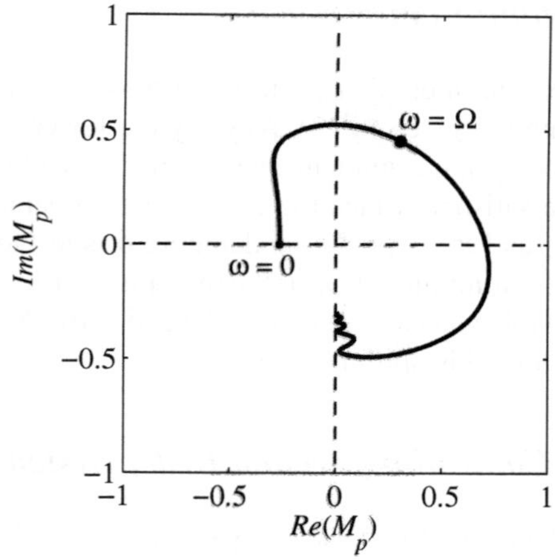

Fig. 3 Argument increment check of pole dominance, Example 1

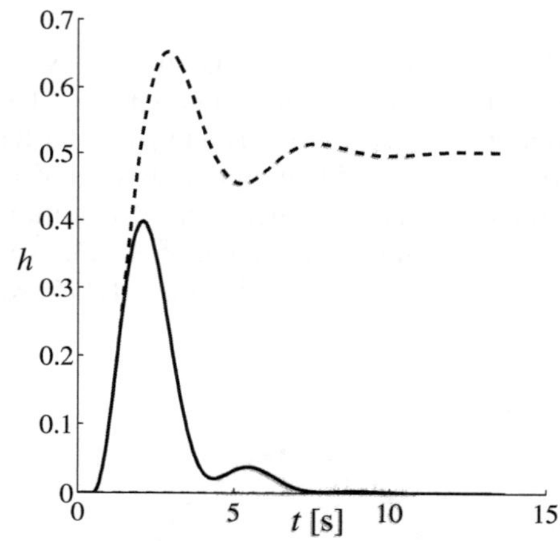

Fig. 4 Disturbance rejection (h) and plant step response (dashed), Example 1

from the initial 180° to -90°, i.e. its total change is -270°. For drawing the contour in Fig. 3 instead of $M(s)$ the following modification is applied

$$M_p(s) = \frac{M(s)}{1+|M(s)|^{1.1}}\,, \tag{25}$$

which does not change $\arg M(s)$ but reduces the module, $|M_p(s)| < |M(s)|$ [14]. This confirms that the whole rest of the system spectrum lies to the left of $p_{1,2,3}$. Due to $\xi = 1.5\beta$ it is even proved that the dominance is as strong that the rest of poles is distributed to the left of the straight line $s = -1.5\beta + j\omega$. At last the control loop response is evaluated for the obtained controller setting in Fig. 4. Comparing this response with that of $\kappa = 1$ in Fig. 2 one can see that the dominance of the poles is very satisfactory since the qualitative agreement of both of them is obvious and no substantial further influence is noticeable in this response.

5.3 Example 2 - Controlling Third Order System

Consider a third-order time delay plant given by the transfer function

$$G(s) = \frac{\exp(-3s)}{(s+1)^2\,(2s+1)} \tag{26}$$

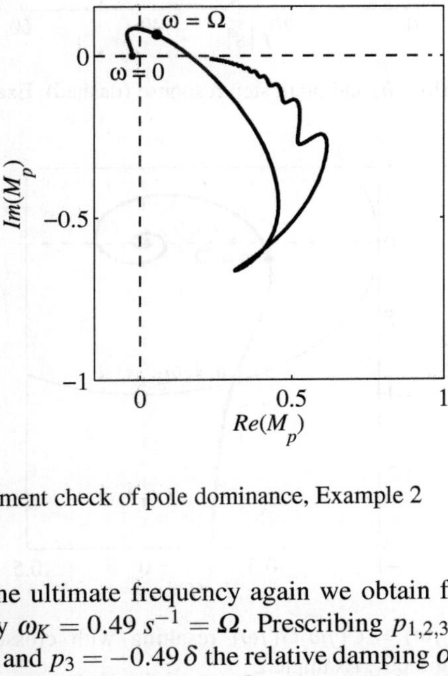

Fig. 5 Argument increment check of pole dominance, Example 2

Beginning with the ultimate frequency again we obtain for the imaginary part of $p_{1,2}$ the frequency $\omega_K = 0.49\ s^{-1} = \Omega$. Prescribing $p_{1,2,3}$ in the same way, i.e. $p_{1,2} = 0.49(-\delta \pm j)$ and $p_3 = -0.49\,\delta$ the relative damping $\alpha/\Omega = \delta$ is optimized. In this case the maximum of K_I is found at $\delta = 0.34$ and the controller setting is then as follows $K_P = 0.920$, $K_I = 0.169\,\mathrm{s}^{-1}$, $K_D = 1.120\,\mathrm{s}$. The dominant position of the prescribed poles is tested by the argument increment with the confirming result in Fig. 5. The dominance of $p_{1,2,3}$ is also confirmed by the disturbance rejection response recorded in Fig. 6. With respect to the third order of (26) the argument increment is -180° now. In Fig. 7 Nyquist plot is recorded with the phase margin equal to 66°. Thus the achieved phase margin is safely higher than usually required value 60° in practice [7].

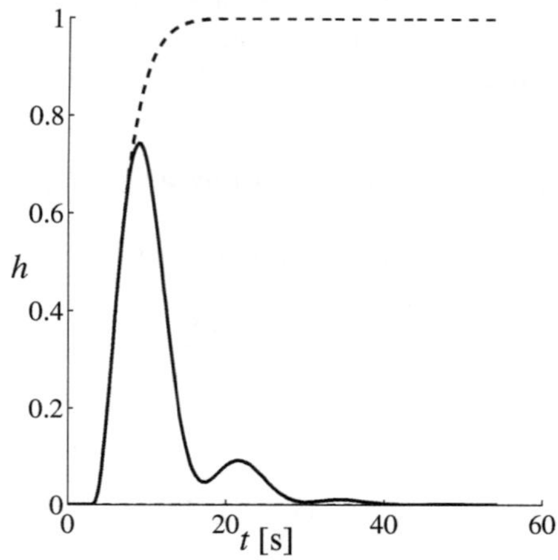

Fig. 6 Disturbance rejection (h) and plant step response (dashed), Example 2

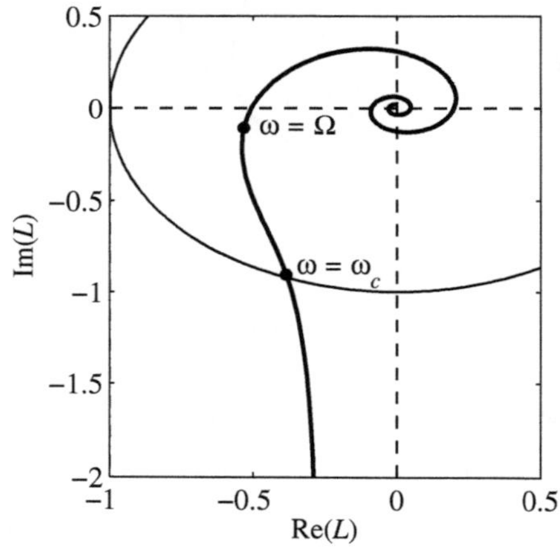

Fig. 7 Nyquist plot $L(j\omega) = C(j\omega)G(j\omega)$ resulting with crossover frequency $\omega_c = 0.201\ s^{-1}$ and phase margin 66°, Example 2

6 Concluding Remarks

The presented method of dominant pole placement is specific by the aim to optimize the disturbance rejection response which is critical for time delay control loops. As any approach using the ultimate gain to identifying the process dynamics the proposed method can be applied only to stable processes both aperiodical and oscillatory. Although the final response is not exactly identical with the nominal response (3) its employment in selecting the prescribed $p_{1,2,3}$ is justified by the checking up on the argument increment. Moreover, due to the selection $\kappa \cong 1$ the degree of the similarity of the response with the ideal pattern (3) can be evaluated by a simple comparison of $h(t)$ and $h_M(t)$ in approaching the time axis after each period of the achieved response. Compared to other methods dealing with dominant pole placement the IAE minimization is based on a simple analytical condition. From the IAE evaluation it turns out that the damping ratio near $\delta \cong 0.35$ is a common optimum value particularly from the point of view of the dominance warranty. A comparison of the IAE criterion graphs for various values of δ and κ led to an important finding that IAE values vary greatly in parts of these graphs where they are irrelevant for controller setting due to loss of dominance of the prescribed poles. However, the IAE graphs unexpectedly concur with each other even for various δ and κ if the prescribed $p_{1,2,3}$ are fully dominant, i.e. determine the control loop response. Due to this finding the values of $\kappa \cong 1.3 \pm 0.1$ and $\delta \cong 0.35 \pm 0.05$ may be recommended universally for any stable plant with an essential delay. This finding makes the method of dominant pole assignment presented here practically uniform in application.

Acknowledgements. The presented research results were supported by The Technology Agency of the Czech Republic under the Competence Centre Project TE01020197, Centre for Applied Cybernetics 3.

References

1. Hwang, S.H., Chang, H.C.: A Theoretical Examination of Closed-Loop Properties and Tuning Methods of Single Loop PI Controllers. Chemical Eng. Sciences 42, 2395–2415 (1987)
2. Persson, P., Åström, K.J.: Dominant Pole Design – a Unified View of PID Controller Tuning. In: Dugard, L., M'Saad, Landau, I.D. (eds.) Adaptive Systems in Control and Signal Processing. Pergamon Press, Oxford (1993)
3. Åström, K.J., Hägglund, T.: PID Controllers Theory Design and Tuning, 2nd edn., Instrument Society of America, Research Triangle Park, North Carolina (1995)
4. Hwang, S.H., Fang, S.M.: Closed Loop Tuning Method Based on Dominant Pole Placement. Chemical Eng. Comm. 136, 45–66 (1995)
5. O'Dwyer, A.: Handbook of PI and PID Controller Tuning Rules. Imperial College Press, London (2003)
6. Wang, Q.G., Zhang, Z., Åström, K.J., Chek, L.S.: Guaranteed Dominant Pole Placement with PID Controllers. Journal of Process Control 19, 349–352 (2009)
7. Tang, W., Wang, Q.G., Ye, Z., Zhang, Z.: PID Tuning for Dominant Poles and Phase Margin. Asian Journal of Control 9, 466–469 (2007)
8. Zítek, P., Fišer, J., Vyhlídal, T.: Ultimate-Frequency based Dominant Pole Placement. In: Proceedings of 9th IFAC Workshop on Time Delay Systems, IFAC 2010, Praha, vol. 1, pp. 87–92 (2010)
9. Zítek, P., Fišer, J., Vyhlídal, T.: Ultimate-Frequency Based Three-Pole Dominant Placement in Delayed PID Control Loop. In: Proceedings of 10th IFAC Workshop on Time Delay Systems, IFAC 2012, Boston, vol. 10, pp. 150–155 (2012)
10. Zítek, P.: Frequency-Domain Synthesis of Hereditary Control Systems via Anisochronic State Space. Int. J. of Control 66, 539–556 (1997)

11. Michiels, W., Vyhlídal, T., Zítek, P.: Control Design for Time-delay Systems Based on Quasi-direct Pole Placement. Journal of Process Control 20, 337–343 (2010)
12. Wang, Q.G., Lee, T.H., Tan, K.K.: Finite Spectrum Assignment Controllers for Time Delay Systems. Springer, London (1995)
13. Vyhlídal, T., Zítek, P.: Mapping Based Algorithm for Large-Scale Computation of Quasi-Polynomial Zeros. IEEE Transactions on Automatic Control 54, 171–177 (2009)
14. Vyhlídal, T., Zítek, P.: Modification of Mikhaylov Criterion for Neutral Time-Delay Systems. IEEE Transactions on Automatic Control 54, 2430–2435 (2009)

Stability of Systems with State Delay Subjected to Digital Control

David Lehotzky and Tamas Insperger

Abstract. Stability of linear delayed systems subjected to digital control is analyzed. These systems can typically be written in the form

$$\dot{\mathbf{x}}(t) = \mathbf{A}\mathbf{x}(t) + \mathbf{B}\mathbf{x}(t-\tau) + \mathbf{C}\mathbf{x}(t_{j-1}), \quad t \in [t_j, t_{j+1}),$$

where $t_j = j\Delta t$ with Δt being the sampling period for the digital controller. The point-delay term $\mathbf{x}(t-\tau)$ is assumed to be inherently present in the governing equation of the uncontrolled system, while the term $\mathbf{x}(t_{j-1})$ is present due to the digital controller. Since the term $\mathbf{x}(t_{j-1})$ can be represented as a term with a piecewise linearly varying time delay, the system is time-periodic at period Δt. The stability analysis for the system is performed using the semi-discretization method. As case studies, the stability charts of the delayed oscillator and the turning process are determined for a digital PD controller.

1 Introduction

Time delays are often inherently present in mechanical systems due to physical interactions between different elements of the system or due to a feedback mechanism. For instance, in wheel shimmy models, the contact between elastic tires and the road is described by a delay-differential equation (DDE) with distributed delay [29]. In car following traffic models, time delay arise due to the reflex delay of the drivers [17]. Machine tool chatter is also modeled by DDEs, where the delay appears due to the regenerative effect as result of the contact of the tool and the workpiece. For simple tool geometry, the regenerative delay can be modeled by a point delay [1], while for more complex tool geometry, such as a milling tool with varying helix angle, the surface regeneration can be described by a distributed

David Lehotzky · Tamas Insperger
Budapest University of Technology and Economics,
Department of Applied Mechanics, Budapest, Hungary
e-mail: `lehotzkydavid@gmail.com`, `insperger@mm.bme.hu`

T. Vyhlídal, J.-F. Lafay, and R. Sipahi (eds.), *Delay Systems*,
Advances in Delays and Dynamics 1,
DOI: 10.1007/978-3-319-01695-5_6,

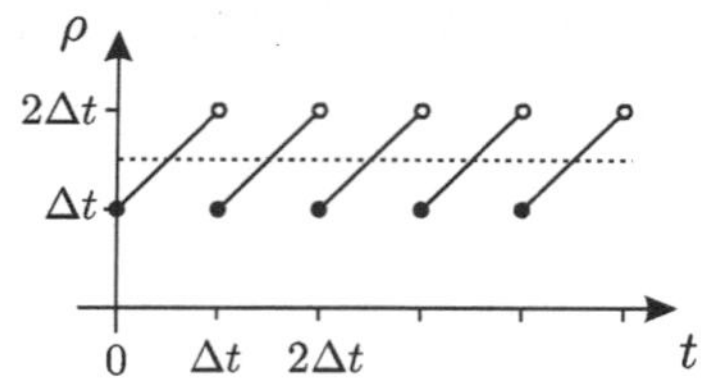

Fig. 1 Representation of the sampling effect as time-varying delay

delay [4]. In these examples, time delays inherently arise due to the structure of the mechanical system and the delayed terms in the governing equations are continuous in time. If these systems are subjected to a digital feedback controller, then discrete-delay terms (i.e., terms with piecewise constant argument) also arise due to the sampling effect [18, 24]. The goal of this chapter is to analyze the stability of systems, where both continuous- and discrete-time delayed terms appears.

Here, we consider Newtonian systems with point delay in the position term subjected to a digital proportional-derivative (PD) controller. The governing equation of such systems can be given in the form

$$\mathbf{M}\ddot{\mathbf{q}}(t)+\mathbf{C}\dot{\mathbf{q}}(t)+\mathbf{K}\mathbf{q}(t)=\mathbf{H}\mathbf{q}(t-\tau)+\mathbf{K}_{\mathrm{p}}\mathbf{q}(t_{j-1})+\mathbf{K}_{\mathrm{d}}\dot{\mathbf{q}}(t_{j-1})\,,\quad t\in\left[t_j,t_{j+1}\right)\,, \tag{1}$$

where $\mathbf{q}\in\mathbb{R}^n$ is the vector of the general coordinates, $\mathbf{M}$, $\mathbf{C}$ and $\mathbf{K}$ are the mass, the damping and the stiffness matrices, $\mathbf{H}$ is a matrix describing the delay effect, τ is the system delay, $\mathbf{K}_{\mathrm{p}}$ and $\mathbf{K}_{\mathrm{d}}$ are the proportional and derivative control matrices, Δt is the sampling step of the digital controller and $t_j=j\Delta t$ are the discrete sampling instants. Thus, the system contains two types of delay terms, the continuous-time point-delay term $\mathbf{q}(t-\tau)$ and the discrete-time delay terms $\mathbf{q}(t_{j-1})$ and $\dot{\mathbf{q}}(t_{j-1})$ with piecewise constant argument over the sampling interval $\left[t_j,t_{j+1}\right)$. Actually, the terms $\mathbf{q}(t_{j-1})$ and $\dot{\mathbf{q}}(t_{j-1})$ can be represented as terms with periodic time delay in the form $\mathbf{q}(t-\rho(t))$ and $\dot{\mathbf{q}}(t-\rho(t))$, where the time delay is a piecewise linear function given as $\rho(t)=t+\Delta t-t_j$, $t\in\left[t_j,t_{j+1}\right)$ (see Fig. 1). According to this interpretation, sampling in the feedback loop presents a parametric excitation in the time delay and the period of the parametric excitation is equal to the sampling period Δt. Consequently, the governing equation is a periodic DDE, and the stability analysis can be performed according to the Floquet theory of DDEs [5, 7]. There exists several numerical methods for the stability analysis of periodic DDEs, the semi-discretization [9, 10], the Chebyshev polynomial approach [2], the spectral element method [11], the method of characteristic matrices [22, 28], Hill's method [12], the full-discretization method [3, 19] or the continuous time approximation [25, 26] can be mentioned as examples. In this chapter, the stability analysis of equation (1) is presented using the semi-discretization method according to [9, 10]. As a new concept, one- and two-point methods with different order of approximations are introduced in the discretization scheme.

2 Semi-discretization

The first-order representation of equation (1) reads

$$\dot{\mathbf{x}}(t) = \mathbf{A}\mathbf{x}(t) + \mathbf{B}\mathbf{x}(t-\tau) + \mathbf{C}\mathbf{x}(t_{j-1}) , \quad t \in [t_j, t_{j+1}) , \tag{2}$$

where

$$\mathbf{x}(t) = \begin{pmatrix} \mathbf{q}(t) \\ \dot{\mathbf{q}}(t) \end{pmatrix}, \mathbf{A} = \begin{pmatrix} \mathbf{0} & \mathbf{I} \\ -\mathbf{M}^{-1}\mathbf{K} & -\mathbf{M}^{-1}\mathbf{C} \end{pmatrix}, \mathbf{B} = \begin{pmatrix} \mathbf{0} & \mathbf{0} \\ \mathbf{H} & \mathbf{0} \end{pmatrix}, \mathbf{C} = \begin{pmatrix} \mathbf{0} & \mathbf{0} \\ \mathbf{K}_\mathrm{p} & \mathbf{K}_\mathrm{d} \end{pmatrix}. \tag{3}$$

Semi-discretization is a numerical technique which can be used for the stability analysis of time-periodic DDEs [9, 10]. The method gives a finite dimensional approximation for the infinite dimensional eigenvalue problem of time-delayed systems. The description presented here is valid for the case when the system delay τ is integer multiple of the sampling period Δt, i.e., when $\kappa = \tau/\Delta t \in \mathbb{Z}$. The semi-discretization is based on the discrete time scale $t_i = ih$, where h is the discretization step determined as $\tau = rh$ and $\Delta t = ph$. Here, r is the delay resolution, p is the period resolution. Clearly, $r/p = \tau/\Delta t = \kappa$. Note that subscript i is used for the discrete time scale of the semi-discretization, while subscript j is used for the discrete time scale $t_j = j\Delta t$ due to the sampling of the controller. In the next two subsections, two types of discretization schemes, the one-point method and the two-point method are detailed.

2.1 One-Point Methods

One-point methods approximate the delayed value of the state variables with values taken from one discrete past time instant. The approximation of equation (1) for the time interval $t \in [t_i, t_{i+1})$ can be given as

$$\dot{\mathbf{x}}(t) = \mathbf{A}\mathbf{x}(t) + \mathbf{D}(t)\mathbf{x}_{i-r} + \mathbf{C}\mathbf{x}_{i-p} , \tag{4}$$

where $\mathbf{D}(t)$ is a weighting matrix which depends on the method and the order of the approximation. Short hand notation is used for $\mathbf{x}(t_{i-r}) = \mathbf{x}_{i-r}$ and respectively for the similar terms. The sketch of the semi-discretization for the case of the zeroth-order one-point method for different steps is shown in Fig. 2 for $r = 20$, $p = 5$ and, consequently, $\kappa = 4$. The initial condition for equation (4) is $\mathbf{x}(t_i) = \mathbf{x}_i$, which provides the continuity of the displacement and velocity functions at time instant $t = t_i$. Using the variation of constants formula, the solution for (4) can be given as

$$\mathbf{x}(t) = \mathrm{e}^{\mathbf{A}(t-t_i)}\mathbf{x}_i + \int_0^{t-t_i} \mathrm{e}^{\mathbf{A}(t-t_i-s)} \left(\mathbf{D}(s)\mathbf{x}_{i-r} + \mathbf{C}\mathbf{x}_{i-p}\right) \mathrm{d}s. \tag{5}$$

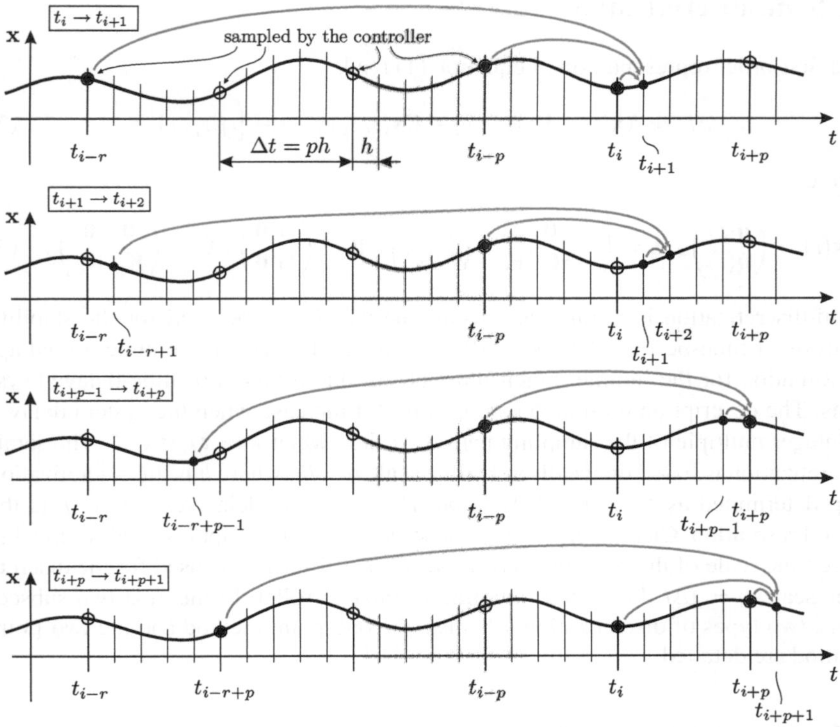

Fig. 2 Sketch of the discretization for the case of zeroth-order one-point method at different steps

Hence the relation between the two end points of the discretization interval is

$$\mathbf{x}_{i+1} = \mathbf{P}\mathbf{x}_i + \mathbf{R}_1\mathbf{x}_{i-r} + \mathbf{R}_C\mathbf{x}_{i-p}, \tag{6}$$

where

$$\mathbf{P} = \mathbf{e}^{\mathbf{A}h}\,, \quad \mathbf{R}_1 = \int_0^h \mathbf{e}^{\mathbf{A}(h-t)}\mathbf{D}(t)\mathrm{d}t\,, \quad \mathbf{R}_C = \int_0^h \mathbf{e}^{\mathbf{A}(h-t)}\mathbf{C}\mathrm{d}t\,. \tag{7}$$

If $\mathbf{A}^{-1}$ exist, then

$$\mathbf{R}_C = -\mathbf{A}^{-1}\left(\mathbf{I} - \mathrm{e}^{\mathbf{A}h}\right)\mathbf{C}\,, \tag{8}$$

where $\mathbf{I}$ denotes the unit matrix. Equation (6) implies the discrete map

$$\mathbf{X}_{i+1} = \mathbf{G}_1\mathbf{X}_i\,, \tag{9}$$

where

$$\mathbf{X}_i = \left(\mathbf{x}_i\ \mathbf{x}_{i-1}\ \ldots\ \mathbf{x}_{i-r}\right)^T \tag{10}$$

is an augmented state vector, and the coefficient matrix for this first step reads

$$\mathbf{G}_1 = \begin{pmatrix} \mathbf{P}\,\mathbf{0} \dots & \mathbf{0}\ \mathbf{R}_C\ \mathbf{0}\ \mathbf{0} \dots & \mathbf{0}\ \mathbf{0}\ \mathbf{R}_1 \\ \mathbf{I}\ \mathbf{0} \dots & \mathbf{0}\ \ \mathbf{0}\ \ \mathbf{0}\ \mathbf{0} \dots & \mathbf{0}\ \mathbf{0}\ \mathbf{0} \\ \mathbf{0}\ \mathbf{I} & \mathbf{0}\ \ \mathbf{0}\ \ \mathbf{0}\ \mathbf{0} & \mathbf{0}\ \mathbf{0}\ \mathbf{0} \\ \vdots & & \vdots \\ \mathbf{0}\ \mathbf{0} & \mathbf{0}\ \ \mathbf{0}\ \ \mathbf{0}\ \mathbf{0} & \mathbf{I}\ \mathbf{0}\ \mathbf{0} \\ \mathbf{0}\ \mathbf{0} \dots & \mathbf{0}\ \ \mathbf{0}\ \ \mathbf{0}\ \mathbf{0} \dots & \mathbf{0}\ \mathbf{I}\ \mathbf{0} \end{pmatrix} . \tag{11}$$

Note that this matrix consists of submatrices of size $2n \times 2n$, namely, $\mathbf{P}$, $\mathbf{R}_C$, $\mathbf{R}_1$, the $2n \times 2n$ unit matrix $\mathbf{I}$ and the $2n \times 2n$ zero matrix $\mathbf{0}$. Matrix $\mathbf{R}_C$ is located at the $(p+1)$th block in the first row of $\mathbf{G}_1$.

Since the control force is constant over the sampling period $[t_i, t_{i+p})$, the approximate differential equation for the second discretization step is

$$\dot{\mathbf{x}}(t) = \mathbf{A}\mathbf{x}(t) + \mathbf{D}(t)\mathbf{x}_{i-r+1} + \mathbf{C}\mathbf{x}_{i-p}\,, \quad t \in [t_{i+1}, t_{i+2})\,. \tag{12}$$

Solving this differential equation similarly to (4), the difference equation between the endpoints of the second discretization is obtained in the form

$$\mathbf{X}_{i+2} = \mathbf{G}_2 \mathbf{X}_{i+1}\,, \tag{13}$$

where the state vector is $\mathbf{X}_i$ is defined as in (10), and the coefficient matrix for the second step reads

$$\mathbf{G}_2 = \begin{pmatrix} \mathbf{P}\,\mathbf{0} \dots & \mathbf{0}\ \mathbf{0}\ \mathbf{R}_C\ \mathbf{0} \dots & \mathbf{0}\ \mathbf{0}\ \mathbf{R}_1 \\ \mathbf{I}\ \mathbf{0} \dots & \mathbf{0}\ \mathbf{0}\ \ \mathbf{0}\ \ \mathbf{0} \dots & \mathbf{0}\ \mathbf{0}\ \mathbf{0} \\ \mathbf{0}\ \mathbf{I} & \mathbf{0}\ \mathbf{0}\ \ \mathbf{0}\ \ \mathbf{0} & \mathbf{0}\ \mathbf{0}\ \mathbf{0} \\ \vdots & & \vdots \\ \mathbf{0}\ \mathbf{0} & \mathbf{0}\ \mathbf{0}\ \ \mathbf{0}\ \ \mathbf{0} & \mathbf{I}\ \mathbf{0}\ \mathbf{0} \\ \mathbf{0}\ \mathbf{0} \dots & \mathbf{0}\ \mathbf{0}\ \ \mathbf{0}\ \ \mathbf{0} \dots & \mathbf{0}\ \mathbf{I}\ \mathbf{0} \end{pmatrix} . \tag{14}$$

Here, matrix $\mathbf{R}_C$ is located at the $(p+2)$th block in the first row of $\mathbf{G}_2$. The only difference between matrices $\mathbf{G}_1$ and $\mathbf{G}_2$ is the location of the sub-matrix $\mathbf{R}_C$. While in $\mathbf{G}_1$, $\mathbf{R}_C$ is located at the $(p+1)$th block in the first row, in $\mathbf{G}_2$, matrix $\mathbf{R}_C$ is located at the $(p+2)$th block in the first row.

For the next discretization interval $[t_{i+2}, t_{i+3})$, matrix $\mathbf{R}_C$ is located at the $(p+3)$th block in the first row, etc. With the induction of this phenomena the structure of the first row of $\mathbf{G}$ is shown in Fig. 3 for different discretization steps.

For the stability analysis of the approximate system (4), the solution should be determined over the period $\Delta t = ph$ of the parametric excitation (i.e., over the principal period). The monodromy mapping for the initial state $\mathbf{X}_i$ is given as

$$\mathbf{X}_{i+p} = \boldsymbol{\Phi}\mathbf{X}_i, \tag{15}$$

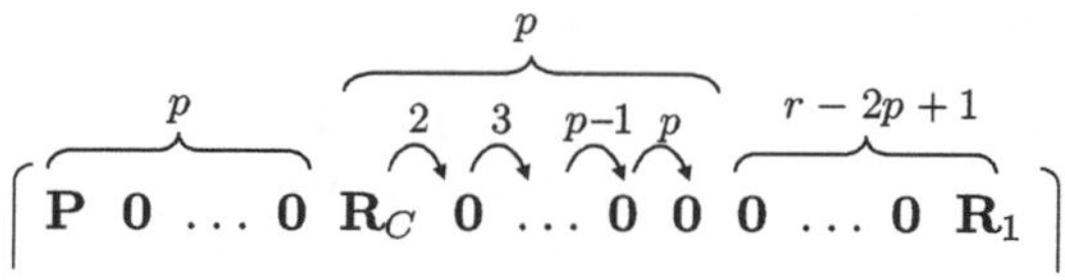

Fig. 3 Top row of **G** matrices for one-point methods

where $\boldsymbol{\Phi} = \mathbf{G}_p \mathbf{G}_{p-1} \ldots \mathbf{G}_2 \mathbf{G}_1$ is the monodromy matrix (Floquet transition matrix). The condition for asymptotic stability is that all eigenvalues of $\boldsymbol{\Phi}$ must be in modulus less then 1, formally

$$|\mu_{\max}| < 1 , \tag{16}$$

where $\mu_{\max} = \max(\mu_i)$ with μ_i, $i = 1, 2, \ldots, (n+1)r$ being the eigenvalues of $\boldsymbol{\Phi}$.

As it was mentioned earlier, semi-discretization of different orders can be represented by the weighting matrix $\mathbf{D}(t)$. In the next points, the zeroth-order and the first-order approximations will be presented for the one-point method.

2.1.1 Zeroth-Order Approximation

This method uses only the discrete vector $\mathbf{q}(t_{i-r})$ of general coordinates to approximate $\mathbf{q}(t-\tau)$ (see Fig. 4). The weighting matrix has the form

$$\mathbf{D} = \begin{pmatrix} \mathbf{0} & \mathbf{0} \\ \mathbf{H} & \mathbf{0} \end{pmatrix} \tag{17}$$

Note that this case corresponds to the standard zeroth-order semi-discretization method given in [10].

2.1.2 First-Order Approximation

This method uses the discrete vector $\mathbf{q}(t_{i-r})$ and its derivative $\dot{\mathbf{q}}(t_{j-r})$ to approximate $\mathbf{q}(t-\tau)$ (see Fig. 4). It can be seen from the structure of the step matrix $\mathbf{G}$ that the derivatives of $\mathbf{q}$ are introduced to the augmented state vector only because the velocity is present in the control force. These derivatives can be used to

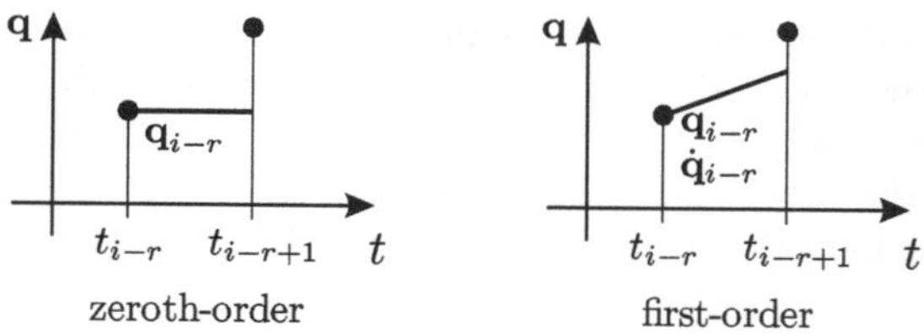

Fig. 4 Sketch of the zeroth- and the first-order one-point methods

give a better approximation for the delayed state variables, which give rise to the first-order approximation, where the delayed term is approximated as $\mathbf{q}(t-\tau) \approx \mathbf{q}(t_{i-r}) + \dot{\mathbf{q}}(t_{i-r})(t-t_{i-r})$. The corresponding weighting matrix reads

$$\mathbf{D}(t) = \begin{pmatrix} \mathbf{0} & \mathbf{0} \\ \mathbf{H} & \mathbf{H}t \end{pmatrix} \tag{18}$$

Note that this first-order approximation is different from the one presented in [10]. Here the first-order approximation of the past state is obtained using $\mathbf{q}(t_{j-r})$ and its derivative $\dot{\mathbf{q}}(t_{j-r})$, while in [10], the first-order approximation is obtained using two subsequent discrete state variables $\mathbf{q}(t_{j-r})$ and $\mathbf{q}(t_{j-r+1})$. This latter case here is called as two-point method.

2.2 Two-Point Methods

Two-point methods take the past values from two subsequent discrete time instants for the approximation of the delayed function. The approximation of equation (2) for the time interval $t \in [t_i, t_{i+1})$ can be given as

$$\dot{\mathbf{x}}(t) = \mathbf{A}\mathbf{x}(t) + \mathbf{D}_1(t)\mathbf{x}_{i-r} + \mathbf{D}_2(t)\mathbf{x}_{i-r+1} + \mathbf{C}\mathbf{x}_{i-p}, \tag{19}$$

where the weighting matrices $\mathbf{D}_1(t)$ and $\mathbf{D}_2(t)$ depend on the order of the approximation and on the weighting between the past values. Similarly to equation (4) the solution of equation (19) can be determined by the variation of constants formula. The relation between the two endpoints of the discretization step is

$$\mathbf{x}_{i+1} = \mathbf{P}\mathbf{x}_i + \mathbf{R}_1\mathbf{x}_{i-r} + \mathbf{R}_2\mathbf{x}_{i-r+1} + \mathbf{R}_C\mathbf{x}_{i-p}, \tag{20}$$

where

$$\begin{aligned} \mathbf{P} &= \mathbf{e}^{\mathbf{A}h}, \quad \mathbf{R}_1 = \int_0^h \mathbf{e}^{\mathbf{A}(h-t)}\mathbf{D}_1(t)\mathrm{d}t, \\ \mathbf{R}_2 &= \int_0^h \mathbf{e}^{\mathbf{A}(h-t)}\mathbf{D}_2(t)\mathrm{d}t, \quad \mathbf{R}_C = \int_0^h \mathbf{e}^{\mathbf{A}(h-t)}\mathbf{C}\mathrm{d}t. \end{aligned} \tag{21}$$

The coefficient matrices $\mathbf{G}$ for two-point methods have similar forms as the ones for the one point methods. The only difference is that one more sub-matrix appears on the right end of the top row. The location of the sub-matrix $\mathbf{R}_C$ for different discretization steps is the same, after each discrete step, this matrix jumps to the right by one, as it is shown in Fig. 5. In the next points, semi-discretization schemes of different orders are presented for the two-point method.

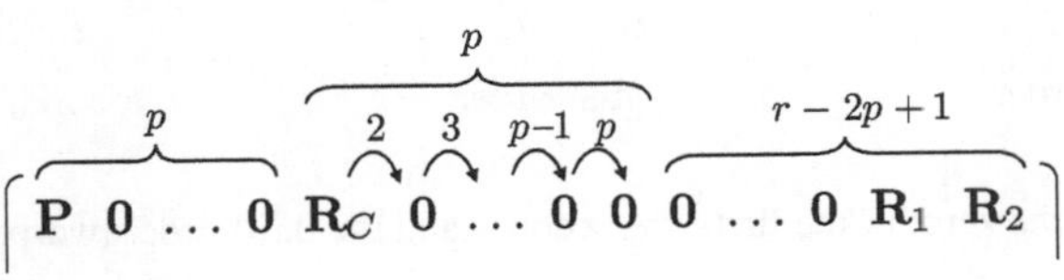

Fig. 5 Top row of **G** matrices for two-point methods

2.2.1 Zeroth-Order Approximation

This method takes the average of the state variable $\mathbf{q}(t)$ at two past time instants, namely at $\mathbf{q}(t_{i-r})$ and $\mathbf{q}(t_{i-r+1})$ to approximate $\mathbf{q}(t-\tau)$ (see Fig. 6). The weighting matrices are

$$\mathbf{D}_1 = \mathbf{D}_2 = \begin{pmatrix} \mathbf{0} & \mathbf{0} \\ \frac{1}{2}\mathbf{H} & \mathbf{0} \end{pmatrix}. \tag{22}$$

Note that this case corresponds to the improved zeroth-order semi-discretization used in [10].

2.2.2 First-Order Approximation

In this method, the delayed term $\mathbf{q}(t-\tau)$ is approximated as a linear function of time using the discrete values $\mathbf{q}(t_{i-r})$ and $\mathbf{q}(t_{i-r+1})$ (see Fig. 6). The weighting matrices are

$$\mathbf{D}_1(t) = \begin{pmatrix} \mathbf{0} & \mathbf{0} \\ (1-t/h)\mathbf{H} & \mathbf{0} \end{pmatrix}, \quad \mathbf{D}_2(t) = \begin{pmatrix} \mathbf{0} & \mathbf{0} \\ t/h\mathbf{H} & \mathbf{0} \end{pmatrix}. \tag{23}$$

Note that this case corresponds to the first-order semi-discretization used in [10].

2.2.3 Second-Order Approximation

This method approximates the state variable values between two past time instants by using not only the past values of the function but also their derivatives. Namely, $\mathbf{q}(t-\tau)$ is approximated by a second-order function using the values $\mathbf{q}(t_{i-r})$, $\mathbf{q}(t_{i-r+1})$ and $\dot{\mathbf{q}}(t_{j-r})$ or $\dot{\mathbf{q}}(t_{j-r+1})$ (see Fig. 6). The second order function is constructed by the linear interpolation between two first order one point approximations at time instants t_{i-r} and t_{i-r+1}. The weighting matrices read

$$\mathbf{D}_1(t) = \begin{pmatrix} \mathbf{0} & \mathbf{0} \\ (1-t/h)\mathbf{H} & t\mathbf{H} \end{pmatrix}, \quad \mathbf{D}_2(t) = \begin{pmatrix} \mathbf{0} & \mathbf{0} \\ t/h\mathbf{H} & (t-h)t/h\mathbf{H} \end{pmatrix}. \tag{24}$$

Note that this discretization concept is different from the ones presented in [10].

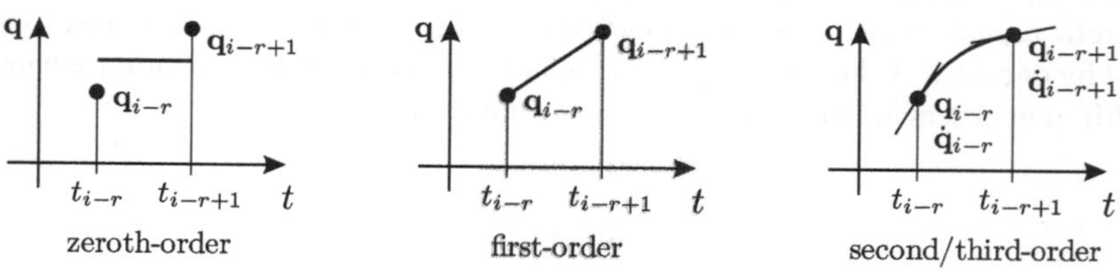

Fig. 6 Sketch of the zeroth-, the first-, the second- and the third-order two-point methods

2.2.4 Third-Order Approximation

In this method, the delayed term $\mathbf{q}(t-\tau)$ is approximated by the discrete values $\mathbf{q}(t_{i-r})$, $\mathbf{q}(t_{i-r+1})$ of the state variables and its derivatives $\dot{\mathbf{q}}(t_{j-r})$ and $\dot{\mathbf{q}}(t_{j-r+1})$ (see Fig. 6). The weighting matrices are

$$\mathbf{D}_1(t) = \begin{pmatrix} \mathbf{0} & \mathbf{0} \\ \left(1-3\left(\frac{t}{h}\right)^2+2\left(\frac{t}{h}\right)^3\right)\mathbf{H} & t\left(1-2\frac{t}{h}+\left(\frac{t}{h}\right)^2\right)\mathbf{H} \end{pmatrix}, \tag{25}$$

$$\mathbf{D}_2(t) = \begin{pmatrix} \mathbf{0} & \mathbf{0} \\ \left(3\left(\frac{t}{h}\right)^2-2\left(\frac{t}{h}\right)^3\right)\mathbf{H} & t\left(-\frac{t}{h}+\left(\frac{t}{h}\right)^2\right)\mathbf{H} \end{pmatrix}. \tag{26}$$

This discretization concept is different from the ones presented in [10].

Comparison of the above methods for different period resolutions shows that the third-order two-point method provides the fastest convergence. Therefore, this method will be used for the for the forthcoming examples.

3 Example: The Delayed Oscillator

Consider first the delayed oscillator subjected to a digital PD controller [13]. The governing equation can be written in the form

$$\ddot{x}(t)+a_1\dot{x}(t)+a_0x(t) = b_0x(t-\tau)-Px(t_{j-1})-D\dot{x}(t_{j-1}), \quad t\in\left[t_j,t_{j+1}\right) \tag{27}$$

where $t_j = j\Delta t$ are the sampling instants for the controller, Δt is the sampling period, P is the proportional gain and D is the derivative gain. The stability chart of this DDE for $P=0$ and $D=0$ is well known in the literature (see the diagrams $P=0$ and $D=0$ in Figures 7 and 8).

The first order representation of equation (27) reads

$$\dot{\mathbf{x}}(t) = \mathbf{A}\mathbf{x}(t)+\mathbf{B}\mathbf{x}(t-\tau)+\mathbf{C}\mathbf{x}(t_{j-1}), \quad t\in\left[t_j,t_{j+1}\right), \tag{28}$$

where

$$\mathbf{x}(t) = \begin{pmatrix} x(t) \\ \dot{x}(t) \end{pmatrix}, \mathbf{A} = \begin{pmatrix} 0 & 1 \\ -a_0 & -a_1 \end{pmatrix}, \mathbf{B} = \begin{pmatrix} 0 & 0 \\ b_0 & 0 \end{pmatrix}, \mathbf{C} = \begin{pmatrix} 0 & 0 \\ -P & -D \end{pmatrix}. \tag{29}$$

Figures 7 and 8 present a series of stability diagrams for different (both negative and positive) proportional and derivative control gains for $\kappa = 2$ and 20. The horizontal and vertical axes are a_0 and b_0 parameters, respectively. The charts were determined by the third-order two-point semi-discretization method. Note that $\kappa = r/p = \tau/\Delta t$, which describes the ratio of the time delay τ and the sampling period Δt. The stability diagrams were obtained numerically by analyzing the eigenvalues of the transition matrix $\boldsymbol{\Phi}$ for a series of fixed parameters.

For large κ values, the sampling period Δt of the digital controller is much smaller than the system delay τ. In these cases, the PD controller practically results

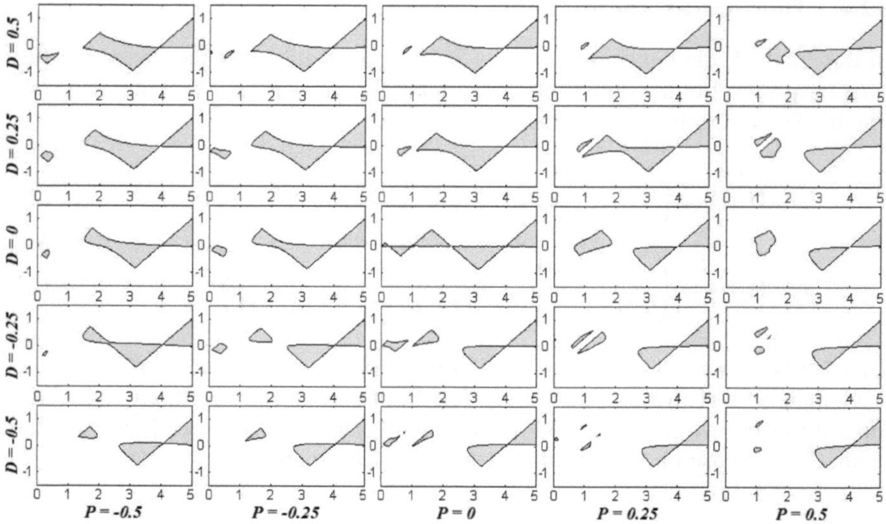

Fig. 7 Stability charts for equation (27) with κ=2, a_1=0, $\tau = 2\pi$ for delay resolution r=40

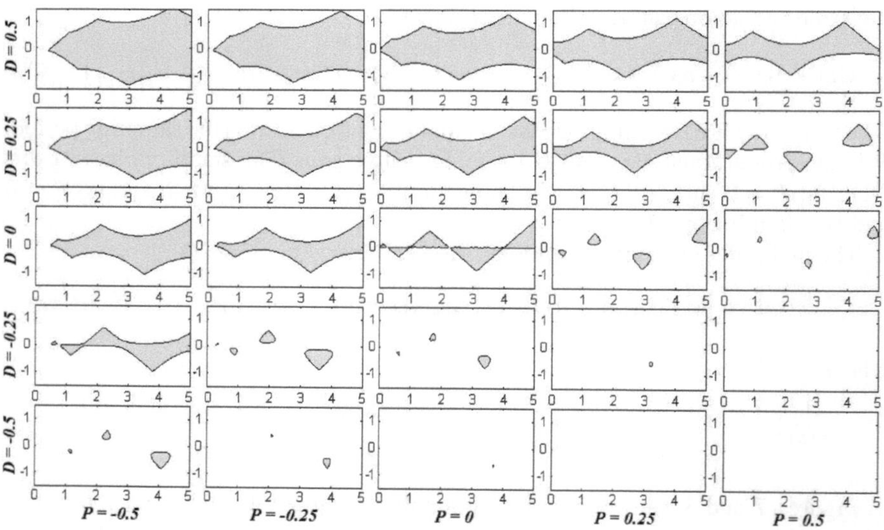

Fig. 8 Stability charts for equation (27) with κ=20, a_1=0, $\tau = 2\pi$ for delay resolution r=40

in an artificial stiffness and a damping in the system, since $a_0 x(t) + P x(t_{j-1}) \approx (a_0 + P)x(t)$ and $a_1 \dot{x}(t) + D\dot{x}(t_{j-1}) \approx (a_1 + D)\dot{x}(t)$ if $t \in [t_j, t_{j+1})$, $t_j = j\Delta t$ and $\Delta t \ll 1$. This tendency can be observed in Fig. 8 (for the case $\kappa = \tau/\Delta t = 20$): positive proportional gains result in a shift of the stability diagram to the left, while positive derivative gains increase the area of the stability domains. An interesting feature in this case is that the stabilizing effect of the positive derivative gains is stronger for negative proportional gains than for positive ones.

For smaller κ values (see Fig. 7), the connection between the control gains and the stability of the system is not so trivial. In these cases, the sampling period Δt of the digital controller and the system delay τ is commensurate, and the combination of the two kind of time delays results in intricate stability charts.

4 Example: Application to Turning Processes

Regenerative machine tool chatter is one of the main limitations of increasing the material removal rate in machining processes [14]. There are several methods and ideas to suppress machine tool chatter, such as the vibration absorber [23], impedance modulation [20], spindle speed variation [21,30] or active control [8,15]. Here, the single degree-of-freedom model of a turning process subjected to a digital PD controller is analyzed. The mechanical model with modal mass m, stiffness k and damping c can be seen in Fig. 9. The linearized governing equation forms as

$$\ddot{\xi}(t)+2\zeta\omega_n\dot{\xi}(t)+(H+\omega_n^2)\xi(t)=H\xi(t-\tau)-k_\mathrm{p}\xi(t_{j-1})-k_\mathrm{d}\dot{\xi}(t_{j-1})\,, \tag{30}$$

where $t\in[t_j,t_{j+1})$, $\xi(t)=x(t)-x_0$ is the displacement around the trivial equilibrium point x_0, $\omega_n=\sqrt{k/m}$ is the undamped natural frequency of the tool, $\zeta=c/(2m\omega_n)$ is the damping ratio of the tool , H is the specific cutting-force coefficient, $Q/m=k_\mathrm{p}\xi(t_{j-1})+k_\mathrm{d}\dot{\xi}(t_{j-1})$ is the specific control force and k_p and k_d are the proportional and derivative control gains [13]. Equation (30) has the same form as (27), hence the stability chart for equation (30) can be analysed by semi-discretization in the same way as for (27). Fig. 10 presents a series of stability diagrams for different (both negative and positive) proportional and derivative control

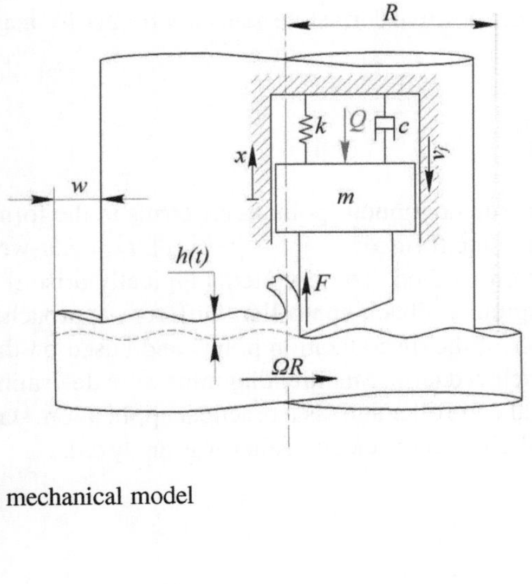

Fig. 9 Sketch of the mechanical model

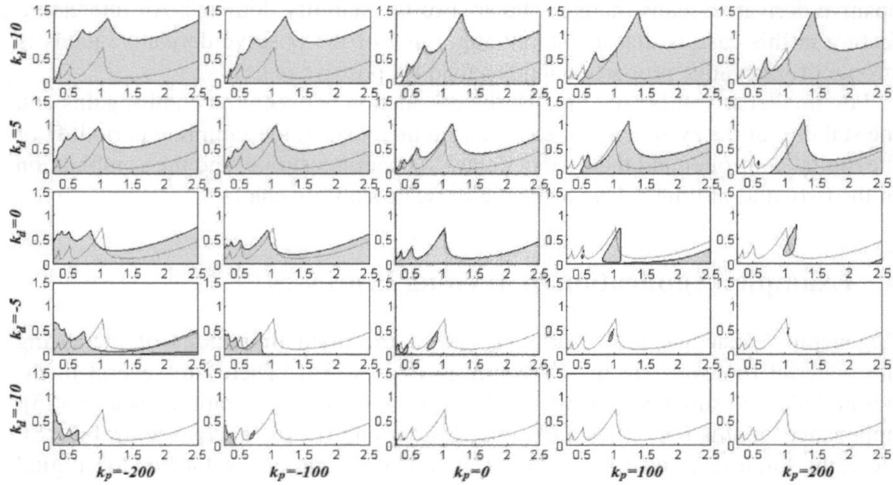

Fig. 10 Stability charts for turning processes with different control parameters for κ=20, r=20 and ζ=0.05

gains for $\kappa = \tau/\Delta t = 20$. The horizontal axis is the dimensionless spindle speed $\Omega/(60 f_n)$, where the spindle speed is $\Omega = 60/\tau$ and the natural frequency of the tool is $f_n = \omega_n/(2\pi)$. The vertical axis is the dimensionless specific cutting-force coefficient H/ω_n^2. In this diagram the exact stability boundaries of the turning process without any control are presented by gray line. The stability diagrams were obtained in the same way as for equation (27). It can be seen that the most important control parameter is the derivative gain k_{d}. Positive derivative gains result in a kind of artificial damping parameter in the system. The effect of the proportional gain k_{p} on the stability is not so significant. Similarly to the delayed oscillator, the stabilizing effect of the positive derivative gains is stronger for negative proportional gains.

5 Conclusions

Dynamical systems with continuous point delay terms in the form $x(t-\tau)$ and discrete delayed terms in the form $x(t_{j-1})$, $t \in [t_j, t_{j+1})$, $t_j = j\Delta t$ were analyzed using the semi-discretization method. These systems typically arise if a delayed system is subjected to a digital feedback controller. Different approaches were presented based on the number of the discretization points and based on the order of the approximation of the delayed term. Stability diagrams were determined for the delayed oscillator with digital controller and, as a practical application, stabilization of turning processes with digital feedback controller was analyzed.

The results related to the delayed oscillator are shown in Figures 7 and 8. The results for the stabilization of the turning process is presented in Fig. 10. The main conclusion is that if the feedback controller is fast enough compared to the time delay in the uncontrolled system, then, since $x(t_{j-1}) \approx x(t)$ and $\dot{x}(t_{j-1}) \approx \dot{x}(t)$ on $t \in [t_j, t_{j+1})$, $t_j = j\Delta t$, positive proportional and derivative gains act as a kind of artificial stiffness and damping in the system. Therefore if $\kappa \gg 1$ then, considering stability, an analogue PD control approximates well the digital PD control. In this case, it was observed that the stabilizing effect of the positive derivative gains is stronger for negative proportional gains. If the sampling period Δt of the feedback controller is commensurate to the system delay τ, then the combination of the two kind of time delays result in an intricate stability picture.

In the equations analyzed in this paper, two types of delays were present: the continuous delay $x(t-\tau)$ and the discrete delay $x(t_{j-1})$. While the continuous delay attributes an infinite dimensional nature to the system, the discrete delay presents a kind of intermittence or discontinuity in the system. This combination of time delays may also be important in human balancing models with reflex delay, where the human motor control is often modeled as a system with discontinuous feedback [6, 16, 27].

Acknowledgements. This work was supported by the Hungarian National Science Foundation under grant OTKA-K105433. The first author was supported by the Rostoczy Foundation Scholarship Program.

References

1. Bachrathy, D., Stepan, G., Turi, J.: State dependent regenerative effect in milling processes. J. Comput. Nonlin. Dyn.–T ASME 6, 041002 (2011)
2. Butcher, E.A., Ma, H., Bueler, E., Averina, V., Szabó, Z.: Stability of time-periodic delay-differential equations via Chebyshev polynomials. Int. J. Numer. Mech. Eng. 59, 895–922 (2004)
3. Ding, Y., Zhu, L.M., Zhang, X.J., Ding, H.: A full-discretization method for prediction of milling stability. Int. J. Mach. Tool Manu. 50, 502–509 (2010)
4. Dombovari, Z., Altintas, Y., Stepan, G.: The effect of serration on mechanics and stability of milling cutters. Int. J. Mach. Tool Manu. 50, 511–520 (2010)
5. Farkas, M.: Periodic motions. Springer, New York (1994)
6. Gawthrop, P., Loram, I., Lakie, M., Gollee, H.: Intermittent control: a computational theory of human control. Biol. Cybern. 104, 31–51 (2011)
7. Hale, J.K.: Theory of functional differential equations. Springer, New York (1977)
8. Hu, Q., Krawcewicz, W., Turi, J.: Stabilization in a state-dependent model of turning processes. SIAM J. Appl. Math. 24, 1–24 (2012)
9. Insperger, T., Stepan, G.: Semi-discretization method for delayed systems. Int. J. Numer. Meth. Eng. 55, 503–518 (2002)
10. Insperger, T., Stepan, G.: Semi-Discretization for Time-Delay Systems. Springer, New York (2011)
11. Khasawneh, F.A., Mann, B.P.: A spectral element approach for the stability of delay systems. Int. J. Numer. Meth. Eng. 87, 566–592 (2011)

12. Lampe, B.P., Rosenwasser, E.N.: Stability investigation for linear periodic time-delayed systems using Fredholm theory. Automat. Rem. Contr. 72(1), 38–60 (2011)
13. Lehotzky, D., Insperger, T.: Stability of delayed oscillators subjected to digital PD control. In: 10th IFAC Workshop on Time Delay Systems, Boston, USA, IFAC-PapersOnline (2012)
14. Mancisidor, I., Zatarain, M., Munoa, J., Dombovari, Z.: Fixed boundaries receptance coupling substructure analysis for tool point dynamics prediction. Adv. Mat. Res. 223, 622–631 (2011)
15. Munoa, J., Mancisidor, I., Loix, N., Uriarte, L.G., Barcena, R., Zatarain, M.: Chatter suppression in ram type travelling column milling machines using a biaxial inertial actuator. CIRP Ann.–Manuf. Techn. 62(1), 407–410 (2013)
16. Milton, J., Townsend, J.L., King, M.A., Ohira, T.: Balancing with positive feedback: the case for discontinuous control. Philos. T. R. Soc. A 367, 1181–1193 (2009)
17. Orosz, G., Stepan, G.: Subcritical Hopf bifurcations in a car-following model with reaction-time delay. P. Roy. Soc. A–Math. Phy. 462, 2643–2670 (2006)
18. Pohlmann, K.C.: Principles of Digital Audio. McGraw-Hill, New York (2000)
19. Quo, Q., Sun, Y., Jiang, Y.: On the accurate calculation of milling stability limits using third-order full-discretization method. Int. J. Mach. Tool Manu. 62, 61–66 (2012)
20. Segalman, D.J., Butcher, E.A.: Suppression of regenerative chatter via impendance modulation. J. Vib. Control 6, 243–256 (2000)
21. Seguy, S., Insperger, T., Arnaud, L., Dessein, G., Peigne, G.: On the stability of high-speed milling with spindle speed variation. Int. J. Adv. Manuf. Tech. 48, 883–895 (2010)
22. Sieber, J., Szalai, R.: Characteristic matrices for linear periodic delay differential equations. SIAM J. Appl. Dyn. Sys. 10(1), 129–147 (2011)
23. Sims, N.D.: Vibration absorbers for chatter suppression: a new analytical tuning methodology. J. Sound Vib. 301, 592–607 (2000)
24. Stepan, G.: Vibrations of machines subjected to digital force control. Int. J. Solids Struct. 38, 2149–2159 (2001)
25. Sun, J.Q.: A method of continuous time approximation of delayed dynamical systems. Commun. Nonlinear Sci. 14, 998–1007 (2009)
26. Sun, J.-Q., Song, B.: Control studies of time-delayed dynamical systems with the method of continuous time approximation. Commun. Nonlinear Sci. Numer. Simulat. 14, 3933–3944 (2009)
27. Suzuki, Y., Nomura, T., Casadio, M., Morasso, P.: Intermittent control with ankle, hip, and mixed strategies during quiet standing: A theoretical proposal based on a double inverted pendulum model. J. Theor. Biol. 310, 55–79 (2012)
28. Szalai, R., Stepan, G., Hogan, S.J.: Continuation of bifurcations in periodic delay-differential equations using characteristic matrices. SIAM J. Sci. Comput. 28, 1301–1317 (2006)
29. Takacs, D., Orosz, G., Stepan, G.: Delay effects in shimmy dynamics of wheels with stretched-string like tyres. Eur. J. Mech. A-Solid 28, 516–525 (2009)
30. Zatarain, M., Bediaga, I., Munoa, J., Lizarralde, R.: Stability of milling processes with continuous spindle speed variation: Analysis in the frequency and time domains, and experimental correlation. CIRP Ann.–Manuf. Techn. 57, 379–384 (2008)

Part II
Networks and Graphs

Control Design for Teleoperation over Unreliable Networks: A Predictor-Based Approach

Alexandre Kruszewski, Bo Zhang, and Jean-Pierre Richard

Abstract. This chapter considers the problem of teleoperation over an unreliable communication network. Roughly, teleoperation applications are two systems communicating trough a communication network, the goal of which is to synchronize some variables. This problem is closely related to network control theory, where not only the stability of the two systems has to be ensured but also some performances and robustness. This chapter proposes a robust control design based on delay-dependent Lyapunov-Krasovskii conditions, where the performances are guaranteed through an H_∞-like optimization.

1 Introduction

Bilateral teleoperation are systems composed with two parts: the master and the slave part. The master part is an human operated low torque robotic arm (haptic interface) used to provide a reference to the slave part and to give some force feedback to the human operator. The salve part is a high torque robotic arm which has to follow the references provided by the master and return back the constraints coming from its environment. The two parts communicate trough a network which introduces in most case a non negligible delay.

Alexandre Kruszewski · Jean-Pierre Richard
LAGIS, CNRS UMR 8219, Laboratoire d'Automatique, Génie Informatique et Signal
Université Lille Nord de France, Ecole Centrale de Lille, 59651 Villeneuve d'Ascq, France

Jean-Pierre Richard
Inria project-team Non-A.
e-mail: {alexandre.kruszewski,jean-pierre.richard}@ec-lille.fr

Bo Zhang
China National Electronics Import & Export Corporation, CEIEC. 17 Fuxing Road, Haidian District, Beijing, China
e-mail: zhangbo829@outlook.com

T. Vyhlídal, J.-F. Lafay, and R. Sipahi (eds.), *Delay Systems*,
Advances in Delays and Dynamics 1,
DOI: 10.1007/978-3-319-01695-5_7,

So from a control point of view, a bilateral teleoperation system is a closed-loop structure with data transmission over a network where the main problem is to design a control structure ensuring the stability and performance (position tracking and force feedback fidelity). These goals can be achieved by modeling the network with time-varying delays for the network part and by considering the non-linear behavior of the robotic arms as time-varying model uncertainties ([7] and the references therein).

Many recent methods have addressed the stability and performance problems of bilateral teleoperation:

• *Passivity-based control* under variable delays: The survey [10] revisits many passivity-based controllers for bilateral teleoperation systems, including scattering and wave variables. Based on the energy and power considerations, time domain passivity control [13], [16] without the transformation of wave variables have also been proposed. Overall, the latest passivity-based results can solve the stabilization problem under time-varying delays, but the system performance is not guaranteed in terms of tracking quality.

• *Non-passive control*: Various control strategies have been proposed for a *non-passive environment* under constant or time-varying delays. The readers can refer to [1], [2] for more details on these methods. However, very few of them focus on perturbations and model uncertainties.

Apart from the stability of teleoperation systems under time-varying delays, there are also two kinds of performances for bilateral teleoperation, which are not that much addressed [14]:

• *Position tracking* (or *position coordination*): The slave robot should follow the motion of the master robot maneuvered by the human operator.

• *Force tracking* (or *force coordination*): The environmental force acting on the slave (when it contacts the external environment) should be accurately and real-time transmitted to the master. This can be achieved by the force-reflecting, in which the human operator feels haptic sensations as if he/she was actually present at the slave side.

[17], [18] introduced a force-reflecting, predictor-based control scheme (a predictor of the master's state is located on the slave side). The control design presented in these papers guarantees the stability and the position/force tracking of the closed-loop system under time-varying delays and uncertainties. These results are based on Lyapunov-Krasovskii functionals (LKF) and H_∞ control techniques [4, 8, 15]. The main results are given as a set of Linear Matrix Inequalities (LMI) to be solved [3]. H_∞ is used to minimize the position tracking error despite the unknown inputs (human operator, environment) and the LKF to ensure the robustness with respect to time varying delays.

In the continuity of this LKF-H_∞ coupling approach, this chapter presents different control structures and discuss about their differences.

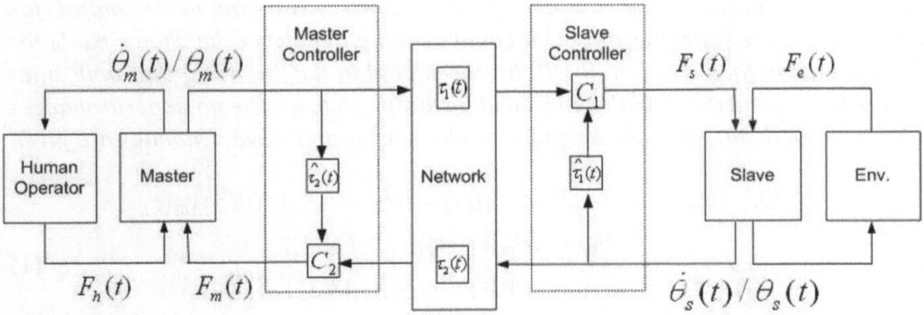

Fig. 1 Bilateral state feedback control scheme

2 A Delay Formulation for Teleoperation Problems

Following [17], it is considered a bilateral state-feedback control scheme. Note that, ignoring the details of master and slave controllers, this scheme corresponds to a general teleoperation system with five entities (human operator, master robot, communication medium, slave robot, environment). In this general scheme [1]:

• Master controller and slave controller are the global controllers we should design so to ensure both the asymptotic stability of the whole system and the position/force tracking between the master and the slave (in Figure 1, C_1 and C_2 are the global controllers that we should design).

• $F_m(t)$ and $F_s(t)$ are the actuated inputs of the master and of the slave.

• $F_h(t)$ and $F_e(t)$ are the forces of the human operator and of the environment. Note that, in an H_∞ context, these forces can be considered as the perturbations in the sense that the global controllers have to minimize their effects on the output tracking error.

• $\tau_1(t)$ (from the master to the slave) and $\tau_2(t)$ (from the slave to the master) are the delays resulting from the communication, access time, and packet loss effects [9]. They are time-varying and asymmetric ($\tau_1(t) \neq \tau_2(t)$).

• $\dot{\theta}_m(t)/\theta_m(t)$ and $\dot{\theta}_s(t)/\theta_s(t)$ are the velocities/positions of the master and the slave.

• The information transferred between the master and the slave can be the positions, the velocities or the estimated forces of the human operator and environment (in Figure 1, the velocities/positions of the master and the slave).

According to the general scheme, the following assumptions are made.

Assumption 1. ***The master and slave robots can be locally controlled so to be modeled by linear systems with uncertainties.*** *It is reasonable to consider linear model for the two sides in many teleoperation situations because most of the used robots are serial ones. These robots are easily linearized by choosing the right controller with a linear reference model, as soon as the trajectories respect the systems boundaries (control saturation, joints not close to a physical limits or singularities ...). The* *use of the calculated torque control (for low friction) or a sliding mode control (for high friction) are examples of how it could be done. Whatever the choice made for this linearizing local control, it will be considered in the following that both sides are disturbed, and the global control will be design so to ensure good performances despite those disturbances/uncertainties. The master and slave dynamics are given by:*

$$\begin{aligned}(\Sigma_m) \quad \dot{x}_m(t) &= ((A_m + \Delta A_m(t)) - (B_m + \Delta B_m(t))K_m^0)x_m(t) \\ &+ (B_m + \Delta B_m(t))(F_m(t) + F_h(t)), \\ (\Sigma_s) \quad \dot{x}_s(t) &= ((A_s + \Delta A_s(t)) - (B_s + \Delta B_s(t))K_s^0)x_s(t) \\ &+ (B_s + \Delta B_s(t))(F_s(t) + F_e(t)),\end{aligned} \tag{1}$$

where $x_m(t) = \dot{\theta}_m(t)$, $x_s(t) = \dot{\theta}_s(t)$, and K_m^0, K_s^0 are the local controllers of the master and slave ensuring the speed stability.

Assumption 2. *The communication delays are bounded:* $\tau_1(t)$, $\tau_2(t) \in [h_1, h_2]$, $h_1 > 0$, $h_2 < \infty$. *The communication delays over unreliable networks (such as the internet or wifi) are not bounded, but in any teleoperation process, a delay limit is chosen to avoid dangerous situations: If the delay goes out of this bound, the teleoperation is stopped in a safe, frozen position.*

Assumption 3. *Master and slave clocks are perfectly synchronized and the data packets include a time-stamp indicating their sending time. This allows the Master (Slave) to compute the Slave-to-Master (Master-to-Slave) delay:* $\hat{\tau}_1(t) = \tau_1(t)$, $\hat{\tau}_2(t) = \tau_2(t)$. *The clocks of both sides can be synchronized. Even if it is impossible to get perfect synchronization, it is possible to get a close enough result to consider it as perfect. It is needed to get an error of synchronization negligible with respect to the delay. For small networks with a low traffic, it could be achieved using a network time protocol (NTP) or a GPS clock [19].*

Assumption 4. *The external forces $F_h(t)$ and $F_e(t)$ are accessible.* *The external forces can be obtained either by using sensors (strain gauge) or a unknown input observer. The choice between these two solutions depends on the degree of fidelity of the models: The result obtained using strain gauges is independent on the model fidelity but can be mechanically hard to adapt. The unknown input observer technique is only suitable if the model is really good and does not present too much friction. Usually, strain gauges are used at the slave side, and observer at the master side.*

3 Force-Reflecting Emulator Control Scheme

3.1 System Description and Problem Formulation

The force-reflecting emulator control scheme is presented in Figure 2. Let us give a description of the control scheme: $F_m(t)$ and $F_s(t)$ are the actuated inputs of the master and the slave; $F_h(t)$ and $F_e(t)$ are the forces of the human operator and environment on the system; $\hat{F}_h(t)$ and $\hat{F}_e(t)$ are the estimations of these two forces,

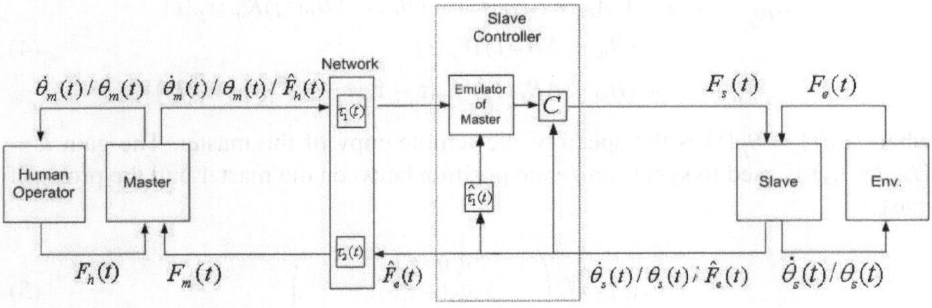

Fig. 2 Force-reflecting emulator control scheme

which can be obtained by adding the perturbation observers in reality; $\dot{\theta}_m(t)/\theta_m(t)$, $\dot{\theta}_s(t)/\theta_s(t)$ are the velocities/positions of the master and the slave.

The communication delays $\tau_1(t)$, $\tau_2(t) \in [h_1, h_2]$, $h_1 \geq 0$. $\hat{\tau}_1(t)$ is the estimated network delay, thanks to time-stamped data packet exchanges using a network time protocol as in [9] between the master and slave, the master and slave clocks are synchronized and $\hat{\tau}_1(t)$ is available at slave's side: $\hat{\tau}_1(t) = \tau_1(t)$.

From the master to slave, the information transferred are the velocity/position of the master and the estimated force $\hat{F}_h(t)$. However, from the slave to the master, only the estimated force $\hat{F}_e(t)$ is transferred, so the force tracking $F_m(t) = \hat{F}_e(t - \tau_2(t))$ is realized if the stability of the whole system is verified.

Note that, in both the master and slave, there exist norm-bounded and time-varying model uncertainties $(\Delta A_m(t), \Delta B_m(t), \Delta A_s(t), \Delta B_s(t))$ as follows:

$$(\Sigma_m) \quad \dot{x}_m(t) = ((A_m + \Delta A_m(t)) - (B_m + \Delta B_m(t))K_m^0)x_m(t) + (B_m + \Delta B_m(t))(F_m(t) + F_h(t)), \tag{2}$$

$$(\Sigma_s) \quad \dot{x}_s(t) = ((A_s + \Delta A_s(t)) - (B_s + \Delta B_s(t))K_s^0)x_s(t) + (B_s + \Delta B_s(t))(F_s(t) + F_e(t)), \tag{3}$$

where $x_m(t) = \dot{\theta}_m(t)$, $x_s(t) = \dot{\theta}_s(t)$. K_m^0 and K_s^0, to be designed later, are the local controllers of the master and slave ensuring the speed stability.

In the slave controller, the emulator of master is like a remote observer of the master, which is introduced at the slave side in order to reduce the impact of the time-varying delays. Thus, the model used for this emulator is the same as the master. Since the master model is a nonlinear one put in the form an uncertain linear model, the remote emulator will follow the same treatment. The main difference between a classical state observer and this structure is in the way states are synchronized. Here, the emulator state is synchronized with a control-like structure. This will provide a more realistic behavior of the emulator state during the convergence stage (acceleration, speed and position trajectory are still consistent).

$$
\begin{aligned}
(\Sigma_p) \quad \dot{x}_p(t) = & ((A_m + \Delta A_p(t)) - (B_m + \Delta B_p(t))K_m^0)x_p(t) \\
& - (B_m + \Delta B_p(t))F_p(t) \\
& + (B_m + \Delta B_p(t))(\hat{F}_e(t - \hat{\tau}_1(t)) + \hat{F}_h(t - \tau_1(t))),
\end{aligned} \tag{4}
$$

where $x_p(t) = \dot{\theta}_p(t)$ is the speed of the remote copy of the master. The gain $L = \begin{pmatrix} L_1 & L_2 & L_3 \end{pmatrix}$ is used to synchronize the position between the master and the proxy of master,

$$
F_p(t) = L \begin{pmatrix} \dot{\theta}_p(t-\hat{\tau}_1(t)) \\ \dot{\theta}_m(t-\tau_1(t)) \\ \theta_p(t-\hat{\tau}_1(t)) - \theta_m(t-\tau_1(t)) \end{pmatrix}. \tag{5}
$$

Next, $K = \begin{pmatrix} K_1 & K_2 & K_3 \end{pmatrix}$ is the gain of the controller C,

$$
F_s(t) = K \begin{pmatrix} \dot{\theta}_s(t) \\ \dot{\theta}_p(t) \\ \theta_s(t) - \theta_p(t) \end{pmatrix}. \tag{6}
$$

In the following, the model uncertainties are considered with the following structure:

$$
\Delta A_i(t) = G_i \Delta(t) D_i, \quad \Delta B_i(t) = H_i \Delta(t) E_i, \tag{7}
$$

where $i = \{m, s, p\}$ and G_i, D_i, H_i, E_i are constant matrices of appropriate dimensions. $\Delta(t)$ is a time-varying matrix satisfying $\Delta(t)^T \Delta(t) \leqslant I$.

The controller design is made by solving three distinct problems and check the global stability/performance index. Note that at the moment, no solution are known to solve everything in a single step because the control gain matrix in the global problem has some strong constraints. For example: the current master state is not available at slave side, which imposes various control gains to be zero. The solving steps are described in the next sections. The conditions provided hereafter are obtained by applying the robustness approach described in [20] on the result of [18].

3.2 Problem 1: Local Controller Design

The local controllers are designed by solving the following LMI conditions for pairs (A, B) given it the systems equations (2) and (3) :

Find $P > 0$, N matrices with suitable dimensions and ρ_A, ρ_B two positive scalars such that [20]:

$$
\begin{pmatrix}
AP + PA^T - N^T B^T - BN & GP & HP & N^T E^T & PD^T \\
* & -\frac{1}{\rho_A} I & 0 & 0 & 0 \\
* & * & -\frac{1}{\rho_B} I & 0 & 0 \\
* & * & * & -\rho_B I & 0 \\
* & * & * & * & -\rho_A I
\end{pmatrix} < 0 \tag{8}
$$

The control gains are given by NP^{-1}. Note that if the local controller performance needs to be improved, it is possible to use pole placement techniques such as D-stability or decay rate techniques.

3.3 Problem 2: Master-Emulator Synchronization

Synchronizing the emulator state to the master state is a control problem where it is considered the following system:

$$\begin{cases} \dot{x}_{mp}(t) = (A+\Delta A(t))x_{mp}(t) \\ \qquad +(B_1+\Delta B_1(t))Lx_{mp}(t-\tau_1(t)) \\ \qquad +(B+\Delta B(t))w_{mp}(t), \\ z_{mp}(t) = Cx_{mp}(t), \end{cases} \tag{9}$$

$$x_{mp}(t) = \begin{pmatrix} \dot{\theta}_p(t) \\ \dot{\theta}_m(t) \\ \theta_p(t)-\theta_m(t) \end{pmatrix}, \quad w_{mp}(t) = \begin{pmatrix} \hat{F}_e(t-\hat{\tau}_1(t))+\hat{F}_h(t-\tau_1(t)) \\ F_m(t)+F_h(t) \end{pmatrix}, \tag{10}$$
$$z_{mp}(t) = \begin{pmatrix} \theta_p(t)-\theta_m(t) \end{pmatrix},$$

$$A = \begin{pmatrix} A_m-B_mK_m^0 & 0 & 0 \\ 0 & A_m-B_mK_m^0 & 0 \\ 1 & -1 & 0 \end{pmatrix},$$
$$\Delta A(t) = \begin{pmatrix} \Delta A_p(t)-\Delta B_p(t)K_m^0 & 0 & 0 \\ 0 & \Delta A_m(t)-\Delta B_m(t)K_m^0 & 0 \\ 0 & 0 & 0 \end{pmatrix},$$
$$B_1(t) = \begin{pmatrix} -B_p(t) \\ 0 \\ 0 \end{pmatrix}, \tag{11}$$
$$\Delta B_1(t) = \begin{pmatrix} -\Delta B_p(t) \\ 0 \\ 0 \end{pmatrix} = H_1\Delta(t)E_1,$$
$$B = \begin{pmatrix} B_m & 0 \\ 0 & B_m \\ 0 & 0 \end{pmatrix} = \begin{pmatrix} B_{mp}^1 & B_{mp}^2 \end{pmatrix},$$
$$\Delta B(t) = \begin{pmatrix} \Delta B_p(t) & 0 \\ 0 & \Delta B_m(t) \\ 0 & 0 \end{pmatrix} = H\Delta(t)E, \quad C = \begin{pmatrix} 0 & 0 & 1 \end{pmatrix}.$$

The synchronization is achieved by designing the emulator control gain L so that the quadratic error between the emulator output and the master output $z_{mp}(t)^2$ is minimized according to the disturbance $w_{mp}(t)^2$ coming from the human interaction and the environment. This system is stable and satisfies $J(w) = \int_0^\infty (z(t)^T z(t) - \gamma^2 w(t)^T w(t))dt < 0$, with the control gains MP_2^{-1} for any time-varying delays $\tau_1(t) \in [h_1, h_2]$ if the following LMI problem is feasible:

Minimize γ with $P > 0$, P_2, M, Q_1, Q_2, R_1, R_2 matrices with suitable dimensions and ξ, ξ_1,ξ_2,ξ_3, ρ_A, ρ_B two positive scalar such that [20]:

$$\Gamma = \begin{pmatrix} \Gamma_{11}^1+\Gamma_{11}^2+\Gamma_{11}^{2\,T} & \Gamma_{12} & \Gamma_{13} & e_1P_2^TC^T & \Gamma_{15} \\ * & \Gamma_{22}^1-\xi P_2-\xi P_2^T & \xi B & 0 & \Gamma_{25} \\ * & * & -\gamma^2 I & 0 & \Gamma_{35} \\ * & * & * & -I & 0 \\ * & * & * & * & \Gamma_{55} \end{pmatrix} < 0, \quad \begin{pmatrix} R_2 & S \\ S^T & R_2 \end{pmatrix} \geqslant 0, \tag{12}$$

$$\Gamma_{11}^2 = \begin{pmatrix} P_2^TA^T & \xi_1P_2^TA^T & \xi_2P_2^TA^T & \xi_3P_2^TA^T \\ -M^TB_1{}^T & -\xi_1M^TB_1{}^T & -\xi_2M^TB_1{}^T & -\xi_3M^TB_1{}^T \\ 0 & 0 & 0 & 0 \\ 0 & 0 & 0 & 0 \end{pmatrix}, \tag{13}$$

$$\Gamma_{12} = e_1 P + \xi \begin{pmatrix} P_2^T A^T \\ -M^T B_1{}^T \\ 0 \\ 0 \end{pmatrix} - \begin{pmatrix} P_2 \\ \xi_1 P_2 \\ \xi_2 P_2 \\ \xi_3 P_2 \end{pmatrix}, \quad \Gamma_{13} = \begin{pmatrix} BP_2 \\ \xi_1 B \\ \xi_2 B \\ \xi_3 B \end{pmatrix}. \tag{14}$$

$$\Gamma_{15}^T = \begin{pmatrix} 0 & DP_2 & 0 & 0 \\ 0 & -E_1 M & 0 & 0 \\ G^T & \xi_1 G^T & \xi_2 G^T & \xi_3 G^T \\ H_1^T & \xi_1 H_1^T & \xi_2 H_1^T & \xi_3 H_1^T \\ 0 & -E_1 M & 0 & 0 \\ DP_2 & 0 & 0 & 0 \\ 0 & 0 & 0 & 0 \\ 0 & 0 & 0 & 0 \\ H^T & \xi_1 H^T & \xi_2 H^T & \xi_3 H^T \\ 0 & 0 & 0 & 0 \\ 0 & 0 & 0 & 0 \\ 0 & 0 & 0 & 0 \end{pmatrix}, \Gamma_{25}^T = \begin{pmatrix} 0 \\ 0 \\ 0 \\ 0 \\ 0 \\ 0 \\ \xi H^T \\ \xi G^T \\ 0 \\ 0 \\ \xi H^T \\ 0 \end{pmatrix}, \Gamma_{35}^T = \begin{pmatrix} 0 \\ 0 \\ 0 \\ 0 \\ 0 \\ 0 \\ 0 \\ 0 \\ 0 \\ EP_2 \\ 0 \\ E \end{pmatrix}, \tag{15}$$

$$\Gamma_{55} = diag\left(\tfrac{-1}{\rho_A} I,\ \tfrac{-1}{\rho_{B_1}} I,\ \rho_A I,\ \rho_{B_1} I,\ \tfrac{-1}{\rho_{B_1}} I,\ \tfrac{-1}{\rho_A} I,\ \rho_{B_1} I,\ \rho_A I,\ \rho_B I,\ \tfrac{-1}{\rho_B} I \right), \tag{16}$$

$$\begin{aligned} e_1 &= col\{I,0,0,0\}, \quad e_2 = col\{0,I,0,0\}, \\ e_3 &= col\{0,0,I,0\}, \quad e_4 = col\{0,0,0,I\}. \end{aligned} \tag{17}$$

$$\begin{aligned} \Gamma_{11}^1 &= e_1 Q_1 e_1^T - e_3 Q_1 e_3^T + e_1 Q_2 e_1^T - e_4 Q_2 e_4^T \\ &- (e_1 - e_3) R_1 (e_1 - e_3)^T \\ &- \begin{pmatrix} e_3 - e_2 & e_2 - e_4 \end{pmatrix} \begin{pmatrix} R_2 & S \\ S^T & R_2 \end{pmatrix} \begin{pmatrix} e_3^T - e_2^T \\ e_2^T - e_4^T \end{pmatrix}, \\ \Gamma_{22}^1 &= h_1^2 R_1 + (h_2 - h_1)^2 R_2. \end{aligned} \tag{18}$$

3.4 Problem 3: Slave-Emulator Synchronization

If the emulator is synchronized with the master, then synchronizing the slave state to the emulator state is a control problem where it is considered the following system:

$$\begin{cases} \dot{x}_{ps}(t) = (A + \Delta A(t)) x_{ps}(t) + (B + \Delta B(t)) w_{ps}(t), \\ z_{ps}(t) = C x_{ps}(t). \end{cases} \tag{19}$$

with:

$$\begin{aligned} x_{ps}(t) &= \begin{pmatrix} \dot{\theta}_s(t) \\ \dot{\theta}_p(t) \\ \theta_s(t) - \theta_p(t) \end{pmatrix}, \quad z_{ps}(t) = \begin{pmatrix} \theta_s(t) - \theta_p(t) \end{pmatrix}, \\ w_{ps}(t) &= \begin{pmatrix} F_e(t) \\ \hat{F}_e(t - \hat{\tau}_1(t)) + \hat{F}_h(t - \tau_1(t)) - F_p(t) \end{pmatrix}, \end{aligned} \tag{20}$$

where:

$$A = \begin{pmatrix} A_s - B_sK_s^0 - B_sK_1 & -B_sK_2 & -B_sK_3 \\ 0 & A_m - B_mK_m^0 & 0 \\ 1 & -1 & 0 \end{pmatrix},$$

$$\Delta A(t) = \begin{pmatrix} \Delta A_s(t) - \Delta B_s(t)K_s^0 - \Delta B_s(t)K_1 & -\Delta B_s(t)K_2 & -\Delta B_s(t)K_3 \\ 0 & \Delta A_p(t) - \Delta B_p(t)K_m^0 & 0 \\ 0 & 0 & 0 \end{pmatrix}, \tag{21}$$

$$B = \begin{pmatrix} B_s & 0 \\ 0 & B_m \\ 0 & 0 \end{pmatrix} = \begin{pmatrix} B_{ps}^1 & B_{ps}^2 \end{pmatrix},$$

$$\Delta B(t) = \begin{pmatrix} \Delta B_s(t) & 0 \\ 0 & \Delta B_p(t) \\ 0 & 0 \end{pmatrix}, \quad C = \begin{pmatrix} 0 & 0 & 1 \end{pmatrix}.$$

The synchronization is achieved by designing the slave control gain K in such a way that the quadratic error between the slave output and the emulator output $z_{ps}(t)^2$ is minimized according to the disturbance $w_{ps}(t)^2$ coming from the environment and the emulator-Master synchronization control. This can be done by solving the following LMI problem:

Minimize γ with $P > 0$, N matrices with suitable dimensions and ρ_A, ρ_B two positive scalar such that:

$$\begin{pmatrix} AP+PA^T-N^TB^T-BN & GP & HP & N^TE^T & PD^T & B & PC \\ * & -\frac{1}{\rho_A}I & 0 & 0 & 0 & 0 & 0 \\ * & * & -\frac{1}{\rho_B}I & 0 & 0 & 0 & 0 \\ * & * & * & -\rho_B I & 0 & 0 & 0 \\ * & * & * & * & -\rho_A I & 0 & 0 \\ * & * & * & * & * & -\gamma^2 I & 0 \\ * & * & * & * & * & * & -I \end{pmatrix} < 0 \tag{22}$$

The control gains are given by NP^{-1}.

3.5 Global Stability and Performance Analysis

The objective of this subsection is to provide the stability and performance analysis by designing the emulator of master and the controller C under time-varying delays and uncertainties. Firstly, the emulator of master, L, is designed by means of Lyapunov-Krasovskii functional, H_∞ control and LMI, so to synchronize the position between the master and the emulator. The objective is to ensure the global stability and tracking performance of the whole system described by:

$$\begin{cases} \dot{x}_{mps}(t) = (A+\Delta A(t))x_{mps}(t) \\ \qquad\qquad +(A_1+\Delta A_1(t))x_{mps}(t-\tau_1(t)) \\ \qquad\qquad +(B+\Delta B(t))w_{mps}(t), \\ z_{mps}(t) = Cx_{mps}(t), \end{cases} \tag{23}$$

with:

$$x_{mps}(t) = \begin{pmatrix} \dot{\theta}_s(t) \\ \dot{\theta}_p(t) \\ \dot{\theta}_m(t) \\ \theta_s(t)-\theta_p(t) \\ \theta_p(t)-\theta_m(t) \end{pmatrix}, w_{mps}(t) = \begin{pmatrix} F_e(t) \\ \hat{F}_e(t-\hat{\tau}_1(t))+\hat{F}_h(t-\tau_1(t)) \\ F_m(t)+F_h(t) \end{pmatrix}, \quad (24)$$

$$z_{mps}(t) = \begin{pmatrix} \theta_s(t)-\theta_p(t) \\ \theta_p(t)-\theta_m(t) \end{pmatrix}.$$

$$A = \begin{pmatrix} A_s-B_sK_s^0-B_sK_1 & -B_sK_2 & 0 & -B_sK_3 & 0 \\ 0 & A_m-B_mK_m^0 & 0 & 0 & 0 \\ 0 & 0 & A_m-B_mK_m^0 & 0 & 0 \\ 1 & -1 & 0 & 0 & 0 \\ 0 & 1 & -1 & 0 & 0 \end{pmatrix},$$

$$\Delta A(t) = \begin{pmatrix} (1,1) & -\Delta B_s(t)K_2 & 0 & -\Delta B_s(t)K_3 & 0 \\ 0 & (2,2) & 0 & 0 & 0 \\ 0 & 0 & (3,3) & 0 & 0 \\ 0 & 0 & 0 & 0 & 0 \\ 0 & 0 & 0 & 0 & 0 \end{pmatrix}, \quad (25)$$

$$\begin{aligned} (1,1) &= \Delta A_s(t) - \Delta B_s(t)K_s^0 - \Delta B_s(t)K_1, \\ (2,2) &= \Delta A_p(t) - \Delta B_p(t)K_m^0, \\ (3,3) &= \Delta A_m(t) - \Delta B_m(t)K_m^0, \end{aligned} \quad (26)$$

$$A_1 = \begin{pmatrix} 0 & 0 & 0 & 0 & 0 \\ 0 & -B_mL_1 & -B_mL_2 & 0 & -B_mL_3 \\ 0 & 0 & 0 & 0 & 0 \\ 0 & 0 & 0 & 0 & 0 \\ 0 & 0 & 0 & 0 & 0 \end{pmatrix} = G\Delta(t)D, \quad (27)$$

$$\Delta A_1(t) = \begin{pmatrix} 0 & 0 & 0 & 0 & 0 \\ 0 & -\Delta B_p(t)L_1 & -\Delta B_p(t)L_2 & 0 & -\Delta B_p(t)L_3 \\ 0 & 0 & 0 & 0 & 0 \\ 0 & 0 & 0 & 0 & 0 \\ 0 & 0 & 0 & 0 & 0 \end{pmatrix} = G_1\Delta(t)D_1, \quad (28)$$

$$B = \begin{pmatrix} B_s & 0 & 0 \\ 0 & B_m & 0 \\ 0 & 0 & B_m \\ 0 & 0 & 0 \\ 0 & 0 & 0 \end{pmatrix}, \quad \Delta B(t) = \begin{pmatrix} \Delta B_s(t) & 0 & 0 \\ 0 & \Delta B_p(t) & 0 \\ 0 & 0 & \Delta B_m(t) \\ 0 & 0 & 0 \\ 0 & 0 & 0 \end{pmatrix}, C = \begin{pmatrix} 0 & 0 & 0 & 1 & 0 \\ 0 & 0 & 0 & 0 & 1 \end{pmatrix}. \quad (29)$$

The goal here is to compute the global ratio between the quadratic tracking error $z_{mps}(t)^2$ and the disturbance $w_{ps}(t)^2$ coming from the environment and the human operator. If the new H_∞ bound is not satisfied, the global approach needs to be rerolled with new performance indices (better bounds, another local controller performance...).

Minimize γ with $P>0$, P_2, P_3, P_4, P_5, M, Q_1, Q_2, R_1, R_2 matrices with suitable dimensions , ρ_A, ρ_{A_1}, ρ_B two positive scalar such that the following conditions are feasible:

$$\Gamma = \begin{pmatrix} \Gamma_{11}^1+\Gamma_{11}^2+\Gamma_{11}^{2\,T} & \Gamma_{12} & \Gamma_{13} & e_1C^T & \Gamma_{15} \\ * & \Gamma_{22}^1-P_3-P_3^T & P_3B & 0 & \Gamma_{25} \\ * & * & -\gamma^2 I & 0 & \Gamma_{35} \\ * & * & * & -I & 0 \\ * & * & * & * & \Gamma_{55} \end{pmatrix} < 0, \quad \begin{pmatrix} R_2 & S \\ S^T & R_2 \end{pmatrix} \geqslant 0, \quad (30)$$

$$\Gamma_{11}^2 = \begin{pmatrix} P_2^T A^T & P_3^T A^T & P_4^T A^T & P_5^T A_1{}^T \\ P_2^T A_1{}^T & P_3^T A_1{}^T & P_4^T A_1{}^T & P_5^T A_1{}^T \\ 0 & 0 & 0 & 0 \\ 0 & 0 & 0 & 0 \end{pmatrix}, \tag{31}$$

$$\Gamma_{12} = e_1 P + \begin{pmatrix} A^T P_3 \\ A_1{}^T P_3 \\ 0 \\ 0 \end{pmatrix} - \begin{pmatrix} P_2^T \\ P_3^T \\ P_4^T \\ P_5^T \end{pmatrix}, \quad \Gamma_{13} = \begin{pmatrix} P_2 B \\ P_3 B \\ P_4 B \\ P_5 B \end{pmatrix}. \tag{32}$$

$$\Gamma_{15}^T = \begin{pmatrix} 0 & D & 0 & 0 \\ 0 & D_1 & 0 & 0 \\ P_2^T G^T & P_3^T G^T & P_4^T G^T & P_5^T G^T \\ P_2^T G_1^T & P_3^T G_1^T & P_4^T G_1^T & P_5^T G_1^T \\ 0 & D_1 & 0 & 0 \\ D & 0 & 0 & 0 \\ 0 & 0 & 0 & 0 \\ 0 & 0 & 0 & 0 \\ P_2^T H^T & P_3^T H^T & P_4^T H^T & P_5^T H^T \\ 0 & 0 & 0 & 0 \\ 0 & 0 & 0 & 0 \\ 0 & 0 & 0 & 0 \end{pmatrix}, \Gamma_{25}^T = \begin{pmatrix} 0 \\ 0 \\ 0 \\ 0 \\ 0 \\ 0 \\ G_1^T \\ G^T \\ 0 \\ 0 \\ \xi H \\ 0 \end{pmatrix}, \Gamma_{35}^T = \begin{pmatrix} 0 \\ 0 \\ 0 \\ 0 \\ 0 \\ 0 \\ 0 \\ 0 \\ 0 \\ EP_3 \\ 0 \\ E \end{pmatrix}, \tag{33}$$

$$\Gamma_{55} = diag\left(\tfrac{-1}{\rho_A} I,\ \tfrac{-1}{\rho_{A_1}} I,\ \rho_A I,\ \rho_{A_1} I,\ \tfrac{-1}{\rho_{A_1}} I,\ \tfrac{-1}{\rho_A} I,\ \rho_{A_1} I,\ \rho_A I,\ \rho_B I,\ \tfrac{-1}{\rho_B} I \right), \tag{34}$$

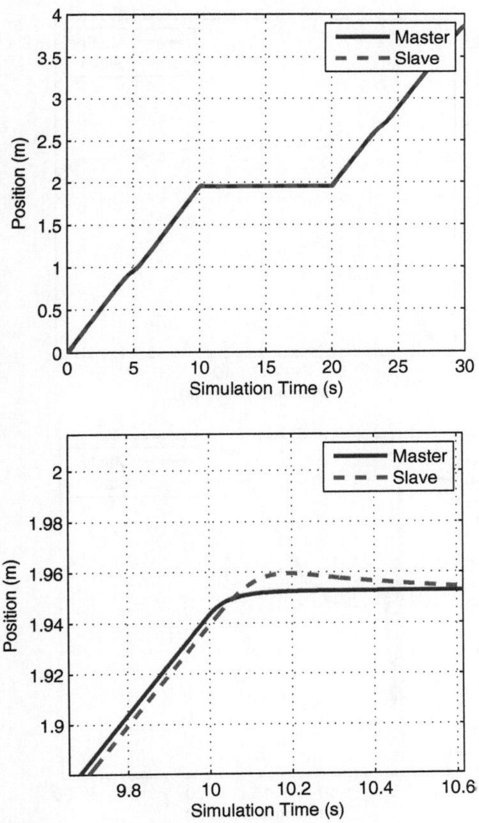

Fig. 3 Position response in abrupt tracking motion

3.6 Tracking in Abrupt Changing Motion

Figure 3 shows the position tracking between the master and slave under time-varying delays and uncertainties, where the human operator ($F_h(t)$) is modeled as the pulse generator.

From the Figure, it can be seen that the method achieves the position tracking, especially at the mutation point (amplified part in Figure 3). Good position convergence between the master and slave has been presented.

3.7 Tracking in Wall Contact Motion

Similarly, the position tracking in wall contact motion is presented Figure 4. Here, the slave is driven to the hard wall with a stiffness of $K_e = 30kN/m$ located at the position $x = 1.0m$.

Based on H_∞ control, the time-varying model uncertainties are handled by the proposed emulator of master with the controller C. The force tracking $F_m(t) = \hat{F}_e(t - \tau_2(t))$ can be seen in the smaller figure of Figure 4.

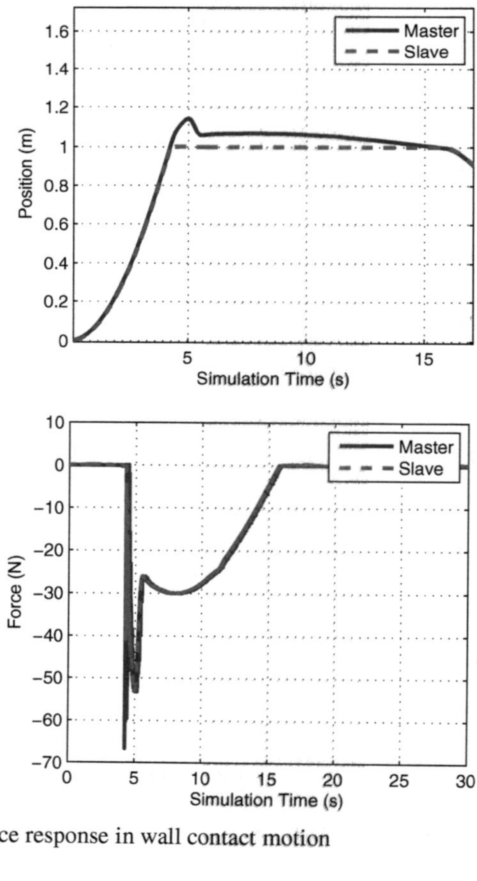

Fig. 4 Position/force response in wall contact motion

4 Conclusions

It was shown how a teleoperation application can be solved by using a time-varying delay framework. The controller proposed is based on predictor-like approaches (the so-called 'emulator') and guarantees some performance in term of force feedback fidelity and position tracking. These results could be enhanced if more information on the environment were available, but also if the design procedure could be achieved globally. This last improvement is still open, since no known result allows for solving this problem in one step.

References

1. Arcara, P., Melchiorri, C.: Control schemes for teleoperation with time delay: A comparative study. Robotics and Autonomous Systems 38(1), 49–64 (2002)
2. Chiasson, J., Loiseau, J.J.: Applications of Time Delay Systems. LNCIS, vol. 352. Springer, Heidelberg (2007)
3. Fridman, E.: A new Lyapunov technique for robust control of systems with uncertain non-small delays. IMA J. Math. Contr. Infor. 23(2), 165–179 (2006)
4. Fridman, E., Niculescu, S.I.: On complete Lyapunov-Krasovskii functional techniques for uncertain systems with fast-varying delays. Int. J. Robust Nonlinear Control 18(3), 364–374 (2008)
5. Fridman, E., Shaked, U.: New bounded real lemma representations for time-delay systems and their applications. IEEE Trans. Automat. Contr. 46(12), 1973–1979 (2001)
6. He, Y., Wu, M., She, J.H., Liu, G.P.: Parameter-dependent Lyapunov functional for stability of time-delay systems with polytopic-type uncertainties. IEEE Trans. Automat. Contr. 49(5), 828–832 (2002)
7. Hokayem, P.F., Spong, M.W.: Bilateral teleoperation: an historical survey. Automatica 42(12), 2035–2057 (2006)
8. Jiang, X., Han, Q.L.: On H_∞ control for linear systems with interval time-varying delay. Automatica 41(12), 2099–2106 (2005)
9. Kruszewski, A., Jiang, W.J., Fridman, E., Richard, J.P., Toguyeni, A.: A switched system approach to exponential stabilization through communication network. IEEE Trans. Contr. Syst. Technol. 20(4), 887–900 (2012)
10. Nuño, E., Basañez, L., Ortega, R.: Passivity-based control for bilateral teleoperation: a tutorial. Automatica 47(3), 485–495 (2011)
11. Park, P., Ko, J.W., Jeong, C.: Reciprocally convex approach to stability of systems with time-varying delays. Automatica 47(1), 235–238 (2011)
12. Richard, J.P.: Time delay systems: An overview of some recent advances and open problems. Automatica 39(10), 1667–1694 (2003)
13. Ryu, J.H., Preusche, C., Hannaford, B., Hirzinger, G.: Time domain passivity control with reference energy following. IEEE Trans. Contr. Syst. Technol. 13(5), 737–742 (2005)
14. Xu, B., Liu, Y.H.: Delay-dependent/delay-independent stability of linear systems with multiple time-varying delays. IEEE Trans. Automat. Contr. 48(4), 697–701 (2003)
15. Xu, S., Lam, J., Zou, Y.: New results on delay-dependent robust H_∞ control for systems with time-varying delays. Automatica 42(2), 343–348 (2006)

16. Ye, Y.Q., Pan, Y.J., Gupta, Y.: Time domain passivity control of teleoperation systems with random asymmetric time delays. In: Proceedings of IEEE Conf. on Decision and Control, and Chinese Control Conference (CDC2009, CCC 2009), Shanghai, China, pp. 7533–7538 (2009)
17. Zhang, B., Kruszewski, A., Richard, J.P.: Tracking improvement based on the proxy control scheme for bilateral teleoperation system under time-varying delays. In: 2011 IEEE Int. Conf. on Emerging Technologies and Factory Automation (ETFA 2011), Toulouse, France (2011)
18. Zhang, B., Kruszewski, A., Richard, J.P.: H_{infty} Robust Control Design for Teleoperation Systems. In: 7th IFAC Symposium on Robust Control Design, Aalborg, Denmark (2011)
19. Jiang, W.J., Kruszewski, A., Richard, J.P., Toguyeni, A.: A Gain Scheduling Strategy for the Control and Estimation of a Remote Robot via Internet. In: Proc. Chinese Control Conference (CCC 2008), Kunming, Yunnan, China (2008)
20. Xie, L., De Souza, C.E., Wang, Y.: Robust control of discrete time uncertain dynamical systems. Automatica 29(4), 1133–1137 (1993)

Graph Laplacian Design of a LTI Consensus System for the Largest Delay Margin: Case Studies*

Wei Qiao and Rifat Sipahi

Abstract. The dynamics of a LTI consensus system with homogeneous inter-agent delays is the focus here where we design, via graph synthesis and our Responsible Eigenvalue (RE) concept, its Graph Laplacian associated with agents' heterogeneous coupling strengths, with the aim to maximize the system's delay margin - the largest delay it can withstand before losing its stability. Over case studies, we present how this calculation can be done.

1 Introduction

Many dynamical systems in biology [11], social networks [1], neural networks [4, 35], vehicular traffic flow [6,29], and supply chains [30] can be seen as systems with coupled agents interacting with each other, some with after effects, i.e., delays [15, 18–20, 32]. In such systems, the network aspect of the system allows coupling the so-called agents, which can be seen as interacting sub-dynamical systems, whereas delay often times arises via the communication medium that facilitates information exchange between the agents. Since in many cases, communication between the agents is necessary, delay effects are inevitable in agents' decision making prepared based on the information received from other agents. Under delays, however, the coupled system can become unstable thereby destroying system functionality. In other words, network connectivity (topology) and time delay together affect the stability of the system [10, 21–24, 27, 33].

Studying the relationship between delays, topology and stability is however challenging since the infinite dimensionality of delay-differential equations cannot be

Wei Qiao · Rifat Sipahi
Northeastern University, Department of Mechanical and Industrial Engineering,
Boston, MA 02115 USA
e-mail: qiao.w@husky.neu.edu, rifat@coe.neu.edu

* Some parts of this chapter are summarized from the authors' work presented at the 2012 IFAC Workshop on Time Delay Systems, Boston, MA.

T. Vyhlídal, J.-F. Lafay, and R. Sipahi (eds.), *Delay Systems*,
Advances in Delays and Dynamics 1,
DOI: 10.1007/978-3-319-01695-5_8, © Springer International Publishing Switzerland 2014

directly correlated to graph properties. In this *reverse problem*, our work starts in 2007 with the work in [29], as well as [28, 30] including a study in 2008 in which *three independent* delays is considered – the first study on this topic with multiple delays, to the best of our knowledge. Moreover, along these lines again, since 2009, the authors have been working on a class of broadly studied LTI consensus system represented by a set of delay-differential equations with homogeneous delays but with heterogenous agent coupling strengths [21, 22, 33, 34]. From these efforts, the *responsible eigenvalue (RE)* concept arose. In summary, the RE concept can be used to calculate the *scalar* quantity the delay margin τ^* of the corresponding *infinite dimensional* delay-differential equation only by checking the *finite* number of eigenvalues of the graph Laplacian of that system.

With the RE concept and inspired by Cartesian product operation on graphs [2], we established some rules with which large scale graphs can be synthesized for the type of delayed systems at hand, while guaranteeing sufficient delay margin [24]. Moreover, we recently proposed some design rules for such systems, while even relaxing their consensus condition, where one can tailor two graphs using Cartesian products and calculate the maximum delay margin possible in the system with the arising large graph [25]. This study also enabled us to design the graph Laplacians, and thereby allowing us to *establish a link to designing multiple heterogeneous agent-coupling strengths all at once*. In this chapter, we revisit these design rules and use them to solve various case studies, in order to demonstrate both the synthesis approach and the numerical technique used to calculate the maximum delay margin based on RE [33, 34].

Notations are standard. We use $\mathbb{C}_+$, $\mathbb{C}_-$, $j\mathbb{R}$ for right half, left half and the imaginary axis of the complex plane, respectively. $\mathbb{R}$ represents the set of real numbers, j is the imaginary unit, and $\lambda_k(\mathbf{A})$ is an eigenvalue of the square matrix $\mathbf{A} \in \mathbb{R}^{n\times n}$, $k = 1,\ldots,n$. Matrices, vectors and sets are denoted by bold face, while scalar entities are with normal font. The vector $\mathbf{v} \in \mathbb{R}^L_+$ defines an L-dimensional vector with positive real entries.

2 Preliminaries

We first present the dynamical system studied in our previous work, briefly explain our RE concept, and next provide background information on Cartesian products, as well as our recent results on “design rules” to calculate the maximum possible delay margin.

2.1 Consensus Dynamics

In [24, 33], we studied the following LTI consensus dynamics:

$$\frac{dx_i(t)}{dt} = \sum_{k=1,k\neq i}^{n} \alpha_{ik}\left[x_k(t-\tau) - \gamma_{ik}x_i(t-\tau)\right], \tag{1}$$

where $x_i(t)$ is the state of agent i, $i = 1, \ldots, n$, $\alpha_{ik} \geq 0$ are the coupling strengths, $\gamma_{ik} = 1$, and $\tau \geq 0$ is the constant time delay. Eq. (1) can be re-written as,

$$\frac{d\mathbf{x}(t)}{dt} = \mathbf{A}\mathbf{x}(t - \tau), \tag{2}$$

where $\mathbf{x}(t) = (x_1(t), \ldots, x_n(t))^T \in \mathbb{R}^n$ is the state vector, and the corresponding graph Laplacian is defined by $\mathscr{L}(G) = -\mathbf{A}$, which is determined by whether or not α_{ik} is zero. Moreover, we assume that $\mathbf{A}$ is diagonalizable.

Eq.(1) and its similar forms have been studied extensively in the context of synchronization [7, 9, 15, 18], traffic flow [3, 6, 29, 31], and autonomous agents [5, 12, 16, 19, 26]. In the previous work, the focus was however on *consensus*, i.e., $\gamma_{ik} = 1$, hence matrix $\mathbf{A}$ was constrained to have zero row-sum. This, along with the assumption that each agent is connected to the other agents with directed links, leads to a single zero eigenvalue $\lambda_1(\mathbf{A}) = 0$, as per the nature of consensus. Moreover, it is common practice to take α_{ik} as non-negative, which then sets the remaining eigenvalues stable, $\lambda_2, \ldots, \lambda_n \in \mathbb{C}_-$, see the cited references. Here, we shall relax these assumptions, following our study in [25] and let $\gamma_{ik} \neq 1$ and $\alpha_{ik} \in \mathbb{R}$ in general. This relaxation indicates that the problem is neither limited to consensus nor to positive coupling strengths, hence system (1) represents a broader class of systems, possibly $\mathbf{A}$ having unstable eigenvalues, $\lambda_k \in \mathbb{C}$.

2.2 Stability, Responsible Eigenvalue (RE), Graph Synthesis

Stability of (1) is determined by the delay parameter τ. In order to find out how large the delay τ can be, while keeping (2) stable, one should study the eigenvalues of (2). In the presence of delay, the dynamics in (2) has *infinitely many eigenvalues*. For this dynamics to be stable, it is necessary and sufficient that all these eigenvalues have *negative* real parts [36]. To be consistent with this stability property, *negative* Laplacian is defined here as $\mathscr{Q}(G) = -\mathscr{L}(G) = \mathbf{A}$.

For $\gamma_{ik} \in \mathbb{R}$ in (1) and so long as $\mathbf{A}$ has stable eigenvalues, there exists an upper bound on the delay value, known as the delay margin τ^*, less than which (2) is stable, see [14, 33] for the special of $\gamma_{ik} = 1$ case. Calculation of τ^* can as a matter of fact be done even without analyzing the infinitely many eigenvalues of (2) as stated in the following lemma:

Lemma 1. *[22, 23, 33] Either one real or one pair of complex conjugate eigenvalue(s) of* $\mathbf{A}$ *determines the delay margin* τ^*.

The eigenvalue of $\mathbf{A}$ that computes τ^* in Lemma 1 is defined as the RE, which was defined in [22, 23, 33, 34] for $\gamma_{ik} = 1$ and $\alpha_{ik} \geq 0$. However, since γ_{ik} and α_{ik} do not affect the main structure of (2), the RE concept holds regardless of what the numerical values of these parameters are. The only slight addition to Lemma 1 for the case of $\gamma_{ik} \neq 1$ and $\alpha_{ik} \in \mathbb{R}$ is to state that the delay margin of (2) does not exist if there exists at least one eigenvalue of $\mathbf{A}$ with a non-negative real part.

With RE, one analytically and precisely computes τ^*. RE also enables a visual tool: the Delay Margin Contour Map (DMCM), see Figure 1 [22,23,33,34]. On DMCM, one superposes all λ_k of $\mathbf{A}$ and identifies the eigenvalue λ_k that resides on the contour with the *smallest contour value*, which is τ^*. This eigenvalue is the RE. We shall use DMCM in the next section to calculate the maximum possible delay margin in (2), which we will synthesize by Cartesian Product graph operations, as defined next.

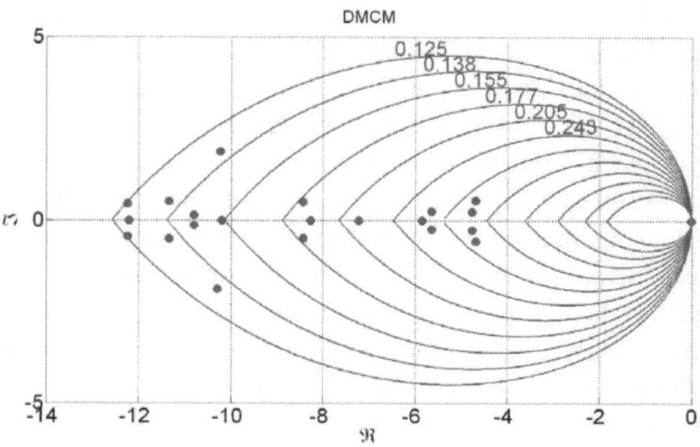

Fig. 1 Eigenvalues (dots) of a representative matrix $\mathbf{A}$ superposed on the Delay Margin Contour Map (DMCM). The label on each contour represents the corresponding τ_k^* value. The RE is the eigenvalue residing on the contour with the smallest contour value, which is $\tau^* = 0.125$ in this example.

Definition 1. [8] The Kronecker sum, $\bigoplus$, is the matrix sum

$$\mathscr{Q}(G_A)\bigoplus\mathscr{Q}(G_B) = \mathscr{Q}(G_A)\bigotimes I_{n_B} + I_{n_A}\bigotimes\mathscr{Q}(G_B), \tag{3}$$

where $\otimes$ denotes the Kronecker product operation, the dimension of $\mathscr{Q}(G_i)$, $i = 1,2$, is n_i, which is also the number of vertices of G_i, and I_q is the $q \times q$ identity matrix. The eigenvalues of $\mathscr{Q}(G_C)$ found from $\mathscr{Q}(G_C) = \mathscr{Q}(G_A)\bigoplus\mathscr{Q}(G_B)$ are all possible sums of the eigenvalues of $\mathscr{Q}(G_A)$ and $\mathscr{Q}(G_B)$ [2, 13].

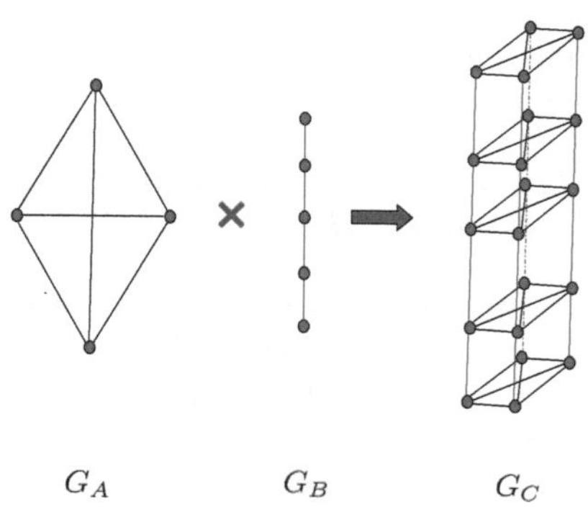

Fig. 2 In this chapter, the Cartesian product of two graphs G_A and G_B is denoted by $G_C = G_A \times G_B$

2.3 Design Rules

Here, we summarize the four design rules we introduced in [25]. These rules lay out the conditions under which the delay margin of the system with graph G_C can be computed as a parameter of one of the eigenvalues of either G_A or G_B. The idea is to exploit the property in Definition 1 to manage the migration of eigenvalues on $\mathbb{C}$, due to Kronecker sum property, and use this in connection with DMCM in Figure 1 to compute the delay margin of the system with graph G_C.

Lemma 2. *[25] Given a stable $\mathscr{Q}(G_A)$ with a RE $\lambda^*(\mathscr{Q}(G_A))$ that is either complex conjugate or real, the largest achievable delay margin $\tau^*_{G_C}$ in G_C can be attained by satisfying the condition $\forall\lambda_k(\mathscr{Q}(G_B)) \in \mathbb{R}$.*

Lemma 2 indicates that in order to build a large-scale system corresponding to G_C with a delay margin $\tau^*_{G_C}$ as large as possible, the first rule is to have $\forall\lambda_k(\mathscr{Q}(G_B)) \in \Re_-$. This is because if one of $\lambda_k(\mathscr{Q}(G_B))$ were complex, the eigenvalue of the system with G_C would migrate diagonally on $\mathbb{C}$, away from the origin, and then due to the specific shape of DMCM contours, the arising delay margin would be guaranteed to be smaller, see contour values in DMCM.

Property 1. Inspecting DMCM in Figure 1, one can find that there exists a threshold on τ^* value for a fixed $\omega = \Im(\lambda_k)$. Moreover, the largest possible τ^* is monotonically decreasing as ω increases for a fixed real part $\sigma = \Re(\lambda_k) < 0$. In other words, for a given system $\mathscr{Q}(G_A)$ with multiple eigenvalues and assuming that the eigenvalues of $\mathscr{Q}(G_C)$, where $G_C = G_A \times G_B$, move horizontally on the complex plane as per the real eigenvalues of $\mathscr{Q}(G_B)$ (see Lemma 2), the maximum possible delay margin τ^* that can be attained in the system with G_C is determined by the eigenvalue of $\mathscr{Q}(G_A)$ with the largest imaginary part [25].

Take the system with G_A as described in (2), where $\mathbf{A} = \mathbf{A}_1$. Let $\mathbf{A}_1$ have more than one eigenvalue in general, and define

$$\sup\{\Re(\lambda)|\lambda = eig(\mathscr{Q}(G_A))\} = c_1.$$

Let another system be defined with graph G_B and $\mathbf{A} = \mathbf{A}_2$ in (2). Assume that $\mathbf{A}_2$ is scalar $\mathbf{A}_2 = \Delta \in \mathbb{R}$, i.e., the corresponding system has only one real eigenvalue, which is Δ. Using Cartesian product, let G_C be obtained as $G_C = G_A \times G_B$. Keeping $\mathbf{A}_1$ fixed, the following design rules hold for the parameter Δ of $\mathscr{Q}(G_B)$, in relation to the maximum delay margin that can be obtained for the system with graph G_C:

Design Rule 1. *If $c_1 > 0$ and thus $\mathbf{A}_1$ is unstable, then for designing a stable delay-free dynamics corresponding to G_C, it is necessary that $\Delta \leq -c_1$.*

Design Rule 2. *For $\mathscr{Q}(G_A)$ and $\mathscr{Q}(G_B)$ stable, a larger $\tau^*_{G_C}$ of G_C exists, for some $\Delta = \Delta^* \in [\Delta_0, 0]$, where $\tau^*(\lambda_k(\mathscr{Q}(G_A)) + \Delta_0) = \tau^*(\lambda_k(\mathscr{Q}(G_A)))$. This $\tau^*_{G_C}$ is computed as $\tau^*(\lambda_k(\mathscr{Q}(G_A)) + \Delta^*)$ corresponding to the eigenvalue $\lambda_k(\mathscr{Q}(G_A)) + \Delta^*$.*

Design Rule 3. *Let the RE of $\mathscr{Q}(G_A)$ reside on the left-hand side of a contour arc in DMCM. Then there exists a $\Delta = \Delta^* \in [0, \Delta_1]$, where $\tau^*(\lambda_k(\mathscr{Q}(G_A)) + \Delta_1) = 0$, for which a larger delay margin $\tau^*_{G_C}$ is computed as $\tau^*(\lambda_k(\mathscr{Q}(G_A)) + \Delta^*)$.*

Design Rule 4. *Given an unstable $\mathscr{Q}(G_A)$, there exists a larger $\tau^*_{G_C}$ of G_C, for some $\Delta = \Delta^* \in [\Delta_0, -c_1]$, where $c_1 > 0$, and Δ_0 satisfies that $\forall(\lambda_k(\mathscr{Q}(G_A)) + \Delta_0)$ lie on the left hand side of a contour arc or on the peak point of a contour arc in DMCM. This $\tau^*_{G_C}$ can be attained as $\tau^*(\lambda_k(\mathscr{Q}(G_A)) + \Delta^*)$.*

3 Case Studies

We *relax the consensus assumption* here by allowing arbitrary entries in $\mathbf{A}$, where $\alpha_{ik} \in \Re$ and $\gamma_{ik} \neq 1$ in (1). That is, the entries of $\mathbf{A}$ can be either negative, zero, or positive, as long as a node is not disconnected from all the remaining nodes. This is a case with heterogeneous inter-agent couplings, where we may in general have both real and complex conjugate eigenvalues $\lambda_k(\mathbf{A})$, some of which may be unstable, see an application of this in [3].

We further note here that, in the sequel, the graphs G_A and G_B are combined together using Cartesian product, giving rise to $G_C = G_A \times G_B$, where $\mathscr{Q}(G_B)$ is first assumed to have a single eigenvalue $\Delta \in \mathbb{R}$. Then the design rules summarized in Section 2 are applied to compute the best Δ selection for which $\tau^*_{G_C}$ can be maximal. Next, pole placement technique [17] is used to increase the dimension of G_B by placing its eigenvalues in the vicinity of Δ, by which $\tau^*_{G_C}$ calculation does not significantly lose accuracy.

3.1 Tailoring Stable $\mathscr{Q}(G_A)$ with Stable $\mathscr{Q}(G_B)$

The system in (2) corresponding to G_A has eigenvalues satisfying $c_1 < 0$. With the knowledge of Lemma 2, we can incrementally add $\Delta \leq 0$ to the eigenvalues of $\mathscr{Q}(G_A)$ where $\Delta \in [\Delta_0, 0]$, and Δ_0 satisfies $\tau^*(\lambda_k(\mathscr{Q}(G_A)) + \Delta_0) = \tau^*(\lambda_k(\mathscr{Q}(G_A)))$. These conditions indicate that $\tau^*_{G_C}$ can be maximized by selecting a location for $\lambda_k(\mathscr{Q}(G_A)) + \Delta$ as $\lambda_k(\mathscr{Q}(G_A))$ migrates horizontally away from the imaginary axis from one point, and *before* reaching another point that is on the same contour in DMCM. This makes sense; delay-margin contour values are larger inside of any of the contours, see Figure 1. If $\Delta_0 = 0$, then the largest achievable delay margin is $\tau^*(\mathscr{Q}(G_A))$ since there is no room for improvement, otherwise, one can refer to the following examples:

Example 1: A stable $\mathscr{Q}(G_A)$ is given with the eigenvalues: $-2 \pm 3.5j, -0.9 \pm 1.3j, -1.5 \pm 3j$. Design Rule 2 renders Figure 3, in which we identify the largest achievable delay margin as $\tau^*_{G_C} = 0.1603$, where $\Delta = \Delta^* = -2.07$. That is, by shifting the real parts of $\lambda_k(\mathscr{Q}(G_A))$ away from the imaginary axis at an amount of 2.07 units, we obtain the delay margin $\tau^*_{G_C} = 0.1603$. In other words, if the eigenvalue Δ of $\mathscr{Q}(G_B)$ is equal to -2.07, then the system in (2) with G_C has a delay margin of $\tau^*_{G_C} = 0.1603$.

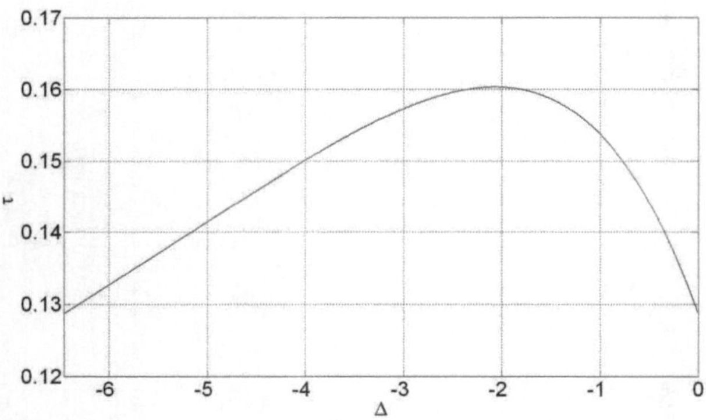

Fig. 3 Example 1: Delay margin τ^* of G_C with respect to the eigenvalue Δ of $\mathcal{Q}(G_B)$

We now design a large dimensional graph G_C corresponding to the dynamics with the largest achievable delay margin. In other words, we also want to increase arbitrarily the system dimension associated with G_C. To achieve this, we can use the well-known pole-placement technique [17] to place all the poles of $\mathcal{Q}(G_B)$ on the real axis around $\Delta^* = -2.07$. This is possible as long as the *place* command in MATLAB yields a feasible solution. To be able to execute the command, we first generate a matrix $V = (v_{ij}) \in \mathbb{R}^{6\times 6}$ where v_{ij} are randomized in $[0,1]$ with uniform distribution, and generate another matrix $W = w_{ij} \in \mathbb{R}^{6\times 1}$, where $w_{ij} = 1$. Next, we select the eigenvalues of $\mathcal{Q}(G_B)$ in close proximity to Δ^*, at $-1.92, -2.02, -2.07, -2.12, -2.22$, see Figure 4. Then, using the *place* command, it becomes possible to compute a matrix $Y \in \mathbb{R}^{1\times 6}$, similar to a control matrix, by which one can then calculate $\mathcal{Q}(G_B) = V - WY$ in $\mathbb{R}^{6\times 6}$. With this six-dimensional matrix $\mathcal{Q}(G_B)$, the delay margin is calculated again, which is found as $\tau^*_{G_C} = 0.1602$. Notice that the tradeoff for eventually utilizing a large dimensional G_B is to have an infinitesimally small decrease in the largest achievable delay margin for the system with G_C, in this case from $\tau^*_{G_C} = 0.1603$ down to 0.1602. If the eigenvalues of $\mathcal{Q}(G_B)$ were significantly spread away from Δ^*, then the largest achievable delay margin would be much smaller than $\tau^*_{G_C} = 0.1602$.

Finally, we note that the above procedure designs the entries of the matrix $\mathcal{Q}(G_B)$, which is formed by the coupling strengths in this system, such that the arising system with $\mathcal{Q}(G_C)$ can have the maximal delay margin. The coupling strengths of the agents in the arising system are related to $\mathcal{Q}(G_C)$, which can be found by utilizing the Kronecker sum defined in (3).

Example 2: A stable $\mathcal{Q}(G_A)$ is given with the eigenvalues: -0.4650, $-1.5307 \pm 2.2208j$, $-3.2104 \pm 1.4406j$, $-4.1838 \pm 0.2521j$, -2.9173. We again use Design Rule 2 which yields Figure 5, where we identify the largest achievable delay margin as $\tau^*_{G_C} = 0.2526$, and $\Delta = \Delta^* = -1.05$. In other words, if $\lambda_k(\mathcal{Q}(G_B)) = -1.05$ holds, then the system in (2) with G_C has a delay margin of $\tau^*_{G_C} = 0.2526$. Larger dimensions of $\mathcal{Q}(G_B)$ can be acquired by pole placement procedure as discussed above in Example 1.

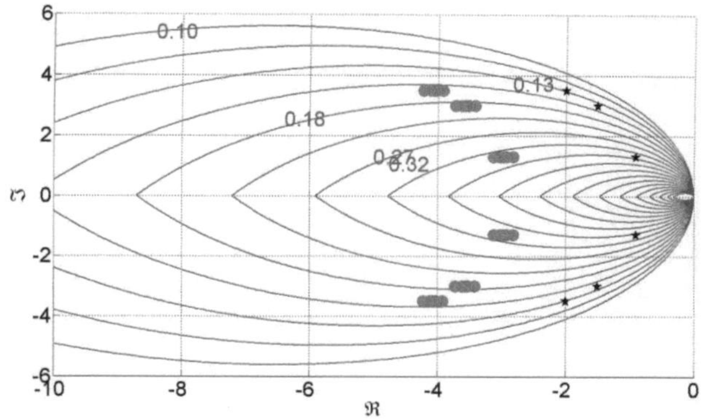

Fig. 4 Example 1: Eigenvalues of $\mathscr{Q}(G_A)$ (stars) and $\mathscr{Q}(G_C)$ (circles)

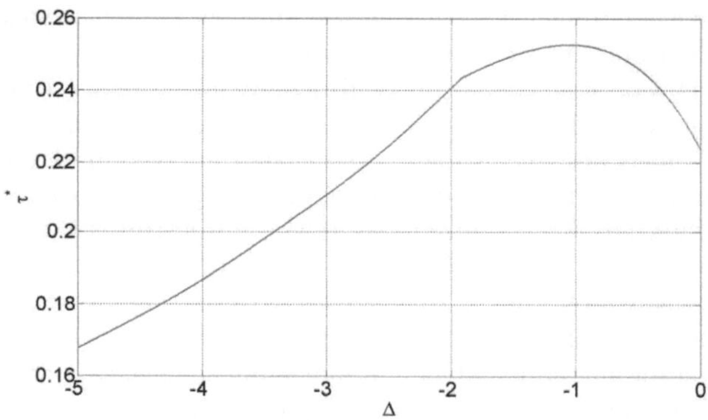

Fig. 5 Example 2: Delay margin τ^* of G_C with respect to the eigenvalue Δ of $\mathscr{Q}(G_B)$

3.2 *Tailoring an Unstable System with a Stable System*

We now analyze the largest achievable delay margin when combining an unstable negative Laplacian and a stable one using Cartesian product. There are two possible cases that fall into this category, either the system corresponding to G_B is unstable (use Design Rule 3), or the designed system corresponding to G_A is unstable (use Design Rule 4). For the latter case, interestingly, due to the unstable eigenvalues of $\mathscr{Q}(G_A)$, we show that there is more room for obtaining a larger achievable delay margin for the system with $\mathscr{Q}(G_C)$.

3.2.1 Case 1: $\mathscr{Q}(G_B)$ Is Unstable

If an eigenvalue of $\mathscr{Q}(G_A)$ resides on the left-hand side of a contour arc in DMCM, it can be seen from Figure 1 that the delay margin decreases monotonically as the eigenvalue migrates away from the imaginary axis while its real part decreases at an amount of $|\Delta|$. In this case, $\Delta_0 = 0$, that is, the largest attainable delay margin is where the eigenvalue already resides, not where it migrates to, as it moves away from the imaginary axis.

Example 3: As shown in Figure 6(a), the delay margin of $\mathscr{Q}(G_C)$ is monotonically decreasing as we decrease Δ from 0 to -5 where $\lambda_k(\mathscr{Q}(G_A))$ are at -4.2731, -7.1597, $-6.1532 \pm 0.5879j$, -6.6820, -6.3555, -5.6477, -4.9577, -5.1771. However, we find that, if Δ is allowed to be *positive*, then by applying Design Rule 3, the delay margin can be increased even more than the case with $\Delta = 0$, and can reach its optimal largest achievable value, which is $\tau^*_{G_C} = 0.5435$, as shown in Figure 6(b). The corresponding Δ^* is $\Delta^* = +4.27$, which is the only eigenvalue of $\mathscr{Q}(G_B)$.

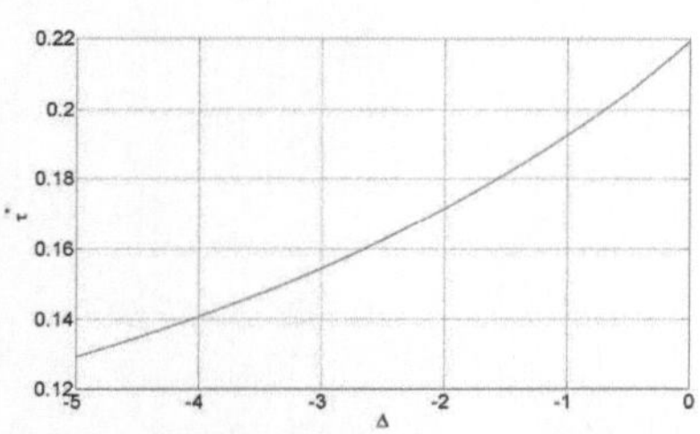

(a) $\tau^*_{G_C}$ vs. the eigenvalue $\Delta < 0$ of $\mathscr{Q}(G_B)$.

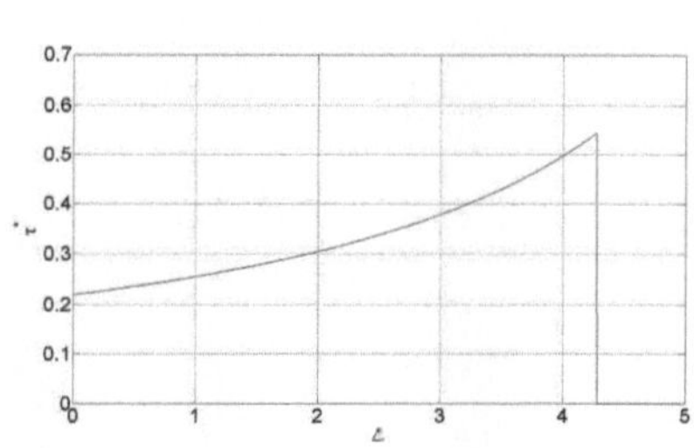

(b) $\tau^*_{G_C}$ vs. the eigenvalue $\Delta > 0$ of $\mathscr{Q}(G_B)$.

Fig. 6 Example 3: Delay margin τ^* of G_C with respect to the eigenvalue Δ of $\mathscr{Q}(G_B)$

3.2.2 Case 2: $\mathscr{Q}(G_A)$ Is Unstable

For the case of $\Re(\lambda_k(\mathscr{Q}(G_A))) > 0$, where the delay margin of the system does not exist since $\mathscr{Q}(G_A)$ is unstable, the design rule is a combination of Design Rule 1 and Design Rule 4.

Example 4: Let $\lambda_k(\mathscr{Q}(G_A))$ be $1.1834 \pm 2.2497j$, $-0.8599 \pm 1.1108j$, 2.7120, $-0.4037 \pm 0.8177j$, -0.6003. By applying Design Rule 4, the largest achievable delay margin is computed as $\tau^*_{G_C} = 0.249$, which corresponds to $\Delta^* = -3.8$, see Figure 7. This indicates that the only eigenvalue of $\mathscr{Q}(G_B)$ is $\Delta^* = -3.8$.

Example 5: Let $\lambda_k(\mathscr{Q}(G_A))$ be $1.3714 \pm 3.3323j$, 1.6620, $-1.5845 \pm 1.8130j$, -2.3915, -1.7505, -1.1474, $-0.0374 \pm 0.3552j$. By applying Design Rule 4, the largest achievable delay margin is computed as $\tau^*_{G_C} = 0.1684$, corresponding to $\Delta^* = -5.25$, see Figure 8.

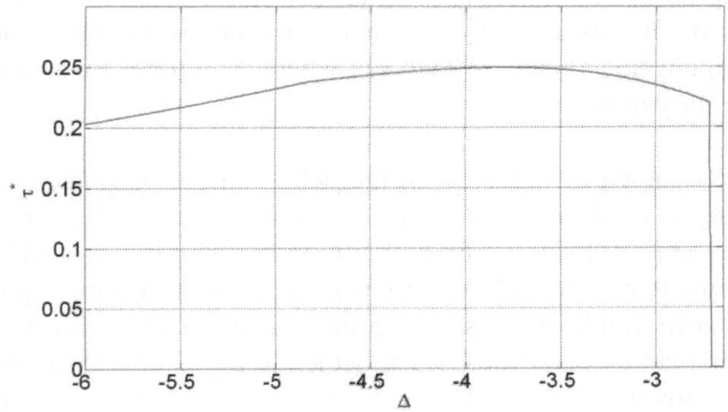

Fig. 7 Example 4: Delay margin τ^* of G_C with respect to the eigenvalue $\Delta < 0$ of $\mathscr{Q}(G_B)$

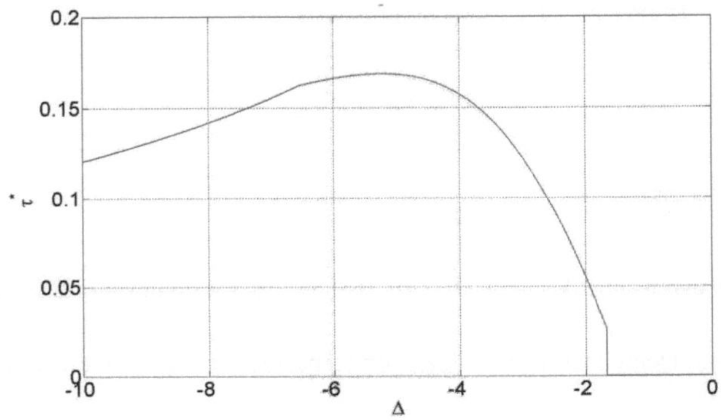

Fig. 8 Example 5: Delay margin τ^* of G_C with respect to the eigenvalue $\Delta < 0$ of $\mathscr{Q}(G_B)$

4 Conclusion

A class of coupled LTI dynamical systems with a time delay representing the delayed communication among the subsystems is studied. A numerical procedure developed earlier is utilized to study several problems, in which we calculate the largest achievable delay margin arising from tailoring, via Cartesian product, stable and/or unstable Laplacians of graphs representing the dynamical system at hand. The calculation becomes possible with the analytical and graphical nature of the authors' previously developed Responsible Eigenvalue concept, and enables us to reveal a relationship with the heterogeneous coupling strengths of multiple agents.

Acknowledgements. This research has been supported in part by the award from the National Science Foundation ECCS 0901442.

References

1. Arenas, A., Díaz-Guilera, A., Kurths, J., Moreno, Y., Zhou, C.: Synchronization in complex networks. Physics Reports 469(3), 93–153 (2008)
2. Atay, F.M., Biyikoglu, T.: Graph operations and synchronization of complex networks. Physical Review E 72(1), 016217 (2005)
3. Bando, M., Hasebe, K., Nakanishi, K., Nakayama, A.: Analysis of optimal velocity model with explicit delay. Physical Review E 58, 5429–5435 (1998)
4. Brandt, S.F., Pelster, A., Wessel, R.: Variational calculation of the limit cycle and its frequency in a two-neuron model with delay. Physical Review E 74(3), 036201 (2006)
5. Fax, J.A., Murray, R.M.: Information flow and cooperative control of vehicle formations. IEEE Transactions on Automatic Control 49(9), 1465–1476 (2004)
6. Helbing, D.: Traffic and related self-driven many-particle systems. Reviews of Modern Physics 73(4), 1067–1141 (2001)
7. Hod, S.: Analytic treatment of the network synchronization problem with time delays. Physical Review Letters 105(20), 208701 (2010)
8. Horn, R.A., Johnson, C.R.: Topics in Matrix Analysis. Cambridge University Press, Cambridge (1994)
9. Hunt, D., Korniss, G., Szymanski, B.K.: Network synchronization in a noisy environment with time delays: Fundamental limits and trade-offs. Physical Review Letters 105(6), 068701 (2010)
10. Jirsa, V.K., Ding, M.: Will a large complex system with time delays be stable? Physical Review Letters 93(7), 070602 (2004)
11. Kadji, H.G.E., Orou, J.B.C., Woafo, P.: Synchronization dynamics in a ring of four mutually coupled biological systems. Communications in Nonlinear Science and Numerical Simulation 13(7), 1361–1372 (2008)
12. Lu, J., Ho, D.W.C., Kurths, J.: Consensus over directed static networks with arbitrary finite communication delays. Physical Review E 80(6), 066121 (2009)
13. Merris, R.: Laplacian matrices of graphs: A survey. Linear Algebra Applications 197-198, 143–176 (1994)
14. Michiels, W., Niculescu, S.-I.: Stability and stabilization of time-delay systems: an eigenvalue-based approach. In: SIAM Advances in Design and Control, Philadelphia, PA, USA (2007)
15. Michiels, W., Nijmeijer, H.: Synchronization of delay-coupled nonlinear oscillators: An approach based on the stability analysis of synchronized equilibria. Chaos: An Interdisciplinary Journal of Nonlinear Science 19(3), 033110 (2009)
16. Nosrati, S., Shafiee, M., Menhaj, M.B.: Synthesis and analysis of robust dynamic linear protocols for dynamic average consensus estimators. Control Theory Applications, IET 3(11), 1499–1516 (2009)
17. Ogata, K.: Morden Control Engineering. Prentice-Hall (2002)
18. Oguchi, T., Nijmeijer, H., Yamamoto, T.: Synchronization in networks of chaotic systems with time-delay coupling. Chaos: An Interdisciplinary Journal of Nonlinear Science 18(3), 037108 (2008)
19. Olfati-Saber, R., Murray, R.M.: Consensus problems in networks of agents with switching topology and time-delays. IEEE Transactions on Automatic Control 49(9), 1520–1533 (2004)
20. Porfiri, M., Roberson, D.G., Stilwell, D.J.: Tracking and formation control of multiple autonomous agents: a two-level consensus approach. Automatica 43(8), 1318–1328 (2007)

21. Qiao, W., Sipahi, R.: Dependence of delay margin on network topology: Single delay case. In: 9th IFAC Workshop on Time Delay Systems, Prague, Czech Republic (2010)
22. Qiao, W., Sipahi, R.: Responsible eigenvalue approach for stability analysis and control design of a single-delay large-scale system with random coupling strengths. In: ASME 3rd Dynamic Systems and Control Conference, Cambridge, MA, USA (2010)
23. Qiao, W., Sipahi, R.: Responsible eigenvalue control for creating autonomy in coupled systems with delays. In: ASME Dynamic Systems and Control Conference, Arlington, VA, USA (2011)
24. Qiao, W., Sipahi, R.: Rules and limitations of building delay-tolerant topologies for coupled systems. Physical Review E 85(1), 016104 (2012)
25. Qiao, W., Sipahi, R.: The Largest achievable delay margin of a class of coupled LTI systems synthesized by graph operations. In: 11th IFAC Workshop on Time Delay Systems, Boston, MA, USA (2012)
26. Ren, W., Beard, R.W., Atkins, E.M.: A survey of consensus problems in multi-agent coordination. In: American Control Conference, Portland, OR, USA (2005)
27. Schöllig, A., Münz, U., Allgöwer, F.: Topology-dependent stability of a network of dynamical systems with communication delays. In: Proceedings of the European Control Conference, Kos, Greece, pp. 1197–1202 (2007)
28. Sipahi, R., Acar, A.: Stability analysis of three-Agent consensus dynamics with fixed topology and three non-identical delays. In: ASME Dynamic Systems and Control Conference, Ann Arbor, Michigan, USA (2008)
29. Sipahi, R., Atay, F.M., Niculescu, S.-I.: Stability of traffic flow behavior with distributed delays modeling the memory effects of the drivers. SIAM Journal on Applied Mathematics 68(3), 738–759 (2007)
30. Sipahi, R., Lämmer, S., Helbing, D., Niculescu, S.-I.: On stability problems of supply networks constrained with transport delay. Journal of Dynamic Systems, Measurement and Control 131(2), 021005 (2009)
31. Sipahi, R., Niculescu, S.-I.: Deterministic time-delayed traffic flow models: a survey. In: Fatihcan M. Atay (ed.), Complex Time-Delay Systems - Theory and Applications, pp. 297–322. Springer, Berlin (2010)
32. Sipahi, R., Niculescu, S.-I., Abdallah, C.T., Michiels, W., Gu, K.: Stability and stabilization of systems with time delay, limitations and opportunities. IEEE Control Systems Magazine 31(1), 38–65 (2011)
33. Sipahi, R., Qiao, W.: Responsible eigenvalue concept for the stability of a class of single-delay consensus dynamics with fixed topology. Control Theory Applications, IET 5(4), 622–629 (2011)
34. Sipahi, R., Qiao, W.: Erratum for 'Responsible eigenvalue concept for the stability of a class of single-delay consensus dynamics with fixed topology 5, 622 (2011)'; Control Theory Applications, IET 6(8), 1154 (2012)
35. Skinner, F.K., Bazzazi, H., Campbell, S.A.: Two-cell to n-cell heterogeneous, inhibitory networks: Precise linking of multistable and coherent properties. Journal of Computational Neuroscience 18(3), 343–352 (2005)
36. Stépán, G.: Retarded dynamical systems: stability and characteristic functions. Pitman Research Notes in Mathematics Series, vol. 210. Longman Scientific & Technical, copublisher John Wiley & Sons, Inc., New York (1989)

Second-Order Leaderless Consensus Protocols with Multiple Communication and Input Delays from Stability Perspective

Rudy Cepeda-Gomez and Nejat Olgac*

Abstract. A leaderless consensus control protocol for double integrators with multiple, rationally-independent time delays is studied in this paper from two intriguing and novel perspectives. First, the crucial stability analysis of time delayed system is performed using a recent technique known as the Cluster Treatment of Characteristic Roots (CTCR). CTCR method is pursued after a block-diagonalization (mode decoupling) transformation on the system. This treatment produces a unique stability outlook for the dynamics in the space of the delays. Furthermore they are non-conservative and exhaustive. Secondly, a much different stability display is created using the Spectral Delay Space as an overture to the CTCR for the determination of the needed potential stability crossing (switching) hypersurfaces in the delay space. Examples are provided to display the effectiveness of this new stability analysis mechanism.

1 Introduction

Within the broad field of cooperative control, the consensus problem for multi-agent systems has received a great deal of attention in recent years. After the work of Olfati-Saber and Murray [1], many researchers have contributed to the knowledge in this area. Some of these studies are limited to the systems with first order agents [1], others focus on second order agent behavior [2], yet others include time delays in the

Rudy Cepeda-Gomez
Universidad Santo Tomas, Bucaramanga, Colombia
e-mail: rudycepedagomez@mail.ustabuca.edu.co

Nejat Olgac
University of Connecticut, Storrs, CT, USA
e-mail: olgac@engr.uconn.edu

* An earlier version of this work was presented at the IFAC TDS 2012 workshop in Boston, MA, June 2012.

T. Vyhlídal, J.-F. Lafay, and R. Sipahi (eds.), *Delay Systems*,
Advances in Delays and Dynamics 1,
DOI: 10.1007/978-3-319-01695-5_9,

communication channels [3,4]. However, very few works such as [6,13,20] offer a practicable procedure for the non-conservative assessment of the stability properties of a consensus system with respect to the delays.

This chapter uses a consensus protocol which is presented by Meng et al. in [16], and applies to it a recent stability analysis methodology [13], which is expanded to handle the case of directed topologies. The agents operating under this protocol are affected by two time delays: a communication delay, which affects only the information coming from other agents, and an input delay, which affects all the state feedback including that of own states of agents. These delays are assumed to be constant and uniform. The stability analysis used in [16] is based on Razhumikin theorem, therefore it is conservative. The new technique performed here, the CTCR, on the contrary, provides non-conservative and exhaustive stability tables.

In this chapter, bold face notation is used for vector quantities, bold capital letters for matrices and italic symbols for scalars.

2 Problem Statement

We focus this study on a group of n agents driven by second order dynamics, $\ddot{x}_j(t) = u_j(t)$. Here $x_j(t)$ represents the scalar position of the agent and $u_j(t)$ the control input. We show a one dimensional case in this text, but the treatment can be easily expanded to higher dimensional dynamics by using the Kronecker product representation [5]. We declare consensus when the agents reach a common position, i.e., when $\lim_{t\to\infty}(x_j(t) - x_k(t)) = 0$ for any j and k. In order to achieve this objective, the members of the group share their positions and velocities with a limited number of neighbors, through one-directional communication channels. The peers from which agent j receives information are called the *informers* of agent j, and this set of $\delta_j < n$ agents is denoted by $\mathcal{N}_j$.

The inter-agent communication topology in this protocol is described by a directed graph with n vertices. We use $\mathbf{A}_\Gamma = [a_{jk}] \in \mathbb{R}^{n\times n}$ to denote the *adjacency matrix* of this graph. Its components are $a_{jk} > 0$ whenever agent k is an informer of agent j and $a_{jk} = 0$ otherwise. The diagonal elements are also taken as zero: $a_{jj} = 0$. Notice that this matrix is, in general, not symmetric. The *in-degree matrix* of the graph is $\boldsymbol{\Delta} = [\Delta_{jk}] \in \mathbb{R}^{n\times n}$, with $\Delta_{jj} = \sum_{k=1}^{n} a_{jk} = \delta_j$ and $\Delta_{jk} = 0$ if $j \neq k$. It is obviously a diagonal matrix.

The control logic followed by the agents is taken from the common literature, with specific parametric selections from [16]:

$$
\begin{aligned}
u_j(t) = &-\frac{1}{\delta_j}\sum_{k=1}^{n} a_{jk}\left(x_j(t-\tau_{in}) - x_k(t-\tau_{in}-\tau_{com})\right) \\
&-\frac{\gamma}{\delta_j}\sum_{k=1}^{n} a_{jk}\left(\dot{x}_j(t-\tau_{in}) - \dot{x}_k(t-\tau_{in}-\tau_{com})\right)
\end{aligned}
\tag{1}
$$

where τ_{in} and τ_{com} are the input and communication delays, respectively, and γ is a positive gain. It is also assumed that each agent has at least one informer, i.e., $\delta_j \neq 0$ for $j = 1, 2, \ldots$. The control logic (1) can be expressed in state space as

$$\begin{aligned}\dot{\mathbf{x}}(t) = \left(\mathbf{I}_n \otimes \begin{bmatrix} 0 & 1 \\ 0 & 0 \end{bmatrix}\right)\mathbf{x}(t) + \left(\mathbf{I}_n \otimes \begin{bmatrix} 0 & 0 \\ -1 & -\gamma \end{bmatrix}\right)\mathbf{x}(t-\tau_1) \\ + \left(\mathbf{C} \otimes \begin{bmatrix} 0 & 0 \\ 1 & \gamma \end{bmatrix}\right)\mathbf{x}(t-\tau_2)\end{aligned} \tag{2}$$

where $\mathbf{x} = [x_1\ \dot{x}_1\ x_2\ \dot{x}_2\ \cdots\ x_n\ \dot{x}_n]^T \in \mathbb{R}^{2n}$ is the state vector. In (2), $\mathbf{I}_n$ represents the identity matrix of order n, $\otimes$ is the Kronecker product [5] and $\mathbf{C} = \boldsymbol{\Delta}^{-1}\mathbf{A}_\Gamma$. The delays have been renamed here, as $\tau_1 = \tau_{in}$ and $\tau_2 = \tau_{in} + \tau_{com}$. The characteristic equation of (2) is a $2n$ degree quasi-polynomial in which the delay terms appear with up to $2n$ degree of commensuracy and with cross-talk terms. Obviously, the complexity of this equation increases rapidly with the number of agents. The only paradigm that provides a non-conservative determination of the stability posture of such systems with respect to the time delays is the Cluster Treatment of Characteristic Roots, CTCR [8,9,12]. The direct deployment of CTCR to (2) in its original form is still very cumbersome as the general problem of multiple time delay systems also known to be NP-hard: it becomes numerically intractable as the order of the characteristic equation increases [22]. We circumvent the complication following the methodology described in [13]. It consists of a factorization procedure followed by the application of CTCR to the simplified system.

Lemma 1. Factorization Property. *The characteristic equation of system* (2) *can always be expressed as the product of a set of second and fourth order factors:*

$$\begin{aligned}&Q(s,\gamma,\tau_1,\tau_2) = \det\left(s\mathbf{I}_{2n} - \mathbf{A} - \mathbf{B}_1 e^{-\tau_1 s} - \mathbf{B}_2 e^{-\tau_2 s}\right) = \\ &\prod_{j=1}^{\ell+m} q_j(s,\gamma,\tau_1,\tau_2,\lambda_j) = \prod_{j=1}^{\ell}\left[s^2 + (\gamma s+1)\left(e^{-\tau_1 s} - \lambda_j e^{-\tau_2 s}\right)\right] \\ &\times \prod_{j=\ell+1}^{\ell+m}\left[s^4 + 2s^2(\gamma s+1)\left(e^{-\tau_1 s} - \Re(\lambda_j)e^{-\tau_2 s}\right) + \right. \\ &\left.(\gamma s+1)^2\left(e^{-2\tau_1 s} - 2\Re(\lambda_j)e^{-(\tau_1+\tau_2)s} + |\lambda_j|^2 e^{-2\tau_2 s}\right)\right]\end{aligned} \tag{3}$$

where $\mathbf{A}$, $\mathbf{B}_1$, *and* $\mathbf{B}_2$ *matrices are self-evident from* (2), λ_j, $j = 1, 2, \cdots, n$ *represent the eigenvalues of* $\mathbf{C}$ *matrix. It is assumed that this matrix has* ℓ *real eigenvalues, denoted by* $j = 1, 2, \cdots, \ell$, *and* m *complex conjugate eigenvalue pairs,* $\left(\lambda_j, \lambda_j^*\right)$, $j = \ell+1, \ell+2, \cdots, m$, $n = \ell + 2n$. *We assume, for simplicity, that each eigenvalue has multiplicity one.*

Proof. Let $\mathbf{T}$ to be the nonsingular similarity transformation matrix that converts $\mathbf{C}$ into its Jordan canonical form: $\boldsymbol{\Lambda} = \mathbf{T}^{-1}\mathbf{C}\mathbf{T}$. The matrix $\boldsymbol{\Lambda} \in \mathbb{R}^{n\times n}$ is block diagonal of the form:

$$\boldsymbol{\Lambda} = \begin{bmatrix} \lambda_1 & 0 & \cdots & 0 & \mathbf{0} & \cdots & \mathbf{0} \\ 0 & \lambda_2 & \cdots & 0 & \mathbf{0} & \cdots & \mathbf{0} \\ 0 & 0 & \cdots & \lambda_\ell & \mathbf{0} & \cdots & \mathbf{0} \\ \mathbf{0} & \mathbf{0} & \cdots & \mathbf{0} & \mathbf{J}_{\ell+1} & \cdots & \mathbf{0} \\ \vdots & \vdots & \vdots & \vdots & \vdots & \ddots & \vdots \\ \mathbf{0} & \mathbf{0} & \cdots & \mathbf{0} & \mathbf{0} & \cdots & \mathbf{J}_{\ell+m} \end{bmatrix} \tag{4}$$

where λ_j, $j = 1, 2, \cdots, \ell$, are the (size 1) Jordan blocks corresponding to the real eigenvalues and

$$\mathbf{J}_j = \begin{bmatrix} \Re(\lambda_j) & -\Im(\lambda_j) \\ \Im(\lambda_j) & \Re(\lambda_j) \end{bmatrix}, \qquad j = \ell+1, \ell+2, \ldots, \ell+m \tag{5}$$

are the 2×2 Jordan blocks corresponding to the complex conjugate eigenvalue pairs. A state transformation $\mathbf{x}(t) = (\mathbf{T} \otimes \mathbf{I}_2)\boldsymbol{\xi}(t)$ in (2) results in:

$$\begin{aligned} \dot{\boldsymbol{\xi}}(t) &= (\mathbf{T}^{-1} \otimes \mathbf{I}_2)\left(\mathbf{I}_n \otimes \begin{bmatrix} 0 & 1 \\ 0 & 0 \end{bmatrix}\right)(\mathbf{T} \otimes \mathbf{I}_2)\boldsymbol{\xi}(t) \\ &+ (\mathbf{T}^{-1} \otimes \mathbf{I}_2)\left(\mathbf{I}_n \otimes \begin{bmatrix} 0 & 0 \\ -1 & -\gamma \end{bmatrix}\right)(\mathbf{T} \otimes \mathbf{I}_2)\boldsymbol{\xi}(t-\tau_1) \\ &+ (\mathbf{T}^{-1} \otimes \mathbf{I}_2)\left(\mathbf{C} \otimes \begin{bmatrix} 0 & 0 \\ 1 & \gamma \end{bmatrix}\right)(\mathbf{T} \otimes \mathbf{I}_2)\boldsymbol{\xi}(t-\tau_2) \end{aligned} \tag{6}$$

A convenient property of the Kronecker product [5] is $(\mathbf{U} \otimes \mathbf{V})(\mathbf{W} \otimes \mathbf{Z}) = \mathbf{UW} \otimes \mathbf{VZ}$, where the matrix pairs $(\mathbf{U}, \mathbf{W})$ and $(\mathbf{V}, \mathbf{Z})$ are of the same dimensions. Using this property, (6) becomes:

$$\begin{aligned} \dot{\boldsymbol{\xi}}(t) &= \left(\mathbf{I}_n \otimes \begin{bmatrix} 0 & 1 \\ 0 & 0 \end{bmatrix}\right)\boldsymbol{\xi}(t) \\ &+ \left(\mathbf{I}_n \otimes \begin{bmatrix} 0 & 0 \\ -1 & -\gamma \end{bmatrix}\right)\boldsymbol{\xi}(t-\tau_1) + \left(\boldsymbol{\Lambda} \otimes \begin{bmatrix} 0 & 0 \\ 1 & \gamma \end{bmatrix}\right)\boldsymbol{\xi}(t-\tau_2) \end{aligned} \tag{7}$$

Since $\mathbf{I}_n$ and $\boldsymbol{\Lambda}$ are diagonal and block diagonal matrices, respectively, (7) is block-diagonalized, thus it can be represented as a set of $\ell + m$ dynamically decoupled subsystems as:

$$\begin{aligned} \dot{\xi}_j(t) &= \begin{bmatrix} 0 & 1 \\ 0 & 0 \end{bmatrix}\xi_j(t) + \begin{bmatrix} 0 & 0 \\ -1 & -\gamma \end{bmatrix}\xi_j(t-\tau_1) \\ &+ \lambda_j \begin{bmatrix} 0 & 0 \\ 1 & \gamma \end{bmatrix}\xi_j(t-\tau_2), \qquad j = 1, 2, \cdots, \ell \end{aligned} \tag{8a}$$

$$\begin{aligned} \dot{\boldsymbol{\xi}}_j(t) &= \left(\mathbf{I}_2 \otimes \begin{bmatrix} 0 & 1 \\ 0 & 0 \end{bmatrix}\right)\boldsymbol{\xi}_j(t) + \left(\mathbf{I}_2 \otimes \begin{bmatrix} 0 & 0 \\ -1 & -\gamma \end{bmatrix}\right)\boldsymbol{\xi}_j(t-\tau_1) \\ &+ \left(\mathbf{J}_j \otimes \begin{bmatrix} 0 & 0 \\ 1 & \gamma \end{bmatrix}\right)\boldsymbol{\xi}_j(t-\tau_2), \qquad j = \ell+1, \ell+2, \ldots, \ell+m \end{aligned} \tag{8b}$$

The characteristic equation of the complete system, therefore becomes a product of $\ell+m$ individual subsystems which are:

$$q_j(s,\gamma,\tau_1,\tau_2,\lambda_j) = s^2 + (\gamma s+1)\left(e^{-\tau_1 s} - \lambda_j e^{-\tau_2 s}\right) = 0 \tag{9a}$$

$$\begin{aligned} q_j(s,\gamma,\tau_1,\tau_2,\lambda_j) = s^4 + 2s^2(\gamma s+1)\left(e^{-\tau_1 s} - \Re(\lambda_j)e^{-\tau_2 s}\right) + \\ (\gamma s+1)^2\left(e^{-2\tau_1 s} - 2\Re(\lambda_j)e^{-(\tau_1+\tau_2)s} + |\lambda_j|^2 e^{-2\tau_2 s}\right) = 0 \end{aligned} \tag{9b}$$

corresponding to (8a) and (8b) respectively. □

Lemma 1 simplifies the problem considerably, by transforming it from a $2n$ order system with time delays of commensuracy degree up to n and delay cross-talk, into ℓ second order and m fourth order systems with highest commensuracy of 2 (e.g., $e^{-2\tau_1 s}$) and single delay cross-talk (e.g., $e^{-(\tau_1+\tau_2)s}$). Notice that, since the only discriminating element from one factor to the other is the eigenvalue λ_j, the stability analysis in the domain of the delays can be performed two times, once for a generic real λ and once for a generic complex λ. These tasks are detailed later. The problem is now considerably simplified to determine the specific eigenvalues of a known matrix **C**, superpose the stability outlook of each factor to obtain the ensemble stability tableau for the system in (3).

We wish to direct the discussion now, to the special features of the eigenvalues of the matrix **C**. From the way this matrix is created, sum of the elements of any row (which are all non-negative) always add up to 1. This property makes **C** a row-stochastic matrix [14]. Using Gershgorin's disk theorem [15], it can be shown that the norm of the eigenvalues of such matrices is always equal to or less than 1. Furthermore, it has been proven [18] that if the topology is connected and has at least one spanning tree, $\lambda = 1$ is one of the eigenvalues of the matrix **C** with multiplicity 1. Then, the corresponding

$$q_j(s,\gamma,\tau_1,\tau_2,\lambda_j) = s^2 + (\gamma s+1)\left(e^{-\tau_1 s} - e^{-\tau_2 s}\right) = 0 \tag{10}$$

is always a factor in the characteristic quasi-polynomial (3). Without loss of generality, we will assign this eigenvalue to the state ξ_1. It can be shown that the normalized eigenvector corresponding to this state is always $\mathbf{t}_1 = 1/\sqrt{n}[1\ 1\ \cdots\ 1]$, and it is selected as the first column of the earlier defined transformation matrix **T**. Factor (10) governs the dynamics of ξ_1, which is proven, in Lemma 2 below, to be nothing other than a weighted average of the positions of the agents. Thus we call ξ_1, the *weighted centroid*, which is topology dependent, since the weights for the computation of ξ_1 arise from the first row of the inverse of the matrix **T**. Furthermore, it is evident that $s=0$ is a stationary root of (10) independent of the delays, τ_1 and τ_2, which implies that the weighted centroid dynamics is at best marginally stable. The other factors of the characteristic equation in (3) are related to the *disagreement dynamics*. When these are stable the agents reach consensus among themselves.

If the communication topology does not have a spanning tree, 1 is a multiple eigenvalue of **C** and equation (10) appears as a factor multiple times within (3). These factors represent the dynamics of the centroids of the subgroups created by the subgraphs that are spanned by a tree. If all the disagreement factors are stable the swarm members within a subgroup reach stationary positions which are generally different. Thus the consensus is not achieved. These facts are stated in the following lemmas.

Lemma 2. Group behavior. *Assume the communication topology has at least one spanning tree. Then, the agents in the group reach a consensus if and only if the factor* (10) *is marginally stable and all the remaining factors of* (3) *are stable. Furthermore, the group consensus value is* $\bar{x} = (1/\sqrt{n})\,\xi_1(t\to\infty) = (1/\sqrt{n})\,\bar{\xi}$, *whereas the other states* $\xi_j(t\to\infty) = 0$ *for* $j = 2, 3, \ldots, n$.

Proof. First, we prove the necessity condition. From the definition of the state, $\boldsymbol{\xi} = [\xi_1(t)\ \xi_2(t)\cdots\xi_n(t)]^T = \mathbf{T}[x_1(t)\ x_2(t)\cdots x_n(t)]$. If consensus is reached, the agents have a common steady state value. This implies $\lim_{t\to\infty} x_j(t) = \bar{x}$, $j = 1, 2, 3, \ldots, n$. Then:

$$\lim_{t\to\infty}[\xi_1(t)\ \xi_2(t)\cdots\xi_n(t)]^T = \bar{x}\mathbf{T}^{-1}[1\,1\cdots 1]^T \tag{11}$$

Since the communication topology is assumed to have a spanning tree, 1 is a simple eigenvalue of $\mathbf{C}$, corresponding to the eigenvector $\mathbf{t}_1 = 1/\sqrt{n}[1\,1\cdots 1]^T$, the first column of the earlier defined transformation matrix $\mathbf{T}$. Since $\mathbf{T}^{-1}[1\,1\cdots 1]^T = \sqrt{n}\mathbf{T}^{-1}\mathbf{t}_1 = \sqrt{n}[1\,0\cdots 0]^T$, equation (11) leads to $\lim_{t\to\infty}\xi_1(t) = \sqrt{n}\bar{x}$, which indicates marginal stability for (10) and $\lim_{t\to\infty}\xi_1(t) = 0$ for $j = 2, 3, \ldots, n$, indicating asymptotic stability in the other factors of (3).

Next the sufficiency condition is proven. If (10) is marginally stable and all the other factors in (3) are stable, the steady state value of $\xi_1(t)$ will be constant whereas the remaining $\xi_j(t)$ will tend to zero as t goes to infinity. Then, $\lim_{t\to\infty}[\xi_1(t)\ \xi_2(t)\cdots\xi_n(t)]^T = \left[\bar{\xi}_1\,0\cdots 0\right]^T$. Going back to x domain using the inverse transformation:

$$\begin{aligned}\lim_{t\to\infty}[x_1(t)\ x_2(t)\cdots x_n(t)]^T &= \mathbf{T}\left[\bar{\xi}_1\,0\cdots 0\right]^T\\ &= \bar{\xi}_1\mathbf{t}_1 = \sqrt{n}\bar{x}\mathbf{t}_1 = [\bar{x}\,\bar{x}\cdots\bar{x}]^T\end{aligned} \tag{12}$$

implying the agents reach consensus. □

Lemma 3. Topologies without spanning trees. *If the given communication topology does not have a spanning tree, the control logic described by* (1) *can not result in consensus.*

The proof of this lemma is omitted here due to space considerations. It can be found in [21].

3 Stability Analysis Using CTCR Paradigm and SDS Domain

The factors given in (9a) and (9b), which dictate the dynamics of the swarm, have the general formations of:

$$g_{11}(s)+g_{12}(s)e^{-\tau_1 s}+g_{13}(s)e^{-\tau_2 s}=0 \tag{13}$$

for (9a), and

$$\begin{aligned} &g_{21}(s)+g_{22}(s)e^{-\tau_1 s}+g_{23}(s)e^{-\tau_2 s} \\ &\qquad +g_{24}(s)e^{-(\tau_1+\tau_2)s}+g_{25}(s)e^{-2\tau_1 s}+g_{26}(s)e^{-2\tau_2 s}=0 \end{aligned} \tag{14}$$

for (9b). In fact, the class of quasi-polynomials given in (13) is a degenerate case of (14). Therefore we will focus the stability treatment of the generic quasi-polynomial (14) with rationally independent time delays. This task is performed deploying a unique methodology, called the Cluster Treatment of Characteristic Roots (CTCR) [12]. The main philosophy behind it is the *clustering* of all possible imaginary root crossings in the (τ_1, τ_2) domain. The method starts with their exhaustive determination. For this we follow a novel approach over a new domain which is called the Spectral Delay Space (SDS). The following paragraphs present some preparatory definitions and key propositions of CTCR paradigm.

Definition 1. *Kernel hypercurves* $\wp_0$: The curves that consist of *all* the points $(\tau_1\ \tau_2) \in \mathbb{R}^{2+}$ exhaustively, which cause an imaginary root $s=\omega i$, $\omega \in \mathbb{R}$ and satisfy the constraint $0<\tau_k\omega<2\pi$ are called the *kernel curves*. The points on this curves contain the smallest delay compositions which correspond to all possible imaginary roots.

Definition 2. *Offspring curves* $\wp$: The curves obtained from the kernel curve by the following pointwise nonlinear transformation:

$$\left\langle \tau_1+\frac{2\pi}{\omega}j_1,\ \tau_2+\frac{2\pi}{\omega}j_2,\ldots,\tau_p+\frac{2\pi}{\omega}j_p \right\rangle,\quad j_k=1,2,\ldots \tag{15}$$

are called the *offspring hypercurves*.

Definition 3. *Root Tendency, RT*: The root tendency indicates the direction of transition of the imaginary root as only one of the delays, τ_j, increases by ε, $0<\varepsilon<<1$, while all the others remain constant:

$$RT\big|_{s=\omega i}^{\tau_j}=\mathbf{sgn}\left[\Re\left(\left.\frac{\partial s}{\partial \tau_j}\right|_{s=\omega i}\right)\right] \tag{16}$$

There are two overarching propositions which support the CTCR paradigm which are stated here from [9] without proof.

Proposition 1. Small number of kernel hypercurves*: The number of kernel hypercurves is manageably small. To be specific for a LTI-TDS of state dimension n, the maximum possible number of kernel hypercurves is n^2 [11].*

Proposition 2. Invariant root tendency property*: Take an imaginary characteristic root, $s = \omega i$, caused by any one of the infinitely many grid points on the kernel and offspring hypercurves in $(\tau_1\ \tau_2) \in \mathbb{R}^{2+}$ defined by the expression* (15)*. The root tendency of these imaginary roots remains invariant from one* offspring hypercurve *to the other when one of the delays is kept constant. That is, the root tendency with respect to the variations of τ_1 (or τ_2) is invariant from the kernel to the corresponding offspring as τ_2 (or τ_1) is fixed.*

Spectral Delay Space (SDS): A new procedure is described in this segment for determining the kernel (and offspring) curves. It is a formalized treatment from a recent thesis work [12, 17]. The procedure is developed on a new domain: SDS. It is defined by the coordinates $\nu_j = \tau_j\omega$ for every point $(\tau_1\ \tau_2)$ on the kernel or the offspring curves. This is a conditional mapping: if a delay set $(\tau_1\ \tau_2)$ creates an imaginary root $s = \omega i$, (i.e., if the point is on the kernel or the offspring curves) then $(\tau_1\omega\ \tau_2\omega)$ forms a point in the SDS. On the contrary, $(\tau_1\ \tau_2)$ points that do not generate an imaginary root have no representation in the SDS.

The main advantage of SDS is that the representation of the kernel curve in the SDS, denoted as $\wp_0^{SDS}$ and called the *building curve*, is confined into a square of edge length 2π. Then, it is only necessary to explore a finite domain to find the representation of the building curves in the SDS. This finite domain is known as the *building block* (BB), i.e., $2\pi \times 2\pi$ squares, as per (15). Another advantage of these coordinates is that the transitions from the building to the *reflection curves* (i.e., the representation of the offspring curves in the SDS) is achieved simply by stacking the copies of the BB as opposed to using the pointwise non-linear transformation (15), which results in an undesirable shape distortion. There are several other intriguing properties of the SDS and BB concepts which can be found in [12].

With these definitions and propositions, we now return to the mentioned preparatory stage of CTCR method. It is the exhaustive determination of all the imaginary roots, $s = \omega i$, for the generic factor of the characteristic equation in (14) within the semi-infinite quadrant of $(\tau_1\ \tau_2) \in \mathbb{R}^{2+}$. We follow the mathematical procedure described in the appendix of [17] which evaluates the building curves. Accordingly, in (14), the exponential terms are replaced by:

$$e^{\tau_k\omega i} = \cos(\nu_k) + i\sin(\nu_k), \quad \nu_k = \tau_k\omega, \quad k = 1,2 \tag{17}$$

and the sine and cosine functions are expressed in terms of half-angle tangent function:

$$\cos(\nu_k) = \frac{1 - z_k^2}{1 + z_k^2}, \quad \sin(\nu_k) = \frac{2z_k}{1 + z_k^2}, \quad z_k = \tan\left(\frac{\nu_k}{2}\right) \tag{18}$$

Equation (14) can now be written as a polynomial in ω with complex coefficients c_k parameterized in z_1 and z_2:

$$q_j(\omega, z_1, z_2) = \sum_{k=0}^{2} c_k(P, D, \lambda_j, z_1, z_2)(\omega i)^k = 0 \tag{19}$$

If there is a solution to (19), both its real and imaginary parts must be zero simultaneously:

$$\Re\left(q_j\left(\omega,z_1,z_2\right)\right) = \sum_{k=0}^{2} f_k\left(z_1, z_2\right)\omega^k = 0 \tag{20a}$$

$$\Im\left(q_j\left(\omega,z_1,z_2\right)\right) = \sum_{k=0}^{2} g_k\left(z_1, z_2\right)\omega^k = 0 \tag{20b}$$

The condition for (20a) and (20b) to have a common root is simply stated using a Sylvester's resultant matrix:

$$\mathbf{M} = \begin{bmatrix} f_2(z_1,z_2) & f_1(z_1,z_2) & f_0(z_1,z_2) & 0 \\ 0 & f_2(z_1,z_2) & f_1(z_1,z_2) & f_0(z_1,z_2) \\ g_2(z_1,z_2) & g_1(z_1,z_2) & g_0(z_1,z_2) & 0 \\ 0 & g_2(z_1,z_2) & g_1(z_1,z_2) & g_0(z_1,z_2) \end{bmatrix} \tag{21}$$

In order for (20) to be satisfied, $\mathbf{M}$ should be singular. This results in the following expression in terms of z_1 and z_2:

$$\det(\mathbf{M}) = F\left(z_1, z_2\right) = F\left(\tan\left(v_1\right), \tan\left(v_2\right)\right) \tag{22}$$

which constitutes a closed form description of the kernel curves in the SDS $(v_1,\ v_2)$, i.e., the building curves. To obtain its graphical depiction, one of the parameters, say v_2, can be scanned in the range of $[0,\ 2\pi]$ and the corresponding v_1 values are calculated again in $[0,\ 2\pi]$. Notice that every point $(v_1,\ v_2)$ on these curves brings an imaginary characteristic root at $\pm\omega i$ which can be evaluated from (20a) or (20b), noting that they share the same imaginary root. That is, we create a continuous sequence of $(v_1,\ v_2,\ \omega)$ sets all along the kernel curves. We then back transform from the $(v_1,\ v_2)$ domain of SDS to the $(\tau_1,\ \tau_2)$ delay space, using the inverse transformation of (17) with the appropriate ω values. This generates the kernel and offspring curves.

The kernel and offspring curves divide the $(\tau_1,\ \tau_2)$ domain in regions of possible stability and instability. To determine these regions we start from the non-delayed system (i.e., $\tau_1 = \tau_2 = 0$). It is trivial to prove that the factors in (3) are stable for the non-delayed case provided that the control gain satisfies the following necessary and sufficient condition:

$$\gamma > \bar{\gamma} = \max_{\lambda_j \neq 1} \sqrt{\frac{\Im\left(\lambda\right)^2}{\left(1-\Re\left(\lambda_j\right)\right)\left(\left(\Re\left(\lambda_j\right)-1\right)^2+\Im\left(\lambda_j\right)^2\right)}} \tag{23}$$

The root tendency invariance property is deployed from one region to the other, resulting in the complete and non-conservative stability outlook of the system. An example of this construction is presented in the following section, for clarity.

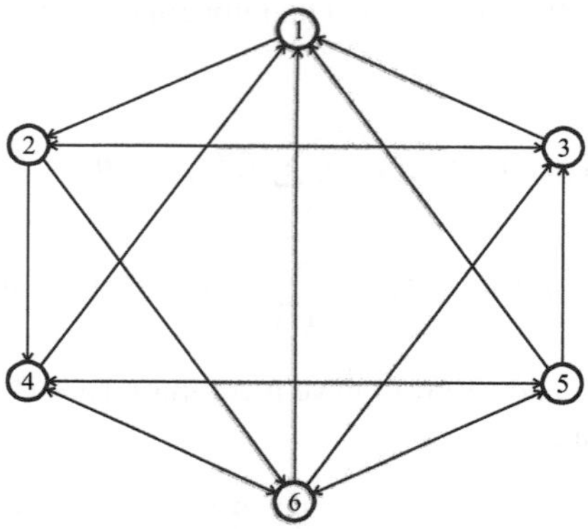

Fig. 1 Communication topology used in the example cases

4 Deployment on a Case Study

We take the communication topology presented in Fig. 1 with the communication weighting factors described in the adjacency matrix:

$$\mathbf{A}_{\Gamma} = \begin{bmatrix} 0 & 0 & 3 & 2.5 & 2 & 2.5 \\ 3 & 0 & 3 & 0 & 2 & 2 \\ 0 & 2 & 0 & 2 & 3 & 3 \\ 1 & 3 & 0 & 0 & 2 & 4 \\ 0 & 2.2 & 0 & 1.8 & 0 & 5 \\ 0 & 6 & 0 & 2 & 2 & 0 \end{bmatrix} \tag{24}$$

The eigenvalues of the $\mathbf{C}$ matrix created from (24) are 1, -0.072, -0.363, -0.286, and -0.140 ± 0.379. With this set of eigenvalues, the minimum control gain that guarantees stability for the non delayed case is $\bar{\gamma} = 0.3281$, defined by (23). A value of $\gamma = 0.5$, which is larger than the lowerbound, is used in the following examples.

After proceeding with the steps presented in the previous section, the building curve (dark) and reflection curves (light) in SDS are obtained. Fig. 2 shows these curves for $\lambda = 1$ only, we do not present the other eigenvalues to avoid overcrowding the figure. The stability switching curves are then transformed to the (τ_1, τ_2) domain, and then to the original (τ_{in}, τ_{com}) space. After superposing the curves obtained, and intersecting their stable regions, the stability map for the complete system is obtained as displayed in Fig. 3.

To validate this stability analysis, we executed several dynamic simulations for specific cases. Figure 4 shows the traces of the agents, using the initial conditions $\mathbf{x}_0 = [-0.4\ 0.5\ 0.7\ 0.4\ 1.2\ 0.3]^T$ and $\dot{\mathbf{x}}_0 = [-0.1\ 0.2\ 0.7\ 0.4 - 0.1\ 0.3]^T$ and a delay combination of $\tau_{in} = 0.03$s and $\tau_{com} = 0.3$s, which is marked by point **a** in Fig. 3. The behavior is stable as declared by the stability map. Furthermore, Fig. 5 shows the traces for delay values selected at point **b** in Fig. 3. This point, outside the shaded region, corresponds to an unstable case, as it is demonstrated by the traces in Fig. 5.

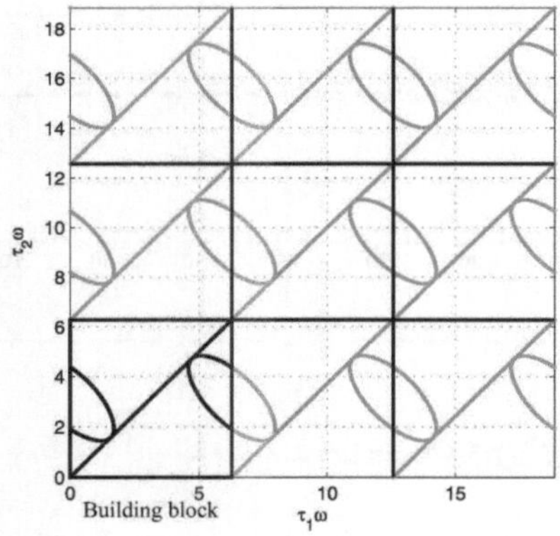

Fig. 2 SDS representation of the stability switching curves created by $\lambda = 1$

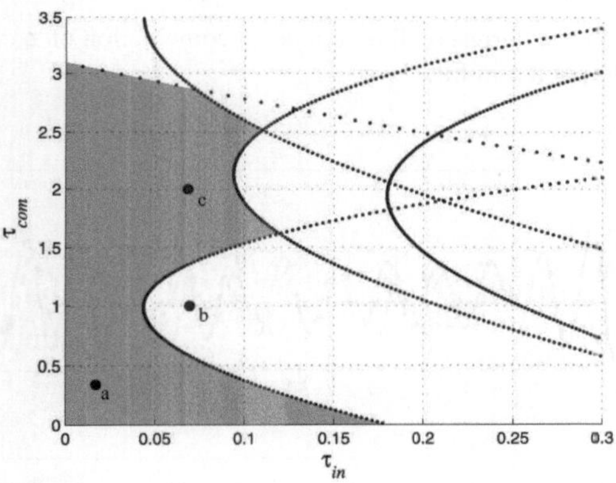

Fig. 3 Stability map in (τ_{in}, τ_{com}) domain for the complete system. The shaded region represents the stable zone.

Certainly, Fig. 3 is an exhaustive and non-conservative stability result, in contrast with those presented by [16] (Theorem 5.1), which provide only a conservative upper bound for the admissible delay. This non-conservative and exhaustive stability map, is made possible by deploying the unique features of the CTCR paradigm. This combination forms the contributory point of this study.

A more intriguing observation that can be made from Fig. 3 is that although the delay combination in point **b** represents an unstable behavior, the stability of the system can be recovered by *increasing* τ_{com} from 1 to 2 seconds, i.e., moving the delay selection to point **c**. This is a counter-intuitive proposition: no one expects an increase in the delay will result in improved performance. The technique of increasing the delays to obtain better tracking results is called *Delay Scheduling*, and it has been successfully used recently for trajectory tracking [23].

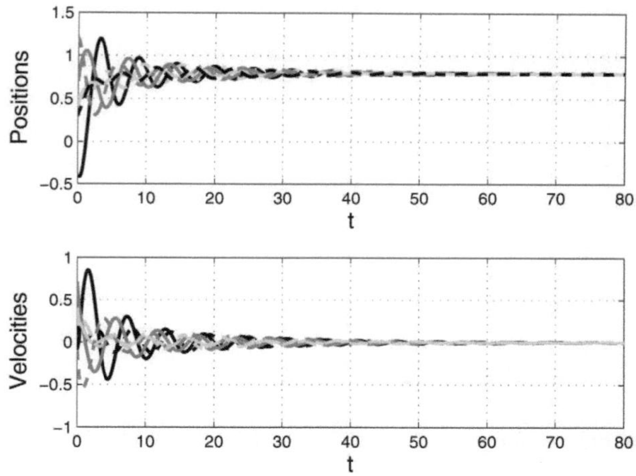

Fig. 4 Traces of the agents corresponding to a delay combination of $\tau_{in} = 0.03$s and $\tau_{com} = 0.3$s, corresponding to point **a** in Fig. 2

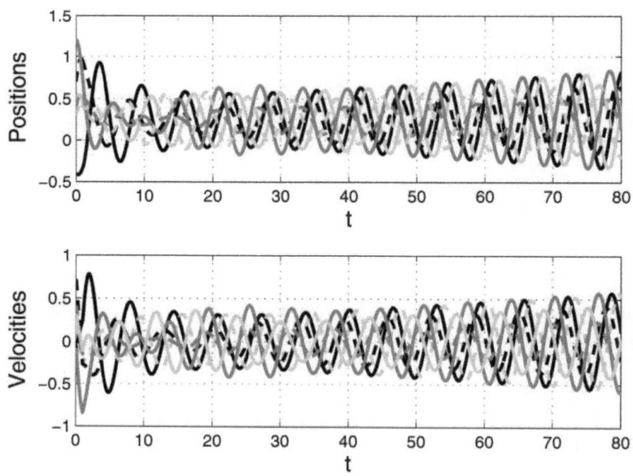

Fig. 5 Traces of the agents corresponding to a delay combination of $\tau_{in} = 0.07$s and $\tau_{com} = 1$s, corresponding to point **b** in Fig. 3

5 Conclusions

We elaborated further on a consensus protocol for a multi-agent system dynamics which has been attracting attention earlier. Interagent communication is governed through a directed network. Their dynamics is, however, influenced by a communication and a separate input delay. These two delays are assumed to be rationally independent. The first delay is on the information coming from the informer agents, whereas the effect of the second one is in both the own state and the state of the informers.

A state-of-the-art and broadly accepted treatment of such complex stability problem is performed using the Razhumikin theorem in some earlier studies. This procedure is replaced here with a novel technique that ultimately decouples the dynamics into a number of second and fourth order subsystems. The CTCR paradigm is then deployed on these primitive subsystems. They are differentiated from each other only by a single scalar parameter. These parameters happen to be the eigenvalues of a certain matrix related to the communication topology. As such, they can be calculated off-line.

The first step in the CTCR establishes an exhaustive display of the potential stability switching hypercurves. For this task, an intriguing domain, Spectral Delay Space, is used here. The efficiency and practicality of the combined deployment is displayed via a case study.

The analysis presented here shows that CTCR is a more powerful tool for the stability analysis of systems with multiple, rationally-independent time delays.

References

1. Olfati-Saber, R., Murray, R.: Consensus problems in networks of agents with switching topology and time-delays. IEEE Transactions on Automatic Control 49(9), 1520–1533 (2004)
2. Ren, W.: On Consensus Algorithms for Double Integrator Dynamics. IEEE Transactions on Automatic Control 53(6), 1503–1509 (2008)
3. Sun, Y., Wang, L.: Consensus problems in networks of agents with double integrator dynamics and time delays. International Journal of Control 82(10), 1937–1945 (2009)
4. Sun, Y., Wang, L.: Consensus of multi-agent systems in directed networks with nonuniform time-varying delay. IEEE Transactions on Automatic Control 54(7), 1607–1613 (2009)
5. Schaefer, R.D.: An Introduction to Nonassociative Algebras. Dover (1996)
6. Cepeda-Gomez, R., Olgac, N.: Exhaustive Stability analysis in a consensus system with time delay and irregular topologies. International Journal of Control 84(4), 746–757 (2011)
7. Zhang, J., Knospe, R., Tsiotras, P.: Stability of time-delay systems: equivalence between Lyapunov and scaled small-gain conditions. IEEE Transactions on Auto 46(3), 482–486 (2001)
8. Olgac, N., Sipahi, R.: An Exact Method for the stability Analysis of Time-Delayed Linear Time invariant Systems. IEEE Transactions on Automatic Control 47(5), 793–797 (2002)
9. Olgac, N., Sipahi, R.: Complete stability robustness of third order LTI multiple time delay systems. Automatica 41(8), 1413–1422 (2005)
10. Olgac, N., Sipahi, R.: An improved procedure in detecting the stability robustness of systems with uncertain delay. IEEE Transactions on Automatic Control 51(7), 1164–1165 (2006)

11. Ergenc, A.F., Olgac, N., Fazelinia, H.: Extended Kronecker summation for cluster treatment of LTI systems with multiple delays. SIAM Journal on Control and Optimization 46(1), 143–155 (2007)
12. Fazelinia, H., Sipahi, R., Olgac, N.: Stability robustness analysis of multiple time-delayed systems using building block concept. IEEE Transactions on Automatic Control 52(5), 799–810 (2007)
13. Cepeda-Gomez, R., Olgac, N.: An Exact Method for the Stability Analysis of Linear Consensus Protocols with Time Delay. IEEE Transactions on Automatic Control 56(7), 1734–1740 (2011)
14. Marcus, M., Minc, H.: A Survey of Matrix Theory and Matrix Inequalities. Dover (1996)
15. Bell, H.E.: Gershgorin's Theorem and the zeros of Polynomials. American Mathematical Monthly 74, 292–295 (1965)
16. Meng, Z., Ren, W., Cao, Y., Zheng, Y.: Leaderless and Leader-Follower Consensus with Communications and Input Delays Under a Directed Network Topology. IEEE Transactions on Systems, Man and Cybernetics - Part B 41(1), 75–88 (2011)
17. Fazelinia, H.: A Novel Stability Analysis of Systems with Multiple Time Delays and Its Application to High Speed Milling Chatter. PhD. Dissertation, University of Connecticut (2007)
18. Agaev, R., Chebotarev, P.: On the spectra of nonsymmetric laplacian matrices. Linear Algebra and its Applications 399, 157–168 (2005)
19. Olgac, N., Vyhlidal, T., Sipahi, R.: A New Perspective in the Stability Assesment of Neutral Systems with Multiple and Cross-Talking Delays. SIAM Journal of Control and Optimization 47(1), 327–344 (2008)
20. Cepeda-Gomez, R., Olgac, N.: Consensus Analysis With Large and Multiple Communication Delays Using Spectral Delay Space (SDS) Concept. International Journal of Control 84(12), 1996–2007 (2011)
21. Cepeda-Gomez, R., Olgac, N.: Formation control based on a consensus protocol under directed communications with two time delays. In: Proceedings of the American Control Conference, pp. 1597–1602 (June 2012)
22. Toker, O., Ozbay, H.: Complexity issues in robust stability of linear delay-differential systems. Mathematics of Control, Signals and Systems 9(4), 386–400 (1996)
23. Olgac, N., Sipahi, R., Ergenc, A.F.: Delay scheduling, an unconventional use of time delay for trajectory tracking. Mechatronics 17(4-5), 199–206 (2007)

Analysis of Gene Regulatory Networks under Positive Feedback

Mehmet Eren Ahsen, Hitay Özbay, and Silviu-Iulian Niculescu

Abstract. In this chapter of the book, a dynamical model of the gene regulatory networks (GRNs) under positive feedback is analyzed. The model considered involve static nonlinearities with negative Schwarzian derivatives, and a time delay in the feedback path. A set of conditions are derived for the global stability of the class of GRNs considered. As a special case, *homogenous* GRNs are also analyzed and an appropriate stability condition is obtained; that depends only on the parameters of the nonlinearity function, which is assumed to be a Hill type function. In particular, conditions leading to bistability of the system are obtained. The results presented here naturally extend to similar classes of *cyclic biological processes* involving time delayed feedback.

1 Introduction

Gene regulation is a tool for the cell to communicate with its environment. Cells respond to certain environmental stimuli by decreasing or increasing the production of certain genes. For example, when grown in a glucose deficient but lactose rich environment, the genes that are responsible for the digestion of lactose is expressed [22]. This way lactose is able to be digested into glucose. In [7] a synthetic oscillatory network have been produced by using gene products, which was very important for the newly emerging field, namely, synthetic biology. By using tools

Mehmet Eren Ahsen
Dept. of Bioengineering, University of Texas at Dallas, Richardson, TX 75080-3021, USA
e-mail: `ahsen@utdallas.edu`

Hitay Özbay
Dept. of Electrical & Electronics Engineering, Bilkent University, 06800, Ankara, Turkey
e-mail: `hitay@bilkent.edu.tr`

Silviu-Iulian Niculescu
LSS-SUPELEC (UMR CNRS 8506), 3 Rue Joliot-Curie, 91192 Gif-sur-Yvette, France
e-mail: `Silviu.Niculescu@lss.supelec.fr`

T. Vyhlídal, J.-F. Lafay, and R. Sipahi (eds.), *Delay Systems*,
Advances in Delays and Dynamics 1,
DOI: 10.1007/978-3-319-01695-5_10, © Springer International Publishing Switzerland 2014

from synthetic biology, it is envisioned that one will be able to use plants as sensor chemicals in order to produce clean and renewable fuels, or even to recognize cancer cells and destroy them [15]. Therefore, accurate modeling and analysis of gene regulatory networks (GRNs) are important for building synthetic networks for specific functions such as oscillatory networks with a predefined period. Another important concept in biology is bistability. Bistability is important in cell functions such as cellular differentiation and apoptosis. Bistability can be generated by adding a positive feedback loop in the circuit and it is a strong network motif especially in developmental circuit [4]. In this work we will give easy conditions that leads to bistability of the homogeneous (GRN), when the nonlinearities are taken as Hill functions.

The present work is based on [2] and concerned with the asymptotic stability of a dynamical model of the GRNs under positive feedback. The model considered involve static nonlinearities with negative Schwarzian derivatives, and a time delay in the feedback path, (see, e.g. [19, 20] for the justification of this model and discussions on the alternative models).

Basically, a GRN can be described as the interaction of DNA segments with themselves and with other biological structures such as the enzymes in the cell. Therefore, it can be thought as an indicator of the genes transcription rates into mRNA, which is used to deliver the coding information required for the protein synthesis, [11]. The model proposed in [6] consists of a set of differential equations in the following form:

$$\begin{cases} \dot{p}_1(t) &= -k_{p1}p_1(t) + f_{p1}(g_m(t-\tau_{gm})) \\ \dot{g}_1(t) &= -k_{g1}g_1(t) + f_{g1}(p_1(t-\tau_{p1})) \\ & \vdots \\ \dot{p}_m(t) &= -k_{pm}p_m(t) + f_{pm}(g_{m-1}(t-\tau_{g_{m-1}})) \\ \dot{g}_m(t) &= -k_{g1}g_m(t) + f_{gm}(p_m(t-\tau_{pm})), \end{cases} \tag{1}$$

where p_i and g_i represent the protein and mRNA concentrations respectively. Models similar to (1) are frequently encountered in the modeling of biological processes such as mitogen-activated protein cascades and circadian rhythm generator [10], [16] and [21]. For instance, in [6], a simplified version of the system (1) is analyzed and a local stability result is given. An explicit computation of the allowable upper bounds on the delay value can be found in [13].

The system (1) under single time-delay and negative feedback has been studied and discussed in [8], where a simple condition for asymptotic stability has been obtained. By using a Hopf bifurcation approach [8] showed the existence of oscillations in some cases. The arguments of [12], [5] are used in [8] to embed the system (1) to a discrete time system. The present work gives and analysis of the GRNs under positive feedback using some of the results of [1]. The techniques similar to the ones used here are employed in [3] for the negative feedback case.

In the rest of this chapter, the functions f_{pi} and f_{gi} are taken as static nonlinearities having negative Schwarzian derivatives. As a special case, *homogenous* GRNs are also analyzed and a sufficient condition is obtained for asymptotic stability of the system; that depends only on the parameters of the nonlinearity function.

The chapter is organized as follows: The problem formulation and some preliminary results are given in the next section. The main results are stated in Section 3. Illustrative examples are given in Section 4, and concluding remarks are made in the last section.

2 Notation, Preliminaries and Problem Formulation

In this section, first, some basic definitions and notations are given. Then, certain properties of Schwarzian derivatives are presented. These are commonly used in the analysis of cyclic nonlinear feedback systems, which are similar to the system considered here, see e.g. [14].

Let a function f be defined from $\mathbb{R}_+$ to $\mathbb{R}_+$. Suppose it is at least three times continuously differentiable. Then, the Schwarzian derivative of the function f, denoted as $Sf(x)$, is given by the following expression (see [17])

$$Sf(x) = \begin{cases} -\infty & \text{if } f'(x) = 0 \\ \dfrac{f'''(x)}{f'(x)} - \dfrac{3}{2}\left(\dfrac{f''(x)}{f'(x)}\right)^2 & \text{if } f'(x) \neq 0\,. \end{cases}$$

The notation f^m is used to denote the function obtained by m compositions of f. For a function f, the point x is a fixed point if $f(x) = x$.

In the sequel, the simplified system shown below (2) is analyzed; this is equivalent to (1), where the delays are lumped into the feedback channel:

$$\begin{cases} \dot{x}_1(t) = -\lambda_1 x_1(t) + g_1(x_2(t)) \\ \dot{x}_2(t) = -\lambda_2 x_2(t) + g_2(x_3(t)) \\ \quad\vdots \\ \dot{x}_n(t) = -\lambda_n x_n(t) + g_n(x_1(t-\tau)). \end{cases} \tag{2}$$

Note the following relation between τ and τ_{gi}, τ_{pi}:

$$\tau = \sum_{i=1}^{m} (\tau_{pi} + \tau_{gi}). \tag{3}$$

Conditions for the asymptotic stability and existence of oscillations regarding the nonlinear time delayed feedback system (2) will be given in Section 3, under the following simplifying assumptions.

Assumption 1. For all $i = 1, 2, ..., n$, the parameters λ_i satisfy $\lambda_i > 0$. □

Assumption 2. For all $i = 1, 2, ..., n$, the nonlinearity functions g_i satisfy:
(i) $g_i(x)$ is a bounded function defined on $\mathbb{R}_+$;
(ii) the derivatives satisfy $g_i'(x) < 0$ or $g_i'(x) > 0$ for all $x \in (0, \infty)$. □

Assumption 2 means that each g_i is a monotone function and takes positive values. The nonlinearity functions have $\mathbb{R}_+$ as their domain since their domain represents biological variables which take positive values. Also note that $g_i'(0) = 0$ is allowed, since it does not violate the monotonicity of g_i. Now define a new function

$$g = (\frac{1}{\lambda_1}g_1) \circ (\frac{1}{\lambda_2}g_2) \circ \ldots \circ (\frac{1}{\lambda_n}g_n). \tag{4}$$

Definition 1. The gene regulatory network is said to be under *positive feedback* if

$$g'(x) > 0 \qquad \forall x \in (0, \infty).$$

Conversely, the gene regulatory network is said to be under *negative feedback* if the above inequality is reversed. In this work, the positive feedback case is studied. For the negative feedback case see [1, 3].

The system is said to be *bistable*, if it has two locally stable equilibrium points. Depending on the initial conditions and external input, the system can switch from one equilibrium point to the other one. The most famous example of bistability is the regulation of *lac* operon in the Bacteria *Escherichia coli*, [22]. When there is enough glucose in the environment, the genes responsible for digesting lactose are not expressed (the low state). When glucose is absent and lactose is present in the environment, the bacteria will express genes responsible for digesting lactose into glucose to meet its need for energy (the high state).

Theorem 1 ([18]). *Consider the system* (2) *under positive feedback. Any solution of* (2) *with any nonnegative initial condition converges to one of its equilibrium points.*

The above result is very important in the sense that the solution does not diverge or show oscillatory behavior. However, the system may have a number of equilibrium points; so, which one is the attractor for a given initial condition is not specified. Moreover, it is important to identify the conditions under which there is a single equilibrium point (or multiple equilibrium points). The present work deals with these issues.

First obvious consequence of Theorem 1 is that when there is single equilibrium point, the system is globally stable (all non-negative initial conditions are brought to the equilibrium point). In the following Corollary a condition for single equilibrium is given.

Corollary 1. *Consider system* (2) *under positive feedback. If the function g defined in* (4) *has a unique fixed point, then the system* (2) *has a unique equilibrium point* x_{eq} *and any solution of the system with a non-negative initial condition will converge to its unique equilibrium point* x_{eq}.

Proof. If g has a unique fixed point, then it is shown in [1] that the system has a unique equilibrium point. The global convergence result follows directly from Theorem 1. □

3 Analysis of the Cyclic Network under Positive Feedback

In the sequel, the system (2) is assumed to be under *positive feedback*. The nonlinearity functions are assumed to have negative Schwarzian derivatives and Assumptions 1 and 2 are satisfied.

3.1 General Conditions for Global Stability

The result below gives a general condition on the existence of *unique equilibrium point* for (2) by using the function g defined in (4).

Proposition 1. *Consider the system* (2) *under positive feedback and assume that g defined in* (4) *has negative Schwarzian derivative. Then, the following results hold.*

(i) *The function g has at most three fixed points.*

(ii) *If $g'(x) < 1$ for all $x \geq 0$, then g has a unique fixed point. In this case, the system defined by* (2) *has a unique equilibrium point x_{eq} which is globally attracting.*

(iii) *If $g'(0) > 1$ then g has a unique positive fixed point.*

Proof. Due to page limitations the proof is omitted here; it can be found in [1]. □

Therefore, if g satisfies conditions *(ii)* or *(iii)* of Proposition 1, then the unique equilibrium point of the system (2) is globally attractive. When the system has three equilibrium points, which one(s) of these is (are) stable equilibrium point(s) has to be determined. Several results are given in the next section along these directions.

3.2 Analysis of Homogenous Gene Regulatory Networks

In this section *homogenous* gene regulatory networks of the form (2) are studied; i.e., it is assumed that $\lambda_i = 1$ and there exists a function f such that

$$g_i(x) = f(x), \qquad \forall i = 1, 2, ..., n.$$

Note that no special structure is assumed for f yet. The following result plays a crucial role in the remaining parts of this section.

Lemma 1. *Let $k(x) : \mathbb{R}_+ \to I \subseteq \mathbb{R}_+$ be a three times continuously differentiable function satisfying $k'(x) > 0$ for all $x \in (0, \infty)$. Let h be defined on $\mathbb{R}_+$ as $h(x) = k^m(x)$. Then, any fixed point of h is a fixed point of k.*

Proof. Suppose that $h(0) = 0$ and $k(0) > 0$ then we have

$$h(0) = k^n(0) > ... > k(k(0)) > k(0) > 0$$

which is contradiction. Therefore, $k(0) = 0$ and 0 is a fixed point of the function k. Let $x > 0$ be a fixed point of the function h and suppose $k(x) \neq x$. Then we have either $x < k(x)$ or $k(x) < x$. If $x < k(x)$, then since k is a strictly increasing function,

$$h(x) = k^n(x) > ... > k(x) > x,$$

which gives us a contradiction. Similarly, if we have $k(x) < x$ then

$$h(x) = k^n(x) < ... < k(x) < x$$

which is again a contradiction. Therefore, we should have $k(x) = x$. Also, it is easy to see that any fixed point x of k is a fixed point of h. Thus we conclude that the functions k and h have the same fixed points. □

Remark 1. The homogenous system is under positive feedback if
(i) $f'(x) > 0$ for all $x \in (0, \infty)$, or
(ii) $f'(x) < 0$ for all $x \in (0, \infty)$ and $n = 2m$ for some positive integer m. □

First consider the case *(ii)* of Remark 1. From linear algebra, every positive number has a unique prime decomposition. Also, it is clear that n is an even integer. Then, one of the following must hold:
(a) $n = 2^l$ for some positive integer l or
(b) $n = 2^{l_1} p_2^{l_2} .... p_n^{l_n}$, where $p_2, p_3, ..., p_n$ are distinct odd primes and $l_i > 0$.
The following result considers the case *(ii)* of Remark 1:

Lemma 2. *Consider the homogenous gene regulatory network* (2) *under positive feedback with $f'(x) < 0$. Moreover, suppose that f has negative Schwarzian derivative. Then, f has a unique fixed point, say $x_0 > 0$, and one of the following holds:*
(a) *We have $n = 2^l$. In this case*

$$g(x) = f^n(x) \tag{5}$$

has the unique fixed point x_0 provided that $|f'(x_0)| < 1$. If $|f'(x_0)| > 1$, then g has exactly three equilibrium points.
(b) *When $n = 2^{l_1} p_2^{l_2} .... p_n^{l_n}$, define $h(x) = f^{(P)}(x)$, where $P = \prod_{i=2}^{n} p_i^{l_i}$. In this case h has a unique fixed point x_0 which is also the unique fixed point of f. If $|f'(x_0)| < 1$ then $|h'(x_0)| < 1$ and g defined in* (5) *has the unique fixed point x_0. If $|f'(x_0)| > 1$, then $|h'(x_0)| > 1$ and g defined in* (5) *has exactly three equilibrium points.*

Proof. Firstly, since f is monotonically decreasing it has a unique fixed point x_0. Suppose $n = 2^l$ and let $g(x) = f^n(x)$. Now, let $h_1(x) = f^{2^{l-1}}(x)$, then $g(x) = h_1(h_1(x))$ and $h_1'(x) > 0$ for all $x \in (0, \infty)$. From Lemma 1 with $m = 2$, it can be concluded that any fixed point x of g is a fixed point of the function h_1. Let $h_2(x) = f^{2^{k-2}}(x)$, then $h_1(x) = h_2(h_2(x))$ and again from Lemma 1, any fixed point of h_1 is a fixed point of h_2. Since $n = 2^l$, g has as many fixed points as h_{l-1}, which is defined as $h_{l-1}(x) = f(f(x))$. If $|f'(x_0)| < 1$ at the unique equilibrium point x_0 of f, it can be concluded that h_{l-1} has a unique equilibrium point. Therefore, from Lemma 1, g has a unique fixed point. Lemma 1 also implies that if $|f'(x_0)| > 1$, then the function $h_{l-1}(x)$ has exactly three fixed points. Therefore, from Lemma 1, the function $g(x)$ has three fixed points.

Now for the second part, consider $n = 2^{l_1} p_2^{l_2} \dots p_n^{l_n}$ and let $P = p_2^{l_2} \dots p_n^{l_n}$ and $h(x) = f^P(x)$. Since P is an odd number, $h'(x) < 0$ for all $x \in (0, \infty)$. It is also known that h has negative Schwarzian derivative by the convolution property of Schwarzian derivatives, see [17]. Therefore, h has a unique fixed point. Since f is decreasing it has a unique fixed point x_0. Also note that

$$h(x_0) = f^P(x_0) = x_0,$$

from which we conclude that the unique fixed point x_0 of f is the unique fixed point of h. Also note that $|h'(x_0)| < 1$ if and only if $|f'(x_0)| < 1$. Similarly, $|h'(x_0)| > 1$ if and only if $|f'(x_0)| > 1$. Note that $g(x) = h^{2^{l_1}}(x)$. Then the rest of the arguments are the same as the proof of the first part. □

Now consider the case where the function f is given as follows:

$$f(x) = \frac{a}{b + x^k}, \qquad a, b > 0, \quad k \in \{1, 2, 3, \dots\}. \tag{6}$$

Note that $f'(x) < 0$. In this case, the following proposition holds.

Proposition 2. *Consider the homogenous gene regulatory network* (2) *under positive feedback with f given as in* (6). *If $k = 1$ or and a, b, k satisfy*

$$\left(\frac{a}{k}\right)^k < \left(\frac{b}{k-1}\right)^{k+1}, \tag{7}$$

then the system has a unique equilibrium point which is globally attractive. Otherwise, the system exhibits bistable behavior.

Proof. Since f has negative derivative, it has a unique fixed point. Let x_0 be the unique fixed point of f. If $k = 1$ or (7) holds, then $|f'(x_0)| < 1$. Otherwise, we have $|f'(x_0)| > 1$. The proof of this fact can be found in [3]. Now, if $|f'(x_0)| < 1$, then from Lemma 2, the system has a unique fixed point which is globally attractive. On the other hand, suppose $|f'(x_0)| > 1$, and let $n = 2m$ be the size of the feedback in the system. Now, let $h(x) = f^m(x)$ and $g(x) = h \circ h(x)$. Then, again from Lemma 2 the function g has three fixed points, one of which is x_0. Let the fixed points of the function g satisfy $x_1 < x_0 < x_2$. We know $h(x_0) = f^m(x_0) = x_0$, since x_0 is a fixed point of f. Now, suppose $h(x_1) \neq x_2$. Then,

$$g(h(x_1)) = h \circ h \circ h(x_1) = h(g(x_1)).$$

Therefore, $h(x_1)$ is a fixed point of the function g. But since $h(x_1) \neq x_2$, then we have $h(x_1) = x_1$. But if m is an odd integer, then $h(x) = f^m(x)$ has the unique fixed point

x_0, but we know $x_0 \neq x_1$, so we get a contradiction. If m is even then, by a similar argument we can show $h(x_1) \neq x_1$ which gives another contradiction. Hence, we have $h(x_1) = x_2$ and $h(x_2) = x_1$. Therefore, $g'(x_1) = g'(x_2)$. We know that $|g'(x_0)| > 1$, so, the equilibrium point associated with it is locally unstable. But from Theorem 1 we know that the solution of the system converges to one of its equilibrium points. Therefore, $|g'(x_1)| = |g'(x_2)| < 1$ and the equilibrium points associated with them are locally stable. Therefore, we have two stable equilibrium points. □

Bistability is an important network motif in biological network models. In fact, the construction of the first synthetic toggle switch in bacteria is considered as one of the milestones in synthetic biology [9]. Therefore, Proposition 2 has implications in biology and this result is illustrated with examples in the next section.

Now consider case *(i)* of Remark 1, where f satisfies

$$f'(x) > 0 \quad \forall x \in (0, \infty). \tag{8}$$

Lemma 3. *Consider the homogenous gene regulatory network* (2) *under positive feedback with the nonlinearity function f satisfying* (8)*. Then, the function $g(x) = f^n(x)$ has as many fixed points as f. In particular, if f has a unique fixed point, then system* (2) *has a unique equilibrium which is globally attractive.*

Proof. Lemma 1 and Proposition 1 gives us the desired result. □

The above results reduce the whole analysis to the investigation of the fixed points of f. If f has a negative Schwarzian derivative then, it has one, two or three fixed points. As an example, consider the following Hill type of functions and try to find some conditions regarding its fixed points:

$$f(x) = \frac{ax^m}{b + x^m} + c, \quad a,\ b,\ c > 0. \tag{9}$$

Note that zero is ruled out as a fixed point by taking the constant c strictly positive. Then $x > 0$ is a fixed point of the function defined in (9) if x is a root of the following polynomial:

$$h(x) = x^{m+1} - (a+c)x^m + bx - bc. \tag{10}$$

Some interesting cases regarding the function (10) may occur. For example, consider the following numerical values: $a = 3.6$, $b = 5$, $m = 2$ and $c = 0.4$, then

$$h(x) = x^{m+1} - (a+c)x^m + bx - bc = (x-1)^2(x-2)$$

which implies that the function f has exactly two fixed points.

Let us now try to find a sufficient condition depending on the parameters a, b, c and m so that the function f defined in (9) has a unique equilibrium point. First note that for arbitrary positive constants a, b, c and m, the following holds:
$h(0) = -bc < 0$. Therefore, if we have

$$h'(x) \geq 0 \qquad \forall x \in \mathbb{R}_+, \tag{11}$$

then h can have at most one positive root so f has a unique fixed point. For $m > 1$,

$$\begin{aligned} h'(x) &= (m+1)x^m - (m)(a+c)x^{m-1} + b \\ &= x^{m-1}((m+1)x - m(a+c)) + b \; = \; h_1(x) + b. \end{aligned}$$

In order to guarantee (11), we should have $h_1(x) \geq -b$ for all $x \in \mathbb{R}_+$. But h_1 takes its minimum at the point y where

$$h_1'(y) = 0. \tag{12}$$

As a result of (12), we get the following equations:

$$\begin{aligned} h_1'(x) \; &= \; (m+1)(m)x^{m-1} - (m)(m-1)(a+c)x^{m-2} \\ &= x^{m-2}(m)(m+1)(x - \frac{m-1}{m+1}(a+c)) \, ; \\ &\Rightarrow h_1'(y) = 0 \Leftrightarrow y = \frac{m-1}{m+1}(a+c) \, ; \\ &\Rightarrow \min_{x \geq 0} h_1(x) = h_1\left(\frac{m-1}{m+1}(a+c)\right) \; = \; -\left(\frac{m-1}{m+1}\right)^{m-1}(a+c)^m. \end{aligned}$$

Combining this with (11) and (12), the following result is obtained:

$$\left(\frac{m-1}{m+1}\right)^{m-1}(a+c)^m \leq b \Rightarrow h_1(x) \geq -b \Rightarrow h'(x) \geq 0.$$

Proposition 3. *Let f be given as a function in the form (9). Then the following hold:*
(i) If $m = 1$, then for any positive a, b and c, the function f has a unique fixed point.
(ii) If $m = 2, 3, \ldots$ and the positive constants a, b and c satisfy

$$\left(\frac{m-1}{m+1}\right)^{m-1}(a+c)^m \leq b,$$

then f has a unique fixed point.

Proof. We already proved the case *(ii)*. For the case where $m = 1$, let a, b and c be arbitrary positive constants. If y is a fixed point of the function f, we have

$$h(y) = y^2 + (b - a - c)y - bc = 0.$$

But h can have at most two roots. Since $h(0) < 0$ and $h(-\infty) = \infty$, the function h has only one positive root; so, f has a unique fixed point. □

It is stated in Theorem 1 that under positive feedback, the solution converges to one of the equilibrium points independent of delay, see also [18]. Therefore, there should always exist at least one equilibrium point which is locally stable. The following result establishes this property.

Proposition 4. *Consider the system (2) under positive feedback, i.e., g defined in (4) satisfies: $g'(x) > 0$ for all $x \in \mathbb{R}_+$. Suppose that g is bounded and continuously differentiable, then g has a fixed point $x_1 \in \mathbb{R}_+$ such that $g'(x_1) \le 1$. Thus, the system is locally stable around the equilibrium point $x_{eq} = (x_1, x_2, ..., x_n)$, where*

$$x_n = g_n(x_1)/\lambda_n, \dots, x_2 = g_2(x_3)/\lambda_2.$$

Proof. Since the function g is bounded, the following supremum is well-defined:

$$a = \sup_{x \in \mathbb{R}_+} (g(x)). \tag{13}$$

It is clear that if x is a fixed point of g, then $x \le a$. Let the set S be defined as

$$S = \{x \in \mathbb{R}_+ : g(x) = x\},$$

then, because of (13), $b = \sup(S)$ exists. Note that since g is bounded and positive, the set S is nonempty. Since $b = \sup(S)$, there exists a sequence $x_i \in S$ such that

$$g(x_i) = x_i \quad \text{and} \quad \lim_{i \to \infty} (x_i) = b.$$

Since g is continuous, we have $g(b) = b$. Suppose that for all fixed points x of g, we have $g'(x) > 1$. Then, $g(b) = b$ and $g'(b) > 1$, but since g bounded then $\exists z > b$ such that $g(z) = z$. But this is contradiction to (13), so there exists some $x_1 \in \mathbb{R}_+$ such that

$$g'(x_1) \le 1. \tag{14}$$

The system has the following linearized transfer function around the equilibrium point x_{eq}:

$$G(s) = \left(1 + \frac{g'(x_1)\prod_{i=1}^{n}(\lambda_i)e^{-\tau s}}{\prod_{i=1}^{n}(s+\lambda_i)}\right)^{-1}.$$

The system is locally stable around x_{eq} if the roots of $G(s)$ are in the left half plane. Combining (14) and the fact that the system is under positive feedback, we can verify the following:

$$0 \le g'(x_1) \le 1.$$

By applying small gain argument, we can see that the system is locally stable independent of the delay value τ. □

4 Examples

The bistable toggle switch in [9] is analyzed here. As in [9], assume that two proteins mutually repress the expression of each other, and the system can be described as:

$$\begin{aligned} \dot{x}_1(t) &= -x_1(t) + f(x_2(t)) \\ \dot{x}_2(t) &= -x_2(t) + f(x_1(t-\tau)), \end{aligned} \tag{15}$$

where the function f is given as follows:

$$f(x) = \frac{a}{b + x^k}, \qquad a, b > 0, \quad k \in \{1, 2, 3, \dots\}. \tag{16}$$

Example 1. Suppose that in (16) the parameters are: $a = 1$, $b = 1$ and $k = 2$. Applying the inequality in Proposition 2, one sees that

$$\left(\frac{a}{k}\right)^k = 0.25 < (\frac{b}{k-1})^{k+1} = 1.$$

The unique fixed point of the function f can be found as $x_0 = 0.6823$. It can be easily verified that the unique equilibrium point of the system is given as $x_{eq} = (0.6823, 0.6823)$. From Proposition 2 we expect that the system has a unique equilibrium point which is stable independent of delay. Simulation results of Figures 1 and 2 illustrate that the solutions of the system converge to x_{eq} under different initial conditions independent of the value of the delay.

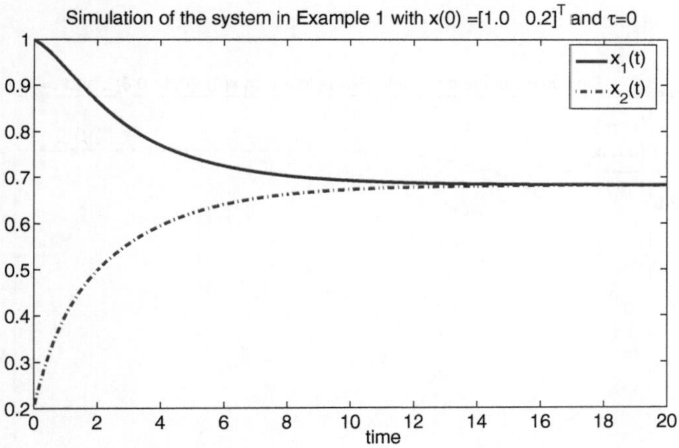

Fig. 1 *Simulation of the system in Example 1 with* $x(0) = [1.0 \ \ 0.2]^{\mathrm{T}}$ *and* $\tau = 0$

Example 2. Now consider the following parameters for f defined in (16): $a = 3$, $b = 1$ and $k = 3$. Again, applying the inequality in Proposition 2, it is observed that $(a/k)^k = 1 > \left((b/(k-1)\right)^{k+1} = 0.0625$.

One can find the unique fixed point of f as $x_0 = 1.164$ and the three equilibrium points of the system can be found as $x_1 = (0.1075, 2.9963)$, $x_2 = (1.164, 1.164)$ and $x_3 = (2.9963, 0.1075)$. From Proposition 2 we expect the system to show a bistable behavior. Simulations for different initial conditions and delay values are illustrated in Figures 3 and 4.

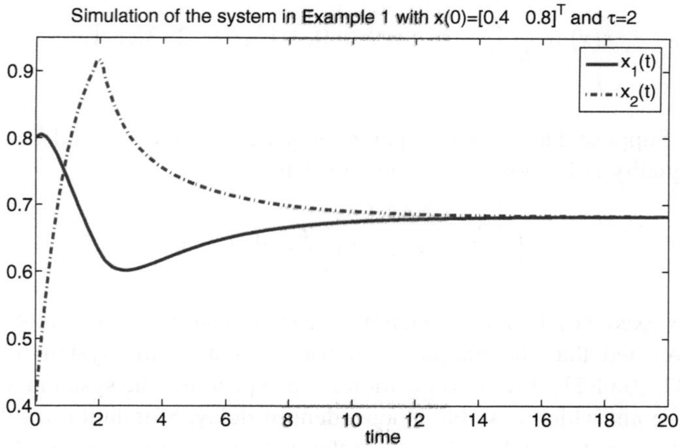

Fig. 2 *Simulation of the system in Example 1 with* $x(0) = [0.4 \quad 0.8]^{\mathrm{T}}$ *and* $\tau = 2$

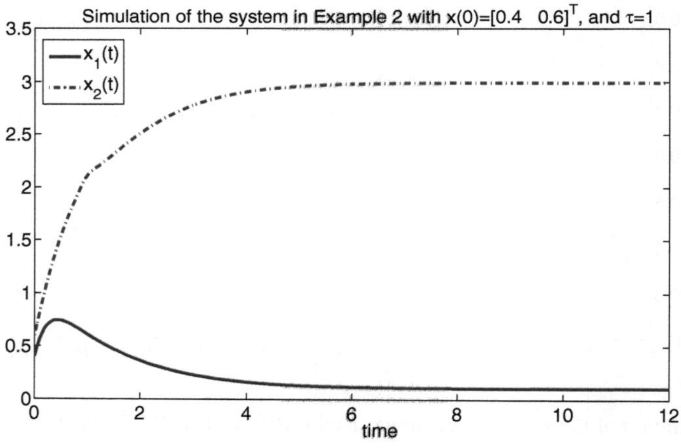

Fig. 3 *Simulation of the system in Example 2 with* $x(0) = [0.4 \quad 0.6]^{\mathrm{T}}$ *and* $\tau = 1.0$

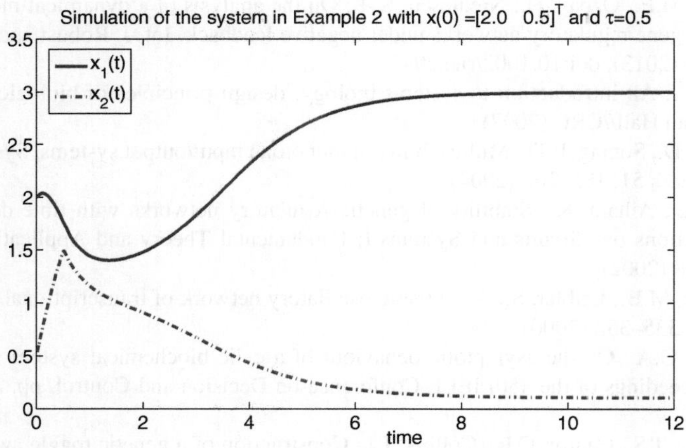

Fig. 4 *Simulation of the system in Example 2 with* $x(0) = [2.0 \quad 0.5]^T$ *and* $\tau = 0.5$

5 Conclusions

In this work gene regulatory networks are modeled as cyclic nonlinear dynamical systems with time delayed feedback. Analysis of this model of the gene regulatory network is done for the positive feedback case, under the assumption that the nonlinearity functions have negative Schwarzian derivatives.

Conditions for the existence of single positive equilibrium point are derived. In such a case the system is globally asymptotically stable independent of delay. In some cases there are more than one equilibrium point. For these situations, it is also shown here how the equilibrium points are computed and how stability of these points are determined. The *homogenous* network under positive feedback is also analyzed and sufficient conditions are derived for asymptotic stability. As a special case, homogenous gene regulatory networks with Hill type of nonlinearities are considered and *sufficient conditions* depending only on the parameters of the nonlinearity, are derived for the asymptotic stability independent of delay.

In this chapter, it is shown that multiple stable equilibrium points may exists for the system (2). An interesting question as a future extension of the current work to find a way to estimate the radius of convergence of each stable equilibrium point.

References

1. Ahsen, M.E.: Analysis of two types of cyclic biological system models with time delays. MS Thesis, Graduate School of Engineering and Sciences, Bilkent University, Ankara, Turkey (July 2011)
2. Ahsen, M.E., Özbay, H., Niculescu, S.-I.: Stability analysis of a dynamical model representing gene regulatory networks. In: Proc. of the 10th IFAC Workshop on Time Delay Systems, Boston, USA, pp. 191–196 (June 2012)

3. Ahsen, M.E., Özbay, H., Niculescu, S.-I.: On the analysis of a dynamical model representing gene regulatory networks under negative feedback. Int. J. Robust and Nonlinear Control (2013), doi:10.1002/rnc.2947
4. Alon, U.: An introduction to systems biology: design principles of biological circuits. Chapman Hall//CRC (2007)
5. Angeli, D., Sontag, E.D.: Multistability in monotone input/output systems. Systems Control Letters 51, 185–202 (2004)
6. Chen, L., Aihara, K.: Stability of genetic regulatory networks with time delay. IEEE Transactions on Circuits and Systems I: Fundamental Theory and Applications 49(5), 602–608 (2002)
7. Elowitz, M.B., Leibler, S.: A synthetic oscillatory network of transcriptional regulators. Nature, 335–338 (2000)
8. Enciso, G.A.: On the asymptotic behaviour of a cylic biochemical system with delay. In: Proceedings of the 45th IEEE Conference on Decision and Control, pp. 2388–2393 (2006)
9. Gardner, T.S., Cantor, C.R., Collins, J.J.: Construction of a genetic toggle switch in Escherichia coli. Nature 403(6767), 339–342 (2000)
10. Goldbeter, A.: Biochemical Oscillations and Cellular Rythms. The Molecular Basis of Periodic and Chaotic Behavior. Cambridge University Press (1996)
11. Levine, M., Davidson, E.H.: Gene regulatory networks for development. Proceedings of the National Academy of Sciences 102(14), 4936–4942 (2005)
12. Liz, E., Pinto, M., Robledo, G., Trofimchuk, S., Tkachenko, V.: Wright type delay differential equations with negative Schwarzian. Discrete and Continuous Dynamical Systems 9(2), 309–321 (2003)
13. Morarescu, C.I., Niculescu, S.-I.: Some remarks on the delay effects on the stability of biochemical networks. In: 16th Mediterranean Conference on Control and Automation, pp. 801–805 (2008)
14. Müller, S., Hofbauer, J., Endler, L., Flamm, C., Widder, S., Schuster, P.: A generalized model of the repressilator. Journal of Mathematical Biology 53, 905–937 (2006)
15. Purnick, P.E.M., Weiss, R.: The second wave of synthetic biology: from modules to systems. Nature Reviews Molecular Cell Biology 10(6) (2009)
16. Scheper, T.O., Klinkenberg, D., Pennartz, C., van Pelt, J.: A mathematical model for the intracellular circadian rhythm generator. The Journal of Neuroscience 19, 40–47 (1999)
17. Sedeghat, H.: Nonlinear Difference Equations. Kluwer Academic Publishers (2003)
18. Smith, H.: Monotone Dynamical Systems: An introduction to the theory of competitive and cooperative systems. American Mathematical Society (2008)
19. Smolen, P., Baxter, D.A., Byrne, J.H.: Modeling transcriptional control in gene networks – Methods, recent results and future directions. Bull. Math. Biol. 62, 247–292 (2000a)
20. Smolen, P., Baxter, D.A., Byrne, J.H.: Mathematical modeling of gene networks. Neuron 26, 567–580 (2000b)
21. Sontag, E.D.: Asymptotic amplitudes and Cauchy gains: a small-gain principle and an application to inhibitory biological feedback. Systems Control Letters 47, 167–179 (2002)
22. Tozeren, A., Byers, S.W.: New biology for engineers and computer scientists. Prentice Hall (2003)

Analysis and Design of Pattern Formation in Networks of Nonlinear Systems with Delayed Couplings

Toshiki Oguchi and Eiichi Uchida

Abstract. In this chapter, we consider formation of oscillatory patterns in networks of identical nonlinear systems with time-delays. First of all, by applying the harmonic balance method, we derive the corresponding harmonic balance equations for networks of identical nonlinear systems with delay couplings. Then, solving the equations by reducing to the stability problem of linear retarded systems, we estimate the oscillation profile such as the frequency, amplitudes and phases of coupled systems. Based on this analysis method, we also develop a design method of networks for nonlinear systems that can achieve prescribed oscillation profiles. The effectiveness of the proposed methods is shown by numerical examples

1 Introduction

In recent years, network systems have attracted attention in applied physics, mathematical biology, social sciences, control theory and interdisciplinary fields. In particular, synchronization and pattern formation of coupled systems have been the subject of intense study. From the standpoint of control engineering, synchronization and pattern formation of coupled systems are important notions to realize decentralized control techniques and bio-mimetic control approaches in increasingly complex applications. For instance, central pattern generators (CPGs), which produce rhythmic movements such as locomotion, breathing and scratching, are considered to consist of a group of neurons. Studies on CPGs have been carried out not only in the understanding of biological behavior but also for the purpose of producing rhythmic locomotion of multi-legged robots.

The behaviors of coupled identical systems and networks have been studied by a large number of researchers in various fields ([3, 6] and so on). Golubitsky *et al.*

Toshiki Oguchi · Eiichi Uchida
Department of Mechanical Engineering, Tokyo Metropolitan University, 1-1, Minami-Osawa, Hachioji-shi, Tokyo 192-0397 Japan
e-mail: t.oguchi@tmu.ac.jp

T. Vyhlídal, J.-F. Lafay, and R. Sipahi (eds.), *Delay Systems*, Advances in Delays and Dynamics 1,
DOI: 10.1007/978-3-319-01695-5_11,

showed the existence of periodic solutions in symmetric coupled systems and the pattern classification of gaits based on the symmetric network structure of the CPG. Pogromsky *et al.* [10, 11] analyzed the synchronization problem in symmetric networks of chaotic systems and derived sufficient conditions for partial synchronization in networks and the existence of periodic solutions. However this research has focused on only the symmetric structure of networks. Iwasaki [4] proposed a systematic approach for the analysis and synthesis of CPGs by applying the multivariable harmonic balance method. The proposed approach has no restriction on the network structure.

On the other hand, research interest on synchronization of coupled systems has shifted to the synchronization problem in delayed networks in recent years [5, 8, 9, 12]. Since time-delays caused by signal transmission affect the behavior of coupled systems in practical situations, it is therefore important to study the effect of time-delay in network systems [13]. In this chapter, we attempt to extend the systematic approach based on the harmonic balance method [4] to delayed network systems.

This chapter is organized as follows. Following this introduction, Section 2 derives harmonic balance equations for networks of identical nonlinear systems with delay couplings, and then by solving the equations based on the stability problem of linear retarded systems, we show an estimation method of the oscillation profiles such as the frequency, amplitudes and phases of coupled systems. In Section 3, based on the estimation method, we also develop a design method of networks of nonlinear systems that can achieve prescribed oscillation profiles. Finally, Section 4 provides the summary of this chapter.

2 Analysis of Oscillatory Patterns in Networks of Nonlinear Systems with Delayed Couplings

2.1 *Nonlinear Network Systems with Delayed Couplings*

We consider n identical nonlinear systems interconnected as follows

$$v_i = \psi_i(q_i), q_i = f(s)u_i, u_i = \sum_{r=1}^{n} \mu_{ir}(s)v_r, \tag{1}$$

where u_i and v_i are the input and output of system i, ψ_i is a static nonlinear function defined by $\psi_i(x) = \tanh(x)$ and μ_{ir} means a transfer function denoting a connection from oscillator r to oscillator i, which is given by $\mu_{ir}(s) = k_{ir}e^{-\tau_m s}$. Here $k_{ir} \in \mathbb{R}$ denotes a coupling gain from system r to system i, and τ_m is a commensurate delay of τ, i.e. $\tau_m = m\tau$ for $m = 1, \ldots, h$. $f(s)$ is a linear time-invariant part of system given by

$$f(s) = \frac{\omega_0}{s + \omega_0}. \tag{2}$$

Then the network of N coupled systems is summarized as the following equations.

$$v = \Psi(q) = (\psi_1, \cdots, \psi_n)^T$$
$$F(s) = \mathrm{diag}\{f(s), \cdots, f(s)\} = F(s)I_n$$

$$q = F(s)M(s)\Psi(q) = \begin{bmatrix} f(s)\mu_{11}(s)\psi_1(q_1) + \cdots + f(s)\mu_{1n}(s)\psi_n(q_n) \\ \vdots \\ f(s)\mu_{n1}(s)\psi_1(q_1) + \cdots + f(s)\mu_{nn}(s)\psi_n(q_n) \end{bmatrix}$$

where $M(s)$ is the transfer matrix whose (i, j) entry is $\mu_{ij}(s)$. For the above system, we deal with the following two problems in a similar way to [4]:

(I) **Analysis problem:** Given a network system with delays, determine whether the coupled systems have oscillatory trajectories, and if so, estimate the oscillation profile (frequency, amplitudes, phase) without actually simulating the differential equations.

(II) **Synthesis problem:** Given a network structure with delays and a desired oscillation profile, determine the coupling strength k_{ir} in the coupling transfer function $\mu_{ir}(s)$ so that the resulting network system achieves the given profile.

2.2 *Analysis of Periodic Solutions*

We assume that system (1) has a periodic solution. The Fourier series expansion of $q_i(t)$ is described by

$$q_i(t) = \sum_{k=0}^{\infty} a_k \sin(\omega k t) + b_k \cos(\omega k t)$$

where $a_k, b_k \in \mathbb{R}^n$ and $\omega \in \mathbb{R}$. Supposing that ω is sufficiently higher than the bandpass frequency of $f(s)$, $q_i(t)$ can be approximated by

$$q_i(t) \simeq \alpha_i \sin(\omega t + \phi_i), i = 1, \ldots, n$$

where α_i and ϕ_i denote the amplitude and phase, respectively. Furthermore, using a describing function κ_i, ψ_i can be approximated by

$$\psi_i(q_i) \simeq \kappa_i(\alpha_i)q_i, \ \kappa_i(x) := \frac{2}{\pi x}\int_0^{\pi} \psi_i(x \sin\theta)\sin\theta d\theta.$$

The describing function $\kappa_i(x)$ is a monotonically decreasing function satisfying $\kappa_i(0) = 1$ and $\kappa_i(\infty) = 0$ since $\psi_i(x) = \tanh(x)$. Using these approximations, we obtain the corresponding multivariable harmonic balance equation for nonlinear delay network systems as follows.

$$(M(j\omega)\mathscr{K}(\alpha) - 1/f(j\omega)I_n)\mathrm{q} = 0 \tag{3}$$

where j denotes the imaginary unit, I_n the $n \times n$ identity matrix, $\mathsf{q}_i := \alpha_i e^{j\phi_i}$ and $\mathscr{K}(\alpha) := \mathrm{diag}(\kappa_i(\alpha_i))$. The triplet (ω, α, ϕ) satisfying equation (3) is the oscillation profile to be solved. In this case, however, the solution is not unique in general, and there may exist infinite number of solutions due to the existence of time-delays. Among the solutions, we have to choose a triplet (ω, α, ϕ) so that the estimated oscillation is stable. Here, replacing $\mathscr{K}(\alpha)$ with a constant matrix $K(\alpha) := \mathscr{K}(|\mathsf{q}|)$, the stability of oscillation can be expected by checking if the following characteristic quasi-polynomial of linear system has a pair of solutions $s = \pm j\omega$ on the imaginary axis and the rest in the open right left plane.

$$\det\Big(M(s)K(\alpha) - 1/f(s)I_n\Big) = 0. \tag{4}$$

If the couplings have no delay, the stability analysis can be reduced to the eigenvalue problem of the constant matrix $MK(\alpha)$ since the matrix $M(s)$ is a constant matrix. However, if the couplings have delays, then the entries of $M(s)$ are not constant but functions of $e^{-s\tau_m}$. Therefore the stability analysis cannot be done in the same way as the delay-free case.

Now we decompose the transfer function matrix $M(s)$ into a constant matrix part and the rest, i.e.

$$M(s) = M_0 + \sum_{m=1}^{h} M_m e^{-s\tau_m} \tag{5}$$

where M_0 and $M_m \in \mathbb{R}^{n\times n}$ for $m = 1, \dots, h$ are constant coupling matrices. Substituting (2) and (5) into the quasi-polynomial (4) and setting $M_c = -I_n + M_0 K(\alpha)$, we obtain

$$\det\left(\sum_{m=1}^{h} M_m e^{-\lambda\omega_0\tau_m} K(\alpha) + M_c - \lambda I_n\right) = 0 \tag{6}$$

where $\lambda := s/\omega_0$. The root λ satisfying equation (6) is identical to the pole of the linear retarded system

$$\dot{x}(t) = M_c x(t) + \sum_{m=1}^{h} M_m K(\alpha) x(t - \tau_m). \tag{7}$$

Therefore, seeking the rightmost characteristic roots of system (7), we can confirm whether the characteristic quasi-polynomial (4) has at least one pair of solutions on the imaginary axis and the rest in the open left half plane. If there exists a solution λ on the imaginary axis, it is a solution of the modified harmonic balance equation given by

$$\left(\sum_{m=1}^{h} M_m e^{-\lambda\omega_0\tau_m} K(\alpha) + M_c - \lambda I_n\right)\mathsf{q} = 0. \tag{8}$$

Assuming that there exists a $\mathsf{q} \in \mathbb{C}^n$ satisfying (8), we can expect that the trajectory of each oscillator $q_i(t)$ has a phase ϕ_i and a frequency ω, i.e. $q_i(t) \simeq \alpha_i \sin(\omega t + \phi_i)$, where a triplet $(\omega, \alpha_i, \phi_i)$ is determined by

$$j\omega/\omega_0 = \lambda, \ \mathsf{q}_i = \alpha_i e^{j\phi_i}, \omega > 0 \tag{9}$$

The analysis on the oscillation profile can be accomplished by finding $\mathsf{q} \in \mathbb{C}^n$ satisfying equation (4) , but $K(\alpha)$ depends on q. Therefore in this paper, extending the algorithm introduced by Iwasaki [4] to delay systems, we propose the following calculation algorithm.

Algorithm:

Step 1. Set the initial values $k = 0$ and $v_k = [1, \ldots, 1]^\top$.

Step 2. Let $K_k = \mathscr{K}(|v_k|)$ and find the rightmost roots λ_k which has the maximum imaginary part and satisfies the following characteristic quasi-polynomial:

$$\det\left(\sum_{m=1}^{h} M_m e^{-\lambda_k \omega_0 m\tau} K_k + M_c - \lambda_k I_n\right) = 0$$

Step 3. Find the corresponding eigenvector x_k to λ_k obtained in Step 2:

$$\left(\sum_{m=1}^{h} M_m e^{-\lambda_k \omega_0 m\tau} K_k + M_c - \lambda_k I_n\right) x_k = 0$$
$$\|x_k\| = 1$$

Step 4. Update the eigenvector

$$v_{k+1} = e^{\sigma_k} y_k, \ \sigma_k := \Re(\lambda_k), \ y_k := x_k x_k^* v_k$$

Step 5. If $\|v_{k+1} - v_k\| \leq \varepsilon$, then the algorithm terminates, and $q := v_{k+1}$ is the solution of equation (8). So, we obtain a triplet (ω, α, ϕ) from (9). Otherwise go to Step 2 and iterate.

Remark 2.1. *Note that in this analysis, network structures are not assumed to be symmetric. Therefore this analysis method is also available for unidirectional network systems.*

Remark 2.2. *For computing the rightmost roots of the characteristic quasi-polynomial, several numerical techniques and some useful softwares like DDE-BIFTOOL [2] are available.*

2.3 Numerical Examples

2.3.1 Example 1

To show the availability of the proposed analysis method for complex network systems, we consider a network with $n = 15$ oscillators shown in Fig. 1. Each vertex denotes a oscillator and the number i inside of each vertex indicates the i-th oscillator. The edge (i, j) represents the existence of coupling from the j-th oscillator to the i-th oscillator.

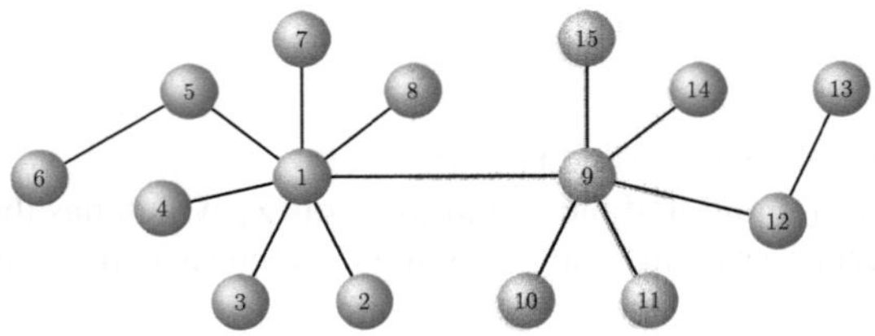

Fig. 1 A scale free network with $n = 15$

Now we assume that all coupling has identical delay τ and the matrix $M(s)$ in (5) is given by $M(s) = M_1 e^{-s\tau}$, where

$$M_1 = \begin{bmatrix} 0 & 1 & 1 & 1 & 1 & 0 & 1 & 1 & 1 & 0 & 0 & 0 & 0 & 0 & 0 \\ -1 & 0 & 0 & 0 & 0 & 0 & 0 & 0 & 0 & 0 & 0 & 0 & 0 & 0 & 0 \\ -1 & 0 & 0 & 0 & 0 & 0 & 0 & 0 & 0 & 0 & 0 & 0 & 0 & 0 & 0 \\ -1 & 0 & 0 & 0 & 0 & 0 & 0 & 0 & 0 & 0 & 0 & 0 & 0 & 0 & 0 \\ -1 & 0 & 0 & 0 & 0 & 1 & 0 & 0 & 0 & 0 & 0 & 0 & 0 & 0 & 0 \\ 0 & 0 & 0 & 0 & -1 & 0 & 0 & 0 & 0 & 0 & 0 & 0 & 0 & 0 & 0 \\ -1 & 0 & 0 & 0 & 0 & 0 & 0 & 0 & 0 & 0 & 0 & 0 & 0 & 0 & 0 \\ -1 & 0 & 0 & 0 & 0 & 0 & 0 & 0 & 0 & 0 & 0 & 0 & 0 & 0 & 0 \\ -1 & 0 & 0 & 0 & 0 & 0 & 0 & 0 & 0 & 1 & 1 & 1 & 0 & 1 & 1 \\ 0 & 0 & 0 & 0 & 0 & 0 & 0 & 0 & -1 & 0 & 0 & 0 & 0 & 0 & 0 \\ 0 & 0 & 0 & 0 & 0 & 0 & 0 & 0 & -1 & 0 & 0 & 0 & 0 & 0 & 0 \\ 0 & 0 & 0 & 0 & 0 & 0 & 0 & 0 & -1 & 0 & 0 & 0 & 1 & 0 & 0 \\ 0 & 0 & 0 & 0 & 0 & 0 & 0 & 0 & 0 & 0 & 0 & -1 & 0 & 0 & 0 \\ 0 & 0 & 0 & 0 & 0 & 0 & 0 & 0 & -1 & 0 & 0 & 0 & 0 & 0 & 0 \\ 0 & 0 & 0 & 0 & 0 & 0 & 0 & 0 & -1 & 0 & 0 & 0 & 0 & 0 & 0 \end{bmatrix}.$$

Table 1 Comparison between the estimated values and the simulation result of the frequency of coupled oscillators for each delay

Delay τ		frequency
0.2	Estimation	2.21
	Simulation	2.22
0.3	Estimation	1.79
	Simulation	1.79

By applying the proposed analysis algorithm for this network system, we estimate the profiles of oscillators for the following two cases: (i) $\tau = 0.2$, (ii) $\tau = 0.3$. Figure 2 shows comparisons between the estimated amplitudes and phases for q_i and those obtained by numerical simulations for $\tau = 0.2$ and $\tau = 0.3$, respectively. Here phase ϕ_i means the phase difference between the first oscillator and the i-th oscillator. In addition, the estimated frequency ω for each delay is summarized in

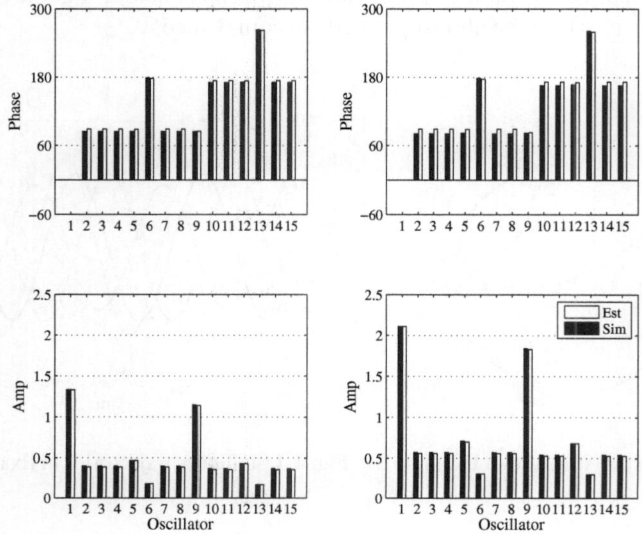

Fig. 2 Estimated and simulated profiles of each oscillator (a) for $\tau = 0.2$ (b) for $\tau = 0.3$

Table 1. These results show that although the oscillations of delay coupled systems depend on the length of time-delay, the proposed method can estimate the profiles successfully.

2.3.2 Example 2

Consider $n = 5$ identical nonlinear systems with delay couplings. The network structure is shown in Fig. 3. Each vertex denotes a oscillator, and the directed edge (i, j) means that there exists a coupling from the i-th oscillator to the j-th oscillator. In particular, the dashed edges mean delay-free couplings and the solid edges are couplings with delay $\tau = 0.3$. Now the transfer function matrix of the coupling $M(s)$ is assumed to be given by $M(s) = M_0 + M_1 e^{-s\tau}$, where

$$M_0 = \begin{bmatrix} 0 & 0 & -1.5991 & 0 & 0 \\ 0 & 0 & 0 & 0 & 0 \\ 0 & 0 & 0 & 0 & 0 \\ -0.6566 & 0 & 0 & 0 & 0 \\ 0 & 0 & 1.6206 & 0 & 0 \end{bmatrix}, M_1 = \begin{bmatrix} 0 & 1.2728 & 0 & 0 & 0 \\ -1.2875 & 0 & -0.5615 & 0 & 0 \\ 0 & -1.7732 & 0 & 0.0875 & 0 \\ 0 & 0 & -1.2973 & 0 & -1.1642 \\ 0 & 0 & 0 & 0.7016 & 0 \end{bmatrix}.$$

The nonzero entries of these matrices are given by random numbers. Figure 4 shows the steady state behaviors of coupled oscillators $q(t)$ and the oscillation of each system is stable. Then, applying the foregoing algorithm, we estimate the profile of each oscillator. The comparison between the estimated values and the simulation results is summarized in Table 2. From Table 2, we can conclude that the estimation of a triplet (ω, α, ϕ) can be almost perfectly accomplished.

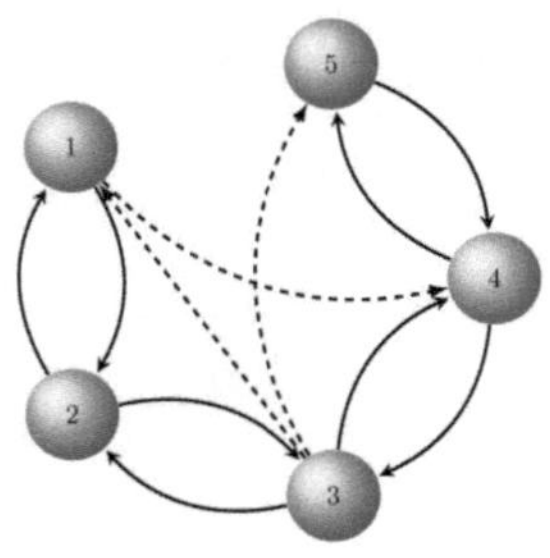

Fig. 3 Network structure in Example 2

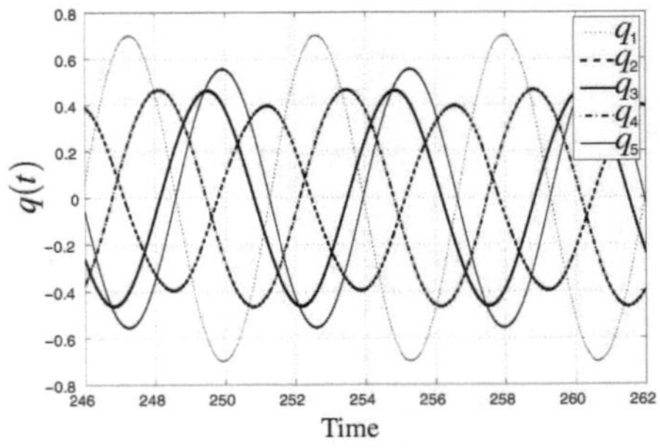

Fig. 4 Oscillatory trajectories in Example 2

Table 2 Estimated and simulated oscillation profiles in Example 2

	q_1		q_2		q_3		q_4		q_5	
	α_1	ϕ_1	α_2	ϕ_2	α_3	ϕ_3	α_4	ϕ_4	α_5	ϕ_5
Estimation	0.73	0.0	0.40	94.0	0.46	-151.0	0.47	-58.1	0.56	180.6
Simulation	0.70	0.0	0.39	95.1	0.46	-151.1	0.47	-57.1	0.56	180.7

3 Synthesis of Networks with Delays

3.1 Design Method

In the previous section, we considered how to estimate the profile of the coupled oscillators under given network systems with delayed couplings. In this section, we consider how to design the coupling gain k_{ij} to achieve given specifications of oscillation profile.

From the discussion in the foregoing section, it is necessary that the coupling gain matrices $M_0, \ldots, M_h$ satisfy equation (4) for a given triplet (ω, α, ϕ). Furthermore, the roots of equation (6) or equivalently the poles of linear retarded system (7) have to be on the imaginary axis and in the left half-plane of the complex plane only.

Based on the above idea, we obtain the following results on the design of coupling gains.

Theorem 3.1 *Let $\omega \in \mathbb{R}$, $\alpha \in \mathbb{R}^n, \phi \in \mathbb{R}^n$ and $m\tau \in \mathbb{R}$ for $m = 1, \ldots, h$ be given, where $\omega > 0$, $\alpha_i > 0$ for $i = 1, \ldots, n$ and $0 \le \tau < 2\pi/\omega$. Furthermore, assume that ϕ is chosen such that there exists at least one ϕ_i satisfying $|\phi_i - \phi_j| \ne k\pi$ for any $j \ne i$ and $k \in \mathbb{Z}$. Then the delay-coupled oscillators are expected to have oscillatory* *behaviors if there exist $M_i \in \mathbb{R}^{n\times n}$ for $i = 0,\ldots,h$ and a positive symmetric matrix P satisfying*

$$\begin{bmatrix}\Re(Aq) & \Im(Aq)\end{bmatrix} = R\Omega \qquad (10)$$

$$B^T P + PB < 0 \qquad (11)$$

where

$$A := \sum_{m=1}^{h} M_m e^{-j\omega\tau_m} K(\alpha) + M_0 K(\alpha) - I$$

$$B := \sum_{m=1}^{h} M_m K(\alpha) + M_0 K(\alpha) - I_n$$

$$R := \begin{bmatrix} c & s\end{bmatrix} \in \mathbb{R}^{n\times 2},\ c_i := \alpha_i \cos\phi_i, s_i := \alpha_i \sin\phi_i$$

$$\Omega := \begin{bmatrix} 0 & \omega/\omega_0 \\ -\omega/\omega_0 & 0 \end{bmatrix}, q = \begin{bmatrix} \alpha_1\cos\phi_1 + j\alpha_1 \sin\phi_1 \\ \vdots \\ \alpha_n \cos\phi_n + j\alpha_n \sin\phi_n \end{bmatrix}$$

Proof. For given ω, α_i, ϕ_i and τ_m, the coupling gain matrices M_i for $i = 0,\ldots,h$ must satisfy the following harmonic balance equation:

$$\left(\sum_{m=1}^{h} M_m e^{-j\omega\tau_m} K(\alpha) + M_c\right) \mathfrak{q} = \lambda \mathfrak{q}$$

where $\lambda = j\omega/\omega_0$ and $M_c = -I_n + M_0 K(\alpha)$. Now if we suppose that the characteristic quasi-polynomial has the characteristic roots at $\pm j\omega$, the following equation equivalently holds:

$$\left(\sum_{m=1}^{h} M_m e^{-j\omega\tau_m} K(\alpha) + M_c\right) \mathfrak{q} = j\frac{\omega}{\omega_0}\mathfrak{q}$$

Paying attention to that $\mathfrak{q}_i = \alpha_i e^{j\phi_i} = \alpha_i \cos\phi_i + j\alpha_i \sin\phi_i$ and rewriting the above equation, we obtain equation (10).

On the other hand, if system (7) with $\tau = 0$, i.e. linear delay-free system

$$\dot{x}(t) = \left(M_c + \sum_{m=1}^{h} A_m\right) x(t)$$

is asymptotically stable, the corresponding matrix pencil Λ is regular. Then Theorem A.1 shows the necessary and sufficient condition for the quasi-polynomial

$$\left| \lambda I_n - M_c - \sum_{m=1}^{h} M_m e^{-\lambda_k \omega_0 m\tau} K_k \right| = 0$$

to have at least one nonzero root on the imaginary axis and the corresponding delay values. From this result, if we choose the minimum value in the delay values, the characteristic quasi-polynomial has at least one pair of roots $\pm j\omega$ on the imaginary axis and the rest on the left half plane. Therefore it is necessary that $\tau < \frac{2\pi}{\omega}$ and there exists a positive definite matrix P satisfying $B^T P + PB < 0$. □

3.2 Examples

In what follows, we consider two network systems with the same network structure given in Figure 5. The first example is a case in which all edges mean delayed coupling and the second one is a case in which a part of edges have time-delays.

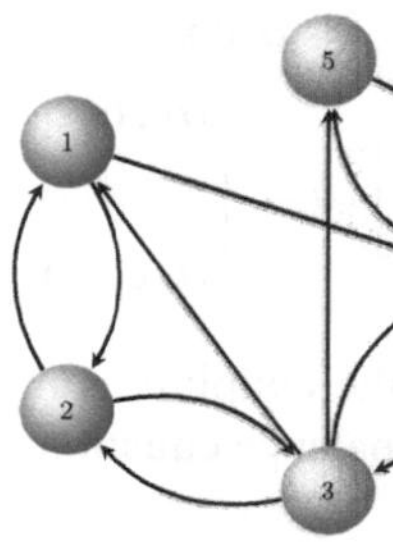

Fig. 5 Network structure in Examples 3 & 4

3.2.1 Example 3

First, we consider a case in which all couplings have the same length of time-delay τ_1. Then we design the coupling gain matrix M_1 so that the coupled oscillators meet the specification given in Table 3.

Table 3 Specifications and simulation results (Example 3)

	Period	α_1	α_2	α_3	α_4	α_5	ϕ_1	ϕ_2	ϕ_3	ϕ_4	ϕ_5
Specification	3.14	0.60	0.80	1.00	1.20	1.40	0	40.0	100.0	170.0	250.0
Case 1)	3.13	0.61	0.78	0.99	1.20	1.42	0	39.7	99.5	168.3	251.4
Case 2)	3.13	0.61	0.78	1.00	1.20	1.42	0	43.4	101.2	169.1	250.1

It should be noted that the structures of M_i have been already fixed due to the network structure. Solving the LMI (11) with respect to M_1 subject to the specified structure and equation (10), we can obtain the following matrix M_1 for $\tau_1 = 0.2$ and 0.3, respectively.

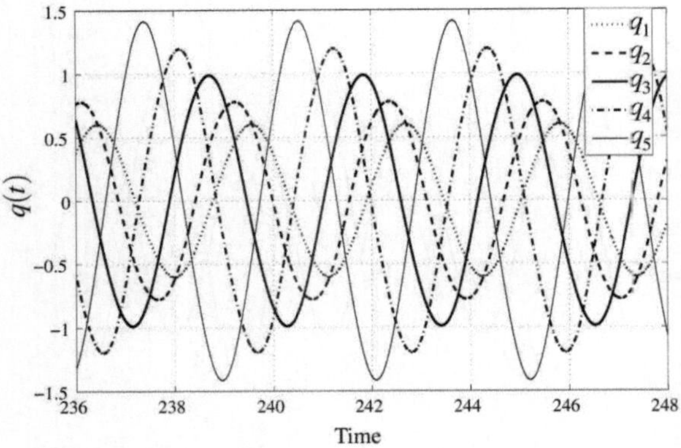

Fig. 6 Behaviors of coupled oscillators ($\tau_1 = 0.2$)

1) Case $\tau_1 = 0.2$:

$$M_1 = \begin{bmatrix} 0 & 0.5258 & 1.3822 & 0 & 0 \\ -1.4632 & 0 & 1.8025 & 0 & 0 \\ 0 & -1.1858 & 0 & 1.785 & 0 \\ 0.3872 & 0 & -1.2293 & 0 & 1.7656 \\ 0 & 0 & -0.9671 & -3.0635 & 0 \end{bmatrix}$$

2) Case $\tau_1 = 0.3$:

$$M_1 = \begin{bmatrix} 0 & 0.0841 & 1.6164 & 0 & 0 \\ -2.0205 & 0 & 1.5026 & 0 & 0 \\ 0 & -1.9642 & 0 & 1.2158 & 0 \\ 0.3392 & 0 & -1.5932 & 0 & 1.5297 \\ 0 & 0 & -0.1546 & -3.4073 & 0 \end{bmatrix}$$

The comparison between the given specification and the simulation results with M_1 mentioned above is summarized in Table 3. Figure 6 shows the behavior of each oscillator in case of $\tau_1 = 0.2$. From these results, we see that the design specifications are almost fulfilled.

3.2.2 Example 4

Consider a case in which the network has both delay-free couplings and time-delay couplings with delay $\tau_1 = 0.2$. As in Example 2, we assume that the edges $\{(3,1),(1,4),(3,5)\}$ are delay free couplings and others time-delay couplings.

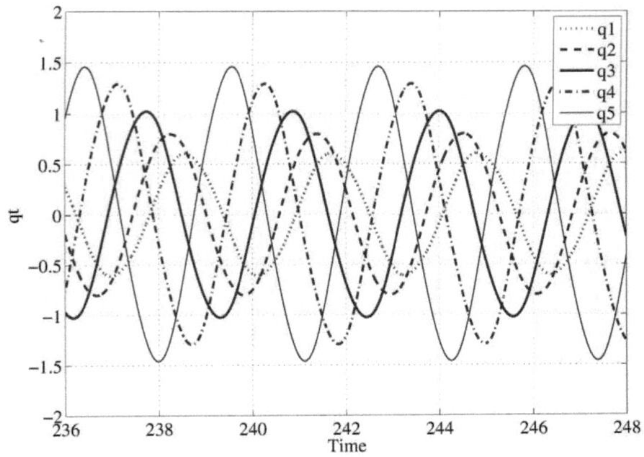

Fig. 7 Behaviors of coupled oscillators (Example 4)

Based on the specifications given in Table 4, the coupling gain matrices M_0 and M_1 are obtained by applying Theorem 3.1 as follows.

$$M_0 = \begin{bmatrix} 0 & 0 & 1.2062 & 0 & 0 \\ 0 & 0 & 0 & 0 & 0 \\ 0 & 0 & 0 & 0 & 0 \\ -0.5250 & 0 & 0 & 0 & 0 \\ 0 & 0 & -1.2412 & 0 & 0 \end{bmatrix}$$

$$M_1 = \begin{bmatrix} 0 & 1.1601 & 0 & 0 & 0 \\ -1.4632 & 0 & 1.8025 & 0 & 0 \\ 0 & -1.1858 & 0 & 1.785 & 0 \\ 0 & 0 & -1.2572 & 0 & 1.7656 \\ 0 & 0 & 0 & -2.6024 & 0 \end{bmatrix}$$

The comparison between the given specification and the design result is summarized in Table 4 and the simulation result is shown in Figure 7. Although the proposed design method contains approximation, the simulation results satisfy almost the given specifications.

Table 4 Specifications and simulation results (Example 4)

	Period	α_1	α_2	α_3	α_4	α_5	ϕ_1	ϕ_2	ϕ_3	ϕ_4	ϕ_5
Specification	3.14	0.60	0.80	1.00	1.20	1.40	0	40.0	100.0	170.0	250.0
Simulation	3.13	0.61	0.80	1.02	1.29	1.46	0	41.3	101.9	172.1	251.9

4 Conclusions

In this chapter, we considered pattern formation in networks of nonlinear systems with delayed couplings. The approach employed in this paper is based on the harmonic balance method and the obtained results are extensions of the results obtained in the reference [4] for delayed networks. In a similar way to the reference [4], we dealt with the analysis problem and the synthesis problem for delayed network systems. Although the proposed method for each problem contains approximations, the validity of the both proposed methods was supported with numerical examples.

Appendix

Regarding to the characteristic roots on the imaginary axis, the following result holds regarding to the stability of linear retarded systems ([1], [7]).

Consider the following linear systems with commensurate delays:

$$\dot{x}(t) = A_0 x(t) + \sum_{i=1}^{m} A_i x(t - i\tau) \tag{12}$$

The corresponding characteristic function is given by

$$p(\lambda;\tau) = \det\Big(\lambda I_n - A_0 - \sum_{i=1}^{m} A_i e^{-\lambda i \tau}\Big)$$

For system (12), the matrix pencil is defined by $\Lambda(z) := zM + N$, where $M, N \in \mathbb{R}^{(2mn^2)\times(2mn^2)}$ are given by

$$M := \begin{bmatrix} I_{n^2} & 0 & \cdots & 0 & 0 \\ 0 & I_{n^2} & \cdots & 0 & 0 \\ & & \ddots & & \\ 0 & 0 & \cdots & I_{n^2} & 0 \\ 0 & 0 & \cdots & 0 & B_m \end{bmatrix}, \quad N := \begin{bmatrix} 0 & -I_{n^2} & 0 \cdots & 0 & \\ 0 & 0 & -I_{n^2} & \cdots & 0 \\ & & & \ddots & \\ 0 & 0 & 0 & \cdots & -I_{n^2} \\ B_{-m} & B_{-m+1} & B_{-m+2} & \cdots & B_{m-1} \end{bmatrix}$$

and B_{-k} for $k = 1, \ldots, m$ and B_i for $i = 1, \ldots, m$ are defined by

$$B_{-k} = I_n \otimes A_k^T, B_i = A_i \otimes I_n, B_0 = A_0 \oplus A_0^T.$$

Here the operators $\otimes$ and $\oplus$ denote the Kronecker product and sum, respectively. Then the following theorem holds ([7]).

Theorem A.1. *Assume that the matrix pencil Λ is regular. Then the quasi-polynomial $p(\lambda;\tau) = 0$ has a crossing root on the imaginary axis for some positive delay value τ if and only if there exists a complex number $z_0 \in \sigma(\Lambda) \cap \mathscr{C}(0,1)$ such that*

$$j\omega_0 \in \sigma\Big(A_0 + \sum_{i=1}^{m} A_i z_o^i\Big)$$

where $\sigma(A)$ denotes a point spectrum of A. Furthermore, for some z_0 satisfying the condition (ii) above, the set of delays corresponding to the induced crossing is given by

$$\mathscr{T}(z_0) = \Big\{\frac{\angle \bar{z}_0}{\omega_0} + \frac{2\pi\ell}{\omega_0} > 0 : j\omega_0 \in \Big(A_0 + \sum_{i=1}^{m} A_i\Big), \ell \in \mathbb{Z}\Big\}.$$

References

1. Chen, J., Gu, G., Nett, C.N.: A new method for computing delay margins for stability of linear delay systems. Systems & Control Letters 26, 107–117 (1995)
2. Engelborghs, K., Luzyaninna, T., Roose, D.: Numerical bifurcation analysis of delay differential equations using DDE-BIFTOOL. ACM Trans. Math. Softw. 28(1), 1–21 (2002)
3. Golubitsky, M., Stewart, I.: The Symmetry Perspective - From Equilibrium to Chaos in Phase Space and Physical Space. Birkhäuser (2004)
4. Iwasaki, T.: Multivariable harmonic balance for central pattern generators. Automatica 44(12), 3061–3069 (2008)
5. Neefs, P., Steur, E., Nijmeijer, H.: Network complexity and synchronous behavior an experimental approach. Int. J. of Neural Systems 20(3), 233–247 (2010)
6. Newman, M., Barabási, A.L., Watts, D.J.: The Structure and Dynamics of Networks. Princeton University Press (2006)
7. Niculescu, S.I., Fu, P., Chen, J.: Stability switches and reversals of linear systems with commensurate delays: A matrix pencil characterization. In: Proceedings of the 16th IFAC World Congress, Prague, Czech Republic, IFAC-PapersOnline, World Congress, vol. 16, Part I, pp. 637–642 (2005)
8. Oguchi, T., Nijmeijer, H., Yamamoto, T.: Synchronization in networks of chaotic systems with time-delay coupling. Chaos 18(3), 037108 (2008)
9. Oguchi, T., Yamamoto, T., Nijmeijer, H.: Synchronization of bidirectionally coupled nonlinear systems with time-varying delay. In: Loiseau, J.J., Michiels, W., Niculescu, S.-I., Sipahi, R. (eds.) Topics in Time Delay Systems. LNCIS, vol. 388, pp. 391–401. Springer, Heidelberg (2009)
10. Pogromsky, A., Glad, T., Nijmeijer, H.: On diffusion driven oscillations in coupled dynamical systems. International Journal of Bifurcation and Chaos 9(4), 629–644 (1999)
11. Pogromsky, A., Santoboni, G., Nijmeijer, H.: Partial synchronization: from symmetry towards stability. Physica D 172(2), 65–87 (2002)
12. Steur, E., Oguchi, T., van Leeuwen, C., Nijmeijer, H.: Partial synchronization in diffusively time-delay coupled oscillator networks. Chaos 22(4), 43144 (2012)
13. Uchida, E., Oguchi, T.: Pattern Formation in Networks of Nonlinear Systems with Delay Couplings. In: Proceedings of the 18th IFAC World Congress 2011, pp. 5118–5123 (2011)

Consensus in Networks of Discrete-Time Multi-agent Systems: Dynamical Topologies and Delays

Wenlian Lu, Fatihcan M. Atay, and Jürgen Jost

Abstract. A stability analysis of general consensus algorithms in discrete-time networks of multi-agents is presented. Here, the networks can have time-varying topologies and delays, as well as nonlinearities. The Hajnal diameter approach is developed for synchronization analysis and sufficient conditions for both consensus at uniform value and synchronization at periodic trajectories are derived, which show how the periods depend on the transmission delay patterns.

1 Introduction

Consensus problems have been recognized to be important in coordination of dynamic agent systems and are widely applied in distributed computing [1], management science [2], flocking/swarming theory [3], distributed control [4], and sensor networks [5]. In these applications, the multi-agent systems need to agree on a common value for a certain quantity of interest that depends on the states of the interests of all agents or is a preassigned value. In this chapter, we consider the following dynamical system of multi-agents:

$$x_i^{t+1} = \phi_i^t\left(x_1^{t-\tau_{i1}(t)}, \ldots, x_m^{t-\tau_{im}(t)}\right),\ i = 1, \ldots, m;\ t \in \mathbb{Z}_{\geq 0}, \tag{1}$$

Wenlian Lu
School of Mathematical Sciences and Centre for Computational Systems Biology,
Fudan University, 200433, Shanghai, China,
Scientific Computing Centre, Department of Computer Science,
The University of Warwick, Coventry CV4 7AL, United Kingdom
e-mail: `wenlian.lu@gmail.com`

Fatihcan M. Atay · Jürgen Jost
Max Planck Institute for Mathematics in the Sciences, 04103 Leipzig, Germany
e-mail: `{fatay,jjost}@mis.mpg.de`

T. Vyhlídal, J.-F. Lafay, and R. Sipahi (eds.), *Delay Systems*,
Advances in Delays and Dynamics 1,
DOI: 10.1007/978-3-319-01695-5_12,

where $x_i^t \in \mathbb{R}$ denotes the state of agent i at time t, $\phi_i^t : \mathbb{R}^m \to \mathbb{R}$ is a differentiable map for each t $\tau_{ij}(t)$ is the time-varying delay from agent j to agent i and $\mathbb{Z}_{\geq 0}$ denotes the discrete time, the set nonnegative integers. We suppose that the delays are uniformly bounded, i.e., $\sup_{i,j,t} \tau_{ij}(t) = \tau_M$ for some finite $\tau_M > 0$.

Let $x^t = [x_1^t, \ldots, x_m^t]^\top \in \mathbb{R}^m$ and $w(t) = [x^{t\top}, x^{t-1\top}, \ldots, x^{t-\tau_M\top}]^\top \in \mathbb{R}^{m(\tau_M+1)}$. We first rewrite (1) in the more abstract form

$$w(t+1) = \Phi^t(w(t)) \tag{2}$$

with $\Phi^t(\cdot) = [\Phi_0^t(\cdot)^\top, \ldots, \Phi_{\tau_M}^t(\cdot)^\top]^\top$, where

$$\begin{cases} \Phi_0^t(w) & = [\phi_1^t(w), \ldots, \phi_m^t(w)]^\top \\ \Phi_\tau^t(w) & = x^{t-\tau+1} \quad \tau \geq 1. \end{cases}$$

We assume that all $\phi_i^t(\cdot)$, $i = 1, \cdots, m$, are $C^{1+\alpha}$ continuous for some $\alpha > 0$ and

$$\phi_i^t(s, \ldots, s) = s \tag{3}$$

for all $s \in D(\subset \mathbb{R})$, i, and t. Eq. (2) is an abstraction and simplification of *consensus algorithm/protocol*, an interaction rule specifying the information communication between each agent and its neighborhood. In the present work, we address the question of consensus when the right-hand side of (2) contains time variations in both couplings and delays.

The condition (3) guarantees that global consensus is a solution of (1). A concept related to consensus, namely *synchronization* [6–8] , indicates that the system's diagonal, i.e. the set

$$\mathscr{S} = \{u \in \mathbb{R}^m : \ u_i = u_j \in \mathbb{R}, \text{ for all } i, j = 1, \ldots, m\}$$

is invariant under the dynamics and asymptotically attracting. Due to fact that the transmission delays $\tau_{ij}(t)$ from agent j to agent i depend on the receiver agent i, the scenario is different from the systems without delays. To specify the argument, let

$$\mathbb{S} = \left\{ w = [w^{0\top}, \cdots, w^{\tau_M\top}]^\top \in \mathbb{R}^{m(\tau_M+1)} : \ w^\tau \in \mathscr{S}, \ \forall \ \tau = 0, 1, \ldots, \tau_M \right\}.$$

Under hypothesis (3), $\mathscr{S}$ may contain subsets that are invariant with respect to (2). However, the more general condition used in [9], namely $\phi_i^t(s, \ldots, s) = \phi(s)$ for some function ϕ independent of index i, does not guarantee that $\mathscr{S}$ contains invariant subsets with respect to Eq. (1).

Actually, the trajectory of system (2), constrained on $\mathbb{S}$, depends on the pattern of the delays. First, let

$$\mathscr{S}_1 = \left\{ w = [w^{0\top}, \ldots, w^{\tau_M\top}]^\top \in \mathbb{S} : \ w_\tau = w_{\tau'}, \text{ for all } \tau, \tau' = 0, \ldots, \tau_M \right\}.$$

Each $s^* = [s, \cdots, s]^\top \in \mathscr{S}$ is an equilibrium of system (2). Next, if

$$P = \gcd\{\tau_{ij}(t) + 1 : i, j = 1, \ldots, m; t \in \mathbb{Z}_{\geq 0}\} > 1, \tag{4}$$

where gcd stands for the greatest common divisor, then the set

$$\mathscr{S}_P = \left\{ w = [w^{0\top}, \ldots, w^{\tau_M \top}]^\top \in \mathbb{S} : w^k = w^{k+P},\ \forall\, k = 0, 1, \ldots, \tau_M - P \right\}$$

consists of invariant periodic solutions of system (2) (with period P). It can be seen that $\mathscr{S}_1$ is a special case of $\mathscr{S}_P$ when $P = 1$. In addition, restricting $\mathbb{S}$ on a local region, for example, the region D where (3) holds, we define

$$\mathscr{S}(C) = \{ u \in \mathbb{R}^m : u_i = u_j \in C, \text{ for all } i, j = 1, \cdots, m \}$$

for some $C \subset \mathbb{R}$. In the same fashion, we define $\mathbb{S}(C)$, $\mathscr{S}_1(C)$ and $\mathscr{S}_P(C)$ as well.

The relationship and difference between consensus and synchronization was presented in [10]. The question we consider is whether the invariant set $\mathscr{S}_0$ or $\mathscr{S}_P$ (according to the delays' gcd) is attracting for dynamical states $[x^t_m, \ldots, x^t_m]$ outside of it, at least locally. First, this question can be translated into synchronization problem as we did in [9]. Then, upon reaching synchronization, hypothesis (3) guarantees that the synchronized trajectory should be an equilibrium or a periodic trajectory (depending of the delay patterns), instead of a general attractor on $\mathbb{S}$.

The motivation for studying (1) (or its abstract form (2)) comes initially from the basic discrete-time consensus algorithm:

$$x_i^{t+1} = \sum_{j=1}^{m} G_{ij} x_j^t,\ i = 1, \cdots, m, \tag{5}$$

where $x_i^t \in \mathbb{R}$ denotes the state variable of the agent i and $G_{ij} \geq 0$ is the nonnegative coupling strength from agent j to agent i and satisfies: $\sum_{j=1}^m G_{ij} = 1$. Define $G = [G_{ij}]_{i,j=1}^m$, which is related to the underlying connecting graph of the system, in the sense that $G_{ij} > 0$ if there is a link from node (agent) j to i and $G_{ij} = 0$ otherwise. It can be seen that G is a stochastic matrix. Then, (5) can be rewritten as

$$x^{t+1} = G x^t, \tag{6}$$

where $x^t = [x_1^t, \ldots, x_m^t]^\top$. Eq. (6) is a general model of the consensus algorithm on a network with fixed topology, which can be a directed graph and may have weights. Additionally, in many real-world applications, the connection structure may change in time, for instance when the agents are moving in physical space. One must then consider time-varying topologies under link failure or creation. Furthermore, delays occur inevitably due to limited information transmission speed. To sum up, the linear model of consensus with transmission delays can be described as

$$x_i^{t+1} = \sum_{j=1}^{m} G_{ij}(t) x_j^{t-\tau_{ij}(t)}, \tag{7}$$

where $\tau_{ij}(t)$, $i,j=1,\cdots,m$, denotes the time-dependent delay from agent j to agent i. We say that a link from j to i is *instantaneous* if $\tau_{ij}(t)\equiv 0$, and *delayed* otherwise. We will associate $G(t)=[G_{ij}(t)]_{i,j=1}^{m}$ with a directed graph sequence (see Sec. 3).

Stability analysis of the consensus in multi-agent networks (the special forms of (7) for discrete-time model) has been intensively investigated in control theory [11–19]. In our recent work [20], we have investigated consensus in dynamic networks and delays under a general stochastic framework, which provides a theoretical method to analyze stability of Eq. (7), and applied the results to analyze consensus in a mobile agent network model [21]. In this paper, we shall address this problem in the context of the general form (1).

The time variation of the connections and delays can be either deterministic or stochastic, which may have a special form, or may be driven by some other dynamical system. Let $\mathscr{Y}=\{\Omega,\mathscr{F},P(\cdot),\theta^t\}$ denote a metric dynamical system, where Ω is the metric state space, $\mathscr{F}$ is the σ-algebra, $P(\cdot)$ is the probability measure, and θ^t is a measure-preserving shift satisfying: $\theta^{t+s}=\theta^t\circ\theta^s$ and $\theta^0=id$, where id denotes the identity map. Then Eq. (7) can be regarded as a random dynamical system (RDS) driven by $\mathscr{Y}$:

$$x_i^{t+1}=\sum_{j=1}^{m}G_{ij}(\theta^t\omega)x_j^{t-\tau_{ij}(\theta^t\omega)},\ i=1,\ldots,m,\ t\in\mathbb{Z}_{\geq 0};$$

or the abstract form (2) can be rewritten as:

$$w^{t+1}=\Phi(w^t,\theta^t\omega),t\in\mathbb{Z}_{\geq 0},\omega\in\Omega. \tag{8}$$

The consensus problem under this scenario is defined in forward and almost-sure sense, i.e., convergence is attained except for a subset of ω of zero probability. For details on random dynamical systems and attractors, we refer the reader to [22].

2 Stability Analysis

In this section we present a linear stability analysis of the invariant sets $\mathscr{S}_1$ and $\mathscr{S}_P$ according to delay patterns. We first consider $\mathscr{S}_1$ in the deterministic time-varying case. We start with a boundedness condition of system (1). The notation $\pi_A(\cdot)$ denotes the orthogonal projection operator from $\mathbb{R}^{m(\tau_M+1)}$ onto a subset A.

B$_1$: *There exists a neighborhood U containing $\mathscr{S}_1(D)$ such that any trajectory $w(t)$ of (2) starting in U is bounded and $\pi_{\mathscr{S}_1}(w(t))\in\mathscr{S}_1(D)$ for all t.*

Due to hypothesis (3), each point $s^*=[s,\ldots,s]^\top\in S$ is an equilibrium of (1). Using the approach in [9] the variational equations of $z(t)=x(t)-s^*$ near an equilibrium point $s^*\in\mathscr{S}$ are

$$z_i^{t+1}=\sum_{j=1}^{m}\frac{\partial\phi_i^t}{\partial x_j}(s^*)z_j^{t-\tau_{ij}(t)},\ i=1,\ldots,m. \tag{9}$$

Hypothesis (3) implies that

$$\sum_{j=1}^{m} \frac{\partial \phi_i^t}{\partial x_j}(s^*) = 1,\ i = 1,\dots,m,$$

for all t and $s^* \in \mathscr{S}(D)$. However, the Jacobian matrix $J(t) = [\frac{\partial \phi_i^t}{\partial x_j}(s^*)]_{i,j=1}^m$ is not necessary a stochastic matrix since some elements may be negative.

With $\tau_M = \sup_{i,j,t} \tau_{ij}(t)$, assumed to be finite as above, partition $J(t)$ into $J_0(t), J_1(t), \dots, J_{\tau_M}(t)$, according to the delays, such that $J(t) = \sum_{\tau=0}^{\tau_M} J_\tau(t)$, and (9) can be rewritten in the general form

$$z^{t+1} = \sum_{\tau=0}^{\tau_M} J_\tau(t) z^\tau, \tag{10}$$

where $z(t) = [z_1(t), \dots, z_m^t]^\top$. Eq. (10) can further be rewritten as

$$y^{t+1} = B(t) y^t,$$

where $y^t = [z^{t\top}, z^{t-1\top}, \dots, z^{t-\tau_M\top}]^\top$ and

$$B(t) = \begin{bmatrix} J_0(t) & J_1(t) & J_2(t) & \cdots & J_{\tau_M}(t) \\ I_m & 0 & 0 & \cdots & 0 \\ 0 & I_m & 0 & \cdots & 0 \\ \vdots & \vdots & \vdots & \ddots & \vdots \\ 0 & 0 & \cdots & I_m & 0 \end{bmatrix}$$

with all row sums equal to 1. To state the main results, we use the concept of the Hajnal diameter introduced in [23, 24]: For a matrix A with row vectors $a_1, \dots, a_m$ and a vector norm $\|\cdot\|$ in $\mathbb{R}^m$, the Hajnal diameter of A is defined by $\operatorname{diam}(A) = \max_{i,j} \|a_i - a_j\|$. The Hajnal diameter of an infinite product of a deterministically time-varying matrix sequence $\{B(t)\}$ is defined as [9]:

$$\operatorname{diam}(B(\cdot)) = \overline{\lim_{T\to\infty}} \sup_{t_0 \ge 0} \left[\operatorname{diam}\left(\prod_{t=t_0}^{t_0+T} B(t)\right) \right]^{1/T}.$$

From Theorem 3.1 in [9], the following result can be concluded.

Theorem 1. *Under the hypothesis* $\mathbf{B}_1$, *if* $\sup_{s^* \in \mathscr{S}_1(D)} \operatorname{diam}(B(\cdot)) < 1$, *then system (1) is (locally) stable with respect to* $\mathscr{S}_1(D)$, *that is, there exists a sufficiently small neighborhood* U *of* $\mathscr{S}_1(D)$ *such that for any initial condition in* U, *the trajectory converges to an equilibrium in* $\mathscr{S}_1(D)$.

In fact, Theorem 3.1 in [9] assumes that there exists an attractor for the system restricted to $\mathscr{S}_1$, which is needed to guarantee that the projection of the trajectory on $\mathscr{S}_1$ are kept in the bounded region defined by the attractor. Here, condition $\mathscr{B}_1$ guarantees that the projection of the trajectory of Eq. (2) with initial condition in U onto $\mathscr{S}_1$ is still in D. So, the proof of Theorem 3.1 in [9] is valid for this theorem, and in addition, this condition also guarantees that hypothesis (3) holds for the system restricted to $\mathscr{S}_1$. When (1) converges to $\mathscr{S}_1(D)$, according to the form of (3), the synchronized trajectory should be an equilibrium. In other words, system (2) reaches consensus, as all agents converge to a uniform value.

If the time-variation is driven by a stochastic process, then the system (1) or (2) becomes the random dynamical system (8). Let

$$V_\lambda^t = \left\{ \omega \in \Omega : \left[\operatorname{diam}\left(\prod_{k=0}^{t} B(\theta^k \omega)\right)\right]^{1/t} < \lambda \right\}.$$

From Theorem 4.3 in [25], we have:

Theorem 2. *Under hypothesis* $\mathbf{B}_1$*, if there exists some* $\lambda \in (0,1)$ *such that* $\sum_{t=0}^{\infty} P(V_\lambda^t) < +\infty$*, then (8) is (locally) stable with respect to* $\mathscr{S}_1(D)$ *in the almost sure sense, that is, for almost every* $\omega \in \Omega$*, there exists a sufficient small neighborhood* $U(\omega)$ *(possibly depending on* ω*) of* $\mathscr{S}_1(D)$ *such that for any initial condition in* U*, the trajectory of (8) converges to an equilibrium in* $\mathscr{S}_1(D)$*.*

We note that the equilibrium depends also on ω. In [25], a sufficient condition was stated in terms of the normal Lyapunov exponent, which was proved to be equivalent to the Hajnal diameter in [9].

We next consider synchronized periodic solutions under condition (4). Note that each $t \in \mathbb{Z}_{\geq 0}$ can be written as $t = kP + l$ for some $k \geq 0$ and $l = 0, \ldots, P-1$. Eq. (1) implies that the state at $t+1$ depends on the states at $t - \tau_{ij}(t)$, $i, j = 1, \ldots, m$. We can write $\tau_{ij}(t) = z_{ij}(t)P - 1$, owing to hypothesis (3). Therefore, $\bmod(t - \tau_{ij}(t), P) = l+1$, which is equal to $\bmod(t+1, P)$ (if $l = P-1$, then $l+1$ equals to 0 in modulus), where $\bmod(a,b)$ denotes the remainder of a divided by b. In other words, hypothesis (4) implies that the state of node at time $t+1$ depends only on those states at the time points that have the same remainder with respect to P. Therefore, after permutation of the $\tau_M + 1$ components in $w^t = [x^{t\top}, \ldots, x^{t-\tau_M\top}]^\top$ such that the time with the same remainder with respect to P are brought together, i.e., $\tilde{w}^t = \left[(\tilde{w}_0^t)^\top, \ldots, (\tilde{w}_{P-1}^t)^\top\right]^\top$ with $\tilde{w}_k^t = \left[(x^{t-k})^\top, (x^{t-P-k})^\top, \ldots, (x^{t-(\tau_M+1)+P-k})^\top\right]^\top$ for all $k = 0, \ldots, P-1$, system (2) has the following block form:

$$\tilde{w}_k^{t+1} = \tilde{\Phi}_k^t\left(\tilde{w}_k^t\right),\ k = 0, \ldots, P-1,\ t \in \mathbb{Z}_{\geq 0},$$

with $\tilde{\Phi}_k^t = \left[\tilde{\Phi}_{k,0}^t, \ldots, \tilde{\Phi}_{k,n}^t\right]^\top$ $(n = (\tau_M + 1)/P - 1)$, and

$$\tilde{\Phi}_{k,z}^t = \begin{cases} [\phi_1^{t-k}(\cdot)^\top, \ldots, \phi_m^{t-k}(\cdot)^\top]^\top & z = 0, \\ x^{t-z\tau_0-k} & z > 0. \end{cases}$$

By linearization, with the same permutation of y^t with that of w^t, we can bring the variational equation near each periodic solution w^* into the form

$$\tilde{y}^{t+1} = \tilde{B}(t)\tilde{y}^t \tag{11}$$

with block-diagonal $\tilde{B}(t)$:

$$\tilde{B}(t) = \mathrm{diag}[\tilde{B}_r(t)]_{r=0}^{P-1}.$$

Thus, after a partition of $\tilde{y} = [\tilde{y}_0^\top, \cdots, \tilde{y}_{P-1}^\top]^\top$, (11) has the block form

$$\tilde{y}_r^{t+1} = \tilde{B}^r(t)\tilde{y}_r^t, \; r = 0, \ldots, P-1. \tag{12}$$

A similar hypothesis to $\mathbf{B}_1$ can be stated as

$\mathbf{B}_2$: *There exists neighborhood U containing $\mathscr{S}_P(D)$ such that any trajectory $w(t)$ of (2) starting in U is bounded and $\pi_{\mathscr{S}_P}(w(t)) \in \mathscr{S}_P(D)$ for all $t \in \mathbb{Z}_{\geq 0}$.*

Then, from Theorem 3.1 in [9], we have

Theorem 3. *Under the hypothesis $\mathbf{B}_2$, if $\sup_{w^* \in \mathscr{S}_P(D)} \max_{r=0,\ldots,P-1} \mathrm{diam}(\tilde{B}^r(\cdot)) < 1$, then system (2) is (locally) stable with respect to $\mathscr{S}_P(D)$, that is, there exists a sufficiently small neighborhood U of $\mathscr{S}_P(D)$ such that from any initial condition in U, the trajectory converges to a periodic trajectory in $\mathscr{S}_P(D)$.*

In a similar fashion as in Theorem 2, if the time-variation is driven by a metric dynamical system $(\Omega, \mathscr{F}, P, \theta^t)$, i. e., Eq. (10) becomes a RDS:

$$\tilde{y}^{t+1} = \tilde{B}(\theta^t \omega)\tilde{y}^t, \tag{13}$$

then letting

$$\tilde{V}_\lambda^t = \left\{ \omega \in \Omega : \max_{r=0,\ldots,P-1} \left[\mathrm{diam}(\prod_{k=0}^{t} \tilde{B}^r(\theta^k \omega)) \right]^{1/t} < \lambda \right\},$$

we can state the following result.

Theorem 4. *Under hypothesis $\mathbf{B}_2$, if there exists some $\lambda \in (0,1)$ such that $\sum_{t=0}^{\infty} P(\tilde{V}_\lambda^t) < +\infty$, then (8) is (locally) stable with respect to $\mathscr{S}_P(D)$ in the almost sure sense, that is, for almost every $\omega \in \Omega$, there exists a sufficient small neighborhood $U(\omega)$ (possibly depending on ω) of $\mathscr{S}_P(D)$ such that for any initial condition in U, the trajectory of (8) converges to a periodic trajectory in $\mathscr{S}_P(D)$.*

Theorems 1 and 2 can be regarded as special cases of Theorems 3 and 4, respectively, when $P = 1$.

Remark 1. Hypotheses $\mathbf{B}_{1,2}$ can be satisfied if system (2) is essentially bounded, i.e., there exists a bounded region $Q \subset \mathbb{R}^{m(\tau_M+1)}$ such that any trajectory enters Q for all $t \geq T$ for a sufficiently large T. Then the set D can be derived by projecting the convex closure of Q onto $\mathscr{S}_1$ or $\mathscr{S}_P$, respectively.

3 Linear Model

Eq. (7) can be regarded as a special case of (1) with $\Phi_i^t = \sum_{j=1}^m G_{ij}(t) x_j^{t-\tau_{ij}^t}$. However, in such a linear model the stability is always global, instead of local for nonlinear systems. In this section, we provide the main results in terms of matrix and graph theories for linear models (7) or (8). The link between stochastic matrices and graphs is an essential feature here.

A stochastic (or simply nonnegative) matrix $A = [a_{ij}]_{i,j=1}^{m} \in \mathbb{R}^{m,m}$ defines a graph $\mathscr{G} = \{\mathscr{V}, \mathscr{E}\}$, where $\mathscr{V} = \{1,\dots,m\}$ denotes the *node (agent) set* with m nodes and $\mathscr{E}$ denotes the *link set* where there exists a directed link from node j to i (i.e., $e(i,j)$ exists) if and only if $a_{ij} > 0$. We denote this graph corresponding to the stochastic matrix A by $\mathscr{G}(A)$. The node i is said to be self-linked if $e(i,i)$ exists, i.e., $a_{ii} > 0$. The node i can *access* the node j, or equivalently, the node j is *accessible* from the node i, if there exists a path from i to j. The graph $\mathscr{G}$ has a *spanning tree* if there exists a node i which can access all other nodes. The graph $\mathscr{G}$ is said to be *strongly connected* if each node is a root. We refer the interested reader to the book [26] for more details. Due to the relationship between nonnegative matrices and graphs, we can call upon and switch between their respective properties as needed. For example, the indecomposability of a nonnegative matrix A is equivalent to that $\mathscr{G}(A)$ has a spanning tree, and the aperiodicity of a graph is associated with the aperiodicity of its corresponding matrix [27]. For a sequence of nonnegative matrices $A(t)$, we can define a graph sequence associated with $A(t)$: $\mathscr{G}(t) = \mathscr{G}(A(t))$. The union of several graphs $\{\mathscr{G}_i,\ i = 1,\dots,p\}$ on the same node set is the union of their link sets.

For a nonnegative matrix A and a given $\delta > 0$, the δ*-matrix* of A, denoted by A^{δ}, is defined as

$$[A^{\delta}]_{ij} = \begin{cases} \delta, \text{ if } A_{ij} \geq \delta; \\ 0, \text{ if } A_{ij} < \delta. \end{cases}$$

The δ*-graph* of A is the directed graph corresponding to the δ-matrix of A. We can then state the following result for the stability of $\mathscr{S}_1$ (noting that in the linear model, $D = \mathbb{R}$).

Theorem 5. *[12] Suppose there exist $\mu > 0$, $L \in \mathbb{Z}_{\geq 0}$, and $\delta > 0$ such that $G^0(\sigma) > \mu I_m$ for all $\sigma \in \Omega$ and the δ-graph of $\sum_{k=n+1}^{n+L} G(k)$ has a spanning tree for all $n \in \mathbb{Z}_{\geq 0}$. Then system (7) is (globally) stable with respect to $\mathscr{S}_1$, i. e., it reaches consensus.*

In fact, with $D = \mathbb{R}$, if the condition in this theorem is satisfied, there exist a sufficiently large integer T and $\lambda \in (0,1)$ such that $\operatorname{diam}\left(\prod_{k=n+1}^{n+T'} G(k)\right) < \lambda^{T'}$ for any $T' > T$. Hence, the conditions in Theorem 1 hold.

We rewrite system (7) in the general form

$$x_i^{t+1} = \sum_{\tau=0}^{\tau_M} \sum_{j=1}^{m} G_{ij}^{\tau}(t) x_j^{t-\tau}, \quad i = 1,\dots,m, \tag{14}$$

by partitioning the inter-links according to delays, as well as in the matrix form

$$x^{t+1} = \sum_{\tau=0}^{\tau_M} G^{\tau}(t) x^{t-\tau}, \tag{15}$$

where $G^{\tau}(t)=[G^{\tau}_{ij}(t)]_{i,j=1}^{m}$. In some cases delays occur at self-links, for example when it takes time for each agent to process its own information. Suppose that the self-linking delay for each node is identical, that is, $\tau_{ii}=P-1>0$. We classify each integer t in the discrete-time set $\mathbb{Z}_{\geq 0}$ (or the whole integer set $\mathbb{Z}$) via $\mathrm{mod}(t+1,P)$ as the quotient group of $(\mathbb{Z}+1)/P$. As a default set-up, we denote $\langle i\rangle_P$ by $\langle i\rangle$. Let $\hat{G}^i(\cdot)=\sum_{j\in\langle i\rangle}G^j(\cdot)$. We have the following result for the stability of $\mathscr{S}_P$.

Theorem 6. *Assume that*

(1) Hypothesis (4) holds for $P>0$;

(2) $\tau_{ii}(t)=P-1$ for all $i=1,\ldots,m$;

(3) $G^{P-1}(t)>\mu I_m$ for some $\mu>0$ and all $t\in\mathbb{Z}_{\geq 0}$.

Suppose further that there exist $L\in\mathbb{Z}_{\geq 0}$ and $\delta>0$ such that the δ-graph of $\sum_{k=n+1}^{n+L}\hat{G}^0(k)$ is strongly connected for all $n\in\mathbb{Z}_{\geq 0}$. Then system (14) is (globally) stable with respect to $\mathscr{S}_P$, i. e., it synchronizes to a P-periodic trajectory.

This theorem can be proved as a consequence from Theorem 3 in a similar fashion as the proof of Theorem 3.4 in [21], but by removing the discussion of randomness, since here we consider deterministic time-variation.

The time-variation can be random, e. g., induced by a stochastic process σ^t. In [20, 21], we considered the case when $\{\sigma^t\}$ is an *adapted stochastic process*: Let $\{A_k\}$ be a stochastic process defined on the basic probability space $\{\Omega,\mathscr{F},P\}$, with the state space Ω, the σ-algebra $\mathscr{F}$, and the probability function $\mathbb{P}$. Let $\{\mathscr{F}^k\}$ be a *filtration*, i. e., a sequence of nondecreasing sub-σ-algebras of $\mathscr{F}$. If A_k is measurable with respect to (w.r.t.) $\mathscr{F}^k$, then the sequence $\{A_k,\mathscr{F}^k\}$ is called an adapted process. Let $\mathbb{E}(\cdot|\mathscr{F}^t)$ denote the conditional expectation with respect to σ-algebra $\mathscr{F}^t$. Then, Eq. (15) becomes

$$x^{t+1}=\sum_{\tau=0}^{\tau_M}G^{\tau}(\sigma^t)x^{t-\tau}. \tag{16}$$

This adapted process can be regarded as a metric dynamical system with invariant probability, $\{\boldsymbol{\Omega},\mathbb{F},\mathbb{P},\theta^t\}$, where $\boldsymbol{\Omega}=\Omega^{\mathbb{Z}_{\geq 0}}$, i. e., each element is the sequence $\{\sigma^t\}_{t\geq 0}$, $\mathbb{F}=\mathscr{F}^{\mathbb{Z}_{\geq 0}}$ is the infinite Cartesian product of $\mathscr{F}$, $\mathbb{P}$ coincides with the intrinsic probability P, and θ^t is the shift map: $\theta\omega=\{\sigma^t\}_{t\geq 1}$. The following results are the stochastic versions of Theorems 5 and 6.

Theorem 7. *[20, 21] Suppose there exist $\mu>0$, $L\in\mathbb{Z}_{\geq 0}$, and $\delta>0$ such that $G^0(\sigma)>\mu I_m$ for all $\sigma\in\Omega$ and the δ-graph of $\mathbb{E}\{\sum_{k=n+1}^{n+L}G(\sigma^k)|\mathscr{F}^n\}$ has a spanning tree for all $n\in\mathbb{Z}_{\geq 0}$ almost surely. Then (16) is stable with respect to $\mathscr{S}_1$ almost surely (i.e. with probability one).*

Theorem 8. *[20, 21] Assume that*

(1) Hypothesis (4) holds;

(2) $\tau_{ii}(t)=P-1$ for all $i=1,\ldots,m$;

(3) $G^{P-1}(t)>\mu I_m$ for some $\mu>0$ and all $t\in\mathbb{Z}_{\geq 0}$.

Suppose further that there exist $L\in\mathbb{Z}_{\geq 0}$ and $\delta>0$ such that the δ-graph of $\mathbb{E}\{\sum_{k=n+1}^{n+L}\hat{G}^0(\sigma^k)|\mathscr{F}^n\}$ is strongly connected for all $n\in\mathbb{Z}_{\geq 0}$ almost surely. Then (16) is stable with respect to $\mathscr{S}_P$ almost surely (i.e. with probability one).

4 Multi-agent Model with Nonlinear Coupling

In this section, we present a stability analysis of a class of nonlinear multi-agent models

$$x_i^{t+1} = \sum_{j=1}^{m} \psi_{ij}^t \left(x_j^{t-\tau_{ij}(t)} - x_i^{t-\tau_{ii}(t)} \right) x_j^{t-\tau_{ij}(t)}, \; i = 1, \ldots, m, \; t \in \mathbb{Z}_{\geq 0}, \tag{17}$$

where $\psi_{ij}^t(\cdot)$ is a (time-dependent) nonlinear function that denotes the coupling strength from agent j to agent i, acting on the difference between the states of the two nodes under the presence of delays. We assume that $\psi_{ij}^t(s)$ is $C^{1+\alpha}$ continuous for some $\alpha > 0$ and attains its maximum value, which is assumed to be nonzero, at $s = 0$. In other words, the coupling strength is maximum when the (delayed) states are equal. Thus, $d\psi_{ij}^t(s)/ds|_{s=0} = 0$ for all $i, j = 1, \ldots, m$ and $t \in \mathbb{Z}_{\geq 0}$. For example, $\psi_{ij}^t(\cdot)$ can be chosen from a class of Gaussian-type kernel functions. In addition, to guarantee that (3) holds, we also assume that $\sum_{j=1}^m \psi_{ij}^t(0) = 1$ for all i and t.

The variational equation near $\mathscr{S}_1$ or $\mathscr{S}_P$ under the assumption (4) is:

$$\delta x_i^{t+1} = \sum_{j=1}^{m} \psi_{ij}^t(0) \delta x_j^{t-\tau_{ij}(t)}, \; i = 1, \cdots, m \;, t \in \mathbb{Z}_{\geq 0}. \tag{18}$$

It has the similar form of (7). Let $\Psi^0(t) = [\psi_{ij}^t(0)]_{i,j=1}^m$ and $\tilde{\Psi}_r^0(t)$ be defined in the same fashion as done in Eqs. (11) and (12). Then we have the following result.

Theorem 9. *Assume all conditions mentioned above for $\psi_{ij}^t(\cdot)$ hold.*

(1) Under hypothesis $\mathbf{B}_1$ for some $D \subset \mathbb{R}$, if $\operatorname{diam}\left(\Psi^0(\cdot)\right) < 1$, *then system (17) is (locally) stable with respect to $\mathscr{S}_1(D)$;*

(2) Under the hypothesis $\mathbf{B}_2$ for some $D \subset \mathbb{R}$, if $\operatorname{diam}\left(\tilde{\Psi}_r^0(\cdot)\right) < 1$ *for all $r = 0, \ldots, P-1$, then system (17) is (locally) stable with respect to $\mathscr{S}_P(D)$.*

In addition, if $\psi_{ij}^t(0)$ are all nonnegative, then $\{\Psi^0(t)\}$ are stochastic matrices, and the conditions in Theorem 9 can be "translated" in terms of graphs associated with the stochastic matrix sequence $\{\Psi^0(t)\}$, namely, into the conditions in Theorems 6 and 7.

When the time-variation is induced by a stochastic process, or generally by a metric dynamical system, the results of Theorems 2 and 4 can be applied to derive sufficient conditions for consensus in the almost surely sense for system (17). Combined with the graph theory used in [20], if $\{\Psi^0(t)\}$ are stochastic matrices, we can derive sufficient conditions for consensus like Theorem 7 and 8. We omit the details due to space constraints.

5 Numerical Examples: Dynamical Networks for Random Waypoint Model

We perform numerical examples to illustrate the results by the "random waypoint" (RWP) model, which is a widely used model in performance evaluation of protocols of ad hoc networks, first introduced in [29]. We use the same set-up of the model as done in [21] to mimic time-varying graph topologies and realize the random waypoint model in a 1000×1000 (m^2) square area, where the agent i moves towards a randomly selected target in this area following the uniform distribution. The velocity of movement is also random, with a uniform distribution in $[10, 20]$ (m/sec). After approaching the target, the agent waits for a random time period following the uniform distribution in $[1, 5]$ (sec). Moreover, each agent's behavior is stochastically independent of the others. The links between agents are generated such that each agent is linked to the agents whose distance is not more than R. We take $R = 120$ (m). There are 50 independent mobile agents in the network, whose location and status of the agents can be modeled as a homogeneous Markov chain [21].

We set up two models of multi-agent systems on the RWP network. The first one is a linear model (stated in the form of (7)):

$$x_i^{t+1} = \frac{1}{\#\mathcal{N}_i^t} \sum_{j \in \mathcal{N}_i(t)} x_j^{t-\tau_{ij}^t}, \ i = 1, \ldots, m, \tag{19}$$

where $\mathcal{N}_i(t)$ denotes the neighborhood of agent i at time t and $\#F$ denotes the number of the elements in a finite set F. The second model is a special case of (17) with coupling functions:

$$\psi_{ij}^t(s) = \frac{1}{\#\mathcal{N}_i(t)} \exp(-\frac{s^2}{2}). \tag{20}$$

It can be verified that all conditions of $\psi_{ij}^t(\cdot)$ in Section 4 are satisfied. We assume that the self-links exist for all nodes. Thus $\mathcal{N}_i(t)$ is always nonempty.

We consider discrete time with a 0.01 (sec) time interval. Each agent operates according to the algorithm (19). Transmission delays exist due to finite information transmission speed and storage buffer. Since the speed of information transmission is typically much higher than the movement of agents, we omit the displacement of the information transmission caused by the movement of agents and define the delays (0.01 sec) as:

$$\tau_{ij}(\sigma^t) = \lambda \lfloor \frac{d_{ij}^t}{v_s} \rfloor + \tau_0, \tag{21}$$

where d_{ij}^t (m) denotes the distance between agents i and j in the two-dimensional space at time t, v_s denotes the transmission speed of information, $\lfloor \cdot \rfloor$ denotes the floor function, i.e., the largest integer less than or equal to its argument, λ is a scaling parameter representing the ratio of the time scale of movement of the agents and that of the information transmission and processing among agents, and τ_0 (0.01 sec) denotes the identical self-linking delay.

Following the arguments in [21], the network has a positive probability of being a complete network, with respect to the stationary probability distribution. This implies that the expectation, with respect to the stationary probability distribution, of the graph topology is a complete graph. Hence, for the case of existence of self-links, the conditions of Theorems 7 and 8 are satisfied. In the absence of self-links, for any initial network graph, there are a path of finite length and a positive probability such that all agents enter a disc with radius less than R. So, the conditional expectation of product of the matrices has a positive probability of being complete. This implies that the conditional expectation is complete. In a similar way, conditions for consensus can be verified for system (17) with (20) as well.

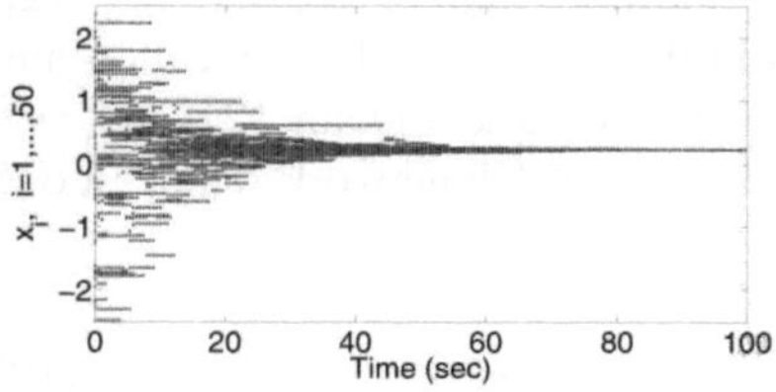

(a) Convergence of (19) with $\lambda = 1$ and $\tau_0 = 0$.

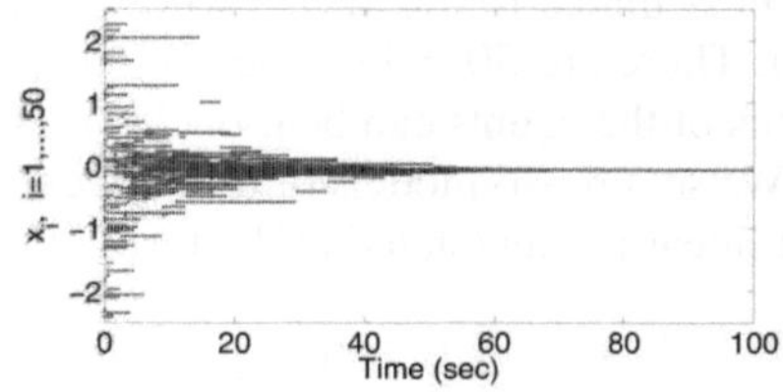

(b) Convergence of (17) with (20), $\lambda = 1$ and $\tau_0 = 0$.

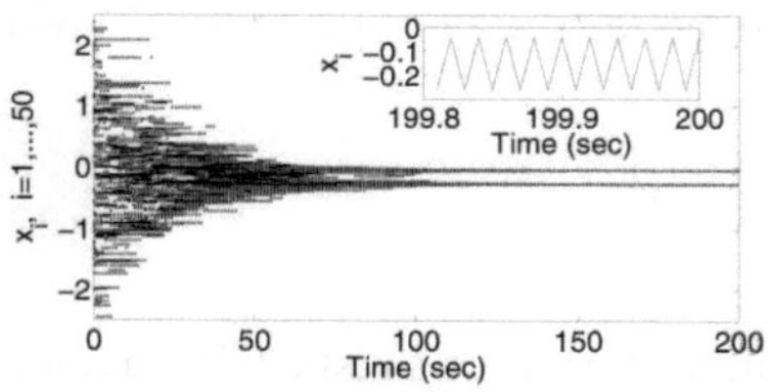

(c) Convergence of (19) with $\lambda = 2$ and $\tau_0 = 1$.

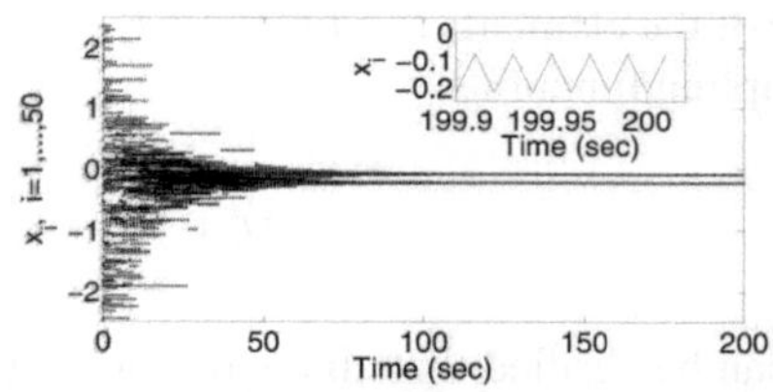

(d) Convergence of (17) with (20), $\lambda = 2$ and $\tau_0 = 1$.

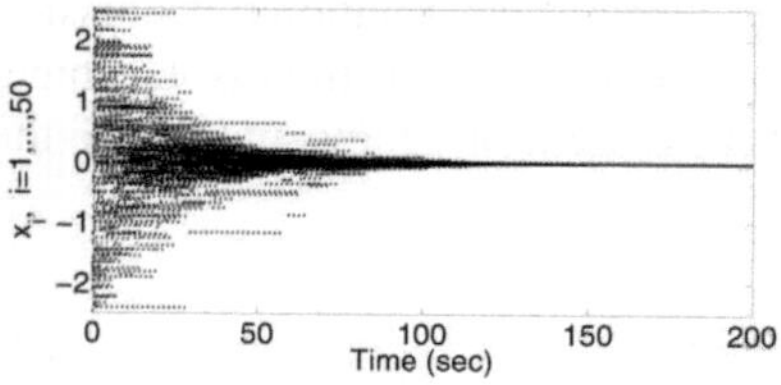

(e) Convergence of (19) with $\lambda = 2$ and $\tau_0 = 2$.

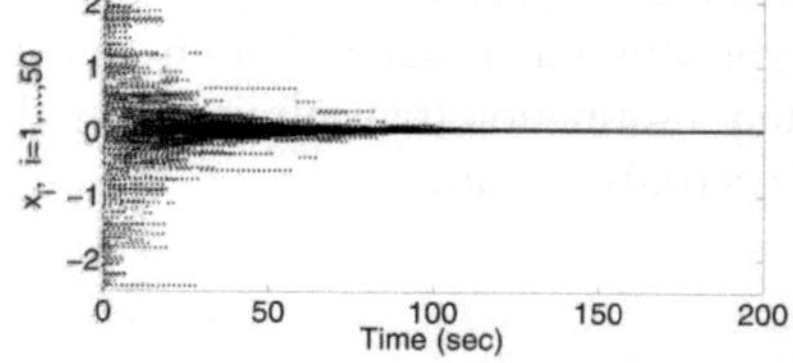

(f) Convergence of (17) with (20), $\lambda = 2$ and $\tau_0 = 2$.

Fig. 1 Convergence dynamics of the multi-agent systems (19) (left column) and (17) with (20) (right column) in RWP networks. The insets show the terminal synchronous orbits.

We fix $v_s = 3000$ (m/sec) and pick different values of τ_0 and λ to illustrate the synchronous or consensus dynamics as mentioned in Theorems 7, 8, and 9.

First, we choose $\lambda = 1$ and $\tau_0 = 0$. Theorem 7 indicates that the multi-agent system (19) reaches consensus. Fig. 1(a) depicts the consensus dynamics of (19) with the delays (21) with respect to $\mathscr{S}_1$. We also observe that system (17) with coupling function (20) reaches consensus, as shown in Fig. 1(b).

We next take $\lambda = 2$ and $\tau_0 = 1$. Thus, the delays can be picked only in the set $\{1,3,5,7,9\}$ and each value in this set can be a possible delay in (21). One can see that $\gcd(\tau_{ij}+1:\ i,j=1,\ldots,m; t\in\mathbb{Z}_{\geq 0}) = 2$. Theorem 8 yields that (19) cannot reach consensus but must instead synchronize to a 2-periodic trajectory. The same conclusion holds also for system (17) with coupling function (20). Fig. 1(c) and 1(d) show the synchronous dynamics of systems (19) and (17) with (20) and the delays (21), $\lambda = 2$, and $\tau_0 = 1$.

Finally, we choose $\lambda = 2$ and $\tau_0 = 2$. Thus, the delays can be picked only from the set $\{2,4,6,8,10\}$. We have $\gcd(\tau_{ij}+1):\ i,j=1,\ldots,m; t\in\mathbb{Z}_{\geq 0}) = 1$. From Theorem 8, similar arguments indicate that (19) reaches consensus, i. e., synchronizes at a periodic trajectory with period $P = 1$. The same conclusion holds for the system (17) with (20). Fig. 1(e) and 1(f) indicate the consensus dynamics with the delays (21), $\lambda = 2$, and $\tau_0 = 2$.

6 Conclusion

We have presented an analysis of consensus problem in discrete-time networks of multi-agent systems, based on our previous results in [9,20,21,25]. Here the model is general, including the linear consensus model as a special example. When the time variation is driven by a metric dynamical system, multi-agent systems become random dynamic systems. Based on a Hajnal diameter approach that we developed for synchronization analysis, we have presented sufficient conditions for both consensus at a uniform value and synchronization at a periodic trajectory, and shown how the periods depend on the transmission delay patterns. As special examples, we have re-derived the stability results for the consensus of the linear model and derived sufficient conditions for the stability of a class of delayed multi-agent systems with nonlinear coupling. To illustrate the theoretical results, we have presented two consensus algorithms in a mobile-agent model under transmission delays.

References

1. Lynch, N.A.: Distributed algorithms. Morgan Kaufmann, San Francisco (1996)
2. DeGroot, M.H.: Reaching a consensus. J. Amer. Statist. Assoc. 69, 118–121 (1974)
3. Vicsek, T., Czirók, A., Ben-Jacob, E., Cohen, I., Shochet, O.: Novel type of phase transition in a system of self-driven particles. Phys. Rev. Lett. 75, 1226–1229 (1995)
4. Fax, J.A., Murray, R.M.: Information flow and cooperative control of vehicle formations. IEEE Trans. Autom. Control 49, 1465–1476 (2004)
5. Olfati-Saber, R., Shamma, J.S.: Consensus filters for sensor networks and distributed sensor fusion. In: 44th IEEE Conference on Decision and Control 2005, and 2005

European Control Conference, CDC-ECC 2005, pp. 6698–6703 (2005)
6. Winfree, A.T.: The Geometry of biological time. Springer, New York (1980)
7. Kuramoto, Y.: Chemical Oscillations, waves, and turbulence. Springer, New York (1984)
8. Pikovsky, A., Rosenblum, M., Kurths, J.: Synchronization: A universal concept in nonlinear sciences. Cambridge University Press (2001)
9. Lu, W., Atay, F.M., Jost, J.: Synchronization of Discrete-Time Dynamical Networks with Time-Varying Couplings. SIAM J. Math. Anal. 39(4), 1231–1259 (2007)
10. Olfati-Saber, R., Fax, J.A., Murray, R.M.: Consensus and cooperation in networked multi-agent systems. Proceedings of the IEEE 95, 215–233 (2007)
11. Olfati-Saber, R., Murray, R.M.: Consensus problems in networks of agents with switching topology and time-delays. IEEE Trans. Autom. Control 49(9), 1520–1533 (2004)
12. Moreau, L.: Stability of multiagent systems with time-dependent communication links. IEEE Trans. Autom. Control 50(2), 169–182 (2005)
13. Cao, M., Morse, A.S., Anderson, B.D.O.: Reaching a consensus in a dynamically changing environment: a graphical approach. SIAM J. Control Optim. 47, 575–600 (2008)
14. Xiao, F., Wang, L.: Asynchronous consensus in continuous-time multi-agent systems with switching topology and time-varying delays. IEEE Trans. Automatic Control 53(8), 1804–1816 (2008)
15. Bliman, P.-A., Ferrari-Trecate, G.: Average consensus problems in networks of agents with delayed communications. Automatica 44(8), 1985–1995 (2008)
16. Michiels, W., Morărescu, C.-I., Niculescu, S.-I.: Consensus problems with distributed delays, with application to traffic flow models. SIAM J. Control Optim. 48, 77–101 (2009)
17. Hatano, Y., Mesbahi, M.: Agreement over random networks. IEEE Trans. Autom. Control 50(11), 1867–1872 (2005)
18. Tahbaz-Salehi, A., Jadbabaie, A.: A necessary and sufficient condition for consensus over random networks. IEEE Trans. Autom. Control 53(3), 791–795 (2008)
19. Wu, C.W.: Synchronization and convergence of linear dynamics in random directed networks. IEEE Trans. Autom. 51(7), 1207–1210 (2006)
20. Lu, W., Atay, F.M., Jost, J.: Consensus and synchronization in discrete-time networks of multi-agents with stochastically switching topologies and time delays. Networks and Heterogeneous Media 6(2), 329–349 (2011)
21. Lu, W., Atay, F.M., Jost, J.: Consensus and Synchronization in Delayed Networks of Mobile Multi-agents. In: The 18th IFAC World Congress, Milano, Italy, August 28 - September 2 (2011)
22. Arnold, L.: Random Dynamical Systems. Springer, Heidelberg (1998)
23. Hajnal, J.: The ergodic properties of non-homogeneous finite Markov chains. Proc. Camb. Phil. Soc. 52, 67–77 (1956)
24. Hajnal, J.: Weak ergodicity in non-homogeneous Markov chains. Proc. Camb. Phil. Soc. 54, 233–246 (1958)
25. He, X., Lu, W., Chen, T.: On transverse stability of random dynamical systems. Discrete and Continuous Dynamic Systems 33(2), 701–721 (2013)
26. Godsil, C., Royle, G.: Algebraic graph theory. Springer, New York (2001)
27. Horn, R.A., Johnson, C.R.: Matrix analysis. Cambridge University Press (1985)
28. Shen, J.: A geometric approach to ergodic non-homogeneous Markov chains. Wavelet Anal. Multi. Meth., LNPAM 212, 341–366 (2000)
29. Johnson, D.B., Maltz, D.A.: Dynamic source routing in ad hoc wireless networks. In: Imielinski, T., Korth, H. (eds.) Mobile Computing, ch. 5, pp. 153–181. Kluwer Academic Publishers (1996)

Part III
Time-Delay and Sampled-Data Systems

Sampled-Data Stabilization under Round-Robin Scheduling

Kun Liu, Emilia Fridman, Laurentiu Hetel, and Jean-Pierre Richard

Abstract. This chapter analyzes the exponential stability of Networked Control Systems (NCSs) with communication constraints, variable sampling intervals and constant delays. We focus on static output feedback controllers for linear systems. The system sensors nodes are supposed to be distributed over a network. The scheduling of sensor information towards the controller is ruled by the classical Round-Robin protocol. We develop a time-delay approach for this problem by presenting the closed-loop system as a switched system with multiple delayed samples. By constructing an appropriate Lyapunov functional, which takes into account the switched system model and the sawtooth delays induced by sampled-data control, we derive the exponential stability conditions in terms of Linear Matrix Inequalities (LMIs). Polytopic uncertainties in the system model can be easily included in the analysis. The efficiency of the method is illustrated on the classical cart-pendulum benchmark problem.

1 Introduction

Networked Control Systems (NCSs) are systems with spatially distributed sensors, actuators and controller nodes which exchange data over a communication data channel. Only one node is allowed to use the communication channel at once. The communication along the data channel is orchestrated by a scheduling rule called protocol. Using such control structures offers several practical advantages: reduced costs, ease of installation and maintenance and increased flexibility. However, from

Kun Liu · Emilia Fridman
School of Electrical Engineering, Tel Aviv University, Tel Aviv 69978, Israel
e-mail: {liukun,emilia}@eng.tau.ac.il

Laurentiu Hetel · Jean-Pierre Richard
University Lille Nord de France, LAGIS, FRE CNRS 3303, Ecole Centrale de Lille, Cite Scientifique, BP 48, 59651 Villeneuve d'Ascq cedex, France
e-mail: {laurentiu.hetel,jean-pierre.richard}@ec-lille.fr

T. Vyhlídal, J.-F. Lafay, and R. Sipahi (eds.), *Delay Systems*,
Advances in Delays and Dynamics 1,
DOI: 10.1007/978-3-319-01695-5_13,

the control theory point of view, it leads to new challenges. Closing the loop over a network introduces undesirable perturbations such as delay, variable sampling intervals, quantization, packet dropouts, scheduling communication constraints, etc. which may affect the system performance and even its stability. It is important in such a configuration to provide a stability certificate that takes into account the network imperfections. For general survey papers we refer to [1, 23, 26]. Recent advancements can be found in [6, 8, 13, 20, 24] for systems with variable sampling intervals, [17] for dealing with the quantization and [2, 9, 14] for control with time delay. Concerning NCSs, three main control approaches have been used: discrete-time models (with integration step), input delay models (together with a Lyapunov-Krasovskii theory) and impulsive/hybrid models.

In the present chapter, we focus on the stabilization of NCSs with communication constraints. We consider a linear (probably, uncertain) system with distributed sensors. The scheduling of sensor information towards the controller is ruled by the classical Round-Robin protocol. The Round-Robin protocol has been considered in [12, 21] (in the framework of hybrid system approach) and in [3, 4] (in the framework of discrete-time systems). In [21], stabilization of nonlinear system based on the impulsive model is studied. However, delays are not included in the analysis. In [12], the authors provide methods for computing the Maximum Allowable Transmission Interval (MATI - i.e. the maximum sampling jitter) and Maximum Allowable Delay (MAD) for which the stability of a nonlinear system is ensured.

In [3], network-based stabilization of Linear Time-Invariant (LTI) systems with Round-Robin protocol and without delay have been considered (see also [4] for delays less than the sampling interval). The analysis is based on the discretization and the equivalent polytopic model at the transmission instants. For LTI systems, discretization-based results are usually less conservative than the general hybrid system-based results. However, discrete-time models do not take into account the system behavior between two transmissions and are not applicable to uncertain systems. Moreover, it is tedious to include large delays in such models and the stability analysis methods may fail when the interval between two transmissions takes small values (see e.g. [8]).

In the present chapter, a direct Lyapunov-Krasovskii approach is developed for stabilization of NCSs with Round-Robin scheduling, constant delays and variable sampling intervals. Discrete-time controllers are considered and the delay may be larger than the sampling interval. We present the closed-loop system as a *switched continuous-time system with multiple delayed samples*. By constructing appropriate Lyapunov functionals, we derive LMIs for the exponentially stability with a given decay rate.

We note that, till recently, only time-independent Lyapunov-Krasovkii functionals (for systems with time-varying delays) were applied to NCSs (see e.g. [7, 9]). These functionals did not take advantage of the sawtooth evolution of the delays induced by sampled-and-hold. The latter drawback was removed in [6], where time-dependent Lyapunov functionals for sampled-data systems were introduced. In some well-studied numerical examples, the results of [6] approach the analytical values of minimum L_2-gain and of maximum sampling interval, preserving the stability.

In the present chapter, we suggest two methods, which extend time-dependent Lyapunov functional constructions developed in [6] and [18] to the switched systems with multiple samples and constant delays, respectively. Our results are applicable to systems with polytopic type uncertainties. The efficiency of the method is illustrated on the classical cart-pendulum example.

Notation: Throughout the chapter the superscript 'T' stands for matrix transposition, $\mathscr{R}^n$ denotes the n dimensional Euclidean space with vector norm $\|\cdot\|$, $\mathscr{R}^{n\times m}$ is the set of all $n\times m$ real matrices, and the notation $P>0$, for $P\in\mathscr{R}^{n\times n}$ means that P is symmetric and positive definite. The symmetric elements of the symmetric matrix will be denoted by $*$. The space of functions $\phi:[a,b]\to\mathscr{R}^n$, which are absolutely continuous on $[a,b]$, have a finite $\lim_{\theta\to b^-}\phi(\theta)$ and have square integrable first order derivatives is denoted by $W[a,b]$ with the norm $\|\phi\|_W = \max_{\theta\in[a,b]}\|\phi(\theta)\| + \left[\int_a^b \|\dot{\phi}(s)\|^2 ds\right]^{\frac{1}{2}}$. $\mathscr{N}$ denotes the set $\{0, 1, 2, 3, \cdots\}$.

2 Problem Formulation

Consider the following system controlled through a network (see Figure 1):

$$\dot{x}(t) = Ax(t) + Bu(t), \tag{1}$$

where $x(t)\in\mathscr{R}^n$ is the state vector, $u(t)\in\mathscr{R}^m$ is the control input, A and B are system matrices with appropriate dimensions. The system presents nodes corresponding to different distributed continuous-time outputs and control computation node situated next to the actuator. For the sake of simplicity, we consider here that the system has two sensor nodes $y^i(t) = C^i x(t)$, $i=1,2$ and we denote $C = \begin{bmatrix} C^1 \\ C^2 \end{bmatrix}$. We let s_k denote the unbounded monotonously increasing sequence of sampling instants, i.e.

$$0 = s_0 < s_1 < \ldots < s_k < \ldots, \quad k\in\mathscr{N}, \ \lim_{k\to\infty} s_k = \infty \tag{2}$$

At each sampling instant s_k, one of the outputs $y^i(t)$ is sampled and transmitted via the network. The choice of the active output node is ruled by a Round-Robin scheduling protocol. The output $y^i(t)$ is transmitted only at the sampling instant $t = s_{2p+i-1}$, $p\in\mathscr{N}$. We assume that the transmission of the information over the network is subject to a constant delay h and that data loss is not possible. We denote by y^i_k, $i=1,2$, the buffers associated to the different system outputs at the controller level. The evolution of these buffers according to the Round-Robin scheduling is given by

$$y^i_k = \begin{cases} y^i(s_k) = C^i x(s_k), & k = 2p+i-1, \\ y^i_{k-1}, & k \neq 2p+i-1, \end{cases} \quad p\in\mathscr{N}. \tag{3}$$

We assume that there exists a matrix $K = \begin{bmatrix} K_1 & K_2 \end{bmatrix}$, $K_1 \in \mathscr{R}^{m \times n_1}$, $K_2 \in \mathscr{R}^{m \times (n-n_1)}$ such that $A + BKC$ is Hurwitz. Then the control law that we consider in this chapter is a static output feedback of the form:

$$u_k = K_1 y_k^1 + K_2 y_k^2, \; k = 1, 2, \ldots \tag{4}$$

and $u_0 = K_1 y_0^1$.

We assume that the controller and the actuator act in an event-driven manner, simultaneously, when data arrive at buffers, at instances $t_k = s_k + h$. Under these hypothesis the control law is piecewise constant with

$$u(t) = u_k, \;\; \forall t \in [t_k, t_{k+1}). \tag{5}$$

The closed-loop system can be presented in the form

$$\begin{aligned} \dot{x}(t) &= Ax(t) + A_1 x(t_k - h) + A_2 x(t_{k-1} - h), \; t \in [t_k, t_{k+1}), \\ \dot{x}(t) &= Ax(t) + A_1 x(t_k - h) + A_2 x(t_{k+1} - h), \; t \in [t_{k+1}, t_{k+2}), \end{aligned} \tag{6}$$

where $k = 2p$, $A_i = BK_iC^i$, $i = 1, 2$, $x(t_{-1} - h) = 0$.

We assume that

$$t_{k+1} - t_k + h \leq \tau_M, \; k \in \mathscr{N} \tag{7}$$

where τ_M denotes the maximum time span between the time $t_k - h$ at which the state is sampled and the time t_{k+1} at which next update arrives at the controller. So we have $t_{k+1} - t_k \leq \tau_M - h$, $k \in \mathscr{N}$.

The objective of the present chapter is to derive exponential stability conditions for system (6) in terms of LMIs via direct Lyapunov method. As a particular case, we will also consider the case of $h = 0$ under the constant/variable sampling. For the constant sampling (as in [20]), our results are less conservative.

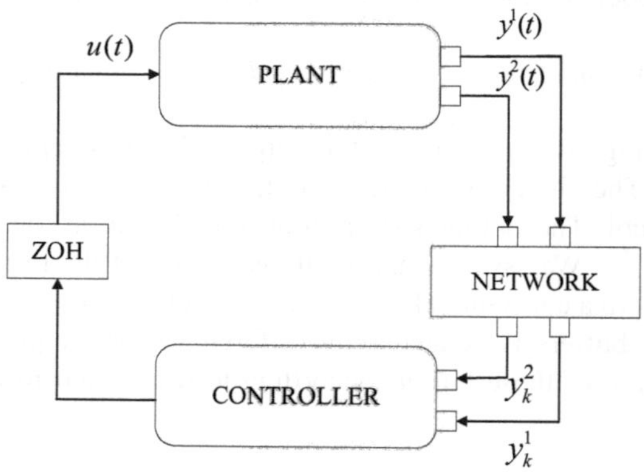

Fig. 1 System architecture

3 Main Results

3.1 *Stability Conditions for NCSs: Variable Sampling and Constant Input/Output Delay*

3.1.1 The First Approach

Extending the constructions of Lyapunov functionals for systems with non-small or interval delay [5], we choose the following Lyapunov functional for exponential stability with the decay rate $\alpha > 0$:

$$V(t,x_t,\dot{x}_t) = \bar{V}(t) = V_0(x_t,\dot{x}_t) + V_1(t,\dot{x}_t) + V_2(t,\dot{x}_t), \tag{8}$$

where

$$\begin{aligned} V_0(x_t,\dot{x}_t) = &[x^T(t)\; x^T(t-h)]P\begin{bmatrix} x(t) \\ x(t-h)\end{bmatrix} \\ &+ \int_{t-h}^{t} e^{2\alpha(s-t)} x^T(s)S_0x(s)ds + h\int_{-h}^{0}\int_{t+\theta}^{t} e^{2\alpha(s-t)}\dot{x}^T(s)R_0\dot{x}(s)dsd\theta \\ &+ \int_{t-h}^{t} e^{2\alpha(s-t)}\dot{x}^T(s)W\dot{x}(s)ds,\; S_0 > 0, R_0 > 0, W > 0, \end{aligned}$$

and

$$\begin{aligned} &V_1(t,\dot{x}_t) = (t_{k+2}-t)\int_{t_k-h}^{t-h} e^{2\alpha(s-t)}\dot{x}^T(s)U_1\dot{x}(s)ds, \quad t\in[t_k,t_{k+2}), \\ &V_2(t,\dot{x}_t) = \begin{cases} (t_{k+1}-t)\int_{t_{k-1}-h}^{t-h} e^{2\alpha(s-t)}\dot{x}^T(s)U_2\dot{x}(s)ds, & t\in[t_k,t_{k+1}), \\ (t_{k+3}-t)\int_{t_{k+1}-h}^{t-h} e^{2\alpha(s-t)}\dot{x}^T(s)U_2\dot{x}(s)ds, & t\in[t_{k+1},t_{k+2}), \end{cases} \\ &U_1 > 0,\; U_2 > 0,\; k = 2p, \end{aligned}$$

and

$$P = \begin{bmatrix} P_1 & P_2 \\ * & P_3 \end{bmatrix} > 0. \tag{9}$$

Note that V_0 is a "nominal" augmented Lyapunov functional for the "nominal" system with a constant delay [15, 22]

$$\dot{x}(t) = Ax(t) + (A_1+A_2)x(t-h). \tag{10}$$

The terms V_1, V_2 are extensions of the time-dependent term introduced in [6] for the sampled-data control. These terms are continuous in time along (6) since

$$\begin{aligned} V_{1|t=t_k^-} &= V_{1|t=t_k^+} = 0,\; V_{1|t=t_{k+1}^-} = V_{1|t=t_{k+1}^+} \geq 0, \\ V_{2|t=t_k^-} &= V_{2|t=t_k^+} \geq 0,\; V_{2|t=t_{k+1}^-} = V_{2|t=t_{k+1}^+} = 0. \end{aligned} \tag{11}$$

By the standard arguments for the switched systems [16,25], the following condition

$$\dot{\bar{V}}(t) + 2\alpha\bar{V}(t) \leq 0 \tag{12}$$

guarantees the exponential stability of (6) with the decay rate α. Therefore, we obtain the following result, the proof of which follows from [19].

Theorem 1. *Given $\tau_M > h \geq 0$, $\alpha > 0$ and K_1, K_2, the system (6) is exponentially stable with the decay rate α, if there exist $n \times n$ matrices $P_1 > 0$, P_2, $P_3 > 0$, $R_0 > 0$, $S_0 > 0$, $W > 0$, $U_i > 0$ $(i = 1,2)$ and Y_{ij}, Z_{ij} $(i = 1,2; j = 1,2,3)$, P_{ij} $(i = 2,3; j = 1,2)$ such that (9) and the following four LMIs are feasible:*

$$\Psi_{11} = \begin{bmatrix} \phi_{11}^1 & \phi_{12}^1 & \phi_{13}^1 & (\tau_M - h)Z_{11}^T & \phi_{15}^1 & \phi_{16}^1 & P_2 \\ * & \phi_{22}^1 & \phi_{23}^1 & (\tau_M - h)Z_{12}^T & \phi_{25}^1 & \phi_{26}^1 & 0 \\ * & * & \phi_{33} & 0 & -Y_{13} & -Z_{13} & P_3 \\ * & * & * & \phi_{55}^2 & 0 & (\tau_M - h)Z_{13} & 0 \\ * & * & * & * & Y_{13} + Y_{13}^T & 0 & 0 \\ * & * & * & * & * & Z_{13} + Z_{13}^T & 0 \\ * & * & * & * & * & * & \phi_{77}^1 \end{bmatrix} < 0, \tag{13}$$

$$\Psi_{12} = \begin{bmatrix} \phi_{11}^1 & \phi_{12}^1 & \phi_{13}^1 & (\tau_M - h)Y_{11}^T & 2(\tau_M - h)Z_{11}^T & \phi_{15}^1 & \phi_{16}^1 & P_2 \\ * & \phi_{22}^1 & \phi_{23}^1 & (\tau_M - h)Y_{12}^T & 2(\tau_M - h)Z_{12}^T & \phi_{25}^1 & \phi_{26}^1 & 0 \\ * & * & \phi_{33} & 0 & 0 & -Y_{13} & -Z_{13} & P_3 \\ * & * & * & \phi_{44}^1 & 0 & (\tau_M - h)Y_{13} & 0 & 0 \\ * & * & * & * & 2\phi_{55}^2 & 0 & 2(\tau_M - h)Z_{13} & 0 \\ * & * & * & * & * & Y_{13} + Y_{13}^T & 0 & 0 \\ * & * & * & * & * & * & Z_{13} + Z_{13}^T & 0 \\ * & * & * & * & * & * & * & \phi_{88}^1 \end{bmatrix} < 0, \tag{14}$$

$$\Psi_{21} = \begin{bmatrix} \phi_{11}^2 & \phi_{12}^2 & \Phi_{13}^2 & (\tau_M - h)Y_{21}^T & \phi_{15}^2 & \phi_{16}^2 & P_2 \\ * & \phi_{22}^2 & \phi_{23}^2 & (\tau_M - h)Y_{22}^T & \phi_{25}^2 & \phi_{26}^2 & 0 \\ * & * & \phi_{33} & 0 & -Y_{23} & -Z_{23} & P_3 \\ * & * & * & \phi_{55}^1 & (\tau_M - h)Y_{23} & 0 & 0 \\ * & * & * & * & Y_{23} + Y_{23}^T & 0 & 0 \\ * & * & * & * & * & Z_{23} + Z_{23}^T & 0 \\ * & * & * & * & * & * & \phi_{77}^2 \end{bmatrix} < 0, \tag{15}$$

$$\Psi_{22} = \begin{bmatrix} \phi_{11}^2 & \phi_{12}^2 & \phi_{13}^2 & 2(\tau_M - h)Y_{21}^T & (\tau_M - h)Z_{21}^T & \phi_{15}^2 & \phi_{16}^2 & P_2 \\ * & \phi_{22}^2 & \phi_{23}^2 & 2(\tau_M - h)Y_{22}^T & (\tau_M - h)Z_{22}^T & \phi_{25}^2 & \phi_{26}^2 & 0 \\ * & * & \phi_{33} & 0 & 0 & -Y_{23} & -Z_{23} & P_3 \\ * & * & * & 2\phi_{55}^1 & 0 & 2(\tau_M - h)Y_{23} & 0 & 0 \\ * & * & * & * & \phi_{44}^2 & 0 & (\tau_M - h)Z_{23} & 0 \\ * & * & * & * & * & Y_{23} + Y_{23}^T & 0 & 0 \\ * & * & * & * & * & * & Z_{23} + Z_{23}^T & 0 \\ * & * & * & * & * & * & * & \phi_{88}^2 \end{bmatrix} < 0, \tag{16}$$

where

$$\phi_{11}^j = A^T P_{2j} + P_{2j}^T A + S_0 - R_0 e^{-2\alpha h} + 2\alpha P_1,$$
$$\phi_{12}^j = P_1 - P_{2j}^T + A^T P_{3j},$$
$$\phi_{13}^j = R_0 e^{-2\alpha h} - Y_{j1}^T - Z_{j1}^T + 2\alpha P_2,$$

$$
\begin{aligned}
&\phi_{15}^j = Y_{j1}^T + P_{2j}^T A_1,\\
&\phi_{16}^j = Z_{j1}^T + P_{2j}^T A_2,\\
&\phi_{22}^j = -P_{3j} - P_{3j}^T + h^2 R_0 + W,\\
&\phi_{23}^j = P_2 - Y_{j2}^T - Z_{j2}^T,\\
&\phi_{25}^j = Y_{j2}^T + P_{3j}^T A_1,\\
&\phi_{26}^j = Z_{j2}^T + P_{3j}^T A_2,\\
&\phi_{33} = -[S_0 + R_0]e^{-2\alpha h} + 2\alpha P_3,\\
&\phi_{44}^j = -(\tau_M - h)U_j e^{-2\alpha\tau_M},\\
&\phi_{55}^j = -(\tau_M - h)U_j e^{-2\alpha(2\tau_M - h)},\\
&\phi_{77}^1 = -[W - 2(\tau_M - h)U_1 - (\tau_M - h)U_2]e^{-2\alpha h},\\
&\phi_{77}^2 = -[W - (\tau_M - h)U_1 - 2(\tau_M - h)U_2]e^{-2\alpha h},\\
&\phi_{88}^j = -[W - (\tau_M - h)U_j]e^{-2\alpha h}, \ j = 1,2.
\end{aligned}
$$

3.1.2 The Second Approach

In this subsection we will adapt to the Round-Robin scheduling a time-dependent Lyapunov functional construction of [18], which is based on the extension of Wirtinger's inequality [10] to the vector case.

Lemma 1. *[18] Let $z(t) \in W[a,b)$ and $z(a) = 0$. Then for any $n \times n$-matrix $R > 0$ the following inequality holds:*

$$\int_a^b z^T(\xi)Rz(\xi)d\xi \le \frac{4(b-a)^2}{\pi^2}\int_a^b \dot{z}^T(\xi)R\dot{z}(\xi)d\xi. \tag{17}$$

We introduce the following *discontinuous in time* Lyapunov functional:

$$V(t,x_t,\dot{x}_t) = \bar{V}_1(t) = V_0(x_t,\dot{x}_t) + V_1(t,x_t,\dot{x}_t) + V_2(t,x_t,\dot{x}_t), \tag{18}$$

where

$$
\begin{aligned}
&V_0(x_t,\dot{x}_t) = x^T(t)Px(t) + \textstyle\int_{t-h}^t x^T(s)S_0x(s)ds + h\int_{-h}^0\int_{t+\theta}^t \dot{x}^T(s)R_0\dot{x}(s)dsd\theta,\\
&V_1(t,x_t,\dot{x}_t) = 4(\tau_M - h)^2 \textstyle\int_{t_k-h}^t \dot{x}^T(s)W_1\dot{x}(s)ds\\
&\qquad -\tfrac{\pi^2}{4}\textstyle\int_{t_k-h}^{t-h}[x(s) - x(t_k-h)]^T W_1[x(s) - x(t_k - h)]ds, \ t \in [t_k,t_{k+2}),
\end{aligned}
$$

$$
V_2(t,x_t,\dot{x}_t) = \begin{cases}
4(\tau_M - h)^2 \int_{t_{k-1}-h}^t \dot{x}^T(s)W_2\dot{x}(s)ds\\
-\frac{\pi^2}{4}\int_{t_{k-1}-h}^{t-h}[x(s)-x(t_{k-1}-h)]^T W_2[x(s)-x(t_{k-1}-h)]ds, \ t \in [t_k,t_{k+1}),\\
4(\tau_M - h)^2 \int_{t_{k+1}-h}^t \dot{x}^T(s)W_2\dot{x}(s)ds\\
-\frac{\pi^2}{4}\int_{t_{k+1}-h}^{t-h}[x(s)-x(t_{k+1}-h)]^T W_2[x(s)-x(t_{k+1}-h)]ds, \ t \in [t_{k+1},t_{k+2}),
\end{cases}
$$

$$W_1 > 0, \ W_2 > 0, \ k = 2p.$$

The terms V_1, V_2 are extensions of the discontinuous constructions of [18]. We note that V_1 can be represented as a sum of the continuous in time term $4(\tau_M - h)^2\int_{t-h}^t \dot{x}^T(s)W_1\dot{x}(s)ds \ge 0$, $t \in [t_k,t_{k+2})$, with the discontinuous (for $t = t_k$) one

$$V_{W1} = 4(\tau_M - h)^2 \int_{t_k-h}^{t-h} \dot{x}^T(s) W_1 \dot{x}(s) ds \\ -\frac{\pi^2}{4} \int_{t_k-h}^{t-h} [x(s) - x(t_k - h)]^T W_1 [x(s) - x(t_k - h)] ds, \quad t \in [t_k, t_{k+2}).$$

Note that $V_{W1|t=t_k} = 0$ and, by the extended Wirtinger's inequality (17), $V_{W1} \geq 0$ for all $t \geq t_0$. Therefore, V_1 does not grow in the jumps.

In a similar way, V_2 can be represented as a sum of the continuous in time term $4(\tau_M - h)^2 \int_{t-h}^{t} \dot{x}^T(s) W_2 \dot{x}(s) ds \geq 0$, with the discontinuous for $t = t_{k+1}$ term

$$V_{W2} = \begin{cases} -\frac{\pi^2}{4} \int_{t_{k-1}-h}^{t-h} [x(s) - x(t_{k-1} - h)]^T W_2 [x(s) - x(t_{k-1} - h)] ds \\ \quad +4(\tau_M - h)^2 \int_{t_{k-1}-h}^{t-h} \dot{x}^T(s) W_2 \dot{x}(s) ds, \quad t \in [t_k, t_{k+1}), \\ -\frac{\pi^2}{4} \int_{t_{k+1}-h}^{t-h} [x(s) - x(t_{k+1} - h)]^T W_2 [x(s) - x(t_{k+1} - h)] ds \\ \quad +4(\tau_M - h)^2 \int_{t_{k+1}-h}^{t-h} \dot{x}^T(s) W_2 \dot{x}(s) ds, \quad t \in [t_{k+1}, t_{k+2}). \end{cases}$$

We have $V_{W2|t=t_{k+1}} = 0$ and, by the extended Wirtinger's inequality (17), $V_{W2} \geq 0$ for all $t \geq t_0$, i.e. V_2 does not grow in the jumps. Therefore, $\bar{V}_1$ does not grow in the jumps: $\lim_{t \to t_k^-} \bar{V}_1(t) \geq \bar{V}_1(t_k)$ and $\lim_{t \to t_{k+1}^-} \bar{V}_1(t) \geq \bar{V}_1(t_{k+1})$ hold.

Theorem 2. *Given $\tau_M > h \geq 0$, the system (6) is asymptotically stable, if there exist $n \times n$ matrices $P > 0$, $R_0 > 0$, $S_0 > 0$, $W_i > 0$ $(i = 1, 2)$, such that the following LMI is feasible:*

$$\Xi = \begin{bmatrix} \Psi_1 & R_0 & PA_1 & PA_2 & A^T W \\ * & \Psi_2 & \frac{\pi^2}{4} W_1 & \frac{\pi^2}{4} W_2 & 0 \\ * & * & -\frac{\pi^2}{4} W_1 & 0 & A_1^T W \\ * & * & * & -\frac{\pi^2}{4} W_2 & A_2^T W \\ * & * & * & * & -W \end{bmatrix} < 0, \tag{19}$$

where

$$\begin{aligned} \Psi_1 &= PA + A^T P + S_0 - R_0, \\ \Psi_2 &= -\frac{\pi^2}{4} W_1 - \frac{\pi^2}{4} W_2 - S_0 - R_0, \\ W &= h^2 R_0 + 4(\tau_M - h)^2 (W_1 + W_2). \end{aligned} \tag{20}$$

Remark 1. Compared to the stability LMI conditions of Theorem 1, the LMI of Theorem 2 is essentially simpler (single LMI of $5n \times 5n$ with fewer decision variables) and is less conservative (see Example below).

Remark 2. Similar to [18], the decay rate of the exponential stability for (6) can be found by changing the variable $\bar{x}(t) = x(t)e^{\alpha t}$ and by applying LMI (19) to the resulting system with polytopic type uncertainty.

3.2 *Stability Conditions for Sample-Data Systems: Constant vs Variable Sampling*

When there is no communication delay, i.e. $h \equiv 0$, the problem for NCSs is reduced to the one for sampled-data systems with scheduling, where the closed-loop system has a form of (6), where $h = 0$.

For $h \to 0$ the conditions of Theorems 1 and 2 become conservative (see Example below). This is different from the stability conditions via conventional Lyapunov functionals for systems with non-small delay $\tau(t) \in [h, \tau_M]$, where for $h \to 0$ the conventional Lyapunov functionals recover the results derived by the corresponding Lyapunov functionals for small delay $[0, \tau_M]$ [11]. Therefore, we will proceed next with exponential stability conditions for (6) with $h = 0$ via time-dependent Lyapunov functionals.

We start with the constant sampling, where $t_{k+1} - t_k = \tau_M$, $k \in \mathcal{N}$. For this case we choose Lyapunov functional of the form

$$\begin{array}{c} V(t, x_t, \dot{x}_t) = x^T(t) P_1 x(t) + \sum_{i=1}^{2} V_i(t, \dot{x}_t) + \sum_{i=1}^{2} V_{Xi}(t, x_t), \\ P_1 > 0, \; t \in [t_k, t_{k+2}), \; k = 2p, \; p \in \mathcal{N}, \end{array} \tag{21}$$

where

$$V_1(t, \dot{x}_t) = (t_{k+2} - t) \int_{t_k}^{t} e^{2\alpha(s-t)} \dot{x}^T(s) U_1 \dot{x}(s) ds,$$

$$V_2(t, \dot{x}_t) = \begin{cases} (t_{k+1} - t) \int_{t_{k-1}}^{t} e^{2\alpha(s-t)} \dot{x}^T(s) U_2 \dot{x}(s) ds, & t \in [t_k, t_{k+1}), \\ (t_{k+3} - t) \int_{t_{k+1}}^{t} e^{2\alpha(s-t)} \dot{x}^T(s) U_2 \dot{x}(s) ds, & t \in [t_{k+1}, t_{k+2}), \end{cases}$$

$$V_{X1}(t, x_t) = (t_{k+2} - t) \xi_0^T(t) \begin{bmatrix} \frac{X+X^T}{2} & -X + X_1 \\ * & -\bar{X}_1 \end{bmatrix} \xi_0(t),$$

$$V_{X2}(t, x_t) = \begin{cases} (t_{k+1} - t) \xi_{-1}^T(t) \begin{bmatrix} \frac{X_2+X_2^T}{2} & -X_2 + X_3 \\ * & -\bar{X}_3 \end{bmatrix} \xi_{-1}(t), & t \in [t_k, t_{k+1}), \\ (t_{k+3} - t) \xi_1^T(t) \begin{bmatrix} \frac{X_2+X_2^T}{2} & -X_2 + X_3 \\ * & -\bar{X}_3 \end{bmatrix} \xi_1(t), & t \in [t_{k+1}, t_{k+2}), \end{cases}$$

with $\xi_i(t) = col\{x(t), x(t_{k+i})\} (i = 0, \pm 1)$, $\bar{X}_1 = X_1 + X_1^T - \frac{X - X^T}{2}$, $\bar{X}_3 = X_3 + X_3^T - \frac{X_2 + X_2^T}{2}$, $U_1 > 0$, $U_2 > 0$, $k = 2p$.

The terms V_i and $V_{Xi} (i = 1, 2)$ are continuous in time along (6) with $h = 0$. The condition $V(t, x_t, \dot{x}_t) \geq \beta \|x(t)\|^2$ holds for $t \in [t_k, t_{k+1})$, $k = 2p$, if

$$\begin{bmatrix} P_1 + \tau_M (X + X^T) + \tau_M \frac{X_2 + X_2^T}{2} & 2\tau_M(-X + X_1) & \tau_M(-X_2 + X_3) \\ * & -2\tau_M \bar{X}_1 & 0 \\ * & * & -\tau_M \bar{X}_3 \end{bmatrix} > 0, \tag{22}$$

$$\begin{bmatrix} P_1 + \tau_M \frac{X + X^T}{2} & \tau_M(-X + X_1) \\ * & -\tau_M \bar{X}_1 \end{bmatrix} > 0, \tag{23}$$

and for $t \in [t_{k+1}, t_{k+2})$, $k = 2p$, if

$$\begin{bmatrix} P_1 + \tau_M (X_2 + X_2^T) + \tau_M \frac{X + X^T}{2} & \tau_M(-X + X_1) & 2\tau_M(-X_2 + X_3) \\ * & -\tau_M \bar{X}_1 & 0 \\ * & * & -2\tau_M \bar{X}_3 \end{bmatrix} > 0, \tag{24}$$

$$\begin{bmatrix} P_1 + \tau_M \frac{X_2+X_2^T}{2} & \tau_M(-X_2+X_3) \\ * & -\tau_M \bar{X}_3 \end{bmatrix} > 0. \tag{25}$$

Lyapunov functional V of (21) with $X = X_i = 0$ $(i = 1,2,3)$ is applicable to systems with variable sampling $t_{k+1} - t_k \leq \tau_M$. Moreover, Lyapunov functional V of (21) with $X = X^T > 0$, $X_2 = X_2^T > 0$ and $X_1 = X_3 = 0$ is applicable to systems with constant sampling. In the latter case the resulting LMIs are convex in τ_M: if they are feasible for τ_M, then they are feasible for any constant $\bar{\tau}_M \in (0, \tau_M]$. Similar to Theorem 1, we arrive to the following result:

Corollary 1. *(i) Given $\tau_M > 0$, $\alpha > 0$ and K_1, K_2, the system (6) with $h = 0$ is exponentially stable with the decay rate α under the constant sampling $t_{k+1} - t_k = \tau_M$, if there exist $n \times n$ matrices $P_1 > 0$, $U_i > 0$, Y_{ij}, Z_{ij} $(i = 1,2; j = 1,2,3)$, P_{ij} $(i = 2,3; j = 1,2)$, X, X_i $(i = 1,2,3)$ such that (22)-(25) and the following LMIs are feasible:*

$$\Sigma_{11} = \begin{bmatrix} \varphi_{11}^1 + X_\alpha^{11} & \varphi_{12}^{11} & \tau_M Z_{11}^T & \varphi_{14}^1 + 2X_{1\alpha} & \varphi_{15}^1 + X_{2\alpha} \\ * & \varphi_{22}^{11} & \tau_M Z_{12}^T & \varphi_{24}^{11} & \varphi_{25}^{11} \\ * & * & -\tau_M U_2 e^{-4\alpha\tau_M} & 0 & \tau_M Z_{13} \\ * & * & * & \varphi_{44}^1 - 4\alpha\tau_M \bar{X}_1 & 0 \\ * & * & * & * & \varphi_{55}^1 - 2\alpha\tau_M \bar{X}_3 \end{bmatrix} < 0,$$

$$\Sigma_{12} = \begin{bmatrix} \varphi_{11}^1 + X_\alpha^{12} & \varphi_{12}^{12} & \tau_M Y_{11}^T & 2\tau_M Z_{11}^T & \varphi_{14}^1 + X_{1\alpha} & \varphi_{15}^1 \\ * & \varphi_{22}^{12} & \tau_M Y_{12}^T & 2\tau_M Z_{12}^T & \varphi_{24}^{12} & \varphi_{25}^{12} \\ * & * & -\tau_M U_1 e^{-2\alpha\tau_M} & 0 & \tau_M Y_{13} & 0 \\ * & * & * & -2\tau_M U_2 e^{-4\alpha\tau_M} & 0 & 2\tau_M Z_{13} \\ * & * & * & * & \varphi_{44}^1 - 2\alpha\tau_M \bar{X}_1 & 0 \\ * & * & * & * & * & \varphi_{55}^1 \end{bmatrix} < 0,$$

$$\Sigma_{21} = \begin{bmatrix} \varphi_{11}^2 + X_\alpha^{21} & \varphi_{12}^{21} & \tau_M Y_{21}^T & \varphi_{14}^2 + X_{1\alpha} & \varphi_{15}^2 + 2X_{2\alpha} \\ * & \varphi_{22}^{21} & \tau_M Y_{22}^T & \varphi_{24}^{21} & \varphi_{25}^{21} \\ * & * & -\tau_M U_1 e^{-4\alpha\tau_M} & \tau_M Y_{23} & 0 \\ * & * & * & \varphi_{44}^2 - 2\alpha\tau_M \bar{X}_1 & 0 \\ * & * & * & * & \varphi_{55}^2 - 4\alpha\tau_M \bar{X}_3 \end{bmatrix} < 0,$$

$$\Sigma_{22} = \begin{bmatrix} \varphi_{11}^2 + X_\alpha^{22} & \varphi_{12}^{22} & 2\tau_M Y_{21}^T & \tau_M Z_{21}^T & \varphi_{14}^2 & \varphi_{15}^2 + X_{2\alpha} \\ * & \varphi_{22}^{22} & 2\tau_M Y_{22}^T & \tau_M Z_{22}^T & \varphi_{24}^{22} & \varphi_{25}^{22} \\ * & * & -2\tau_M U_1 e^{-4\alpha\tau_M} & 0 & 2\tau_M Y_{23} & 0 \\ * & * & * & -\tau_M U_2 e^{-2\alpha\tau_M} & 0 & \tau_M Z_{23} \\ * & * & * & * & \varphi_{44}^2 & 0 \\ * & * & * & * & * & \varphi_{55}^2 - 2\alpha\tau_M \bar{X}_3 \end{bmatrix} < 0,$$

where

$$\begin{aligned}
\varphi_{11}^{j} &= A^T P_{2j} + P_{2j}^T A - Y_{j1} - Y_{j1}^T - Z_{j1} - Z_{j1}^T + 2\alpha P_1 - \tfrac{X+X^T}{2} - \tfrac{X_2+X_2^T}{2},\\
\varphi_{12}^{11} &= P_1 - P_{21}^T + A^T P_{31} - Y_{12} - Z_{12} + \tau_M(X+X^T) + \tau_M \tfrac{X_2+X_2^T}{2},\\
\varphi_{12}^{12} &= P_1 - P_{21}^T + A^T P_{31} - Y_{12} - Z_{12} + \tau_M \tfrac{X+X^T}{2},\\
\varphi_{12}^{21} &= P_1 - P_{22}^T + A^T P_{32} - Y_{22} - Z_{22} + \tau_M(X_2+X_2^T) + \tau_M \tfrac{X+X^T}{2},\\
\varphi_{12}^{22} &= P_1 - P_{22}^T + A^T P_{32} - Y_{22} - Z_{22} + \tau_M \tfrac{X_2+X_2^T}{2},\\
\varphi_{14}^{j} &= Y_{j1}^T + P_{2j}^T A_1 - Y_{j3} + X - X_1,\\
\varphi_{15}^{j} &= Z_{j1}^T + P_{2j}^T A_2 - Z_{j3} + X_2 - X_3,\\
\varphi_{22}^{11} &= -P_{31} - P_{31}^T + 2\tau_M U_1 + \tau_M U_2,\\
\varphi_{22}^{12} &= -P_{31} - P_{31}^T + \tau_M U_1,\\
\varphi_{22}^{21} &= -P_{32} - P_{32}^T + \tau_M U_1 + 2\tau_M U_2,\\
\varphi_{22}^{22} &= -P_{32} - P_{32}^T + \tau_M U_2,\\
\varphi_{24}^{11} &= Y_{12}^T + P_{31}^T A_1 + 2\tau_M(-X+X_1),\\
\varphi_{24}^{12} &= Y_{12}^T + P_{31}^T A_1 + \tau_M(-X+X_1),\\
\varphi_{24}^{21} &= Y_{22}^T + P_{32}^T A_1 + \tau_M(-X+X_1),\\
\varphi_{24}^{22} &= Y_{22}^T + P_{32}^T A_1,\\
\varphi_{25}^{11} &= Z_{12}^T + P_{31}^T A_2 + \tau_M(-X_2+X_3),\\
\varphi_{25}^{12} &= Z_{12}^T + P_{31}^T A_2,\\
\varphi_{25}^{21} &= Z_{22}^T + P_{32}^T A_2 + 2\tau_M(-X_2+X_3),\\
\varphi_{25}^{22} &= Z_{22}^T + P_{32}^T A_2 + \tau_M(-X_2+X_3),\\
\varphi_{44}^{j} &= Y_{j3} + Y_{j3}^T + \bar{X}_1,\\
\varphi_{55}^{j} &= Z_{j3} + Z_{j3}^T + \bar{X}_3,\\
X_{\alpha}^{11} &= 2\alpha\tau_M(X+X^T) + \alpha\tau_M(X_2+X_2^T),\\
X_{\alpha}^{12} &= \alpha\tau_M(X+X^T),\\
X_{\alpha}^{21} &= \alpha\tau_M(X+X^T) + 2\alpha\tau_M(X_2+X_2^T),\\
X_{\alpha}^{22} &= \alpha\tau_M(X_2+X_2^T),\\
X_{1\alpha} &= 2\alpha\tau_M(-X+X_1),\\
X_{2\alpha} &= 2\alpha\tau_M(-X_2+X_3),\ j=1,2.
\end{aligned}$$

(ii) If the above LMIs are feasible for $X = X_i = 0$ $(i = 1,2,3)$, then the system (6) with $h = 0$ is exponentially stable with the decay rate α under the variable sampling $t_{k+1} - t_k \le \tau_M$.

(iii) If the above LMIs are feasible for $X = X^T > 0$, $X_2 = X_2^T > 0$ and $X_1 = X_3 = 0$, then the system (6) with $h = 0$ is exponentially stable with the decay rate α under the constant sampling $t_{k+1} - t_k = \bar{\tau}_M \in (0, \tau_M]$.

Remark 3. LMIs of Theorems 1, 2 and Corollary 1 are affine in the system matrices. Therefore, in the case of system matrices from the uncertain time-varying polytope

$$\Xi = \textstyle\sum_{j=1}^{N} \mu_j(t)\Xi_j,\ \ 0 \le \mu_j(t) \le 1,\ \ \sum_{j=1}^{N} \mu_j(t) = 1,\ \ \Xi_j = \left[A^{(j)}\ B^{(j)}\right],$$

one have to solve these LMIs simultaneously for all the N vertices Ξ_j, applying the same decision matrices.

4 Example

Consider the following linearized model of the inverted pendulum on a cart:

$$\begin{bmatrix} \dot{x} \\ \ddot{x} \\ \dot{\theta} \\ \ddot{\theta} \end{bmatrix} = \begin{bmatrix} 0 & 1 & 0 & 0 \\ 0 & 0 & \frac{-mg}{M} & 0 \\ 0 & 0 & 0 & 1 \\ 0 & 0 & \frac{(M+m)g}{Ml} & 0 \end{bmatrix} \begin{bmatrix} x \\ \dot{x} \\ \theta \\ \dot{\theta} \end{bmatrix} + \begin{bmatrix} 0 \\ \frac{a}{M} \\ 0 \\ \frac{-a}{Ml} \end{bmatrix} u \tag{26}$$

with $M = 3.9249\text{kg}$, $m = 0.2047\text{kg}$, $l = 0.2302\text{m}$, $g = 9.81\text{N/kg}$, $a = 25.3\text{N/V}$. In the model, x and θ represent cart position coordinate and pendulum angle from vertical, respectively.

Ideally, the pendulum can be stabilized by a state feedback $u(t) = K\left[x\ \dot{x}\ \theta\ \dot{\theta}\right]^T$ with the gain

$$K = [5.825\ 5.883\ 24.941\ 5.140],$$

which leads to the closed-loop system eigenvalues $\{-100, -2+2i,\ -2-2i,\ -2\}$. In practice the variables θ, $\dot{\theta}$ and x, $\dot{x}$ are not accessible simultaneously. We consider

$$C^1 = \begin{bmatrix} 1\,0\,0\,0 \\ 0\,1\,0\,0 \end{bmatrix},\ C^2 = \begin{bmatrix} 0\,0\,1\,0 \\ 0\,0\,0\,1 \end{bmatrix}.$$

The applied control is obtained from the following blocks of K

$$K_1 = \left[5.825\ 5.883\right],\ K_2 = \left[24.941\ 5.140\right].$$

For the values of h given in Table 1, applying Theorem 1 with $\alpha = 0$ and Theorem 2, we find the maximum values of τ_M that preserve the asymptotic stability (see Table 1).

Table 1 Example: Max. value of τ_M for constant delay h

$\tau_M \backslash h$	1.0×10^{-3}	2.0×10^{-3}	3.0×10^{-3}	4.0×10^{-3}
Theorem 1	3.7×10^{-3}	4.5×10^{-3}	5.2×10^{-3}	6.0×10^{-3}
Theorem 2	4.9×10^{-3}	5.5×10^{-3}	6.1×10^{-3}	6.8×10^{-3}

For $h = 0$, by applying Corollary 1 with $\alpha = 0$, the system remains stable for all variable samplings up to 5.3×10^{-3} and for all constant samplings up to 6.4×10^{-3}. Using the classical discretization-based model for the case of constant sampling, it can be shown that the system is stable up to 6.8×10^{-3}, which is close to our result. Moreover, by Corollary 1 with $\alpha = 0$ and $X = X^T > 0$, $X_2 = X_2^T > 0$, $X_1 = X_3 = 0$, the system remains stable for all constant samplings up to 6.2×10^{-3}. Application of Theorems 1 and 2 for $h \to 0$ leads to essentially smaller variable samplings up to 3.0×10^{-3} and 4.3×10^{-3}, respectively.

Choosing further $\tau_M = 5.0 \times 10^{-3}$, $h = 3.0 \times 10^{-3}$ and applying Theorem 1, we find that the system is exponentially stable with the decay rate $\alpha = 1.50$. Considering next $h = 0$, $\tau_M = 5.0 \times 10^{-3}$ and applying Corollary 1, the system is exponentially stable with a greater decay rate $\alpha = 1.86$ for the constant sampling $t_{k+1} - t_k = 5.0 \times 10^{-3}$. For the variable sampling, the decay rate is reduced to 1.39. By Corollary 1 with $X = X^T > 0, X_2 = X_2^T > 0$ and $X_1 = X_3 = 0$, the system is exponentially stable with decay rate $\alpha = 1.68$ for all constant samplings up to 5.0×10^{-3}.

5 Conclusions

In this chapter, a time-delay approach has been introduced for the exponential stability of NCSs with Round-Robin scheduling, constant input/output delays and variable sampling intervals. By constructing appropriate time-dependent Lyapunov functionals, that take into account the switched system model and the sawtooth delays induced by sampled-data control, we derive the exponential stability conditions in terms of LMIs. In the absence of input and output delays, our results for the constant sampling intervals are less conservative than those for the variable ones.

Acknowledgements. This work was supported by Israel Science Foundation (grant No 754/10).

References

1. Antsaklis, P., Baillieul, J.: Special issue on technology of networked control systems. Proceedings of the IEEE 95(1), 5–8 (2007)
2. Cloosterman, M.B.G., Hetel, L., van de Wouw, N., Heemels, W.P.M.H., Daafouz, J., Nijmeijer, H.: Controller synthesis for networked control systems. Automatica 46(10), 1584–1594 (2010)
3. Donkers, M.C.F., Hetel, L., Heemels, W.P.M.H., van de Wouw, N., Steinbuch, M.: Stability Analysis of Networked Control Systems Using a Switched Linear Systems Approach. In: Majumdar, R., Tabuada, P. (eds.) HSCC 2009. LNCS, vol. 5469, pp. 150–164. Springer, Heidelberg (2009)
4. Donkers, M.C.F., Heemels, W.P.M.H., van de Wouw, N., Hetel, L.: Stability analysis of networked control systems using a switched linear systems approach. IEEE Transactions on Automatic Control 56(9), 2101–2115 (2011)
5. Fridman, E.: A new Lyapunov technique for robust control of systems with uncertain non-small delays. IMA Journal of Mathematical Control and Information 23, 165–179 (2006)
6. Fridman, E.: A refined input delay approach to sampled-data control. Automatica 46, 421–427 (2010)
7. Fridman, E., Seuret, A., Richard, J.P.: Robust sampled-data stabilization of linear systems: an input delay approach. Automatica 40, 1441–1446 (2004)
8. Fujioka, H.: A discrete-time approach to stability analysis of systems with aperiodic sample-and-hold devices. IEEE Transactions on Automatic Control 54(10), 2440–2445 (2009)

9. Gao, H., Chen, T., Lam, J.: A new system approach to network-based control. Automatica 44(1), 39–52 (2008)
10. Hardy, G., Littlewood, J., Polya, G.: Inequalities. Cambridge University Press, Cambridge (1934)
11. He, Y., Wang, Q.G., Lin, C., Wu, M.: Delay-range-dependent stability for systems with time-varying delay. Automatica 43, 371–376 (2007)
12. Heemels, W.P.M.H., Teel, A.R., van de Wouw, N., Nesic, D.: Networked control systems with communication constraints: tradeoffs between transmission intervals, delays and performance. IEEE Transactions on Automatic Control 55(8), 1781–1796 (2010)
13. Hetel, L., Daafouz, J., Iung, C.: Analysis and control of LTI and switched systems in digital loops via an event-based modeling. International Journal of Control 81(7), 1125–1138 (2008)
14. Jiang, W., Fridman, E., Kruszewski, A., Richard, J.P.: Switching controller for stabilization of linear systems with switched time-varying delays. In: Proceedings of the 48th IEEE Conference on Decision and Control, Shanghai, China (2009)
15. Kolmanovskii, V.B., Richard, J.P.: Stability of some linear systems with delays. IEEE Transactions on Automatic Control 44(5), 984–989 (1999)
16. Liberzon, D.: Switching in systems and control. Systems and Control: Foundation and Applications. Birkhauser, Boston (2003)
17. Liberzon, D.: Quantization, time delays, and nonlinear stabilization. IEEE Transactions on Automatic Control 51(7), 1190–1195 (2006)
18. Liu, K., Fridman, E.: Wirtinger's inequality and Lyapunov-based sampled-data stabilization. Automatica 48(1), 102–108 (2012)
19. Liu, K., Fridman, E., Hetel, L., Richard, J.P.: Sampled-data stabilization via Round-Robin scheduling: a direct Lyapunov-Krasovskii approach. In: Proceedings of the 18th World Congress of the International Federation of Automatic Control, Milano, Italy (2011)
20. Naghshtabriz, P., Hespanha, J., Teel, A.R.: Exponential stability of impulsive systems with application to uncertain sampled-data systems. Systems & Control Letters 57, 378–385 (2008)
21. Nesic, D., Teel, A.R.: Input-output stability properties of networked control systems. IEEE Transactions on Automatic Control 49(10), 1650–1667 (2004)
22. Park, P.G.: A delay-dependent stability criterion for systems with uncertain time-invariant delays. IEEE Transactions on Automatic Control 44(4), 876–877 (1999)
23. Richard, J.P., Divoux, T.: Systèmes Commandés en réseau. Hermes-Lavoisier, IC2, Systemes Automatises (2007)
24. Seuret, A.: A novel stability analysis of linear systems under asynchronous samplings. Automatica 48(1), 177–182 (2012)
25. Sun, X.M., Zhao, J., Hill, D.: Stability and L_2-gain analysis for switched delay systems: a delay-dependent method. Automatica 42, 1769–1774 (2006)
26. Zhang, W., Branicky, M.S., Phillips, S.M.: Stability of networked control systems. IEEE Control Systems Magazine 21(1), 84–99 (2001)

Structure of Discrete Systems with Variable Nonlocal Behavior

Erik I. Verriest

Abstract. We clarify, for discrete time systems, two notions of a system with time variant delay, and explore their structural properties. The ultimate purpose is to shed some light on the potential ill-posedness associated with problems with rapidly increasing delay (delay increment greater than one system update step). The main approach is to consider the time delay system using an extension of the state space. For the case of periodically varying delays, this problem becomes a special case of the lifting for periodic systems. Structural problems, such as stability and reachability, can then be investigated using time-invariant theory. For instance, the property that the quotient space of reachable periodic delay systems modulo state space similarity is a smooth manifold, is inherited. A special reflecto-difference equation is analyzed in more detail as an example of a system with unbounded nonlocal behavior.

1 Introduction: Behavioral Approach

In line with recent investigations on continuous time systems with time variable delay [11], especially in view of inconsistencies arising where the delay rate exceeds one, we report similar problems in the discrete time case. Parts of this chapter appeared in preliminary form in preprint and on-line [12]. The present version delves in more detail. We present our ideas in the behavioral approach towards discrete time systems as expounded in [7]. A brief synopsis, together with some new definitions follows. Let $\mathbb{T}$ denote the time set. For discrete systems $\mathbb{T} = \mathbb{Z}$. Let $\mathbb{W}$ be the set where the signals assume their values, say for some $n \geq 1$, $\mathbb{W} = \mathbb{R}^n$. As usual, $\mathbb{W}^{\mathbb{T}}$ denotes the set of all maps from $\mathbb{T}$ to $\mathbb{W}$. A dynamical system Σ is defined as a triple $\Sigma = (\mathbb{T}, \mathbb{W}, \mathscr{B})$, where $\mathscr{B}$ is called the behavior, which is an appropriately restricted subset of $\mathbb{W}^{\mathbb{T}}$. We define the evaluation functional σ_k operating on se-

Erik I. Verriest
Georgia Institute of Technology, Atlanta, GA 30332-0250, USA
e-mail: erik.verriest@ece.gatech.edu

T. Vyhlídal, J.-F. Lafay, and R. Sipahi (eds.), *Delay Systems*,
Advances in Delays and Dynamics 1,
DOI: 10.1007/978-3-319-01695-5_14,

quences by $\sigma_k(w) = w_k$. The (left-) shift operator $\mathbf{T}$ is defined by $\sigma_k(\mathbf{T}w) = \sigma_{k+1}w$. Hence $\mathbf{T}^n$ corresponds with an n-step shift to the left.

The dynamical system $\Sigma = (\mathbb{T}, \mathbb{W}, \mathscr{B})$ is said to be linear if $\mathbb{W}$ is a vector space over $\mathbb{R}$ or $\mathbb{C}$, and the behavior $\mathscr{B}$ is a linear subspace of $\mathbb{W}^{\mathbb{T}}$. The dynamical system $\Sigma = (\mathbb{T}, \mathbb{W}, \mathscr{B})$ is said to be shift invariant if $w \in \mathscr{B}$ implies $\mathbf{T}^\tau w \in \mathscr{B}$ for all $\tau \in \mathbb{T}$.

A sequence in $\mathbb{W}^{\mathbb{T}}$ is said to be ℓ_2-locally bounded if $\sum_{i=m}^{p} \|w_i\|^2 < \infty$ for all finite $m < p$. The space of all locally bounded sequences in the ℓ_2 norm is denoted ℓ_2^{loc}. This space is a separable Hilbert space. Any bounded operator defined everywhere on a separable Hilbert space $\mathscr{H}_1$ and mapping to a separable Hilbert space $\mathscr{H}_2$ admits a matrix representation which uniquely defines this operator. Hence, without loss of generality, any discrete time behavior can be represented by a infinite matrix. A behavioral description restricts the trajectories that are allowed by the system. Thus, a global (timeless) viewpoint of the behavior can be characterized by the nullspace of some infinite matrix. For causal systems, this matrix is upper triangular (using the convention that scrolling the vector down the signal vector means going back in time). For a time invariant system, the matrix is (block)-Toeplitz. Such a global matrix approach is taken in [2] and other work referenced therein.

A behavior is called *autonomous* if for all $w^{(1)}, w^{(2)} \in \mathscr{B}$ the condition $w_k^{(1)} = w_k^{(2)}$ for $k \leq 0$ implies $w_k^{(1)} = w_k^{(2)}$ for all k. For an autonomous system, the future is entirely determined by its past.

The notion of *controllability* is an important concept in the behavioral theory, as in all of system theory. Let $\mathscr{B}$ be the behavior of a linear time invariant system. This system is called controllable if for any two trajectories $w^{(1)}$ and $w^{(2)}$ in $\mathscr{B}$, there exists a $0 \leq \tau \in \mathbb{Z}$ and a trajectory $w \in \mathscr{B}$ such that

$$\sigma_k(w) = \begin{cases} \sigma_k(w^{(1)}) & k \leq 0 \\ \sigma_k(\mathbf{T}^{-\tau} w^{(2)}) & k \geq \tau \end{cases},$$

i.e., one can switch from one trajectory to the other, with perhaps a delay of τ steps. Note that an autonomous system cannot get of a trajectory once you are on it. Hence an autonomous system is not controllable.

We define some new operators, which will be useful in the sequel: Let $\Pi : \mathbb{R}^{\mathbb{Z}} \to \mathbb{R}^{\mathbb{Z}}$ denote the future-time projection operator characterized by

$$\sigma_k \Pi w = \begin{cases} \sigma_k w \text{ if } k \geq 0 \\ 0 \quad \text{if } k < 0. \end{cases}$$

This induces a mapping to the one sided sequences $\widehat{\Pi} : \mathbb{R}^{\mathbb{Z}} \to \mathbb{R}^{\mathbb{Z}_+}$ such that $\widehat{\Pi}\{w_k\}_{k=-\infty}^{\infty} = \{w_k\}_{k=0}^{\infty}$, i.e., the past gets forgotten under application of $\widehat{\Pi}$. Introduce also $\mathbf{R}$, the *parity* or *reflection* operator, defined by: $\mathbf{R}\{w_k\}_{k=-\infty}^{\infty} = \{w_{-k}\}_{k=-\infty}^{\infty}$.

Denote by δ_ℓ the *unit pulse* sequences with $\sigma_k \delta_\ell = 1$ if $k = \ell$ and 0 else, then the following commutation relations are readily established

$$\begin{aligned} \Pi \mathbf{T} &= \mathbf{T} \Pi - \delta_{-1} \sigma_0 \\ \mathbf{R}\mathbf{T} &= \mathbf{T}^{-1} \mathbf{R}, \end{aligned}$$

and by induction on the first, for all $k > 0$

$$\Pi \mathbf{T}^k = \mathbf{T}^k \Pi - \delta_{-k}\sigma_0 - \delta_{-(k-1)}\sigma_1 - \dots - \delta_{-1}\sigma_{k-1}.$$

2 Discrete Delay System

In this paper we are interested in a simple interconnected system (Figure 1) described locally (in time) by

$$x_{k+1} = Ax_k + z_k + bu_k \tag{1}$$

$$z_k = Bx_{k-n}. \tag{2}$$

Let us first consider the case where the delay, n, is fixed. For simplicity, we consider the single input case. The vectors x_k and z_k are in $\mathbb{R}^m$, hence A and B are in $\mathbb{R}^{m\times m}$. Furthermore, we assume that at time k, the variables x_k and u_k are available (full observability).

We first make a preliminary observation. If $\operatorname{rank}(B) < m$, not all components of the m-vector x_k need to be memorized. Indeed, a simple similarity transformation (in $Gl_m(\mathbb{R})$), $x \mapsto [\xi^\top, \eta^\top]^\top$ may recast the system as

$$\xi_{k+1} = A_1\xi_k + A_2\eta_k + B_1\eta_{k-n} + b_1u_k \tag{3}$$

$$\eta_{k+1} = A_3\xi_k + A_4\eta_k + B_2\eta_{k-n} + b_2u_k \tag{4}$$

where now $\dim \xi_k = m - \operatorname{rank} B$, $\dim \eta_k = \operatorname{rank} B$ and $[B_1^\top, B_2^\top] \in \mathbb{R}^{\operatorname{rank}(B)\times m}$. It is obvious from these equations that in order to evolve the system forward in time, from k onward, we will need accessibility to the variables ξ_k, and $\eta_{k-n}, \eta_{k-n+1}, \dots, \eta_k$ This means: $(m - \operatorname{rank} B) + (n+1)\operatorname{rank} B = m + n\operatorname{rank} B$ real variables suffice to predict the future behavior, if also the future inputs, $u_k, u_{k+1}, \dots,$ are known. We call this sufficient information a "state" of the system at time k, as this is indeed a sufficient statistic replacing full knowledge of the past behavior.

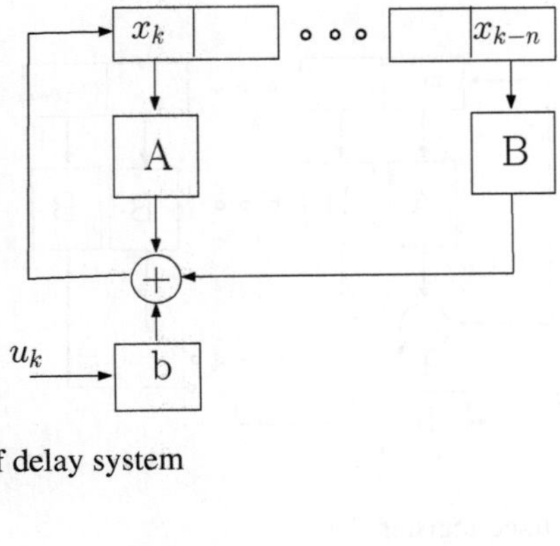

Fig. 1 Representation of delay system

Figure 1 gives a realization of such a delay system for the case when B has full rank. The shift register on the top is the memory device, and operates by right shifting the content from each cell (here an m-vector) to the one on its right at each clock pulse. The data in the rightmost cell is subsequently lost, and the new item generated by the adder is loaded in the leftmost cell. We observe that the true state space of such a system is in general a vector space of dimension larger than m.

3 Time-Variant Delay

Let now the delay in this model at step k depend on k, then one might think that $m + n(k)\mathrm{rank}B$ variables would suffice to characterize future behavior, and worse, associate a time-varying state space to the system. However, this is nonsense. A state space must itself have a fixed structure, otherwise one cannot talk about evolution or trajectories (or rather sequences in this discrete case) in it. Even more to the point, without trajectories it cannot be clear what would be meant by stability or convergence. In what follows, two causal interpretations of discrete systems with time-variant delays are presented For notational and display purposes, we assume that the minimal delay is at least 1. The other case can easily be handled as well.

3.1 Causal Models

In the first interpretation, we envision that each memory cell connects to a copy of the B-matrix, but a switching device chooses which one to connect to the adder. Only one connection is made at each step. This is shown in Figure 2. Here $n_{\max}$ is the maximal anticipated delay, but this could in principle be unbounded. In this form, the buffer has a fixed size. Hence, the state space dimension is constant, but this may not be a *minimal* state space. Effectively, the system is a special case of a *distributed* time delay system, but with time variant delay distribution. This is the discrete time equivalent of the *lossless causalization* discussed for continuous time delay systems in [8, 10]. The fact that the buffer length depends on future memory

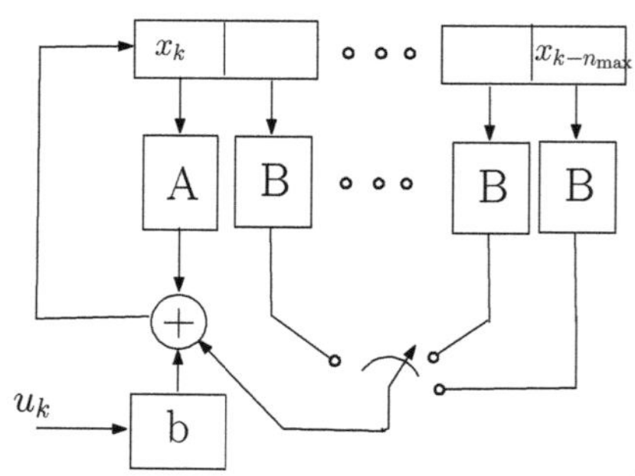

Fig. 2 Interpretation with fixed register

sizes makes this system noncausal in a certain sense. In many cases this required (buffer size) information may be known a priori, but there are situations, such as with state dependent buffer size (see [4]), where this cannot be anticipated. This problem can be resolved by defining the *information structure* of the model. Guided by the notion of a *filtration* in the theory of stochastic processes, this was presented in [10].

The second interpretation considers at each step k a system as in figure 1 but with $n = n(k)$, thus with k-dependent buffer size (see Figure 3). At each step, the input for the B-matrix is taken from the rightmost register in the buffer. In particular, this implies that if $n_{k+1} \geq n_k + 1$, no information is available to store in the added cells on the right, except for the leftmost of these cells, where thus $x_{k-n} = x_{(k+1)-(n+1)}$ will be stored. On the other hand, if $n_{k+1} < n_k$, the buffer gets shortened, and the information that was stored in the deleted cells is now *irretrievably lost*. No anticipation of the future delays is needed here. For this reason, this system causalization was called the *forgetful causalization* in [8, 10]. Of course both causalization models will differ in their outcome if the delay is allowed to increase by *more than one step* at a time, and it makes no sense to determine which is right or wrong. We just point out that the physics behind both models is different, and therefore which one is appropriate depends on the phenomenon to be studied. We caution that a blind use of a mathematical model may lead to inconsistencies (see [11]).

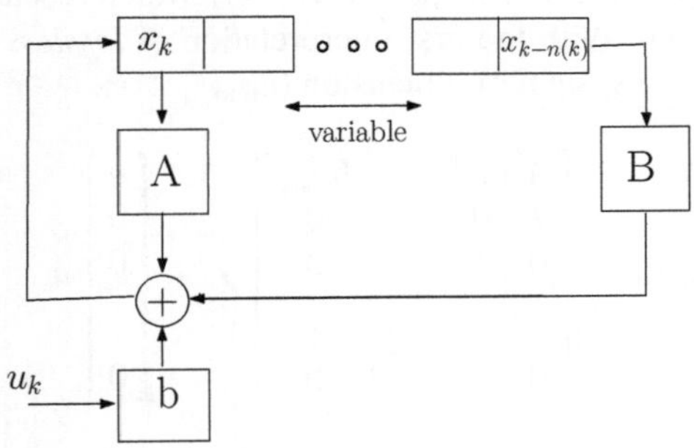

Fig. 3 Interpretation with variable length register

3.2 *State Space and Trajectories*

The main difference in the causal interpretations given above is that for the first, the memory size stays constant, but not everything stored in memory may be used at any time. This version of the time-varying delay systems may be lifted to a large dimensional time-invariant system, but perhaps at the expense of resulting in a non-minimal model. With the second interpretation, the memory size changes with k. As mentioned, this creates a problem with the conventional notion of a state space. This problem can be resolved by casting the model in the class of multi-mode, multi-dimensional (M^3D) systems. In [9] we studied the continuous version of this problem, and discussed their structure and aspects of optimal control further in [13].

Characterize each physically different buffer size as a mode of the system. The state space of such a M^3D system is then represented by a *sheaf* over a discrete space (the mode set). In [9, 13], the state space was defined as a fiber bundle, which is in error as the fibers cannot be isomorphic if modes have different dimensions. If B has full rank (m), the stalk over j is a vector space of dimension $(n_j+1)m$ where n_j is the delay in phase j. In general, the dimension of the j-th stalk is $d_j = n_j \operatorname{rank} B + m$. The stalks correspond to the state space restricted to the different modes. In addition, at the switches, the state transition from one buffer to the next is simply given by projection (to the lower dimensional stalk) or embedding (in the higher dimensional stalk), where empty cells are loaded with the zero state (the equilibrium). Hence the *inter-mode* is specified by sections in the sheaf. In contrast, the *intra-mode* behavior is defined by the affine structure on the stalks. We emphasize that a sheaf structure is necessary as we want a *stationary* structure for the state space. If that were not the case we could not reasonably talk about trajectories in the state space.

4 Periodic Time Delay System

For a periodically varying delay, the previous models simplify considerably. Let the system have delay sequence $n(k); k = 1,2,\ldots,N$ which repeats periodically. Let $n_{\max}$ be their maximum. With the first interpretation of *lossless causalization*, the system can be lifted to a system of dimension $(n_{\max}+1)m$.

$$\chi_{k+1} = \begin{bmatrix} A & \beta_1^{(k)} & \cdots & & \beta_{n_{\max}}^{(k)} \\ I & 0 & & & 0 \\ 0 & I & & & 0 \\ \vdots & & \ddots & & \vdots \\ 0 & \cdots & \cdots & I & 0 \end{bmatrix} \chi_k + \begin{bmatrix} b \\ 0 \\ 0 \\ \vdots \\ 0 \end{bmatrix} u_k. \tag{5}$$

where only one of the $\beta_i^{(k)}$ in the top block row equals B at any time, the other entries being zero. Denote this compactly by

$$\chi_{k+1} = \mathscr{A}_{[k]}\chi_k + b^d u_k, \tag{6}$$

where the subscript $[k]$ denotes $[k] = ((k-1)\bmod N)+1$. This determines an N-period system of fixed state space dimension. Its stability is completely determined by the monodromy matrix $F = \mathscr{A}_{[N]}\mathscr{A}_{[N-1]}\cdots\mathscr{A}_{[1]}$, and its reachability properties by the monodromy systems (F_i, G_i) with $F_i = \mathscr{A}_{[i-1]}\mathscr{A}_{[i-2]}\cdots\mathscr{A}_{[1]}\mathscr{A}_{[N]}\cdots\mathscr{A}_{[i]}$, and $G_i = [b^d, \mathscr{A}_{[i-1]}b^d, \mathscr{A}_{[i-1]}\mathscr{A}_{[i-2]}b^d, \ldots, \mathscr{A}_{[i-1]}\mathscr{A}_{[i-2]}\cdots\mathscr{A}_{[i-N+1]}b^d]$. The monodromy system (F_i, G_i) is the system that describes how all inputs over one past period contribute to χ_i. If the i-th monodromy system is reachable, then the full memory state χ_i can be made arbitrary, and thus as a special case, can be zeroed. The time-invariant nature of the monodromy systems implies that if this is possible at all, it will require at most $Nm(n_{\max}+1)$ steps. However it may still be possible to control to zero, even if the monodromy system is not reachable: Reachability implies controllability (accessibility of the zero state), but is not implied by it in the discrete time case [5, p. 100].

With *forgetfull causalization*, the individual mode systems have dimension $(n_k + 1)m$. Let χ_k denote the entire contents of the shift register, by concatenating the state in each cell. Thus $\chi_k^\top = [x_k^\top, x_{k-1}^\top, \dots, x_{k-n_k}^\top]$. Represent the system in phase k by

$$\chi_{k+1}^- = \begin{bmatrix} A & 0 & \cdots & 0 & B \\ I & 0 & & & 0 \\ 0 & I & & & 0 \\ \vdots & & \ddots & & \vdots \\ 0 & \cdots & \cdots & I & 0 \end{bmatrix} \chi_k^+ + \begin{bmatrix} b \\ 0 \\ 0 \\ \vdots \\ 0 \end{bmatrix} u_k \tag{7}$$

which we shall denote as

$$\chi_{k+1}^- = A_{[k]}\chi_k^+ + b_{[k]}u_k, \tag{8}$$

with $[k]$ as before. In addition, if the delay switches from $n(k)$ at step k to $n(k+1)$ at step $k+1$, this transition involves a mode change, which induces a change in the dimension of the delay system representation. This requires additional transitions

$$\chi_k^+ = [\,I\ 0\,]\,\chi_k^-, \quad \text{if } n_k < n_{k-1} \tag{9}$$

$$\chi_k^+ = \begin{bmatrix} I \\ 0 \end{bmatrix} \chi_k^-, \quad \text{if } n_k > n_{k-1}. \tag{10}$$

These matrices are respectively a projection and an embedding. Denote this transition matrix from k to $k+1$ by $S_{k+1,k}$. Combining, we set $A_k = S_{k+1,k}A_{[k]}$ as the total transition from the $m(n_k+1)$-dimensional state at k to the $m(n_{[k+1]}+1)$-dimensional state at $k+1$, and likewise, let $S_{k+1,k}b_{[k]} = b_k$ be the effect of the input u_k to χ_{k+1}. In general, these A_i are no longer square matrices.

Every N-periodic system is therefore completely described by a periodic sequence of matrices

$$(A_k, B_k)_N = ((A_1, B_1), (A_2, B_2), \dots, (A_N, B_N)) \tag{11}$$

Again noting that $\dim \chi_k = m(n_{[k]} + 1)$, we see that the j-th phase matrix A_j is in $\mathbb{R}^{m(n_{j+1}+1)\times m(n_j+1)}$. The system $(A_k, B_k)_N$ is now purely periodic, and the structure of such systems was described in [3, 14].

With every N-periodic system $(A_k, B_k)_N$ a time-invariant linear system is associated as follows. Let (A_e, B_e) denote the *extended system* defined as follows:

$$A_e := \begin{bmatrix} 0 & \dots & 0 & A_N \\ A_1 & \dots & 0 & 0 \\ \vdots & \ddots & \vdots & \vdots \\ 0 & \dots & A_{N-1} & 0 \end{bmatrix} \tag{12}$$

$$B_e := \begin{bmatrix} b_N & \dots & \dots & 0 \\ \vdots & b_1 & & \vdots \\ \vdots & & \ddots & \vdots \\ 0 & \dots & \dots & b_{N-1} \end{bmatrix}. \tag{13}$$

This extended system has dimensions $n_e = m \sum_{k=1}^{N} (n_k + 1)$ and $m_e = N$. If we use the time-varying state space transformations and input transformations as in [6]

$$(x_k) \longmapsto (z_k) \quad , \quad (u_k) \longmapsto (v_k)$$

then the system with periodically varying delay is equivalent to the extended time-invariant dynamics

$$z_{k+1} = A_e z_k + B_e v_k \quad , \quad k \in \mathbb{Z}.$$

We shall now relate the reachability properties of the periodic system to the extended system. But first we need to define what these concepts mean for systems with variable dimensions.

Definitions
(i) The periodic system is *reachable at phase* j if the state x_j can be made arbitrary by applying suitable inputs prior to j, regardless the initial state.
(ii) The periodic system is *reachable*, iff it is reachable at all phases $j = 1, \dots, N$.

In the literature a stronger notion of reachability is defined for systems of constant dimension (n), sometimes referred to as unform reachability. This notion is equivalent to reachability in n steps. In [1, p. 115] a characterization of uniform reachability is given via the existence of a periodic rational canonical form. In [3], we proved the extension:

Theorem 1. *The periodic system is reachable if and only if the extended system is reachable.*

Various invariant theoretic characterizations of reachability are given in [3], and it was shown that the quotient space of reachable periodic systems modulo state space equivalence is a smooth manifold. An embedding into a Grassman manifold can be given, extending the well-known Kalman embedding of reachable pairs. The analysis of observability proceeds by duality. Hence all of this will carry over for discrete periodic delay systems. A simple scalar example was shown in [12].

5 Reflecto-difference Equation

So far we focused on systems with finite delay. In this section, we consider the behavior

$$\mathscr{B} = \{w \mid \mathbf{T}w = Aw + B\mathbf{R}w\},$$

as an example of a system with unbounded delay (and advance).

It corresponds to the system with time varying delay, $n(k) = 2k$, for $k \geq 0$, $w_{k+1} = Aw_k + Bw_{k-2k}$. However, for $k > 0$ it poses an additional restriction on the behavior restricted to $[-|k|, |k|]$. How is one to interpret such a relation involving the shift and the parity operator? The continuous time equivalent is somewhat simpler and was analyzed in [11]. The idea is to resolve the behavior in an even and an odd sequence, with respect to $k = 0$. Then the doubly infinite sequence w is completely characterized by the one sided sequences $\widehat{\Pi}\mathbf{E}w$ and $\widehat{\Pi}\mathbf{Od}w$, where also $\sigma_0\mathbf{Od}w = 0$. The even and odd part of a sequence w are respectively defined as

$$\mathbf{E}w = \frac{1}{2}(w + \mathbf{R}w) \tag{14}$$

$$\mathbf{Od}w = \frac{1}{2}(w - \mathbf{R}w), \tag{15}$$

and note that, since $\mathbf{RT} = \mathbf{T}^{-1}\mathbf{R}$, it holds for all k and x that

$$\begin{aligned}
\sigma_k\mathbf{E}(\mathbf{TE}x) &= \frac{1}{2}(\sigma_k(\mathbf{TE}x) + \sigma_k(\mathbf{RTE}x)) \\
&= \frac{1}{2}(\sigma_k(\mathbf{TE}x) + \sigma_k(\mathbf{T}^{-1}\mathbf{RE}x)) \\
&= \frac{1}{2}(\sigma_k(\mathbf{TE}x) + \sigma_k(\mathbf{T}^{-1}\mathbf{E}x)) \\
&= \frac{1}{2}(\sigma_{k+1}(\mathbf{E}x) + \sigma_{k-1}(\mathbf{E}x)) \\
&= \frac{1}{2}(\sigma_{k+1} + \sigma_{k-1})\mathbf{E}x.
\end{aligned} \tag{16}$$

Consequently, the operator identity $\mathbf{E\,T\,E} = \frac{1}{2}\left(\mathbf{T} + \mathbf{T}^{-1}\right)\mathbf{E}$ follows. Likewise, the following relations are established

$$\mathbf{E\,T\,Od} = \frac{1}{2}\left(\mathbf{T} - \mathbf{T}^{-1}\right)\mathbf{Od} \tag{17}$$

$$\mathbf{Od\,T\,Od} = \frac{1}{2}\left(\mathbf{T} + \mathbf{T}^{-1}\right)\mathbf{Od} \tag{18}$$

$$\mathbf{Od\,T\,E} = \frac{1}{2}\left(\mathbf{T} - \mathbf{T}^{-1}\right)\mathbf{E} \tag{19}$$

These identities are needed since we note that $\mathbf{TE}x$ is neither odd nor even in general. Combining (16) and (17) with the system behavior, we find:

$$(\mathbf{T} + \mathbf{T}^{-1})\,\mathbf{E} + (\mathbf{T} - \mathbf{T}^{-1})\,\mathbf{Od} = 2(A + B)\,\mathbf{E}. \tag{20}$$

Likewise, equations (18) and (19) give with the system behavior

$$(\mathbf{T}-\mathbf{T}^{-1})\,\mathbf{E}+(\mathbf{T}+\mathbf{T}^{-1})\,\mathbf{Od}=2(A-B)\,\mathbf{Od}. \tag{21}$$

Evaluation of the latter at $k=0$ is not informative since $\sigma_0\mathbf{Od}x=0$ and $\sigma_0\mathbf{TE}x=\sigma_0\mathbf{T}^{-1}\mathbf{E}x$. On the other hand the evaluation of (20) at 0 yields the relation

$$\sigma_1\,\mathbf{E}x+\sigma_1\mathbf{Od}x=(A+B)\,\sigma_0\mathbf{E}x. \tag{22}$$

Unlike the situation for one sided sequences, the operator $\mathbf{T}$ is *invertible* on $\mathbb{R}^{\mathbb{Z}}$. Equivalently, $\ker\mathbf{T}=0$, and $\mathbf{T}w=0$ implies $w=0$. This implies further that with a polynomial representation in $\mathbb{R}[\xi,\xi^{-1}]$, besides the usual elementary row and column operations of elimination theory, another operation: *multiplication of any row or column by a monomial* ξ^k, $k\in\mathbb{Z}$ is allowed as well. Hence (20) and (21) may be combined as

$$\begin{bmatrix}\mathbf{T}^2-2(A+B)\mathbf{T}+I & \mathbf{T}^2-I\\ \mathbf{T}^2-I & \mathbf{T}^2-2(A-B)\mathbf{T}+I\end{bmatrix}\begin{bmatrix}\mathbf{E}\\ \mathbf{Od}\end{bmatrix}=0. \tag{23}$$

This can be unimodularly reduced to

$$\begin{bmatrix}(A+B)\mathbf{T}-I & I-(A-B)\mathbf{T}\\ \mathbf{T}^2-I & \mathbf{T}^2-2(A-B)\mathbf{T}+I\end{bmatrix}\begin{bmatrix}\mathbf{E}\\ \mathbf{Od}\end{bmatrix}=0. \tag{24}$$

Specific forms for A and B allow then further unimodular reduction. For simplicity, we consider in what follows the case of a scalar reflecto-difference equation, and thus set $A=a$, $B=b$. We consider three cases:
Case 1: $a=b=0$: This is the trivial case, having only the null solution $w=0$ as consistent solution in $\mathbb{R}^{\mathbb{Z}}$.
Case 2: $a=-b\neq 0$: The system (24) is unimodularly equivalent to

$$\begin{bmatrix}1 & 2a\mathbf{T}-1\\ 0 & \mathbf{T}^2-\frac{1}{a}\mathbf{T}+1\end{bmatrix}\begin{bmatrix}\mathbf{E}\\ \mathbf{Od}\end{bmatrix}=0. \tag{25}$$

Note that the general solution of $(\mathbf{T}^2-\frac{1}{a}\mathbf{T}+1)x=0$ is $x_k=\alpha\lambda^k+\beta\lambda^{-k}$, where $\lambda=\frac{1}{2a}\left(1-\sqrt{1-4a^2}\right)$. Hence the space of *odd* solutions is restricted to the one-parameter set $\mathbf{Od}x_k=\alpha(\lambda^k-\lambda^{-k})$. The even part follows from

$$\mathbf{E}x=(1-2a\mathbf{T})\mathbf{Od}x,$$

which gives then the one dimensional solution set (α is free).

$$\sigma_k\mathbf{E}x=\alpha\sqrt{1-4a^2}\,(\lambda^k+\lambda^{-k}).$$

Combining, the solutions to the scalar reflecto-difference equation are in this case

$$w_k = 2\alpha a(\lambda^{k-1} - \lambda^{1-k}) = \frac{\alpha}{(2a)^{k-2}}\left[(1-\sqrt{1-4a^2})^{k-1} - (1+\sqrt{1-4a^2})^{k-1}\right].$$

Note that all solutions satisfy $w_1 = 0$.
Case 3: $a+b \neq 0$: This case is notationally quite involved, and we shall omit the details here. If $a \neq 0$, $a+b \neq 0$ and $(a+b)^2 - 1 \neq 0$, then the array(24) is equivalent to

$$\begin{bmatrix} 1 & \frac{2a(a+b)\mathbf{T}^2+2(b^3+ab^2-a^2b-a^3)\mathbf{T}+(a+b)^2-1}{1-(a+b)^2} \\ 0 & \mathbf{T}^2+\frac{b^2-a^2-1}{a}\mathbf{T}+1 \end{bmatrix}\begin{bmatrix} \mathbf{E} \\ \mathbf{Od} \end{bmatrix} = 0. \tag{26}$$

The odd part is determined by the roots of $\lambda^2 + \frac{b^2-a^2-1}{a}\lambda + 1$, whose product is one, so that the odd solution always exists. The even part is determined by the first equation of (26), giving upon reconstitution a one dimensional solution set.
The excluded case $a = 0$ is rather special. For instance with $b = 1$, the array in (24) reduces to

$$\begin{bmatrix} \mathbf{T}-1 & I+\mathbf{T} \\ 0 & 0 \end{bmatrix}\begin{bmatrix} \mathbf{E} \\ \mathbf{Od} \end{bmatrix} = 0. \tag{27}$$

This implies that $\Pi\mathbf{Od}w$ may be chosen completely arbitrary, and then $\Pi\mathbf{E}w$ and hence w follow. This gives infinite dimensionality to the solution set. For instance, the choice $\sigma_k\mathbf{Od}w = k$ yields $w(k) = k^2 - k + 1$. On the other hand, for $a = 0, b = 2$, the array (24) reduces to the identity matrix, so that $\mathbf{Od}w = 0$ and $\mathbf{E}w = 0$ giving only the trivial solution.

Adding the relations (20) and (21) and separating odd and even parts gives

$$(\mathbf{E}x)_{k+1} - (A+B)(\mathbf{E}x)_k = -[(\mathbf{Od}x)_{k+1} - (A-B)(\mathbf{Od}x)_k].$$

Let $-u_k$ be the common value in the above equation, then we can write, for $k \geq 0$

$$(\mathbf{E}x)_{k+1} = (A+B)(\mathbf{E}x)_k - u_k \tag{28}$$

$$(\mathbf{Od}x)_{k+1} = (A-B)(\mathbf{Od}x)_k + u_k. \tag{29}$$

The solutions are readily obtained, noting that $\mathbf{Od}x_0 = 0$ and $\mathbf{E}x_0 = x_0$.

$$(\mathbf{E}x)_k = (A+B)^k x_0 - \sum_{i=0}^{k-1}(A+B)^i u_{k-i-1} \tag{30}$$

$$(\mathbf{Od}x)_k = \sum_{i=0}^{k-1}(A-B)^i u_{k-i-1}. \tag{31}$$

Note however that this is not a reachable system if u_k is considered as a generating input. Indeed, the PBH-test yields

$$\text{rank}\begin{bmatrix} zI-(A+B) & & -I \\ & zI-(A-B) & I \end{bmatrix} = \text{rank}\begin{bmatrix} zI-A & B & 0 \\ 0 & 0 & I \end{bmatrix},$$

so that the Odd/Even system (28-29) is only reachable iff the pair (A,B) is reachable. Even and odd parts of x_k can be uniquely determined from $\{x_k\}, k \geq 0$ also if and only if the pair (A,B) is observable, as easily shown by the dual PBH

$$\operatorname{rank}\begin{bmatrix} zI-(A+B) & \\ & zI-(A-B) \\ I & I \end{bmatrix} = \operatorname{rank}\begin{bmatrix} 0 & zI-A \\ 0 & B \\ I & 0 \end{bmatrix}.$$

Clearly, the odd and even parts are related through the sequence $\{u_k\}$. Subtracting equation (21) from (20) gives, after stepping up the index by one, and substituting the relations (28) and (29):

$$[I-(A+B)^2](\mathbf{E}x)_k + 2Au_k = [I-(A-B)^2](\mathbf{Od}x)_k. \tag{32}$$

This implies that u_0 may not be chosen freely, since for $k=0$, one gets the relation

$$[I-(A+B)^2]x_0 + 2Au_0 = 0.$$

In fact other restrictions apply. Unlike the continuous time case, where solutions can be shown to exist for arbitrary x_0 (see [11]), the discrete time situation is quite different. For instance, for $A=0$, it is easily seen that the given behavior should imply $x_{k+1} = Bx_{-k}$ for $k \geq 0$ and $x_{-k+1} = Bx_k$ for $k \geq 1$. But this implies $x_k = B^2 x_k$ for all k. Thus either all x_k are eigenvectors of B, corresponding to an eigenvalue ± 1, or $x_k \equiv 0$ if B does not have such eigenvalues.

In the other extreme, where A has full rank, it is easily shown that there is no freedom whatsoever to choose a sequence u_k, if the behavior is specified for all integers. Indeed, it follows that, if we set $x_{-k} = y_k$ for $k \geq 0$, then the system defining behavioral equation specifies that for $k \geq 1$ also $x_{(-k)+1} = Ax_{-k} + Bx_k$, which implies, for $k' = k-1$

$$x_{-k'} = Ax_{-(k'+1)} + Bx_{k'+1}.$$

where now $k' \geq 0$. Thus the given behavior is equivalent to the one restricted for $k \geq 0$ and given by

$$\begin{aligned} x_{k+1} &= Ax_k + By_k \\ y_k &= Ay_{k+1} + Bx_{k+1} \end{aligned}$$

with initial condition $[x_0^\top, y_0^\top]^\top = [x_0^\top, x_0^\top]^\top$. In turn this determines an autonomous generalized system

$$\begin{bmatrix} I & 0 \\ B & A \end{bmatrix}\begin{bmatrix} x_{k+1} \\ y_{k+1} \end{bmatrix} = \begin{bmatrix} A & B \\ 0 & I \end{bmatrix}\begin{bmatrix} x_k \\ y_k \end{bmatrix} \tag{33}$$

Full rankness of A implies invertibility of $\begin{bmatrix} A & B \\ 0 & I \end{bmatrix}$, hence $[x_k^\top, y_k^\top]$ is completely specified by the initial condition x_0.

If one relaxes the behavior, and only requires satisfaction for nonnegative k, thus

$$\mathscr{Y} = \{x \mid \widehat{\Pi}\mathbf{T}x = A\widehat{\Pi}x + B\widehat{\Pi}\mathbf{R}x\},$$

then all behaviors are represented by

$$\mathscr{Y} = \{x \mid \widehat{\Pi}\mathbf{T}x = A\widehat{\Pi}x + B\widehat{\Pi}u,\ \mathbf{R}x = u\}.$$

If $u \in \ell_2$ then $\widehat{\Pi}x \in \ell_2$ if A is a Schur-Cohn (discrete time stable) matrix. Consequently, $x \in \ell_2$ for all ℓ_2 sequences u. Thus infinitely many trajectories are consistent with this behavior if the state space for the system at time zero is chosen as $\mathbb{R}^m$, consistent with $x_1 = (A+B)x_0$, and noncausal behavior is allowed.

The growing buffer interpretation in the forgetful causalization makes the system behavior equivalent to $x_{k+1} = Ax_k$, since at no time $k > 0$, the value of x_k had been preserved. Hence all trajectories of the system are of the form $x_k = A^{k-1}(A+B)x_0$ for $k > 0$. If A is Schur-Cohn stable, the system is stable.

If the infinite past constitutes the state space, then for time zero (lossless causalization), there is no freedom in the choice of u, and the evolution of the system is uniquely defined. Again, if A is Schur-Cohn stable, any ℓ_2 initial condition will give an ℓ_2 bounded solution, and thus $x_k \to 0$, as $k \to \infty$.

6 Conclusions

Two approaches (lossless and forgetful causalization) for modeling discrete delay systems with time varying delay were given. Which one should be used depends on the physical nature of the memory or information storage and structure (e.g. foresight in future delay) in the system. For systems with periodic time variation, it was shown that with either system interpretation, the model may be embedded in large dimensional periodic system, and thus the results on the structural properties (parameterization and canonical forms) are inherited from the latter.

We have illustrated that the reflecto-difference equation, unlike its continuous time counterpart, may not possess a nontrivial solution. If however the description is limited to positive time, then nontrivial solutions exist, and we have contrasted the noncausal behavior with the lossless and forgetful causalization.

References

1. Bittanti, S., Colaneri, P.: Periodic Systems. Springer (2009)
2. Dewilde, P.M., van der Veen, A.J.: Time-varying systems and computations. Kluwer (1998)
3. Helmke, U., Verriest, E.I.: Structure and parameterization of periodic systems. Math. Control Signals Syst. 23(1-3), 67–99 (2011)
4. Michiels, W., Verriest, E.I.: A look at fast varying and state dependent delays from a systems theory point of view, TW Report No. 586. Department of Computer Science, Katholieke Universiteit Leuven, Belgium (March 2011)

5. Kailath, T.: Linear Systems. Prentice-Hall (1980)
6. Park, B.P., Verriest, E.I.: Canonical forms on discrete linear periodically time-varying systems and a control application. In: Proc. 28th Conference on Decision and Control, Tampa, FL, pp. 1220–1225 (1989)
7. Polderman, J.W., Willems, J.C.: Introducton to Mathematical Systems Theory. A Behavioral Approach. Springer (1998)
8. Verriest, E.I.: Causal behavior of switched delay systems as multi-mode multi-dimensional systems. In: Proc. 8th IFAC Workshop on Time-Delay Systems, Sinaia, Romania (2009)
9. Verriest, E.I.: Multi-mode multi-dimensional systems with application to switched systems with delay. In: Proceedings of the 48th Conference on Decision and Control, Shangai, PRC, pp. 3958–3963 (2009)
10. Verriest, E.I.: Well-posedness of problems involving systems with time varying delays. In: Proceedings of the 18th International Symposium on Mathematical Theory of Networks and Systems, Budapest, Hungary, pp. 1203–1210 (2010)
11. Verriest, E.I.: Inconsistencies in systems with time-varying delays and their resolution. IMA Journal of Mathematical Control and Information 28, 147–162 (2011)
12. Verriest, E.I.: Structure of Discrete Systems with Switched Delay. In: Preprints of the 18th IFAC World Congress, Milano, Italy, August 28-September 2 (August 28, 2011), `www.IFAC-PapersOnLine.net`
13. Verriest, E.I.: Pseudo-Continuous Multi-Dimensional Multi-Mode systems: Behavior, Structure and Optimal Control. Journal Discrete Event Dynamical Systems 22(1), 27–59 (2012)
14. Verriest, E.I.: Periodic Systems: A Sequence Algebra Approach to Realization, Paramaterization and Topology. International Journal of Control (in press, 2013)

Decentralized Robustification of Interconnected Time-Delay Systems Based on Integral Input-to-State Stability

Hiroshi Ito, Pierdomenico Pepe, and Zhong-Ping Jiang

Abstract. This article deals with interconnected systems described by retarded nonlinear equations with discontinuous right-hand side. The problem of feedback control redesign to achieve ISS (input-to-state stability) and iISS (integral input-to-state stability) with respect to additive disturbances acting on each subsystem is solved. It is shown that it is possible to design a decentralized controller accomplishing the robustification whenever a small-gain condition is satisfied.

1 Introduction

Lyapunov redesign is an important strategy in nonlinear control, which allows us to enhance system properties by additional feedback compensation exploiting the knowledge of a Lyapunov function. The ISS feedback control redesign was introduced by [14], for finite dimensional nonlinear systems. This methodology allows to attenuate the actuator disturbance in terms of ISS. It has been extended to different classes of systems with time-delays in [10], [11] and [12]. In particular, in [12], systems described by nonlinear functional differential equations with discontinuous right-hand side are considered, and the saturation problem of the input magnitude is

Hiroshi Ito
Department of Systems Design and Informatics, Kyushu Institute of Technology, Iizuka, Fukuoka 820-8502, Japan
e-mail: `hiroshi@ces.kyutech.ac.jp`

Pierdomenico Pepe
Information Engineering, Computer Science and Mathematics, University of L'Aquila, 67040 Poggio di Roio, L'Aquila, Italy
e-mail: `pierdomenico.pepe@univaq.it`

Zhong-Ping Jiang
Department of Electrical and Computer Engineering,
Polytechnic Institute of New York University, Brooklyn, NY 11201, USA
e-mail: `zjiang@poly.edu`

T. Vyhlídal, J.-F. Lafay, and R. Sipahi (eds.), *Delay Systems*,
Advances in Delays and Dynamics 1,
DOI: 10.1007/978-3-319-01695-5_15,

addressed, in both an iISS and an ISS fashion. The problem of iISS (ISS) feedback control redesign is based on the knowledge of a Lyapunov-Krasovskii functional which needs to be constructed a priori for the unforced (disturbance equal to zero) system. In [5], the construction of Lyapunov-Krasovskii functionals is addressed for ISS and iISS of interconnected systems under a small-gain condition. The small-gain condition allows to split the problem of finding an overall Lyapunov-Krasovskii functional for the whole system, into the problems of finding Lyapunov-Krasovskii functionals for each subsystem. This renders the original problem much easier. The reader can refer to [7] for an application of a similar, but different small-gain characterization to stabilization of a chemostat.

This article shows, for a class of nonlinear retarded interconnected systems, that it is possible, under a small-gain condition, to achieve the iISS (ISS) feedback control redesign by means of decentralized controllers. That is, the redesign allows us to attenuate the effect of disturbances acting on each subsystem by means of a feedback from the state of each subsystem itself. For this aim, we exploit the Lyapunov-Krasovskii functional which is proved to exist when the global asymptotic stability of the disturbance-free overall closed-loop is secured via a small-gain condition in [5]. We cover multiple discrete as well as distributed time-delays, and the maps describing the dynamics are allowed to be discontinuous. A preliminary version of this article has been presented in [6].

Notations. The symbol $\mathbb{R}$ denotes the set of real numbers $(-\infty,+\infty)$. $\overline{\mathbb{R}}$ denotes the extended real line $[-\infty,+\infty]$. We also use $\mathbb{R}_+ := [0,+\infty)$ and $\overline{\mathbb{R}}_+ := [0,+\infty]$. For a positive integer n, $\mathbb{R}^n$ denotes the n-dimensional Euclidean space with norm $|\cdot|$. A function $v : \mathbb{R}_+ \to \mathbb{R}^m$, with positive integer m, is said to be essentially bounded if $\operatorname{ess\,sup}_{t\geq 0} |v(t)| < \infty$. For given times $0 \leq T_1 < T_2$, we indicate with $v_{[T_1,T_2)} : \mathbb{R}_+ \to \mathbb{R}^m$ the function given by $v_{[T_1,T_2)}(t) = v(t)$ for all $t \in [T_1,T_2)$ and $= 0$ elsewhere. The function v is said to be locally essentially bounded if, for any $T > 0$, $v_{[0,T)}$ is essentially bounded. The essential supremum norm is indicated with the symbol $\|\cdot\|_\infty$. For a positive integer n and a positive real Δ: $\mathscr{C}_n$ denotes the space of continuous functions mapping $[-\Delta,0]$ into $\mathbb{R}^n$; $\mathscr{Q}_n$ denotes the space of bounded, continuous, except at a finite number of points, and right-continuous functions mapping $[-\Delta,0)$ into $\mathbb{R}^n$. For $\phi \in \mathscr{C}_n$, $\phi_{[-\Delta,0)}$ is the function in $\mathscr{Q}_n$ defined as $\phi_{[-\Delta,0)}(\tau) = \phi(\tau)$, $\tau \in [-\Delta,0)$. For a function $x : [-\Delta,c) \to \mathbb{R}^n$, with $0 < c \leq +\infty$, for any real $t \in [0,c)$, x_t is the function in $\mathscr{C}_n$ defined as $x_t(\tau) = x(t+\tau)$, $\tau \in [-\Delta,0]$. For given positive integers n,m, a map $f : \mathscr{C}_n \to \mathbb{R}^{n\times m}$ is said to be Lipschitz on bounded sets if, for any positive real q there exists a positive real L_q such that, for any $\phi_1,\phi_2 \in \mathscr{C}_n$ satisfying $\|\phi_i\|_\infty \leq q$, $i = 1,2$, the inequality holds $|f(\phi_1) - f(\phi_2)| \leq L_q\|\phi_1 - \phi_2\|_\infty$. A function $\gamma : \mathbb{R}_+ \to \mathbb{R}_+$ is said to be: of class $\mathscr{P}$ if it is continuous, zero at zero, and positive at any positive real; of class $\mathscr{K}$ if it is of class $\mathscr{P}$ and strictly increasing; of class $\mathscr{K}_\infty$ if it is of class $\mathscr{K}$ and it is unbounded; of class $\mathscr{L}$ if it is continuous and it monotonically decreases to zero as its argument tends to $+\infty$. A function $\beta : \mathbb{R}_+ \times \mathbb{R}_+ \to \mathbb{R}_+$ is of class $\mathscr{KL}$ if $\beta(\cdot,t)$ is of class $\mathscr{K}$ for each $t \geq 0$ and $\beta(s,\cdot)$ is of class $\mathscr{L}$ for each $s \geq 0$. The symbols $\vee$ and $\wedge$ denote logical sum and logical

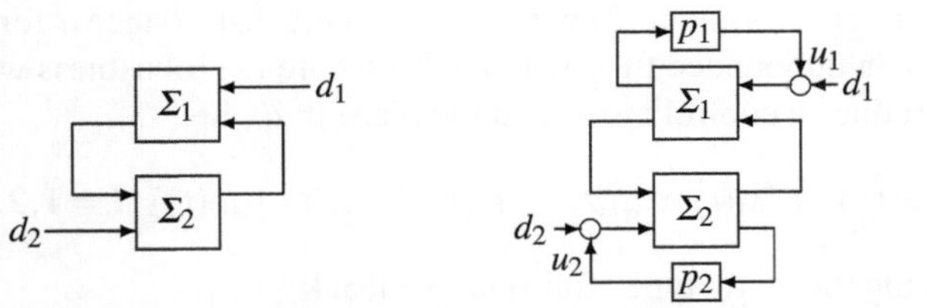

(a) Interconnected subsystem (b) Decentralized compensation

Fig. 1 Decentralized robustification with respect to disturbances

product, respectively. For $x \in \mathbb{R}$, $\tanh(x) = (e^x - e^{-x})/(e^x + e^{-x})$. For $x \in \mathbb{R} \setminus \{0\}$, $\mathrm{sgn}(x) = x/|x|$ and $\mathrm{sgn}(0) = 0$. Proofs are omitted due to the space limitation.

2 Idea and Issues to Be Solved

Decentralized Robustification. Consider a finite-dimensional dynamical system Σ consisting of two subsystems Σ_1 and Σ_2, and suppose that the trivial solution $x = 0$ of the overall system Σ is globally asymptotically stable (GAS)[1]. If the GAS property is characterized by a Lyapunov function in a desirable form, we can secure robustness of the system Σ against the additional disturbances d_1 and d_2 shown in Fig. 1(a) by introducing decentralized compensators. Such decentralized robustification is to insert local feedback inputs in the places where the disturbances come in as shown in Fig. 1(b). To illustrate this idea, let subsystems Σ_1 and Σ_2 be

$$\dot{x}_i(t) = f_i(x_i(t), x_{3-i}(t)), \quad i = 1, 2 \tag{1}$$

and define $x = [x_1^T, x_2^T]^T$ and $f = [f_1^T, f_2^T]^T$. Let $V(x)$ be a Lyapunov function describing the GAS of Σ, i.e., $V(x)$ is a C^1 function satisfying

$$\underline{\alpha}(|x|) \le V(x) \le \overline{\alpha}(|x|), \quad \dot{V}(t) \le -\alpha(|x(t)|) \tag{2}$$

along the trajectories of (1) for some $\alpha \in \mathscr{P}$ and $\underline{\alpha}, \overline{\alpha} \in \mathscr{K}_\infty$. To assess robustness of the interconnected system Σ, we consider the disturbances d_1 and d_2 as

$$\dot{x}_i(t) = f_i(x_i(t), x_{3-i}(t)) + g_i(x_i(t)) d_i(t), \quad i = 1, 2. \tag{3}$$

depicted in Fig. 1(a). Then along the trajectories of (3) with $d = [d_1^T, d_2^T]^T$, we have

$$\dot{V} = L_f V(x) + L g V(x) d \le -\alpha(|x|) + L g V(x) d, \tag{4}$$

where $g = [g_1^T, g_2^T]^T$. A bounded α can yield a fair stability margin for GAS of the original system (1) without disturbance. However, we cannot derive either the ISS

[1] For brevity, a system without input is said to be GAS if an equilibrium of the system is GAS.

or iISS property of the system (3) with respect to the disturbance d for the bounded α if $LgV(x)$ is an unbounded function of x. To secure the robustness with respect to d, we can introduce a control input u_i at the place of d_i, i.e.,

$$\dot{x}_i(t) = f_i(x_i(t), x_{3-i}(t)) + g_i(x_i(t))(d_i(t) + u_i(t)) \ , i = 1, 2. \tag{5}$$

Indeed, applying the "L_gV-type" full state feedback

$$u(t) = [u_1^T(t), u_2^T(t)]^T = -a(LgV(x(t)))^T \tag{6}$$

with a real number $a > 0$ to (5) we obtain

$$\dot{V} = L_f V(x) - a(LgV(x))(LgV(x))^T + LgV(x)d \leq -\alpha(|x|) + \frac{1}{4a}|d|^2 \tag{7}$$

along the trajectories of (5) with the help of Young's inequality. The disadvantage of using (6) is its centralized structure. Since $L_gV(x)$ in (6) usually contain both x_1 and x_2, the control input $u_i(t)$ of subsystem i is based not only on the local state x_i, but also on the state x_{3-i} of the other subsystem $3-i$. To make the robustifying compensation decentralized, instead of $V(x)$, we consider C^1 functions $V_i(x_i)$ which only contain local information for $i = 1, 2$. Applying the local version

$$u_i(t) = -a_i(Lg_iV_i(x_i(t)))^T, \quad i = 1, 2 \tag{8}$$

with a real number $a_i > 0$ to (5) we obtain

$$\begin{aligned} \dot{V}_i &= L_{f_i}V_i(x_i, x_{3-i}) - a_i(Lg_iV_i(x_i))(Lg_iV_i(x_i))^T + Lg_iV_i(x_i)d_i \\ &\leq L_{f_i}V_i(x_i, x_{3-i}) + \frac{1}{4a_i}|d_i|^2, \quad i = 1, 2. \end{aligned} \tag{9}$$

At this point, the property (9) does not give us information about robustness of the overall system (5) with the decentralized state feedback (8). Indeed, it is true in general that $a(LgV(x)) = [a_1(Lg_1V_1(x_1)), a_2(Lg_2V_2(x_2))]$ does not hold for any choice of positive constants a, a_1 and a_2. If the function $V(x)$ fulfilling (2) happens to be in the form of $V(x) = V_1(x_1) + V_2(x_2)$, then the property (9) implies (7) for the choice $a = a_1 = a_2$. The larger the feedback gain a_i is, the stronger the robustness with respect to d is. However, general nonlinear systems consisting of (1) often disallow any V in the form of $V(x) = V_1(x_1) + V_2(x_2)$ to accomplish (2) even if the equilibrium $x = 0$ is GAS. In this way, the feedback input u_i which uses only local state x_i as in Fig. 1(b) achieves the desired robustness of the overall system Σ with respect to the disturbance d only if the construction of the Lyapunov function $V(x)$ and the selection of the local feedback control laws $u_i(x_i)$ are judiciously coordinated. Therefore, it is significantly useful to derive a condition under which such a desirable pair V and u can be constructed, and to provide the formulas of V and u.

iISS. If $\alpha \in \mathcal{K}_\infty$ holds in (2), the property (7) implies ISS of the system Σ with respect to the disturbance d. In the case of $\alpha \in \mathcal{P} \setminus \mathcal{K}_\infty$, the system Σ is iISS with respect to d. It is, however, not guaranteed to be ISS. The existence of $V(x)$ satisfying (2) ensures the existence of another C^1 function $V(x)$ satisfying (2) with a class $\mathscr{K}_\infty$ function α. Indeed, replacing $V(x)$ by $F(V(x))$ with an appropriate C^1 function $F : \mathbb{R}_+ \to \mathbb{R}_+$ always allows us to achieve (2) with $\alpha \in \mathscr{K}_\infty$. Here, it is important to notice that this transformation into $\alpha \in \mathscr{K}_\infty$ via redefinition of $V(x)$ does not preserve the decentralized structure of robustifying controllers. In fact, the redefinition of $V(x)$ yields the "L_gV-type" feedback (6) as

$$u(t) = -a(LgF(V(x(t))))^T = [u_1^T(x(t)), u_2^T(x(t))]^T \tag{10}$$

in which $u_i = u_i^T(x_i)$ does not hold true in general. The transformation by F results in the centralized feedback $u_i = u_i^T(x_1,x_2)$, $i = 1,2$, even if the original V is in the form of $V(x) = V_1(x_1) + V_2(x_2)$. In addition, there are a lot of GAS systems for which no matter how we choose C^1 functions $F_1, F_2 : \mathbb{R}_+ \to \mathbb{R}_+$, the composite function $V(x) = F_1(V_1(x_1)) + F_2(V_2(x_2))$ never achieves (2) with $\alpha \in \mathscr{K}_\infty$. Such examples are found in the iISS framework (see e.g. [2]). Hence, it is unreasonable to expect that the interconnected system Σ achieves ISS with respect to the disturbance d. In this way, allowing $\alpha \notin \mathscr{K}_\infty$ is imperative to avoid unreasonably stringent constraints on systems Σ_i, and it is quite useful to develop a method of achieving iISS and including ISS as a special case.

Limitation of Input Magnitude. The control laws (6) and (8) are unbounded unless a strong constraint is imposed on the system Σ. In practical situations where the magnitude of control input is limited, the laws (6) and (8) need to be implemented with saturation functions. Then the property (7) does not hold true. From (5) it is obvious that, if $|d_i(t)|$ becomes larger than the upper limit of $|u_i(t)|$ an actuator can generate, such a control input cannot enhance the robustness against d. However, the upper bound of $|d_i(t)|$ is known and it is smaller than the actuator limitation, the robustness of Σ should be enhanced by applying appropriately saturated control input $u_i(t)$. Therefore, it is practically important to clarify how robust the system Σ can be by judiciously designing robustifying controllers meeting the input constraints.

***LgV* in the Presence of Delays and Discontinuities.** In addition to the inevitability of time delays in dynamical systems, discontinuity in the right-hand side often arises in practical models of control and sliding mode control laws. Such delays and discontinuities need to be incorporated into the right-hand side of (1), (3) or (5). Moreover, the map V needs to be extended to a functional in order to characterize the behavior of systems with delays whose solutions are defined as the evolution of segments defined on the delay interval along the time axis. In (6) and (8), the symbol LgV indicated the Lie derivative of the C^1 function V along g, i.e., $LgV = \frac{\partial V}{\partial x} g(x)$. When V is a functional, this definition is inapplicable. Furthermore, the relation between LgV and the estimation of the solutions $x(t)$ to the system Σ is not immediate at all for time-delay discontinuous right-hand side systems. It is necessary to redefine LgV in accordance with a feasible estimation of the behavior of Σ subject to time-delays and discontinuities.

An Approach. To address the above issues, we take an new approach based on

- Invariantly differential functionals to characterize the robustification in the form of LgV;
- A sum-type construction of a Lyapunov-Krasovskii functional to obtain V leading to the decentralized robustification;
- An iISS small-gain condition to formulate the robustification in the iISS framework in the presence of actuator limitations.

These tools have been investigated and developed very recently in [4, 5, 12]. This article demonstrates how successfully the problem of decentralized robustification for time-delay discontinuous right-hand side systems can be solved.

3 Invariantly Differentiable Functionals

This article borrows the definition of invariant differentiable functionals from [8] (see Definitions 2.2.1, 2.5.2 in Chapter 2). In the subsequent sections, we will assume that Lyapunov-Krasovskii functionals are invariantly differentiable. The formalism used in [8] is slightly modified here for the purpose of formalism uniformity throughout this article. For any given $x \in \mathbb{R}^n$, $\phi \in \mathcal{Q}_n$ and any continuous function $\mathcal{Y} : [0,\Delta] \to \mathbb{R}^n$ with $\mathcal{Y}(0) = x$, let $\psi_h^{(x,\phi,\mathcal{Y})} \in \mathcal{Q}_n$, $h \in [0,\Delta)$, be defined as

$$\begin{aligned} &\psi_0^{(x,\phi,\mathcal{Y})} = \phi \\ &\psi_h^{(x,\phi,\mathcal{Y})}(s) = \begin{cases} \phi(s+h), & s \in [-\Delta,-h) \\ \mathcal{Y}(s+h), & s \in [-h,0) \end{cases} \quad \text{for } h > 0. \end{aligned} \tag{11}$$

Definition 1. (see [8]) A functional $V : \mathbb{R}^n \times \mathcal{Q}_n \to \mathbb{R}_+$ is said to be invariantly differentiable if, at any point $(x,\phi) \in \mathbb{R}^n \times \mathcal{Q}_n$:
i) for any continuous function $\mathcal{Y} : [0,\Delta] \to \mathbb{R}^n$ with $\mathcal{Y}(0) = x$, there exists the finite right-hand derivative $\partial V\left(x, \psi_h^{(x,\phi,\mathcal{Y})}\right)/\partial h\Big|_{h=0}$ and such derivative is invariant with respect to the function $\mathcal{Y}$;
ii) there exists the finite derivative $\partial V(x,\phi)/\partial x$; iii) for any $z \in \mathbb{R}^n$, for any continuous function $\mathcal{Y} : [0,\Delta] \to \mathbb{R}^n$ with $\mathcal{Y}(0) = x$, for any $h \in [0,\Delta)$,

$$\begin{aligned} &V\left(x+z, \psi_h^{(x,\phi,\mathcal{Y})}\right) - V(x,\phi) = \\ &\qquad \frac{\partial V(x,\phi)}{\partial x} z + \left.\frac{\partial V\left(x, \psi_\ell^{(x,\phi,\mathcal{Y})}\right)}{\partial \ell}\right|_{\ell=0} h + o\left(\sqrt{|z|^2+h^2}\right) \end{aligned} \tag{12}$$

with $\lim_{s\to 0^+} o(\sqrt{s})/\sqrt{s} = 0$.

The first two terms in (12) serve as a differential of $V(x,\phi)$, and they are independent of $\mathcal{Y}$ defining the increment of the second argument ϕ of the functional $V(x,\phi)$. As explained in [12, 16], due to the invariant differentiability, we can define an appropriate derivative by which we can estimate the behavior of the trajectories of time-delay discontinuous right-hand side systems with a locally Lipschitz functional $V : \mathbb{R}^n \times \mathscr{Q}_n \to \mathbb{R}_+$ as in the classical Lyapunov theory for ordinary differential equations. Lemma 6 in [12] provides a tool to rescale invariantly differentiable functionals, which helps us evaluate robustness of interconnected systems effectively by means of invariantly differentiable functionals.

4 Interconnected Time-Delay Systems with Discontinuous Right-Hand Side

Consider an interconnected system Σ described by the following functional differential equations with discontinuous right-hand side

$$\Sigma \begin{cases} \Sigma_1 : \dot{x}_1(t) = f_1(x_{1,t}, x_{2,t}) + g_1(x_{1,t})(u_1(t) + d_1(t)) \\ \Sigma_2 : \dot{x}_2(t) = f_2(x_{2,t}, x_{1,t}) + g_2(x_{2,t})(u_2(t) + d_2(t)) \end{cases} \tag{13}$$
$$x_{1,0} = \xi_{1,0}, \quad x_{2,0} = \xi_{2,0},$$

where, for $i = 1,2$, $x_i(t) \in \mathbb{R}^{n_i}$; $d_i(t) \in \mathbb{R}^{m_i}$ is a disturbance adding to the control input (measurable, locally essentially bounded); n_i and m_i are positive integers. For $t \in \mathbb{R}_+$, $x_{i,t} : [-\Delta, 0] \to \mathbb{R}^{n_i}$ denotes the function $x_{i,t}(\tau) = x_i(t+\tau)$, where $\Delta > 0$ is the maximum involved delay. Suppose that $\xi_{i,0} \in \mathscr{C}_{n_i}$. The locally bounded maps $f_i : \mathscr{C}_{n_i} \times \mathscr{C}_{n_{3-i}} \to \mathbb{R}^{n_i}$ are continuous with respect to the second argument, and are allowed to be discontinuous with respect to the first argument, the maps $g_i : \mathscr{C}_{n_i} \to \mathbb{R}^{n_i \times m_i}$ are assumed to be Lipschitz on bounded sets. We combine vectors as $x(t) = [x_1(t)^T, x_2(t)^T]^T \in \mathbb{R}^n$, $n = n_1 + n_2$, $u(t) = [u_1(t)^T, u_2(t)^T]^T \in \mathbb{R}^m$, $d(t) = [d_1(t)^T, d_2(t)^T]^T \in \mathbb{R}^m$, $m = m_1 + m_2$, $\xi_0 = [\xi_{1,0}^T, \xi_{2,0}^T]^T \in \mathscr{C}_n$, $f() = [f_1()^T, f_2()^T]^T$, $\phi = [\phi_1^T, \phi_2^T]^T \in \mathscr{C}_n$ and $g() = [g_1()^T, g_2()^T]^T$. We define x_t as done for its i-th component $x_{i,t}$. It is assumed that $f_i(0,0) = 0$, $i = 1,2$. We use semi-norms $\|\cdot\|_{a,i} : \mathscr{C}_{n_i} \to \mathbb{R}_+$ and $\|\cdot\|_a : \mathscr{C}_n \to \mathbb{R}_+$, $i = 1,2$, respectively, for which there exist class $\mathscr{K}^\infty$ functions $\underline{\gamma}_{a,i}$, $\overline{\gamma}_{a,i}$, $\underline{\gamma}_a$ and $\overline{\gamma}_a$ such that

$$\underline{\gamma}_{a,i}(|\phi_i(0)|) \le \|\phi_i\|_{a,i} \le \overline{\gamma}_{a,i}(\|\phi_i\|_\infty), \quad \forall \phi_i \in \mathscr{C}_{n_i} \tag{14}$$

$$\underline{\gamma}_a(|\phi(0)|) \le \|\phi\|_a \le \overline{\gamma}_a(\|\phi\|_\infty), \qquad \forall \phi \in \mathscr{C}_n \,. \tag{15}$$

The retarded inclusions corresponding to Σ represented by (13) are given by

$$\begin{aligned} &\dot{x}_1(t) \in \Psi_1(x_{1,t}, x_{2,t}, u_1(t) + d_1(t)), && t \ge 0, \ a.e., \\ &\dot{x}_2(t) \in \Psi_2(x_{2,t}, x_{1,t}, u_2(t) + d_2(t)), && t \ge 0, \ a.e., \\ &x(\tau) = \xi_0(\tau), \qquad \tau \in [-\Delta, 0], && \xi_0 \in \mathscr{C}_n, \end{aligned} \tag{16}$$

where, for $(\phi_i, \phi_{3-i}, v) \in \mathscr{C}_{n_i} \times \mathscr{C}_{n_{3-i}} \times \mathbb{R}^{m_i}$, $\Psi_i(\phi_i, \phi_{3-i}, v)$ is the set given by

$$\Psi_i(\phi_i, \phi_{3-i}, v) = \{\xi_i + g_i(\phi_i)v, \ \ \xi_i \in F_i[f_i](\phi_i, \phi_{3-i})\}, \tag{17}$$

and $F_i[f_i](\phi_i,\phi_{3-i})$ is the convex closure of all limit values of the map f_i at the point (ϕ_i,ϕ_{3-i}). We introduce here the following standard assumption on the maps f_i of subsystems in (13): For each $(\phi_i,\phi_{3-i}) \in \mathscr{C}_{n_i} \times \mathscr{C}_{n_{3-i}}$, the set $F_i[f_i](\phi_i,\phi_{3-i})$ is assumed to be compact in $\mathbb{R}^{n_i}$; for each bounded set $W \in \mathscr{C}_{n_i} \times \mathscr{C}_{n_{3-i}}$, the set $\cup_{(\phi_i,\phi_{3-i})\in W} F_i[f_i](\phi_i,\phi_{3-i})$ is assumed to be bounded; the multimap $(\phi_i,\phi_{3-i}) \to F_i[f_i](\phi_i,\phi_{3-i})$ is assumed to satisfy the Carathéodory conditions (see Sections 4.2, 4.3, pp. 121-126, in [9]).

For the system (13), as in [12], we consider situations where essential bounds of the disturbance $d(t)$ are known in the following sense:

$$\underline{d}_{i,j} \leq \operatorname{ess}\inf_{t\in\mathbb{R}_+} d_{i,j}(t), \quad \overline{d}_{i,j} \geq \operatorname{ess}\sup_{t\in\mathbb{R}_+} d_{i,j}(t). \tag{18}$$

Here, $\underline{d}_{i,j}, \overline{d}_{i,j} \in \overline{\mathbb{R}}$, $i=1,2$, $j=1,2,...,m_i$ satisfying $\underline{d}_{i,j} \leq 0 \leq \overline{d}_{i,j}$ are given *a priori*. Note that, when do not have any *a priori* knowledge of the disturbance magnitude at Σ_i, we let $-\underline{d}_{i,j} = \overline{d}_{i,j} = \infty$, $i=1,2$ for $j=1,2,...,m_i$. The notions of ISS and iISS with the essential bounds are defined as follows:

Definition 2. The system (13) with $u(t) \equiv 0$ is said to be input-to-state stable (ISS) with respect to d with the essential bounds (18) if there exist a $\mathscr{KL}$ function β and a $\mathscr{K}$ function γ such that, for any initial state ξ_0 and any measurable, locally essentially bounded input d satisfying (18), any corresponding solution in the sense of (16) exists for all $t \geq 0$ and furthermore it satisfies

$$|x(t)| \leq \beta\left(\|\xi_0\|_\infty, t\right) + \gamma\left(\|d_{[0,t)}\|_\infty\right). \tag{19}$$

Definition 3. The system (13) with $u(t) \equiv 0$ is said to be integral input-to-state stable (iISS) with respect to d with the essential bounds (18) if there exist a $\mathscr{K}_\infty$ function χ, a $\mathscr{KL}$ function β and a $\mathscr{K}$ function γ such that, for any initial state ξ_0 and any measurable, locally essentially bounded input d satisfying (18), any corresponding solution in the sense of (16) exists for all $t \geq 0$ and furthermore it satisfies

$$\chi(|x(t)|) \leq \beta\left(\|\xi_0\|_\infty, t\right) + \int_0^t \gamma(|d(\tau)|)\,d\tau. \tag{20}$$

It is stressed that, in the situation where we take $-\underline{d}_{i,j} = \overline{d}_{i,j} = \infty$ for $i=1,2$ and $j=1,2,...,m_i$, the above definition reduces to the standard definitions of ISS and iISS without any bounds of the disturbance d (see [1, 13–15]). For example, the system (13) with $u(t) \equiv 0$ is ISS with respect to d with the essential bounds (18) for $-\underline{d}_{i,j} = \overline{d}_{i,j} = \infty$, $i=1,2$, $j=1,2,...,m_i$, if and only if the system (13) with $u(t) \equiv 0$ is ISS with respect to d. This equivalence also holds in the iISS case.

The following assumption is imposed on each unforced ($u_i(t) = d_i(t) \equiv 0$) subsystem in (13): For each subsystem Σ_i ($i=1,2$) defined in (13) with $u_i(t) = d_i(t) \equiv 0$, we assume the existence of a Locally Lipschitz invariantly differentiable functional $V_i : \mathbb{R}^{n_i} \times \mathscr{Q}_{n_i} \to \mathbb{R}_+$ such that

$$\underline{\alpha}_i(\|\phi_i\|_{a,i}) \le V_i(\phi_i(0),(\phi_i)_{[-\Delta,0)}) \le \overline{\alpha}_i(\|\phi_i\|_{a,i}), \tag{21}$$
$$D^+V_i(\phi_i,\phi_{3-i}) \le \rho_i(\phi_i,\phi_{3-i}), \quad \forall\, \phi_j \in \mathscr{C}_j, j=1,2 \tag{22}$$

hold, where $\underline{\alpha}_i, \overline{\alpha}_i$ are $\mathscr{K}_\infty$ functions and $\rho_i : \mathscr{C}_{n_i} \times \mathscr{C}_{n_{3-i}} \to \mathbb{R}$ is a continuous functional given by

$$\rho_i(\phi_i,\phi_{3-i}) = -\alpha_i(\|\phi_i\|_{a,i}) + \sigma_{i,0}(\|\phi_{3-i}\|_{a,i}) + \sum_{j=1}^{h} \sigma_{i,j}\left(\underline{\gamma}_{a,3-i}|\phi_{3-i}(-\Delta_j|)\right)$$
$$+ \sum_{j=h+1}^{h+h_d} \int_{-\Delta_j}^{0} \sigma_{i,j}\left(\underline{\gamma}_{a,3-i}|\phi_{3-i}(\tau)|\right) d\tau. \tag{23}$$

Here, h and h_d are non-negative integers, α_i and $\sigma_{i,j}$ are class $\mathscr{K}$ functions, and $\Delta_j \in (0,\Delta]$ for $j=0,1,\dots,h+h_d$. The left hand side of (22) is defined with

$$D^+V_i(\phi_i,\phi_{3-i}) = \sup_{\xi_i \in F_i[f_i](\phi_i,\phi_{3-i})} \left.\frac{\partial V_i(x_i,\phi_i)}{\partial x_i}\right|_{x_i=\phi_i(0)} \xi_i + \left.\frac{\partial V_i(\phi_i(0),\phi_{i,h})}{\partial h}\right|_{h=0} \tag{24}$$

$$\phi_{i,h}(s) = \begin{cases} \phi_i(s+h), & s \in [-\Delta,-h) \\ \phi_i(0), & s \in [-h,0] \end{cases} \quad \text{for } h \in [0,\Delta). \tag{25}$$

5 Decentralized iISS and ISS Feedback Redesign

We introduce a few notations and definitions. Define an operator $\alpha_i^{\ominus}: \overline{\mathbb{R}}_+ \to \overline{\mathbb{R}}_+$ as

$$\alpha_i^{\ominus}(s) = \sup\{v \in \mathbb{R}_+ : s \ge \alpha_i(v)\}. \tag{26}$$

Thus, we have $\alpha_i^{\ominus}(s) = \infty$ for $s \ge \lim_{\tau\to\infty}\alpha_i(\tau)$, and $\alpha_i^{\ominus}(s) = \alpha_i^{-1}(s)$ elsewhere. For a class $\mathscr{K}$ function $\omega : \mathbb{R}_+ \to \mathbb{R}_+$, this article uses the extension ω: $\overline{\mathbb{R}}_+ \to \overline{\mathbb{R}}_+$ defined as

$$\omega(s) := \sup_{v \in \{w \in \mathbb{R}_+ : w \le s\}} \omega(v).$$

The reader may refer to [3] for the benefit of these extended operators. We define the following set $\mathscr{D}(\underline{w},\overline{w})$ of continuous functions:

Definition 4. Given $-\infty \le \underline{w} < 0 < \overline{w} \le +\infty$, a function $\omega : \mathbb{R} \to \mathbb{R}$ is said to belong to $\mathscr{D}(\underline{w},\overline{w})$ if it is a strictly increasing and locally Lipschitz function such that $\omega(0)=0 \wedge \{\lim_{s\to-\infty}\omega(s) < \underline{w} \vee \lim_{s\to-\infty}\omega(s) = -\infty\} \wedge \{\overline{w} < \lim_{s\to+\infty}\omega(s) \vee \lim_{s\to+\infty}\omega(s) = +\infty\}$.

For a mapping $\omega \in \mathscr{D}(\underline{w},\overline{w})$ from $\mathbb{R}$ onto $(a,b) \subseteq \mathbb{R}$, the inverse of ω is a strictly increasing continuous function denoted by $\omega^{-1} : (a,b) \to \mathbb{R}$. For $\omega \in \mathscr{D}(\underline{w},\overline{w})$, the function $\omega^{-1}(s)s : (\underline{w},\overline{w}) \to \mathbb{R}$ is locally Lipschitz. For each $i=1,2$, let N_i be

$$N_i = \sum_{j=0}^{h+h_d} \mathrm{sgn}(\sigma_{i,j}(1)) \,, \tag{27}$$

which describes the number of non-zero functions among $\sigma_{i,0}$, ..., $\sigma_{i,h+h_d}$, in (23). The following achieves decentralized robustification under a small-gain condition.

Theorem 1. *Define* $\sigma_i \in \mathscr{K}$, $i = 1,2$, *by*

$$\sigma_i(s) = N_i \max\left\{ \max_{j=0,1,\ldots,h} \sigma_{i,j}(s), \max_{j=h+1,\ldots,h+h_d} \Delta_j \sigma_{i,j}(s) \right\}. \tag{28}$$

Suppose that there exist $c_i > 1$, $i = 1,2$, *such that*

$$c_1\sigma_1 \circ \underline{\alpha}_2^{-1} \circ \overline{\alpha}_2 \circ \alpha_2^{\ominus} \circ c_2\sigma_2(s) \le \alpha_1 \circ \overline{\alpha}_1^{-1} \circ \underline{\alpha}_1(s), \quad \forall s \in \mathbb{R}_+ \tag{29}$$

holds. Pick $\tau_i, \mu_i > 0$ *and* $\varphi \ge 0$ *such that*

$$1 < \tau_i < \frac{c_i}{1+\mu_i}, \quad \left(\frac{\tau_i(1+\mu_i)}{c_i} \right)^{\varphi} \le \tau_i - 1, \quad i = 1,2 \tag{30}$$

are satisfied. Define class $\mathscr{K}$ *functions* λ_i, $i = 1,2$, *by*

$$\lambda_i(s) = \left[\frac{1}{\tau_i} \alpha_i(\overline{\alpha}_i^{-1}(s)) \right]^{\varphi} \left[(1+\mu_i)\sigma_{3-i}(\underline{\alpha}_i^{-1}(s)) \right]^{\varphi+1}. \tag{31}$$

Assume that the mapping

$$\begin{aligned} h_i(\phi_i) &= [h_{i,1}(\phi_i), h_{i,2}(\phi_i), \ldots, h_{i,m_i}(\phi_i)] \\ &= \lambda_i(V_i(\phi_i(0), (\phi_i)_{[-\Delta,0)})) \cdot \left. \frac{\partial V_i(x_i, (\phi_i)_{[-\Delta,0)})}{\partial x_i} \right|_{x_i=\phi_i(0)} g_i(\phi_i) \end{aligned} \tag{32}$$

from $\mathscr{C}_{n_i}$ *into* $\mathbb{R}^{m_i}$ *is Lipschitz on bounded sets for* $i = 1,2$. *Define*

$$p_i(\phi_i) = -[Y_{i,1}(h_{i,1}(\phi_i)), Y_{i,2}(h_{i,2}(\phi_i)), \cdots, Y_{i,m_i}(h_{i,m_i}(\phi_i))]^T \tag{33}$$

for $Y_{i,j} \in \mathscr{D}(\underline{d}_{i,j}, \overline{d}_{i,j})$, $i = 1,2$, $j = 1,2,\ldots,m_i$. *Then the decentralized feedback control laws* ($i = 1,2$)

$$u_i(t) = p_i(x_{i,t}) \tag{34}$$

render the closed-loop system consisting of (13) *and* (34) *iISS with respect to the disturbance d with the essential bounds* (18). *Moreover, if* $\lim_{s\to\infty} \alpha_i(s) = \infty$ *holds true for* $i = 1,2$, *then the closed-loop system is ISS with respect to the disturbance d with the essential bounds* (18).

For any $c_i > 1$, there always exist $\tau_i, \mu_i > 0$ and $\varphi \ge 0$ fulfilling (30). Note that Theorem 1 establishes ISS and iISS even if $-\underline{d}_{i,j} = \overline{d}_{i,j} = \infty$ for $i = 1,2$, $j = 1,2,\ldots,m_i$. In such a case, $Y_{i,j}$'s are required to be unbounded and the magnitude of the robustifying inputs $u_i(t)$ become large arbitrarily for arbitrarily large disturbances $d_{i,j}(t)$. If time delays reside only in communication channels, the mappings V_i are functions which do not involve any terms for time delays. In such cases, equations (32), (33) and (34) yield the compensations $u_i(t)$ which are delay free. Theorem 1 is established by making use of the functional $V : \mathbb{R}^n \times \mathscr{Q}_n \to \mathbb{R}_+$:

$$\begin{aligned} V(\phi(0),(\phi)_{[-\Delta,0)}) = &\sum_{i=1}^{2} \int_0^{V_i(\phi_i(0),(\phi_i)_{[-\Delta,0)})} \lambda_i(s)ds \\ &+ \sum_{j=1}^{h} \int_{-\Delta_j}^{0} F_{i,j}(\tau)\tilde{\sigma}_{i,j}\left(\underline{\gamma}_{a,3-i}(|\phi_{3-i}(\tau)|)\right)d\tau \\ &+ \sum_{j=h+1}^{h+h_d} \int_{-\Delta_j}^{0} F_{i,j}(\tau)\int_{\tau}^{0} \tilde{\sigma}_{i,j}\left(\underline{\gamma}_{a,3-i}(|\phi_{3-i}(\theta)|)\right)d\theta d\tau. \end{aligned} \tag{35}$$

where, for $i = 1,2$ and $j = 1,2,...,h+h_d$, the continuous functions $F_{i,j} : [-\Delta_j, 0] \to \mathbb{R}$ and the functions $\tilde{\sigma}_{i,j} \in \mathscr{K}$ are given by

$$F_{i,j}(\tau) = \frac{-\tau}{\Delta_j} + (1+\mu_i)\frac{\tau+\Delta_j}{\Delta_j}, \quad \tilde{\sigma}_{i,j}(s) = \lambda_i(\theta_{i,j}(s))\sigma_{i,j}(s)$$

$$\theta_{i,j}(s) = \begin{cases} \overline{\alpha}_i \circ \alpha_i^{\ominus} \circ N_i\tau_i\sigma_{i,j}(s) & , j = 0,1,...,h \\ \overline{\alpha}_i \circ \alpha_i^{\ominus} \circ N_i\tau_i\Delta_j\sigma_{i,j}(s) & , j = h+1, h+2, ..., h+h_d. \end{cases}$$

Let $\underline{\alpha}$, $\overline{\alpha} \in \mathscr{K}_\infty$ be such that $\underline{\alpha}(\|\phi\|_a) \le V(\phi(0),(\phi)_{[-\Delta,0)}) \le \overline{\alpha}(\|\phi\|_a)$. The functional V in (35) plays the role of a Lyapunov-Krasovskii functional to estimate the influence of the disturbance d on the resulting system as follows:

Corollary 1. *Suppose that all the assumptions in Theorem 1 are fulfilled. Then the closed-loop system consisting of* (13) *and* (34) *satisfies*

$$D^+V(\phi,d) \le -\alpha(\|\phi\|_a) + \sigma(|d|), \tag{36}$$

where $\alpha \in \mathscr{K}$ is given by (37)*, and σ is any class $\mathscr{K}$ function satisfying* (38)*:*

$$\begin{aligned} \alpha(s) = \min_{\{\phi=[\phi_1^T,\phi_2^T]^T \in \mathscr{C}_n : s=\|\phi\|_a\}} &\left\{ \sum_{i=1}^{n}\sum_{j=1}^{h} \frac{\mu_i}{\Delta_j}\int_{-\Delta_j}^{0} \tilde{\sigma}_{i,j}\left(\underline{\gamma}_{a,3-i}(|\phi_{3-i}(\tau)|)\right)d\tau + \right. \\ &\sum_{j=h+1}^{h+h_d} \frac{\mu_i}{\Delta_j}\int_{-\Delta_j}^{0}\int_{\tau}^{0} \tilde{\sigma}_{i,j}\left(\underline{\gamma}_{a,3-i}(|\phi_{3-i}(\theta)|)\right)d\theta d\tau + \\ &\left. \frac{\left(1-\frac{\tau_i}{c_i}\right)(\tau_i-1)}{\tau_i}\lambda_i(V_i(\phi_i(0),(\phi_i)_{[-\Delta,0)}))[\alpha_i \circ \overline{\alpha}_i^{-1}(V_i(\phi_i(0),(\phi_i)_{[-\Delta,0)}))] \right\} \end{aligned} \tag{37}$$

$$\sigma(s) \ge \sup_{\{d \in \mathbb{R}^m : s \ge |d|,\ d_{i,j} \in (\underline{d}_{i,j}, \overline{d}_{i,j})\}} \sum_{i=1}^{2}\sum_{j=1}^{m_i} Y_{i,j}^{-1}(d_{i,j})d_{i,j}. \tag{38}$$

Furthermore, a pair of χ *and* γ *in* (20) *is given by* $\chi(s) = \underline{\alpha} \circ \underline{\gamma}_a(s)$ *and* $\gamma(s) = 2\sigma(s)$. *Moreover, if* $\lim_{s\to\infty} \alpha_i(s) = \infty$ *holds for* $i = 1,2$, *a function* γ *satisfying* (19) *is* $\gamma(s) = \underline{\gamma}_a^{-1} \circ \underline{\alpha}^{-1} \circ \overline{\alpha} \circ \alpha^{-1}(2\sigma(s))$.

Equation (31) is a special case of the more general formula of λ_i presented in [4]. The free parameters in [4] allow us to replace (31) by the one presented in [5].

6 An Example

Consider the interconnection of two scalar subsystems:

$$\dot{x}_1(t) = -\frac{\operatorname{sgn}(x_1(t))}{1+|x_1(t)|} + \frac{\gamma_1}{1+|x_1(t)|} x_2(t-\Delta) + \cos(x_1(t))(u_1(t)+d_1(t)) \tag{39}$$
$$\dot{x}_2(t) = -x_2(t)(2+\operatorname{sgn}(x_2(t)-1)) + \gamma_2 \frac{x_1(t-\Delta)}{1+|x_1(t-\Delta)|} + x_2(t)(u_2(t)+d_2(t)),$$

where $\Delta > 0$ is a channel delay, $\gamma_i \in \mathbb{R}$, $i = 1,2$, are interaction parameters. Choose $V_i(\phi_i(0), (\phi_i)_{[-\Delta,0)}) = \phi_i(0)^2$, $i = 1,2$. For $u_i(t) \equiv 0$, $d_i(t) \equiv 0$, we obtain

$$D^+V_1 \le -\frac{2|\phi_1(0)|}{1+|\phi_1(0)|} + 2|\gamma_1||\phi_2(-\Delta)|,\ D^+V_2 \le -|\phi_2(0)|^2 + \gamma_2^2 \left(\frac{|\phi_1(-\Delta)|}{1+|\phi_1(-\Delta|)}\right)^2.$$

If $|\gamma_1\gamma_2| < 1$, (29) is satisfied. For example, in the case of $|\gamma_1| = |\gamma_2| = 1/2$, the formula (31) gives $\lambda_1(s) = \frac{1}{4}(\sqrt{s}/(1+\sqrt{s}))^2$ and $\lambda_2(s) = \sqrt{s}$ for $\varphi = 0$ and $\tau_i(1+\mu_i) = 17/8$, $i = 1,2$. Assume that $|d_1(t)| \le 2$ and $|d_2(t)| \le 7$ hold for all $t \ge 0$. Setting $\underline{d}_{1,1} = \overline{d}_{1,1} = 2$ and $\underline{d}_{2,1} = \overline{d}_{2,1} = 7$, we can choose $Y_{1,1}(s) = 3\tanh(s)$ and $Y_{2,1}(s) = 10\tanh(s)$. Thus, equations (32)-(34) yield

$$u_1(t) = -3\tanh(\lambda_1(x_1^2(t))2x_1(t)\cos(x_1(t)))$$
$$u_2(t) = -10\tanh(\lambda_2(x_2^2(t))2x_2^2(t)) \tag{40}$$

which achieve iISS of the overall system (39) with respect to $d_i(t)$, $i = 1,2$. Figure 2(a) illustrates the effectiveness of (40) for (39) with $\xi_0(\tau) = [-1,3]^T$, $\tau \in [-\Delta, 0]$, $\Delta = 2$ and $\gamma_1 = \gamma_2 = 0.5$ in the presence of $d = [2\cos(2t), 4+3\cos(4t)]^T$ which satisfies $|d_1(t)| \le 2$ and $|d_2(t)| \le 7$ for all $t \ge 0$. Compared with Fig.2(c), the local feedback laws (40) with the input magnitude limitations significantly improve robustness with respect to the disturbance d. If no limitations of input magnitude are necessary, Theorem 1 yields the unbounded local feedback laws

$$u_1(t) = -3\lambda_1(x_1^2(t))2x_1(t)\cos(x_1(t)),$$
$$u_2(t) = -10\lambda_2(x_2^2(t))2x_2^2(t) \tag{41}$$

which produce state trajectories shown in Fig.2(b). The robustness achieved by the bounded control (40) is almost identical to the robustness achieved by the unbounded control (41). For $d = [7\cos(2t), 9+11\cos(4t)]^T$ exceeding the upper

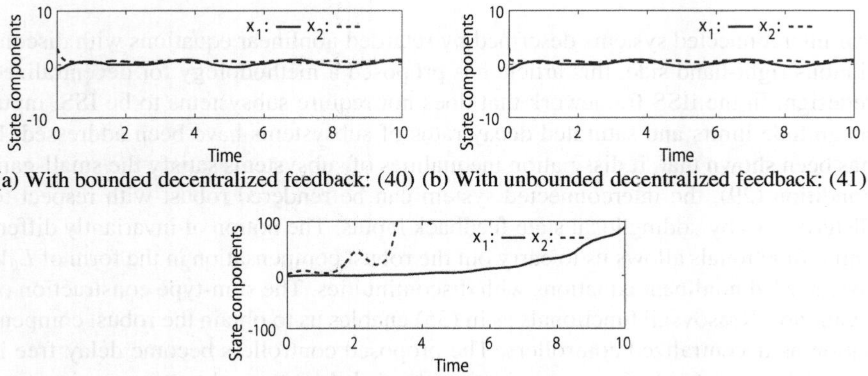

(a) With bounded decentralized feedback: (40) (b) With unbounded decentralized feedback: (41)

(c) Without feedback: $u_i(t) \equiv 0, i = 1,2$

Fig. 2 State transition $x(t) = [x_1(t), x_2(t)]^T$ of (39) with $d = [2\cos(2t), 4 + 3\cos(4t)]^T$

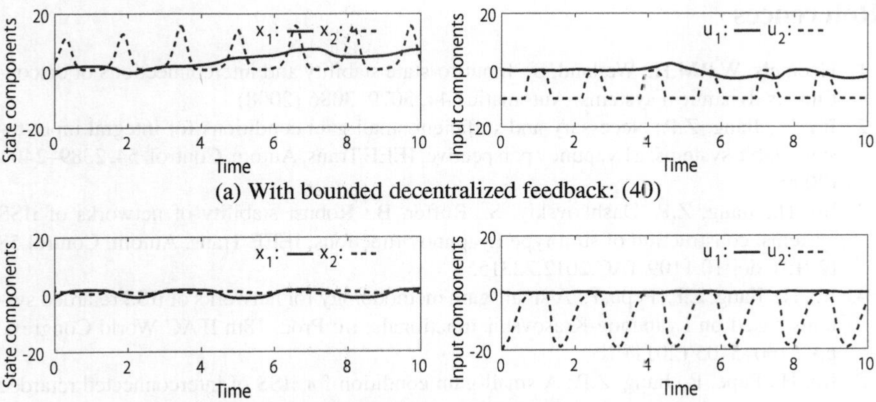

(a) With bounded decentralized feedback: (40)

(b) With unbounded decentralized feedback: (41)

Fig. 3 State transition $x(t) = [x_1(t), x_2(t)]^T$ of (39) with $d = [7\cos(2t), 9 + 11\cos(4t)]^T$

bounds, the trajectories with the bounded laws (40) and the unbounded laws (41) are plotted in Figs.3(a) and (b), respectively. The parameters Δ, γ_1, γ_2 and ξ_0 are the same as those used in Figs. 2(a), (b) and (c). The control inputs (40) fulfill the magnitude constraints $|u_1(t)| \leq 3$ and $|u_2(t)| \leq 10$ for all $t \geq 0$. However, they cannot ensure the robustness the larger control inputs (41) can attain. In the case of no control inputs, the simulation exhibited a vertical increase of x_2 at $t = 1.80$.

7 Conclusions

For interconnected systems described by retarded nonlinear equations with discontinuous right-hand side, this article has proposed a methodology for decentralized redesign. In the iISS framework that does not require subsystems to be ISS, input magnitude limits and saturated decay rates of subsystems have been addressed. It has been shown that, if dissipation inequalities of subsystems satisfy the small-gain condition (29), the interconnected system can be rendered robust with respect to disturbances by adding local state feedback inputs. The notion of invariantly differential functionals allows us to carry out the robust compensation in the form of L_gV for retarded nonlinear equations with discontinuities. The sum-type construction of Lyapunov-Krasovskii functionals as in (35) enables us to obtain the robust compensation as decentralized controllers. The proposed controllers become delay free if time delays exist only in communication channels between subsystems.

Acknowledgements. The work of P. Pepe is supported by MIUR PRIN 2009 and DEWS. The work of Z.P. Jiang has been supported by NSF grant ECCS-1230040.

References

1. Heemels, W.P.M.H., Weiland, S.: Input-to-state stability and interconnections of discontinuous dynamical systems. Automatica 44, 3079–3086 (2008)
2. Ito, H., Jiang, Z.P.: Necessary and sufficient small gain conditions for integral input-to-state stable systems: a Lyapunov perspective. IEEE Trans. Autom. Control. 54, 2389–2404 (2009)
3. Ito, H., Jiang, Z.P., Dashkovskiy, S., Rüffer, B.: Robust stability of networks of iISS systems: construction of sum-type Lyapunov functions. IEEE Trans. Autom. Control 58 (2013), doi:10.1109/TAC.2012.2231552
4. Ito, H., Jiang, Z.P., Pepe, P.: A small-gain methodology for networks of iISS retarded systems based on Lyapunov-Krasovskii functionals. In: Proc. 18th IFAC World Congress, pp. 5100–5105 (2011)
5. Ito, H., Pepe, P., Jiang, Z.P.: A small-gain condition for iISS of interconnected retarded systems based on Lyapunov-Krasovskii functionals. Automatica 46, 1646–1656 (2010)
6. Ito, H., Pepe, P., Jiang, Z.P.: Decentralized iISS robustification of interconnected time-delay systems: a small-gain approach. In: Proc. 10th IFAC Workshop on Time Delay Systems, pp. 219–224 (2012)
7. Karafyllis, I., Jiang, Z.P.: A new small-gain theorem with an application to the stabilization of the chemostat. Int. J. Robust and Nonlinear Contr. 22, 1602–1630 (2012)
8. Kim, A.V.: Functional differential equations, application of i-smooth calculus. Kluwer, Dordrecht (1999)
9. Kolmanovskii, V., Myshkis, A.: Introduction to the theory and applications of functional differential equations. Kluwer, Dordrecht (1999)
10. Pepe, P.: Input-to-state stabilization of stabilizable, time-delay, control affine, nonlinear systems. IEEE Trans. Automat. Contr. 54, 1688–1693 (2009)
11. Pepe, P.: On the actuator disturbance attenuation for systems described by neutral equations. IMA J. Mathematical Control and Information 28, 163–181 (2011)

12. Pepe, P., Ito, H.: On saturation, discontinuities and delays in iISS and ISS feedback control redesign. IEEE Trans. Automat. Contr. 57, 1125–1140 (2012)
13. Pepe, P., Jiang, Z.P.: A Lyapunov-Krasovskii methodology for ISS and iISS of time-delay systems. Systems & Contr. Letters 55, 1006–1014 (2006)
14. Sontag, E.D.: Smooth stabilization implies coprime factorization. IEEE Trans. Automat. Contr. 34, 435–443 (1989)
15. Sontag, E.D.: Comments on integral variants of ISS. Systems & Contr. Letters 34, 93–100 (1998)
16. Surkov, A.V.: On the stability of functional-differential inclusions using invariantly differentiable Lyapunov functionals. Differential Equations 43, 1079–1087 (2007)

Global Stability Analysis of Nonlinear Sampled-Data Systems Using Convex Methods

Matthew M. Peet and Alexandre Seuret

Abstract. In this chapter, we consider the problem of global stability of nonlinear sampled-data systems. Sampled-data systems are a form of hybrid model which arises when discrete measurements and updates are used to control continuous-time plants. In this paper, we use a recently introduced Lyapunov approach to derive stability conditions for both the case of fixed sampling period (synchronous) and the case of a time-varying sampling period (asynchronous). This approach requires the existence of a Lyapunov function which decreases over each sampling interval. To enforce this constraint, we use a form of slack variable which exists over the sampling period, may depend on the sampling period, and allows the Lyapunov function to be temporarily increasing. The resulting conditions are enforced using a new method of convex optimization of polynomial variables known as Sum-of-Squares. We use several numerical examples to illustrate this approach.

1 Introduction to the Problem of Stability of Sampled-Data Systems

Consider an aircraft in combat being remotely piloted by an operator. Directed energy or some other form of electronic warfare is used to deny portions of the electromagnetic spectrum and thus reduce the communication bandwidth between vehicle and operator. The change in bandwidth restricts the rate at which information can be transmitted to the vehicle. The question we ask is what is the minimum rate of transfer of information the aircraft can tolerate before it becomes unstable. This situation

Matthew M. Peet
Arizona State University, P.O. Box 876106, Tempe, AZ 85287
e-mail: `mpeet@asu.edu`

Alexandre Seuret
CNRS, LAAS, 7 avenue du Colonel Roche, 31077 Toulouse, France
Univ. de Toulouse, LAAS, F-31400 Toulouse, France
e-mail: `aseuret@laas.fr`

T. Vyhlídal, J.-F. Lafay, and R. Sipahi (eds.), *Delay Systems*,
Advances in Delays and Dynamics 1,
DOI: 10.1007/978-3-319-01695-5_16,

is similar to the use of electronic countermeasures on an active radar-guided missile. In both cases, there is a set of continuous-time dynamics representing the dynamics of the controlled system. These continuous dynamics are regulated by continuous real-time feedback using digital sensing and actuation. During normal operation, the controller is updated continuously and so the digitization of the controller does not affect the dynamics of the closed-loop system. When interference occurs, however, the update rate of the controller may be sparse or unpredictable. In this case, the system becomes neither discrete nor continuous, but rather a special type of hybrid system referred to as a *Sampled-Data* system, modeled as

$$\dot{x}(t) = f(x(t), x(t_k)) \qquad \text{for } t \in [t_k, t_k + T_k], \quad k = 1, \cdots, \infty.$$

where $t_{k+1} = t_k + T_k$ for all k and T_k is the sampling period which may be constant or may depend on k. Typically, systems of this form arise when the dynamics depend on external updates - often through the use of a controller so that $f(x(t), x(t_k)) = f^*(x(t), u(t))$ with $u(t) = k(x(t_k))$. The sampling period T_k may be thought of as the time between updates from an external controller. In the scenarios described above, T_k would vary with k - the so-called 'asynchronous' case. However, there are certain situations when T_k may not vary from update to update - such as when the controller is implemented using an A/D converter with step-times. We refer to this situation as the 'synchronous' case.

Linear Sampled-Data systems have been well-studied in the literature [1–4], including work on nonlinear systems in [5, 6]. One popular approach has been to regard the system in continuous-time and use a discontinuous, time-varying delay to represent the hybrid part of the dynamics [7]. Unfortunately, this approach has not been completely successful, as the understanding of nonlinear systems with time-varying delay is itself a difficult problem. An alternative approach has been to regard the system in discrete time [8–10], where the update law is given by the solution map of the continuous-time system over a period T_k. For a linear system, this solution map is well-defined using matrix exponentials. For nonlinear systems, it can be approximated over bounded intervals using methods such as the extended Picard iteration [11]. The difficulty with this approach is that the update law is different for every sampling period - meaning that although the approach may work well for a fixed sampling period, for unknown and time-varying sampling periods, one has to verify stability over a family of potential solutions. Even in the linear case, this means verification of stability with parametric uncertainty which enters through the exponential. If we have a nonlinear sampled-data system, then even if the vector field is polynomial, the extended Picard iteration yields a polynomial approximation to the solution map - meaning we must test stability of a complicated polynomial vector field with parametric uncertainty - an NP-hard problem.

In this chapter, we consider the use of a new Lyapunov-based approach to stability analysis of sampled-data systems. Specifically, we rely on a Lyapunov result which states that while Lyapunov functions must experience a net decrease over the sampling period, it may be instantaneously increasing [12]. This constraint can be implemented in a Lyapunov context through the use of 'spacing functions' - functions which are required to vanish at the endpoints of the sampling period. The main idea behind these functions is that instead of requiring negativity of the Lyapunov function over the entire sampling interval, we only require the sum of the Lyapunov function and the spacing function to be decreasing for all time. The inspiration for this approach came from the previous work on spacing functions for Lyapunov-Krasovskii functionals for stability of time-delay systems in [13]. In [14], we considered the use of this approach for construction of quadratic Lyapunov functions for linear sampled-data systems in both the synchronous and asynchronous cases. The contribution of this chapter is to show how this approach can be extended to prove global stability of nonlinear sampled-data systems.

This chapter is organized as follows. In Section 2, we introduce the sampled-data system model and define our concepts of stability. We then give the Lyapunov theorem whose conditions we will test. We then introduce the Sum-of-Squares approach to optimization of polynomial variables, including the use of Positivstellensatz results to enforce local positivity. In Section 2, we show how the Sum-of-Squares framework can be used to enforce the stability conditions of Section 2. Finally, in Section 4, we apply the results of the chapter to several cases of nonlinear stability analysis in both the synchronous and asynchronous cases.

2 Background

In this section we will first describe the Lyapunov theorem we will use and discuss the conditions that a Lyapunov function must satisfy. Following this, we will briefly discuss the computational framework we will use to enforce the conditions of the Lyapunov theorem. Specifically, we will give background on optimization of polynomials using the Sum-of-Squares methodology (SOS).

2.1 Sampled-Data Systems

In this chapter, we consider the stability of solutions of equations of the form

$$\begin{aligned} \dot{x}(t) &= f(x(t), x(t_k)) \qquad \text{for } t \in [t_k, t_k + T_k], \quad k = 1, \cdots, \infty. \\ x(t) &= x_0 \end{aligned} \tag{1}$$

where $t_0 = 0$, $t_{k+1} = t_k + T_k$ for all $k \geq 0$ and T_k is the sampling period which may be constant or may depend on k. We assume that T_k satisfies some upper bound $T_k \leq T_{\max}$ for all k. When it exists, we define the continuous-time flow-map $\Gamma(s)$ to be any function which satisfies $\frac{d}{ds}\Gamma(s)z = f(\Gamma(s)z, z)$ for all $s \in [0, T_{\max}]$ and $\Gamma(0)z = z$. If Γ exists, then the sampled-data system can be reduced to a discrete-time system as $x_{k+1} = \Gamma(T_k)x_k$. For the linear sampled-data system

$$\dot{x}(t) = A_0 x(t) + A_1 x(t_k),$$

we have the explicit solution

$$\Gamma(s)z = \left(e^{A_0 s} + \int_0^s e^{A_0(s-\theta)} A_1 d\theta\right) z.$$

For a nonlinear system, the solution map Γ is difficult to find - although it may be approximated using such methods as Picard iteration.

Definition 1. We say the Sampled-Data System (1) is globally exponentially stable if there exist positive constants K, γ such that for any initial condition $x_0 \in \mathbb{R}^n$, and any x satisfying (1), we have $\|x(t)\| \leq K\|x_0\|e^{-\gamma t}$ for all $t \geq 0$.

For a synchronous linear sampled-data system with period T_k, global exponential stability is equivalent to $\rho\left(e^{A_0 s T_k} + \int_0^{T_k} e^{A_0(s-\theta)} A_1 d\theta\right) < 1$.

Definition 2. We say the Sampled-Data System (1) is locally exponentially stable on domain X if there exist positive constants K, γ such that for any initial condition $x_0 \in X$, and any x satisfying (1), we have $x(t) \in X$ and $\|x(t)\| \leq K\|x_0\|e^{-\gamma t}$ for all $t \geq 0$.

2.2 A Lyapunov Theorem

In this theorem, we assume global existence and continuity of solutions.

Notation: For a given solution, x, of System (1), define the function $x_k(s) = \Gamma(s)x(t_k)$ for $s \in [0, T_k]$. Associated with $x_k \in \mathscr{C}[0, T_k]$, we denote the supremum norm $\|x_k\|_\infty = \sup_{s \in [0,\, T_{\max}]} \|x_k(s)\|$.

Theorem 1. *[12] Suppose $V : \mathbb{R}^n \to \mathbb{R}^+$ is continuously differentiable and*

$$\mu_1\|x\|^2 \leq V(x) \leq \mu_2\|x\|^2, \quad \text{for all } x \in \mathbb{R}^n. \tag{2}$$

for positive scalars μ_1, μ_2 with $\mu_1 > \mu_2 > 0$. Then for any positive constants $\alpha, T_{\min}$ and $T_{\max}$ such that $T_k := t_{k+1} - t_k \in [T_{\min}, T_{\max}]$ for all $k \in \mathbb{N}$, the following are equivalent.

(i) There exists positive constants ε, α such that for all solutions x of Equation (1)*, and for all $k \geq 0$,*

$$V(x(t_{k+1})) < e^{-2\alpha T_k} V(x(t_k)) - \varepsilon\|x(t_k)\|^2.$$

(ii) There exists a positive constants δ and α and continuously differentiable functions $Q_k : [0, T_k] \times \mathscr{C}[0, T_k] \to \mathbb{R}$ which satisfy the following for all $k \geq 0$.

$$Q_k(T_k, z) = e^{-2\alpha T_k} Q_k(0, z) \quad \text{for all } z \in \mathscr{C}[0, t_k] \tag{3}$$

and such that for all solutions of Equation (1)*, and for all $t \in [t_k, t_{k+1}]$*

$$\frac{d}{dt}[V(x(t)) + Q_k(t - t_k, x_k)] + 2\alpha V(x(t)) + 2\alpha Q_k(t - t_k, x_k) < -\delta\|x_k\|_\infty. \tag{4}$$

Moreover, if either of these statements is satisfied, then System (1) *is globally exponentially stable about the origin with decay rate* $\gamma = \alpha$.

Note that the function Q is an operator on an infinite-dimensional vector space. Parametrization of a dense subspace of such operators is impossible using digital computation. However, in this paper, we avoid this difficulty by choosing the operator Q to have the form of $Q(s,z) = F(s,z(0),z(T_k),z(s))$. This choice for the structure of Q comes from the proof of Theorem 1 and is non-conservative.

2.3 Sum-of-Squares Optimization

Theorem 1 reduces the question of global exponential stability of sampled-data systems to the existence of a Lyapunov function V and a piecewise-continuous 'spacing function' Q, which jointly satisfy certain pointwise constraints. Specifically, using the structure $Q_k(s,z) = F_k(s,z(0),z(T_k),z(s))$, we require

$$\begin{aligned}Q_k(T_k,z) &= F_k(T_k,z(0),z(T_k),z(T_k))\\ &= e^{-2\alpha T_k} F_k(0,z(0),z(T_k),z(0))\\ &= e^{-2\alpha T_k} Q_k(0,z)\end{aligned}$$

and

$$\begin{aligned}&\frac{\mathrm{d}}{\mathrm{d}t}[V(x(t)) + Q_k(t-t_k,x_k)] + 2\alpha V(x(t)) + 2\alpha Q_k(t-t_k,x_k)\\ &= \nabla V(x)^T f(x(t),x(t_k)) + \nabla_x F_k(t-t_k,x(t),x(t_{k+1}),x(t_k))^T f(x)\\ &\quad + \frac{d}{dt} F_k(t-t_k,x(t),x(t_{k+1}),x(t_k)) + 2\alpha V(x(t))\\ &\quad + 2\alpha F_k(t-t_k,x(t),x(t_{k+1}),x(t_k)) < -\delta \|x(t)\|\end{aligned}$$

for all $x(t),x(t_{k+1}),x(t_k) \in \mathbb{R}^n$ and $t - t_k \in [0,T_k]$.

To find the functions F_k and V and enforce these constraints, we must optimize functional variables subject to positivity constraints. While this is a very difficult form of optimization, there has been recent progress in this area through the use of sum-of-squares variables. Specifically, we assume the functions F and V are polynomials of bounded degree. The vector space of polynomials of bounded degree is finite dimensional and can be represented using e.g. a set of monomial basis functions. Specifically, if we define the vector of monomials in variables x of degree d or less as $Z_d(x)$, then we can assume that F_k has the form

$$F_k(s,x,y,z) = c^T Z_d(s,x,y,z)$$

for some vector $c \in \mathbb{R}^n$. To enforce the positivity constraints, we assume that any positive polynomial, h can be represented as the sum of squared polynomials as

$$h(x) = \sum_i g_i(x)^2.$$

While this assumption is somewhat conservative, the conservatism is not significant, as Sum-of-Squares polynomials are known to be dense in the set of positive polynomials. The key advantage to requiring positive polynomials to be sum-of-squares is that the set of sum-of-squares polynomials of bounded degree is precisely parameterized by the set of positive semidefinite matrices with size corresponding to the degree of the polynomials. That is, a polynomial $y(x) = c^T Z_{2d}(x)$ is SOS if and only if

$$y(x) = c^T Z_{2d}(x) = Z_d(x)^T Q Z_d(x)$$

for some positive semidefinite matrix Q and where recall Z_d is the vector of monomials in variables x of degree d or less. Thus the constraint that y be a SOS polynomial is equivalent to a set of linear equality constraints between the variables c and Q, as well as the constraint that $Q \geq 0$. Thus optimization of SOS polynomials is actually a form of semidefinite programming - for which we have efficient numerical algorithms and implementations - e.g. [15, 16].

Notation: We denote the constraint that a polynomial p be Sum-of-Squares as $p \in \Sigma_s$.

While polynomials which are SOS will always be globally positive, we occasionally would like to search for polynomials which are only positive on a subset of $\mathbb{R}^n$. This is typically accomplished through the use of SOS multipliers, formalized through certain 'Positivstellensatz' results.

Lemma 1. *Suppose that there exists polynomials t_i and SOS polynomial $s_i \in \Sigma_s$ such that*

$$v(x) = s_0(x) + \sum_i s_i(x) g_i(x) + \sum_j t_i(x) h_j(x)$$

Then $v(x) \geq 0$ for any $x \in X := \{x \in \mathbb{R}^n : g_i(x) \geq 0, h_i(x) = 0\}$.

Thus if we can represent the subset of interest X as a semialgebraic set, then we can enforce positivity on this set using SOS and polynomial variables. Note that v in Lemma 1 is not itself a Sum-of-Squares.

As an example, if we wish to enforce positivity of $F_k(s,x,y,z)$ on the interval $s \in [0, T_k]$, then we can search for *SOS* functions s_0, s_1 such that

$$F_k(s,x,y,z) = s_0(s,x,y,z) + s_1(s,x,y,z) g(s)$$

where $g(s) = s(T_k - s)$. This function g was chosen because $s \in [0, T_k]$ if any only if $g(s) \geq 0$. Positivstellensatz results [17–19] give conditions under which Lemma 1 is not conservative.

Polynomial positivity and Sum-of-Squares have been studied for some time. For additional information, we refer the reader to the references [13, 20–23].

3 Main Results

Now that we have described our approach, the main results of the paper follow directly. We will describe both the synchronous and asynchronous cases and consider global exponential stability. Note that in the following theorems, we restrict Q to have the structure

$$Q_k(s,z) = F_k(s,x(0),x(s)).$$

That is, there is no dependence on $x(T_k)$. This was done in order to be consistent with our approach to linear Sampled-Data systems described in [14] and also to reduce the computational complexity of the stability conditions. This restriction may, however, introduce additional conservatism and should be considered carefully by the user.

3.1 The Synchronous Case

We first consider stability in the 'synchronous' case - that is, when $T_i = T_j$ for all $i, j > 0$. In this case, the updates to the state occur after regular intervals. As we have argued before, this case is often unrealistic. However, there exist certain scenarios where this model is relevant - such as in the case of an A/D converter. Synchronous sampled-data systems are well-represented by conversion to a discrete-time system as the resulting state update law

$$x_{k+1} = f(x_k)$$

will not depend on k. However, as we mentioned before, derivation and stability analysis of the resulting nonlinear discrete-time system are still difficult problems. For this reason and others, the method we outline in this section will not rely on conversion to discrete-time, but will use SOS programming to perform global exponential stability analysis while retaining the full hybrid model of the dynamics.

Theorem 2. *Suppose there exist polynomials V, F, s_0, and s_1 such that*

$$V(x) - \mu_1\|x\|^2 \in \Sigma_s \tag{5}$$

$$\nabla V(z)^T f(z,x) + \nabla_z F(t,x,z)^T f(z,x) + \frac{d}{dt}F(t,x,z) + 2\alpha V(z) + 2\alpha F(t,x,z)$$
$$= -s_0(t,x,z) - s_1(t,x,z)t(T-t) \tag{6}$$

$$F(T,x,y) = e^{-2\alpha T} F(0,x,x) \tag{7}$$

then if $T_k = T$ for all $k > 0$, System 1 is globally exponentially stable.

Proof. Using V as given and $Q_k(t,z) = F(t,z(0),z(t))$ for all $k > 0$, we first get from Condition (5) that

$$V(x(t)) - \mu_1\|x(t)\|^2 \geq 0$$

and hence

$$V(x(t)) \geq \mu_1\|x(t)\|^2.$$

Furthermore, since V is a polynomial, it is upper bounded by some function $\mu_2\|x\|^p$ for sufficiently large μ_2 and p.

Next, we see that from Condition (7),

$$\begin{aligned} Q_k(T_k,z) &= F(T,z(0),z(T)) \\ &= e^{-2\alpha T}F(0,z(0),z(0)) \\ &= e^{-2\alpha T_k}Q_k(0,z). \end{aligned}$$

Finally, we have from Condition (6) and Lemma 1 that

$$\begin{aligned} &\frac{\mathrm{d}}{\mathrm{d}t}[V(x(t))+Q_k(t-t_k,x_k)]+2\alpha V(x(t))+2\alpha Q_k(t-t_k,x_k) \\ &= \nabla V(x(t))^T f(x(t),x(t_k))+\nabla_3 F(t,x(t_k),x(t))^T f(x(t),x(t_k)) \\ &\qquad +\nabla_1 F(t,x(t_k),x(t))+2\alpha V(x(t))+2\alpha F(t,x(t_k),x(t)) \\ &\leq 0 \end{aligned}$$

for all $s \in [0,T]$. Thus the conditions for exponential stability in Theorem 1 are satisfied. We conclude that System (1) is stable if $T = T_k$ for all $k > 0$.

3.2 *The Asynchronous Case*

In this Subsection, we consider the case when the sampling period is time-varying, yet is known to lie within some interval $[T_{\min}, T_{\max}]$. To illustrate, suppose that during a Denial-of-Service attack the rate of controller updates is reduced, but still does not drop below the rate of one packet per second. Thus implies a maximum sampling period of $T_{\max} = 1s$. However, it is possible and even likely that the duration between most of the updates during and after the attack may be significantly less that this $T_{\max}$. Hence, there is also a minimum sample time determined to be either $T_{\min} = 0$ or possibly to be the communication delay between controller and system if the application is tele-operation.

To address the problem where we have $T_k \in [T_{\min}.T_{\max}]$, we allow the 'spacing function' F to vary with T_k. This is allowable since the spacing function is not part of the storage function, V.

Note that we do not allow V to be a function of T_k. The restriction that V not vary with k is similar to the Quadratic Stability condition for general classes of switched systems. However, while quadratic stability is known to be conservative for general classes of hybrid system, for sampled-data systems it is not known whether quadratic stability is conservative.

Theorem 3. *Suppose there exist constant α and polynomials V, F, s_0, and s_1 such that*

$$V(x) - \mu_1 \|x\|^2 \in \Sigma_s \tag{8}$$

$$\begin{aligned}
&\nabla V(z)^T f(z,x) + \nabla_z F(t,x,z,T)^T f(z,x) + \frac{d}{dt} F(t,x,z,T) + 2\alpha V(z) \\
&\qquad\qquad + 2\alpha F(t,x,z,T) \\
&= -s_0(t,x,z,T) - s_1(t,x,z,T)t(T-t) - s_2(t,x,z,T)(T-T_{\min})(T_{\max}-T)
\end{aligned} \tag{9}$$

$$F(T,x,y,T) = e^{-2\alpha T_{\max}} F(0,x,x) \tag{10}$$

then if $T_k \in [T_{\min}, T_{\max}]$ for all $k > 0$, System (1) *is globally exponentially stable.*

Proof. The proof is similar to the synchronous case. We use $V(x)$ as given and define $Q_k(t,z) = F(t,z(0),z(t),T_k)$ for all $k > 0$. From Condition (8) we have that

$$V(x(t)) \geq \mu_1 \|x(t)\|^2.$$

As before, since V is a polynomial, it is upper bounded by some function $\mu_2 \|x\|^p$ for sufficiently large μ_2 and p.

Next, we see that from Condition (10),

$$\begin{aligned}
Q_k(T_k,z) &= F(T_k,z(0),z(T_k),T_k) \\
&= e^{-2\alpha T_k} F(0,z(0),z(0),T_k) \\
&= e^{-2\alpha T_k} Q_k(0,z).
\end{aligned}$$

Finally, we have from Condition (9) and Lemma 1 that

$$\begin{aligned}
&\frac{\mathrm{d}}{\mathrm{d}t}[V(x(t)) + Q_k(t-t_k,x_k)] + 2\alpha V(x(t)) + 2\alpha Q_k(t-t_k,x_k) \\
&= \nabla V(x(t))^T f(x(t),x(t_k)) + \nabla_3 F(t,x(t_k),x(t),T_k)^T f(x(t),x(t_k)) \\
&\qquad + \nabla_1 F(t,x(t_k),x(t),T_k) + 2\alpha V(x(t)) + 2\alpha F(t,x(t_k),x(t),T_k) \\
&\leq 0
\end{aligned}$$

for all $s \in [0,T_k]$ and $T_k \in [T_{\min}, T_{\max}]$. Thus the conditions for exponential stability in Theorem 1 are satisfied. We conclude that System (1) is stable if $T_k \in [T_{\min}, T_{\max}]$ for all $k > 0$.

4 Numerical Examples

To verify the algorithms described above, we performed global stability analysis on a set of nonlinear sampled-data systems. In the examples considered here, we let $\alpha \cong 0$, meaning that we are not interested in finding exponential rates of decay. For a study of estimating exponential rates of decay as a function of sampling period for linear systems, we refer to [14].

4.1 Example 1: 1-D Nonlinear Dynamical System

For our first set of numerical examples, we consider the class of 1-D nonlinear dynamical systems parameterized by

$$\dot{x}(t) = f(x(t)) = ax(t)^3 + bx(t)^2 + cx(t)$$

where we assume that the $u(t) = cx(t)$ term represents negative feedback. Without sampling, we know this system is globally stable if and only if $x(t)f(x(t)) > 0$ for all $x \neq 0$. It can be shown that this condition is satisfied if and only if $a < 0$ and $c < \frac{b^2}{4a}$. For this example, we initially chose $a = -1$, $b = 2$ and $c = -1.1$. Then, we used a sampled signal for the term $u(t) = cx(t_k)$ to get the following dynamics.

$$\dot{x}(t) = -x(t)^3 + 2x(t)^2 - 1.1x(t_k).$$

In Table 1, we list the maximum verifiably globally stable sampling period for this system as a function of the polynomial degree used for the variables V, F and s_1. These results were obtained using the conditions of Theorem 2 implemented using SOSTOOLS coupled with SeDuMi. Due to the known potential for numerical inaccuracies, all solutions were verified a-posteriori using SOS and via simulation. The resulting Lyapunov function and function F are illustrated over a single sampling period in Figure 1. The evolution of the system can be seen over multiple sampling periods in Figure 2.

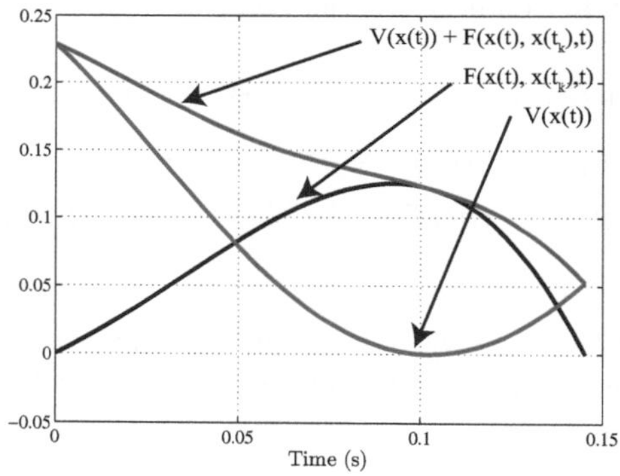

Fig. 1 Evolution of V and F over one sampling period for Numerical Example 1

4.2 Example 2: Controlled Model of a Jet Engine

In our second example, we consider a controlled model of a jet engine with dynamics

$$\dot{x}(t) = -y(t_k) - \frac{3}{2}x(t)^2 - \frac{1}{2}x(t)^3$$
$$\dot{y}(t) = -y(t) + x(t)$$

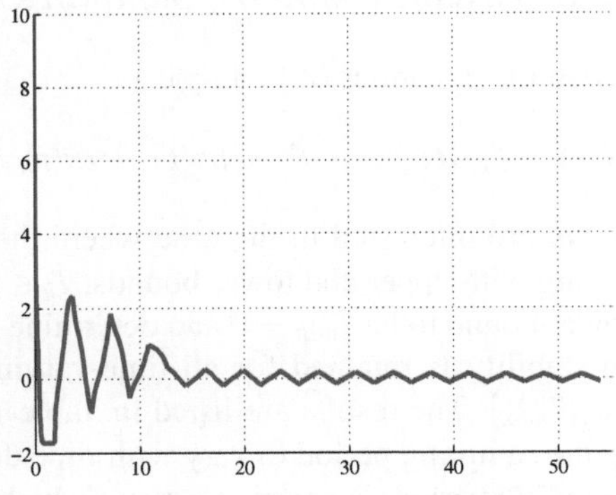

Fig. 2 Evolution of $x(t)$ over 30 sampling periods for Numerical Example 1 with $T_s = 1.8$

We consider the case where the negative feedback to the first state is provided using a sampled-data controller. When $T_s = 0$, this system is known to be globally stable. Figure 3 illustrates the trajectories of this system plotted against the level set of one such Lyapunov function for $T_s = .4$.

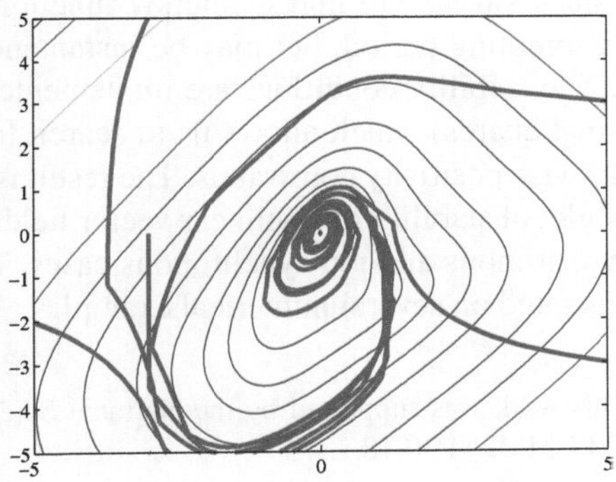

Fig. 3 Level Sets of a Lyapunov function for Example 2, with multiple trajectories simulated over 30 sampling periods

Table 1 Maximum allowable sampling period T_s for Examples 1, 2, and 3 with $T_1 = 0$

Degree	$N = 2$	$N = 4$	$N = 6$	$N = 8$	$N = 10$
Example 1: Maximum Synchronous T_s	∅	0.7901	1.5449	1.8192	1.8411
Example 2: Maximum Synchronous T_s	∅	.171	.4599	N/A	N/A
Example 3: Maximum Asynchronous T_s	∅	.7891	1.542	N/A	N/A

4.3 Example 3: 1-D System, Unknown Sampling Period

In this example, we revisit the dynamics of Example 1.

$$\dot{x}(t) = f(x(t)) = ax(t)^3 + bx(t)^2 + cx(t)$$

However, in this case, we are interested in the case where the sampling period is unknown and time-varying with upper and lower bounds, $T_k \in [T_{\min}, T_{\max}]$. Specifically, we choose the lower bound to be $T_{\min} = 0$ and determine the maximum upper bound $T_{\max}$ for which stability is retained for all time-varying sampling periods which satisfy $T_k \in [T_{\min}, T_{\max}]$. The results are listed in Table 1. As we can see, in this example, allowing the sampling period to vary with time does not significantly affect the maximum sampling period - a surprising result which indicates that using a Lyapunov function V which does not depend on T_k may not be conservative.

5 Conclusion

In this chapter, we have studied the question of global stability of nonlinear sampled-data systems in both the synchronous and asynchronous cases. These systems arise through the use of digitized sensing and actuation to control continuous-time dynamics where the controller updates may be irregular. Our approach has been to exploit a new type of slack variable to find Lyapunov functions which experience net decrease over each sampling period, but may be instantaneously increasing at certain points in time. The stability conditions are implemented using a new form of optimization (Sum-of-Squares) which allows us to search for polynomial functions which satisfy pointwise positivity constraints. The result is a convex algorithm which is able to assess global stability of nonlinear vector fields with sampled-data signals in both the asynchronous and the synchronous cases. The effectiveness of the algorithm is demonstrated on several numerical examples.

Acknowledgements. This work was supported by the National Science Foundation under Grants CMMI 110036 and CMMI 1151018.

References

1. Chen, T., Francis, B.: Optimal sampled-data control systems. Springer, Berlin (1995)
2. Fridman, E., Seuret, A., Richard, J.-P.: Robust sampled-data stabilization of linear systems: An input delay approach. Automatica 40(8), 1141–1446 (2004)
3. Fujioka, H.: Stability analysis of systems with aperiodic sample-and-hold devices. Automatica 45(3), 771–775 (2009)
4. Zhang, W., Branicky, M., Phillips, S.: Stability of networked control systems. IEEE Control Systems Magazine (21) (2001)
5. Jury, E., Lee, B.: On the stability of a certain class of nonlinear sampled-data systems. IEEE Transactions on Automatic Control 9(1), 51–61 (1964)

6. Zaccarian, L., Teel, A.R., Nešić, D.: On finite gain lp stability of nonlinear sampled-data systems. Systems & Control Letters 49(3), 201–212 (2003)
7. Mikheev, Y., Sobolev, V., Fridman, E.: Asymptotic analysis of digital control systems. Automation and Remote Control 49(9), 1175–1180 (1988)
8. Suh, Y.: Stability and stabilization of nonuniform sampling systems. Automatica 44(12), 3222–3226 (2008)
9. Oishi, Y., Fujioka, H.: Stability and stabilization of aperiodic sampled-data control systems: An approach using robust linear matrix inequalities. In: Joint 48th IEEE Conference on Decision and Control and 28th Chinese Control Conference, December 16-18 (2009)
10. Hetel, L., Daafouz, J., Iung, C.: Stabilization of arbitrary switched linear systems with unknown time-varying delays. IEEE Transactions on Automatic Control 51(10), 1668–1674 (2006)
11. Peet, M., Papachristodoulou, A.: A converse sum of squares Lyapunov result with a degree bound. IEEE Transactions on Automatic Control 57(9) (2012)
12. Seuret, A.: A novel stability analysis of linear systems under asynchronous samplings. Automatica 48(1), 177–182 (2012)
13. Peet, M.M., Papachristodoulou, A., Lall, S.: Positive forms and stability of linear time-delay systems. SIAM Journal on Control and Optimization 47(6) (2009)
14. Seuret, A., Peet, M.: Stability analysis of sample-data systems using sum-of-squares. IEEE Transactions on Automatic Control 58(6) (2013)
15. Sturm, J.F.: Using SeDuMi 1.02, a Matlab Toolbox for optimization over symmetric cones. Optimization Methods and Software 11-12, 625–653 (1999)
16. Borchers, B.: CSDP, a C library for semidefinite programming. Optimization Methods and Software 11(1-4), 613–623 (1999)
17. Stengle, G.: A nullstellensatz and a positivstellensatz in semialgebraic geometry. Mathematische Annalen 207, 87–97 (1973)
18. Schmüdgen, C.: The K-moment problem for compact semi-algebraic sets. Mathematische Annalen 289(2), 203–206 (1991)
19. Putinar, M.: Positive polynomials on compact semi-algebraic sets. Indiana Univ. Math. J. 42(3), 969–984 (1993)
20. Reznick, B.: Some concrete aspects of Hilbert's 17th problem. Contemporary Mathematics 253, 251–272 (2000)
21. Parrilo, P.A.: Structured semidefinite programs and semialgebraic geometry methods in robustness and optimization. Ph.D. dissertation, California Institute of Technology (2000)
22. Lasserre, J.B.: A sum of squares approximation of nonnegative polynomials. SIAM Journal of Optimization 16(3), 751–765 (2006)
23. Chesi, G., Garulli, A., Tesi, A., Vincino, A.: Polynomially parameter-dependent Lyapunov functions for robust stability of polytopic systems: An lmi Approach 50(3), 365–370 (2005)

DDE Model-Based Control of Glycemia via Sub-cutaneous Insulin Administration

Pasquale Palumbo, Pierdomenico Pepe, Simona Panunzi, and Andrea De Gaetano

Abstract. Plasma glucose regulation is commonly attained in Type 1 Diabetes Mellitus (T1DM) patients, as well as in advanced Type 2 Diabetes Mellitus (T2DM), by means of Sub-Cutaneous (SC) insulin administration. In order to study this extremely common and relevant clinical problem from a theoretical point of view, a Delay Differential Equation (DDE) model of the glucose-insulin system has been considered. The model extends a previous DDE model, already used for glucose control, by endowing it with a SC Insulin compartment and by introducing modifications regarding insulin-independent glucose uptake and Hepatic Glucose Output (HGO). Pancreatic insulin release (non-negligible in T2DM) is considered, in order for the control method to address both T1DM and T2DM. The method of exact input/output feedback linearization and stabilization is used, to ensure the local convergence of the tracking error to zero. Numerical simulations show the effectiveness of the proposed approach.

1 Introduction

Diabetes Mellitus comprises a group of metabolic diseases characterized by hyperglycemia. The chronic hyperglycemia of diabetes is associated with long-term damage, dysfunction, and failure of different organs, especially eyes, kidneys, nerves,

Pasquale Palumbo · Simona Panunzi · Andrea De Gaetano
BioMatLab, Istituto di Analisi dei Sistemi ed Informatica "A. Ruberti",
Consiglio Nazionale delle Ricerche (IASI-CNR), Largo A. Gemelli 8, 00168, Rome, Italy
e-mail: pasquale.palumbo@iasi.cnr.it,
{simona.panunzi,andrea.degaetano}@biomatematica.it

Pierdomenico Pepe
Dipartimento di Ingegneria e Scienze dell'Informazione, e Matematica,
University of L'Aquila
e-mail: pierdomenico.pepe@univaq.it

T. Vyhlídal, J.-F. Lafay, and R. Sipahi (eds.), *Delay Systems*,
Advances in Delays and Dynamics 1,
DOI: 10.1007/978-3-319-01695-5_17, © Springer International Publishing Switzerland 2014

heart, and blood vessels. Patients with diabetes have an increased incidence of atherosclerotic cardiovascular, peripheral vascular, and cerebrovascular diseases.

In a healthy person, blood glucose is maintained between 3.9 mmol/L and 6.9 mmol/L by means of a complex control system which ensures a balance between glucose entering the bloodstream after liver gluconeogenesis and intestinal absorption following meals, and glucose uptake from the peripheral tissues. This balance is regulated mainly by insulin, a hormone produced by the β-cells of the pancreas when stimulated by rising level of plasma glycemia: insulin enhances glucose uptake in muscles and adipose tissues and promotes the storage of excess circulating glucose in the liver.

A pathological increase in blood glucose concentration (hyperglycemia) results from defects in insulin secretion, insulin action, or both. In case of an absolute deficiency of insulin secretion, caused by an autoimmune destruction of the pancreatic β cells, Type 1 Diabetes Mellitus (T1DM) occurs: these patients require exogenous insulin administration for survival. On the other hand, in case of hyperglycemia caused by a combination of resistance to insulin action and inadequate compensatory insulin secretory response, Type 2 Diabetes Mellitus (T2DM) occurs: these patients usually have a relative (rather than absolute) insulin deficiency, increased levels of circulating glucose, and may or may not need supplemental insulin therapy.

Exogenous insulin administration is a basic procedure to cope with any malfunctioning of the endogenous insulin feedback action (in T1DM only exogenous insulin is available, while in T2DM exogenous insulin complements pancreatic production of the hormone). Control of glycemia by means of subcutaneous insulin injections, with the dose adjusted on the basis of capillary plasma glucose concentration measurements, is by far the most widespread insulin therapy, since the dose is habitually administered by the patients themselves (see [1] and references therein). In order to design closed-loop control strategies in this case, insulin absorption from the subcutaneous depot needs to be considered.

This note proposes to synthesize a closed-loop insulin administration according to a model-based approach. The advantages of a model-based approach are evident since, by using a glucose/insulin model of the subject, the control problem may be treated mathematically and optimal strategies may be determined. Clearly, the more accurate the model, the more effective is the control law. Different approaches have been proposed in the literature, based on nonlinear models such as the Minimal Model [2], or more exhaustive compartmental models [6, 11, 17, 34]: see, e.g., papers on Model Predictive Control, [10, 18, 29], on Parametric Programming, [7], on H_∞ control [4, 13, 14, 30, 33]. It has to be stressed that most of these approaches are based on the approximation of the original nonlinear model, provided by linearization, discretization and model reduction (balanced truncation). An excellent review of the available models presently adopted for blood glucose regulation as well as the closed loop control methodologies and technical devices (blood glucose sensors and insulin pumps) may be found in [3] and references therein.

Differently from previously mentioned model-based approaches, which use nonlinear Ordinary Differential Equation (ODE) models, the one presented here uses a nonlinear Delay Differential Equation (DDE) model to describe the glucose/insulin regulatory system. Such a model consists of a slight modification of a previous model already published, [23, 28]. In fact, it may be considered as an extension of [28], modeling in more details the insulin-independent glucose uptake and the Hepatic Glucose Output (HGO) in the glucose kinetics. On the other hand, the insulin kinetics is not changed, with an explicit apparent delay allowing a better representation of pancreatic Insulin Delivery Rate (IDR) (e.g. [16, 19] and references therein). In fact, DDE models have been recently exploited with the aim of glucose control in [24, 26, 27], where an intra-venous insulin administration was designed to track a desired plasma glycemia. In that case the DDE model approach allowed to synthesize a closed-loop control for virtual T2DM patients, where the IDR is not absent, differently from T1DM. In this chapter we aim to achieve the same goal of tracking a desired glucose reference, by means of subcutaneous (instead of intra-venous) infusions. Therefore a linear model of the subcutaneous insulin absorption has been considered, in order to design the closed loop (see [15, 21, 35] to have a comprehensive review of the many different models of insulin absorption). The model of insulin absorption here adopted refers to [32] with no insulin degradation at the injection site. It has been recently analyzed in the papers of [35] and [5], and it has been exploited with the aim of glucose control in [10], whose notation we use in the present work.

The proposed control law is based on recent results on differential geometry for time-delay systems (see [8], [9], [20], and [22]). An exactly linearized input/output map is first obtained by a nonlinear inner feedback which makes use of the state variables at the current and at delayed time. Tracking of the output (blood glucose concentration) is then achieved by means of an outer feedback on the exactly linear input/output map. No approximations have been used in this contribute. The control law is obtained without linearizing the system equations (i.e. without first order approximations). The control law here provided is meant to work also in case of large deviations from the desired final level, and not only for small deviations.

Preliminary results have been presented in [25], where a different DDE model has been used [28].

Numerical simulations show the effectiveness of the proposed control law, and encourage further developments involving insulin infusion therapies more and more constrained to real frameworks.

2 The Glucose-Insulin Model

Denote $G(t)$, [mM], $I(t)$, [pM], plasma glycemia and insulinemia, respectively, and S_1 [pmol], S_2 [pmol] the insulin mass in the accessible and not-accessible subcutaneous depot, respectively. The model considered consists of a single discrete-delay differential equation system:

$$
\begin{aligned}
\frac{dG(t)}{dt} &= -T_{xg}\frac{G(t)}{G(t)+\widetilde{G}} - K_{xgi}G(t)I(t) + \frac{T_{ghmax}}{V_G}e^{-\lambda G(t)I(t)},\\
\frac{dI(t)}{dt} &= -K_{xi}I(t) + \frac{T_{iGmax}}{V_I}f\big(G(t-\tau_g)\big) + \frac{1}{V_I t_{max,I}}S_2(t),\\
\frac{dS_2(t)}{dt} &= \frac{1}{t_{max,I}}S_1(t) - \frac{1}{t_{max,I}}S_2(t),\\
\frac{dS_1(t)}{dt} &= -\frac{1}{t_{max,I}}S_1(t) + u(t)
\end{aligned}
\tag{1}
$$

where

- T_{xg}, [mM/min], is the maximal insulin-independent rate constant for glucose brain uptake;
- $\widetilde{G}$, [mM], is the glycemia at which the insulin-independent rate is half of its maximal value
- K_{xgi}, [min^{-1} pM^{-1}], is the rate of glucose uptake by tissues (insulin-dependent) per pM of plasma insulin concentration;
- T_{ghmax}, [min^{-1}(mmol/kgBW)], is the maximal hepatic glucose output at zero glycemia, zero insulinemia;
- V_G, [L/kgBW], is the apparent distribution volume for glucose;
- λ, [$\text{mM}^{-1}\text{pM}^{-1}$], is the rate constant for hepatic glucose output decrease with increase of glycemia and insulinemia
- K_{xi}, [min^{-1}], is the apparent first-order disappearance rate constant for insulin;
- T_{iGmax}, [min^{-1}(pmol/kgBW)], is the maximal rate of second-phase insulin release;
- V_I, [L/kgBW], is the apparent distribution volume for insulin;
- τ_g, [min], is the apparent delay with which the pancreas varies secondary insulin release in response to varying plasma glucose concentrations;
- $t_{max,I}$, [min], is the time-to-maximum insulin absorption;
- $u(t)$, [pM/min], is the subcutaneous insulin delivery rate, i.e. the control input.

The nonlinear function $f(\cdot)$ models the pancreas Insulin Delivery Rate as:

$$
f(G) = \frac{\left(\frac{G}{G^*}\right)^\gamma}{1+\left(\frac{G}{G^*}\right)^\gamma},
\tag{2}
$$

where γ is the progressivity with which the pancreas reacts to circulating glucose concentrations and G^* [mM] is the glycemia at which the insulin release is half of its maximal rate.

Besides the presence of the subcutaneous insulin compartment, the model differs from the DDE model [23,28] already used in glucose control [24,26,27] for the glucose kinetics in (1). More in details, modifications concern the insulin-independent glucose uptake, mainly due to the brain and the nerve cells, and the Hepatic Glucose Output (HGO). As far as the first term, except very low cases of hypoglycemia, it may well be approximated as a constant term ($\simeq -T_{xg}$) as, indeed, it has been done in the original DDE model [23,28]. As far as the HGO, instead to be considered constant (like in [23,28]), it is made dependent on circulating plasma glucose and insulin, motivated by the fact that liver glucose production is suppressed and glycogen synthesis is enhanced in the presence of high plasma glucose and insulin concentrations. In the present formulation this indirect relationship has been represented by a decreasing exponential net glucose production for increasing glycemia and insulinemia. Both these modifications have been considered also in a model of the euglycemic hyperinsulinemic clamp [31], within a different mathematical framework (stochastic instead of deterministic as the present one).

3 The Feedback Control Law

We make use here of the elementary theory of nonlinear feedback (see [12]) for time-delay systems (see [8,9,20,22]).

Let $G_{ref}(t)$ be the desired glucose reference signal to be tracked, which we assume to be smooth and bounded. Let

$$x(t) = \begin{bmatrix} x_1(t) \\ x_2(t) \\ x_3(t) \\ x_4(t) \end{bmatrix} = \begin{bmatrix} G(t) \\ I(t) \\ S_2(t) \\ S_1(t) \end{bmatrix}, \qquad y(t) = G(t) - G_{ref}(t) \tag{3}$$

and

$$z(t) = \begin{bmatrix} z_1(t) \\ z_2(t) \\ z_3(t) \\ z_4(t) \end{bmatrix} = \begin{bmatrix} y(t) \\ y^{(1)}(t) \\ y^{(2)}(t) \\ y^{(3)}(t) \end{bmatrix} = \begin{bmatrix} G(t) - G_{ref}(t) \\ \dot{G}(t) - \dot{G}_{ref}(t) \\ \ddot{G}(t) - \ddot{G}_{ref}(t) \\ G^{(iii)}(t) - G^{(iii)}_{ref}(t) \end{bmatrix} \tag{4}$$

with, by suitably exploiting the equations (1):

$$\begin{aligned} \ddot{G}(t) = & -T_{xg}\widetilde{G}\frac{\dot{G}(t)}{\left(G(t)+\widetilde{G}\right)^2} \\ & -\left(K_{xgi} + \lambda\frac{T_{ghmax}}{V_G}e^{-\lambda G(t)I(t)}\right)\left(\dot{G}(t)I(t) + G(t)\dot{I}(t)\right) \end{aligned} \tag{5}$$

$$\begin{aligned} G^{(iii)}(t) = & -T_{xg}\widetilde{G}\frac{(G(t)+\widetilde{G})\ddot{G}(t) - 2\dot{G}(t)^2}{\left(G(t)+\widetilde{G}\right)^3} \\ & +\lambda^2\frac{T_{iGmax}}{V_G}e^{-\lambda G(t)I(t)}\left(\dot{G}(t)I(t)+G(t)\dot{I}(t)\right)^2 \\ & -\left(K_{xgi}+\lambda\frac{T_{iGmax}}{V_G}e^{-\lambda G(t)I(t)}\right)\left(\ddot{G}(t)I(t)+2\dot{G}(t)\dot{I}(t)+G(t)\ddot{I}(t)\right) \end{aligned} \tag{6}$$

$$\ddot{I}(t) = -K_{xi}\dot{I}(t) + \frac{T_{iGmax}}{V_I} \cdot \left.\frac{df(\alpha)}{d\alpha}\right|_{\alpha=G(t-\tau_g)} \cdot \left.\frac{dG(\theta)}{d\theta}\right|_{\theta=t-\tau_g} + \frac{1}{V_I t_{max,I}}\dot{S}_2(t) \tag{7}$$

Notice that $z(t)$ does not depend of the input $u(t)$.
The derivative of $z_4(t)$ is as follows

$$\dot{z}_4(t) = \mu(*) - \left(K_{xgi} + \lambda \frac{T_{ghmax}}{V_G} e^{-\lambda G(t)I(t)}\right) \frac{G(t)}{V_I t_{max,I}^2} u(t) \tag{8}$$

with

$$\begin{aligned}\mu(*) = & -T_{xg}\frac{\widetilde{G}}{(G+\widetilde{G})^4}\left((G(t)+\widetilde{G})^2 G^{(iii)}(t) - 4\dot{G}(t)\ddot{G}(t)(G(t)+\widetilde{G}) + 6\dot{G}(t)^3\right)\\ & + \frac{\lambda^2 T_{ghmax}}{V_G} e^{-\lambda G(t)I(t)}\left(\dot{G}(t)I(t)+G(t)\dot{I}(t)\right)\\ & \cdot\left(-\lambda\left(\dot{G}(t)I(t)+G(t)\dot{I}(t)\right)^2 + 3(\ddot{G}(t)I(t)+2\dot{G}(t)\dot{I}(t)+G(t)\ddot{I}(t))\right)\\ & - \left(K_{xgi} + \lambda\frac{T_{ghmax}}{V_G}e^{-\lambda G(t)I(t)}\right)\\ & \cdot (G^{(iii)}(t)I(t) + 3\ddot{G}(t)\dot{I}(t) + 3\dot{G}(t)\ddot{I}(t) + \beta(*)G(t)) - G_{ref}^{(iv)}(t)\end{aligned} \tag{9}$$

and

$$\begin{aligned}\beta(*) = & -K_{xi}\ddot{I}(t) + \frac{T_{iGmax}}{V_I}\left(\left.\frac{d^2 f(\alpha)}{d\alpha^2}\right|_{\alpha=G(t-\tau_g)}\left(\left.\frac{dG(\theta)}{d\theta}\right|_{\theta=t-\tau_g}\right)^2\right.\\ & \left.\left.\frac{df(\alpha)}{d\alpha}\right|_{\alpha=G(t-\tau_g)}\left.\frac{d^2G(\theta)}{d\theta^2}\right|_{\theta=t-\tau_g}\right) + \frac{1}{V_I t_{max,I}^3}(S_2(t) - 2S_1(t))\end{aligned} \tag{10}$$

Without loss of generality we assume that, for $\theta \in [-\tau_g, 0]$, $\frac{dG(\theta)}{d\theta} = 0$.

Notice that, according to (1), it comes that $\left.\frac{d^2G(\theta)}{d\theta^2}\right|_{\theta=t-\tau_g}$ depends of the delayed state components $G(t-\tau_g)$ and $G(t-2\tau_g)$.

From (5), (6), (8), it follows that the dynamics of vector $z(t)$ is given by:

$$\dot{z}(t) = A_b z(t) + B_b\left(\mu(*) - \left(K_{xgi} + \lambda\frac{T_{ghmax}}{V_G}e^{-\lambda G(t)I(t)}\right)\frac{G(t)}{V_I t_{max,I}^2}u(t)\right) \tag{11}$$

where $\mu(*)$, according to (8) is a suitable function of:

$$G(t),\ I(t),\ G(t-\tau_g),\ S_2(t),\ S_1(t),\ \left.\frac{d^iG(\theta)}{d\theta^i}\right|_{\theta=t-\tau_g},\quad i=1,2 \tag{12}$$

and of $G_{ref}^{(j)}(t)$, $j = 0, 1, \ldots, 4$. Thus, the (inner) feedback control law

$$u(t) = \frac{\mu(*) - v(t)}{\left(K_{xgi} + \lambda \frac{T_{ghmax}}{V_G} e^{-\lambda G(t) I(t)}\right) \frac{G(t)}{V_I t_{max,I}^2}}, \tag{13}$$

where $v(t)$ is a new (outer input), yields the following linear equation

$$\dot{z}(t) = A_b z(t) + B_b v(t), \tag{14}$$

with A_b and B_b a Brunowskii pair, given as follows:

$$A_b = \begin{bmatrix} 0 & 1 & 0 & 0 \\ 0 & 0 & 1 & 0 \\ 0 & 0 & 0 & 1 \\ 0 & 0 & 0 & 0 \end{bmatrix}, \quad B_b = \begin{bmatrix} 0 \\ 0 \\ 0 \\ 1 \end{bmatrix}. \tag{15}$$

Finally, by choosing the new input $v(t)$ as the (outer) feedback

$$v(t) = \Gamma z(t), \tag{16}$$

with Γ a suitable row vector in $\mathbb{R}^4$, the following equation is obtained,

$$\dot{z}(t) = (A_b + B_b \Gamma) z(t) \tag{17}$$

Thus, by designing Γ such that $A_b + B_b \Gamma$ is Hurwitz (this is possible since A_b, B_b is a controllable pair), we get that $z(t)$ goes to zero exponentially, which returns the glucose to converge to the desired reference signal exponentially. From a mathematical point of view, the control law (13), (16) can always be computed, since the variable $G(t)$ (at the denominator of (13)) never vanishes: indeed, as it is required from basic assumptions on the qualitative behavior of the solutions, the glucose dynamics is strictly positive whatever are chosen the initial conditions in the positive orthant (see [23]). It follows that, from a mathematical point of view, the control law (13), (16) can be used with any physically meaningful initial conditions. As well, the equation (17) holds with any physically meaningful initial conditions.

Remark 1. In general, in the application of the elementary theory of nonlinear feedback for systems with time-delays, further dynamics, given by continuous time difference equations, must be taken into account (see [8,9]), even if the relative degree is full (as in our case). In this case, no unstable zero dynamics occur, since the relationship between the variable $x(t)$ and the variable $z(t)$ does not involve any further dynamics.

4 Simulation Results

Simulations have been carried out on a virtual patient identified by the following parameters:

$$
\begin{array}{llll}
G_b = 10.66 & I_b = 49.29 & T_{iGmax} = 0.236 & T_{xg} = 0.0145 \\
\gamma = 3.205 & G^* = 9 & \tau_g = 24 & \lambda = 7.9 \cdot 10^{-3} \\
V_G = 0.187 & K_{xi} = 1.211 \cdot 10^{-2} & T_{ghmax} = 0.194 & t_{max,I} = 55 \\
V_I = 0.25 & K_{xgi} = 3.11 \cdot 10^{-5} & \widetilde{G} = 0.1 &
\end{array}
\tag{18}
$$

Parameters $t_{max,I}$ and λ have been taken from [10] and [31], respectively. As far as T_{xg} it is supposed to be about 0.19 mmol/min. In fact considering a daily intake of 2000 Kcal corresponding to about a daily intake of 500 gr of glucose and supposing that the brain consumption is the 10% of the total glucose amount, the brain consumption can be estimated at 0.0145 mM/min (L = 0.19 · 70 Kg). The other parameters refer to a Type 2 diabetic patient, with a substantial degree of insulin resistance, see also [25].

In order to regulate the resulting hyperglycemia down to a safe level, we choose matrix Γ such that the closed loop matrix $A + B\Gamma$ has eigenvalues -0.0464, -0.0512, -0.0496, -0.0480. The reference signal is chosen such to obtain the plasma glycemia decreasing exponentially from the value of 10.66 to the new value 4.7:

$$
G_{\text{ref}}(t) = 4.7 + (10.66 - 4.7) \cdot \exp(-0.01t). \tag{19}
$$

The subject is supposed to be at rest before the experiment begins, which means that the initial state is given by $G_0(\tau) = G_b$, $I_0(\tau) = I_b$ for $\tau \in [-\tau_g, 0]$. Initial conditions for the subcutaneous depot are $S_1(0) = 0$ and $S_2(0) = 0$.

As it clearly appears from Fig.1, a reasonably low plasma glycemia (i.e. < 6mM) is reached within the first four hours of simulation.

Fig.2 shows the plasma insulin concentration, Fig.3 reports the insulin infusion rate to be administered, according to the proposed control law and Fig.4 reports the insulin in the subcutaneous depots.

Remark 2. While the control input cannot be negative, the proposed approach does not consider, from a theoretical point of view, saturation problems for the control law. We have taken into account this fact in the simulations: whenever the designed control law becomes negative, a zero control input is given to the system. Note that we have chosen the control parameters in order not to allow such a drawback, as it appears from Fig.3.

Remark 3. As a final remark, we point out that the present chapter is a key starting point for further, and more significant results. Indeed, the proposed control law is based on the complete knowledge of the state vector, which means available

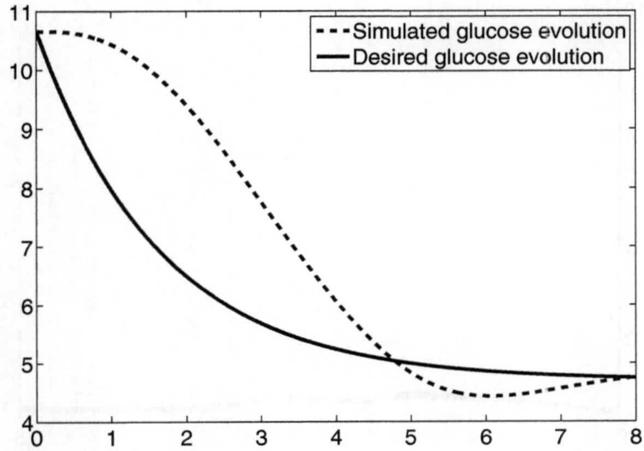

Fig. 1 Plasma glycemia [mM], compared with the desired glucose reference; time is in hours

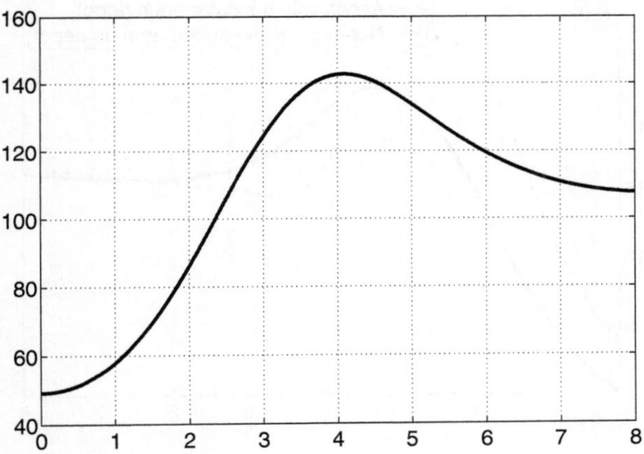

Fig. 2 Plasma insulinemia [pM]; time is in hours

measurements from both glycemia and insulinemia, as well as from the subcutaneous depot. Of course, such assumptions are far to be reliable, especially for what concerns the subcutaneous depot. The next step, which is a work in progress of the authors, will be to design the feedback control by using only plasma glucose measurements, by suitably exploiting a state observer for time-delay systems, as it has already been presented in [27] for the case of intra-venous insulin administration.

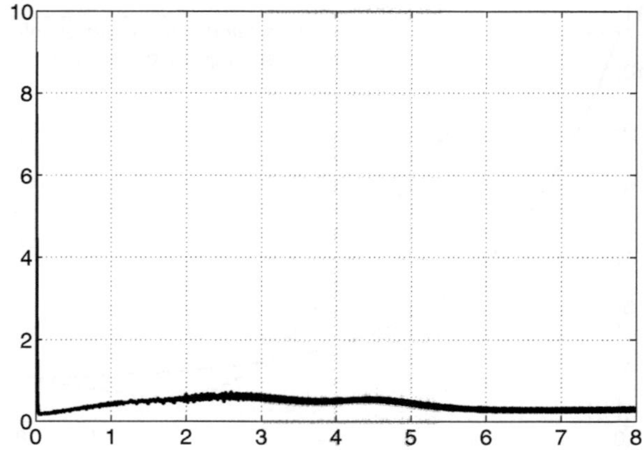

Fig. 3 Insulin infusion rate $u(t)$, [pmol/min]; time is in hours

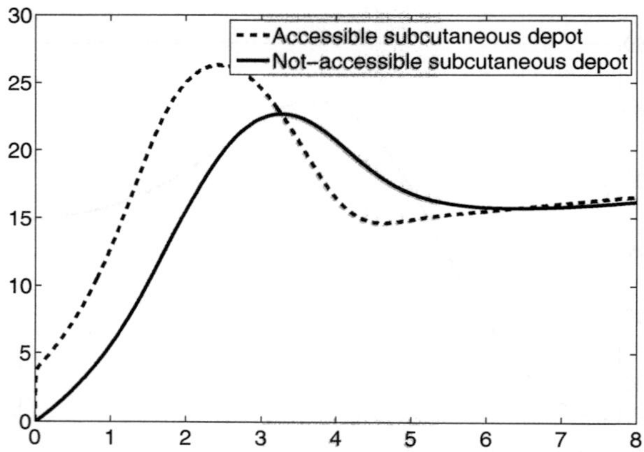

Fig. 4 Insulin mass in the sc depots [pmol]; time is in hours

5 Conclusions

A closed-loop control law has been proposed for the glucose-insulin system, with sub-cutaneous administration of insulin. A DDE-model based approach has been adopted, with the mathematical model given by the extension of a recent DDE-model. Asymptotic tracking of a desired time evolution for the blood glucose concentration is achieved by means of this nonlinear control law. The method of exact input/output feedback linearization for nonlinear time-delay systems is used. No linear approximations are used. Numerical simulations show the good performance of the proposed control law.

References

1. Bellazzi, R., Nucci, G., Cobelli, C.: The subcutaneous route to insulin-dependent diabetes therapy. IEEE Engineering in Medicine and Biology 20, 54–64 (2001)
2. Bergman, R.N., Ider, Y.Z., Bowden, C.R., Cobelli, C.: Quantitative estimation of Insulin sensitivity. Am. Journal on Physiology 236, 667–677 (1979)
3. Chee, F., Fernando, T.: Closed-loop control of blood glucose. Springer, Heidelberg (2007)
4. Chee, F., Savkin, A.V., Fernando, T.L., Nahavandi, S.: Optimal H_∞ insulin injection control for blood glucose regulation in diabetic patients. IEEE Trans. on Biomedical Engineering 52, 1625–1631 (2005)
5. Clausen, W.H.O., De Gaetano, A., Vølund, A.: Within-patient variation of the pharmacokinetics of subcutaneously injected biphasic insulin aspart as assessed by compartmental modelling. Diabetologia 49, 2030–2038 (2006)
6. Dalla Man, C., Rizza, R.A., Cobelli, C.: Meal simulation model of the glucose-insulin system. IEEE Trans. Biomedic. Eng. 54, 1740–1749 (2007)
7. Dua, P., Doyle, P.F.J., Pistikopoulos, E.N.: Model-based blood glucose control for Type 1 diabetes via parametric programming. IEEE Transactions on Biomedical Engineering 53, 1478–1491 (2006)
8. Germani, A., Manes, C., Pepe, P.: Local Asymptotic Stability for Nonlinear State Feedback Delay Systems. Kybernetika 36, 31–42 (2000)
9. Germani, A., Manes, C., Pepe, P.: Input-Output Linearization with Delay Cancellation for Nonlinear Delay Systems: the Problem of the Internal Stability. International Journal of Robust and Nonlinear Control 13(9), 909–937 (2003)
10. Hovorka, R., Canonico, V., Chassin, L.J., Haueter, U., Massi-Benedetti, M., Federici, M.O., Pieber, T.R., Schaller, H.C., Schaupp, L., Vering, T., Wilinska, M.E.: Nonlinear model predictive control of glucose concentration in subjects with type I diabetes. Physiological Measurements 25, 905–920 (2004)
11. Hovorka, R., Shojaee-Moradie, F., Carroll, P.V., Chassin, L.J., Gowrie, I.J., Jackson, N.C., Tudor, R.S., Umpleby, A.M., Jones, R.H.: Partitioning glucose distribution/transport, disposal and endogenous production during IVGTT. Am. J. Physiol. Endocrinol. Metab. 282, 992–1007 (2007)
12. Isidori, A.: Nonlinear Control Systems. Springer, New York (1995)
13. Kovács, L., Kulcsár, B., György, A., Benyó, Z.: Robust servo control of a novel type 1 diabetic model. Optimal Control Applications and Methods 32, 215–238 (2011)
14. Kovács, L., Benyó, B., Bokor, J., Benyó, Z.: Induced L_2-norm minimization of glucose-insulin system for Type I diabetic patients. Computer Methods and Programs in Biomedicine 102, 105–118 (2011)
15. Li, J., Johnson, J.D.: Mathematical models of subcutaneous injection of insulin analogues: a mini-review. Discr. Cont. Dynamical Syst. 12, 401–414 (2009)
16. Li, J., Kuang, Y., Li, B.: Analysis of IVGTT glucose-insulin interaction models with time-delay. Discrete and Continuous Dynam. Systems - B 1, 103–124 (2001)
17. Liu, W., Tang, F.: Modeling a simplified regulatory system of blood glucose at molecular levels. J. Theoretical Biology 252, 608–620 (2008)
18. Magni, L., Raimondo, D.M., Dalla Man, C., De Nicolao, G., Kovatchev, B., Cobelli, C.: Model predictive control of glucose concentration in type I diabetic patients: An in silico trial. Biomedical Signal Processing and Control 4, 338–346 (2009)
19. Makroglou, A., Li, J., Kuang, Y.: Mathematical models and software tools for the glucose-insulin regulatory system and diabetes: an overview. Applied Numerical Mathematics 56, 559–573 (2006)

20. Marquez-Martinez, L.A., Moog, C.H.: Input-output feedback linearization of time-delay systems. IEEE Transactions on Automatic Control 49(5), 781–785 (2004)
21. Nucci, G., Cobelli, C.: Models of subcutaneous insulin kinetics. A critical review. Comp. Methods and Programs in Biomed. 62, 249–257 (2000)
22. Oguchi, T., Watanabe, A., Nakamizo, T.: Input-Output Linearization of Retarded Nonlinear Systems by Using an Extension of Lie Derivative. International Journal of Control 75(8), 582–590 (2002)
23. Palumbo, P., Panunzi, S., De Gaetano, A.: Qualitative behavior of a family of delay differential models of the glucose insulin system. Discrete and Continuous Dynamical Systems - Series B 7, 399–424 (2007)
24. Palumbo, P., Pepe, P., Panunzi, S., De Gaetano, A.: Robust Closed-Loop Control of Plasma Glycemia: a Discrete-Delay Model Approach. Discrete and Continuous Dynamical Systems Series B 12, 455–468 (2009)
25. Palumbo, P., Pepe, P., Panunzi, S., De Gaetano, A.: Glucose control by subcutaneous insulin administration: a DDE modeling approach. In: Proc. IFAC World Congress, Milan, Italy, pp. 1471–1476 (2011)
26. Palumbo, P., Pizzichelli, G., Panunzi, S., Pepe, P., De Gaetano, A.: Tests on a virtual patient for an observer-based, closed-loop control of plasma glycemia. In: 50th Conf. on Decision and Control and European Control Conference (CDC 2011), Orlando, Florida, pp. 6936–6941 (2011)
27. Palumbo, P., Pepe, P., Panunzi, S., De Gaetano, A.: Time-delay model-based control of the glucose-insulin system, by means of a state observer. Eur. J. Control 18(6), 591–606 (2012)
28. Panunzi, S., Palumbo, P., De Gaetano, A.: A discrete single delay model for the Intra-Venous Glucose Tolerance Test. Theoretical Biology and Medical Modelling 4(35), 1–16 (2007)
29. Parker, R.S., Doyle III, F.J., Peppas, N.A.: A model-based algorithm for blood glucose control in type I diabetic patients. IEEE Trans. on Biomedical Engineering 46, 148–157 (1999)
30. Parker, R.S., Doyle III, F.J., Ward, J.H., Peppas, N.A.: Robust H_∞ glucose control in diabetes using a physiological model. AIChE Journal 46, 2537–2549 (2000)
31. Picchini, U., Ditlevsen, S., De Gaetano, A.: Modeling the euglycemic hyperinsulinemic clamp by stochastic differential equations. J. Mathematical Biology 53, 771–796 (2006)
32. Puckett, W.R., Lightfoot, E.N.: A model for multiple subcutaneous insulin injections developed from individual diabetic patient data. Am. J. Physiol. 269, E1115–E1124 (1995)
33. Ruiz-Velázquez, E., Femat, R., Campos-Delgado, D.U.: Blood glucose control for type I diabetes mellitus: a robust H^∞ tracking problem. Control Engineering Practice 12, 1179–1195 (2004)
34. Sorensen, J.T., Colton, C.K., Hillman, R.S., Soeldner, J.S.: Use of a physiologic pharmacokinetic model of glucose homeostasis for assesment of performance requirements for improved insulin therapies. Diabetes Care 5, 148–157 (1982)
35. Wilinska, M.E., Chassin, L.J., Schaller, H.C., Schaupp, L., Pieber, T.R., Hovorka, R.: Insulin kinetics in Type-1 diabetes: continuous and bolus delivery of rapid acting insulin. IEEE Trans. Biom. Eng. 52, 3–12 (2005)

Part IV
Computational and Software Tools

Eigenvalue Based Algorithms and Software for the Design of Fixed-Order Stabilizing Controllers for Interconnected Systems with Time-Delays*

Wim Michiels and Suat Gumussoy

Abstract. An eigenvalue based framework is developed for the stability analysis and stabilization of coupled systems with time-delays, which are naturally described by delay differential algebraic equations. The spectral properties of these equations are analyzed and their stability properties are studied, taking into account the effect of small delay perturbations. Subsequently, numerical methods for stability assessment and for designing stabilizing controllers with a prescribed structure or order, based on a direct optimization approach, are briefly addressed. The effectiveness of the approach is illustrated with a software demo. The paper concludes by pointing out the similarities with the computation and optimization of $\mathscr{H}_\infty$ norms.

1 Introduction

We consider the stability analysis and stabilization of systems described by delay differential algebraic equations (DDAEs), also called descriptor systems [3], of the form

$$E\dot{x}(t) = A_0 x(t) + \sum_{i=1}^{m} A_i x(t-\tau_i), \;\; x(t) \in \mathbb{R}^n, \tag{1}$$

where E is allowed to be singular. The time-delays $\tau_i, i = 1,\dots,m$, satisfy

$$0 < \tau_1 < \tau_2 < \dots < \tau_m$$

and the capital letters are real-valued matrices of appropriate dimensions.

Wim Michiels
Department of Computer Science, KU Leuven, Celestijnenlaan 200A,
3001 Heverlee, Belgium
e-mail: Wim.Michiels@cs.kuleuven.be

Suat Gumussoy
MathWorks, 3 Apple Hill Drive, Natick, MA, 01760, USA
e-mail: Suat.Gumussoy@mathworks.com

* This work is based on and the extension of the Authors' conference paper [7].

T. Vyhlídal, J.-F. Lafay, and R. Sipahi (eds.), *Delay Systems*,
Advances in Delays and Dynamics 1,
DOI: 10.1007/978-3-319-01695-5_18,

The motivation for the system description (1) in the context of designing controllers lies in its generality in modeling interconnected systems. For instance, the feedback interconnection of the system

$$\begin{cases} \dot{z}(t) = \sum F_i z(t-r_i) + \sum G_i u(t-r_i) \\ y(t) = \sum H_i x(t-r_i) + \sum L_i u(t-r_i) \end{cases} \tag{2}$$

and the controller

$$\begin{cases} \dot{z}_c(t) = \sum \hat{F}_i z_c(t-s_i) + \sum \hat{G}_i y(t-s_i) \\ u(t) = \sum \hat{H}_i z_c(t-s_i) + \sum \hat{L}_i y(t-s_i) \end{cases} \tag{3}$$

can be directly brought in the form (1), where

$$x = [z^T \; z_c^T \; u^T \; y^T], \;\; \{\tau_1, \ldots, \tau_m\} = \{r_i\} \cup \{s_i\}.$$

In this way no elimination of inputs and outputs is required, which may even not be possible in the presence of delays [4]. Another favorable property is the linear dependence of the matrices of the closed-loop system on the elements of the matrices of the controller. The increase in the number of equations, on the contrary, is a minor problem in most applications because the delay difference equations or algebraic constraints are related to inputs and outputs, as illustrated above, and the number of inputs and outputs is usually much smaller than the number of state variables. Finally, we note that also neutral systems can be dealt with in this framework, by introducing slack variables. The neutral equation

$$\frac{d}{dt}\left(z(t) + \sum_{i=1}^{m} G_i z(t-\tau_i)\right) = \sum_{i=0}^{m} H_i z(t-\tau_i) \tag{4}$$

can namely be rewritten as

$$\begin{cases} \dot{v}(t) &= \sum_{i=0}^{m} H_i z(t-\tau_i) \\ 0 &= -v(t) + z(t) + \sum_{i=1}^{m} G_i z(t-\tau_i) \end{cases}, \tag{5}$$

where v is the slack variable. Clearly (5) is of the form (1), if we set $x(t) = [v(t)^T \; z(t)^T]^T$.

The stability analysis of the null solution of (1) in this work is based on a spectrum determined growth property of the solutions, which allows us to infer stability information from the location of the characteristic roots. For instance, exponential stability will be related to a strictly negative spectral abscissa (the supremum of the real parts of the characteristic roots). As we shall see, the spectral abscissa of (1) may not be a continuous function of the delays. Moreover, this may lead to a situation where infinitesimal delay perturbations destabilize an exponentially stable system. These properties are similar to the spectral properties of neutral equations. Since in a practical control design the robustness of stability against infinitesimal changes of parameters is a prerequisite, we will define the concept of strong stability, inspired by the common terminology for neutral equations [5], and we will introduce the notion of the robust spectral abscissa, which explicitly takes small parametric perturbations into account. We will also provide explicit conditions and expressions that eventually lead to numerical algorithms.

Numerical algorithms for the computation of characteristic roots and the robust spectral abscissa are outlined, and subsequently applied to the design of stabilizing controllers. Similarly to [12], a direct optimization approach towards stabilization is taken, based on minimizing the (robust) spectral abscissa as a function of the parameters of the controller. In the example (2)-(3) these parameters may correspond to elements of the controller matrices. In this way stabilization is achieved on the moment that the objective function becomes strictly negative. This approach allows us to design stabilizing controllers with a prescribed structure or order (dimension). It is also possible to fix elements of the controller matrices, allowing to impose additional structure, e.g., a proportional-integral-derivative (PID)-like structure, or sparsity.

After a software demo of the stabilization algorithms we point out how the computational and optimization of $\mathscr{H}_\infty$ norm leads to similar problems as well as similar solutions and algorithms.

2 Preliminaries and Assumptions

Let matrix E in (1) satisfy

$$\text{rank}(E) = n - \nu,$$

with $1 \leq \nu < n$, and let the columns of matrix $U \in \mathbb{R}^{n\times\nu}$, respectively $V \in \mathbb{R}^{n\times\nu}$, be a (minimal) basis for the right, respectively left nullspace of E, which implies

$$U^T E = 0, \;\; EV = 0. \tag{6}$$

Throughout the paper we make the following assumption.

Assumption 1. *The matrix $U^T A_0 V$ is nonsingular.*

The equations (1) can be separated into coupled delay differential and delay difference equations. When we define

$$\mathbf{U} = \left[U^\perp \; U\right], \;\; \mathbf{V} = \left[V^\perp \; V\right],$$

a pre-multiplication of (1) with $\mathbf{U}^T$ and the substitution

$$x = \mathbf{V}\; [x_1^T \; x_2^T]^T,$$

with $x_1(t) \in \mathbb{R}^{n-\nu}$ and $x_2(t) \in \mathbb{R}^\nu$, yield the coupled equations

$$\begin{array}{l} E^{(11)}\dot{x}_1(t) = \sum_{i=0}^m A_i^{(11)} x_1(t-\tau_i) + \sum_{i=0}^m A_i^{(12)} x_2(t-\tau_i), \\ 0 = A_0^{(22)} x_2(t) + \sum_{i=1}^m A_i^{(22)} x_2(t-\tau_i) + \sum_{i=0}^m A_i^{(21)} x_1(t-\tau_i), \end{array} \tag{7}$$

where

$$E^{(11)} = U^{\perp T} E V^{\perp} \tag{8}$$

and

$$\begin{aligned} A_i^{(11)} &= U^{\perp T} A_i V^{\perp}, A_i^{(12)} = U^{\perp T} A_i V, \\ A_i^{(21)} &= U^T A_i V^{\perp}, \quad A_i^{(22)} = U^T A_i V, \quad i = 0, \ldots, m. \end{aligned} \tag{9}$$

In (7) matrix $E^{(11)}$ is invertible, following from

$$\mathrm{rank}(E^{(11)}) = \mathrm{rank}(\mathbf{U}^T E \mathbf{V}) = \mathrm{rank}(E) = n - \nu,$$

and matrix $A_0^{(22)}$ is invertible as well, induced by Assumption 1.

3 Spectral Properties and Stability

3.1 Exponential Stability

Stability conditions for the zero solution of (1) can be expressed in terms of the position of the *characteristic roots*, i.e., the roots of the equation

$$\det \Delta(\lambda) = 0, \tag{10}$$

where Δ is the characteristic matrix, $\Delta(\lambda) := \lambda E - A_0 - \sum_{i=1}^m A_i e^{-\lambda \tau_i}$. In particular, we have the following result.

Proposition 1. *The null solution of (1) is exponentially stable if and only if $c < 0$, where c is the* spectral abscissa, $c := \sup\{\Re(\lambda) : \det \Delta(\lambda) = 0\}$.

3.2 Continuity of the Spectral Abscissa and Strong Stability

We discuss the dependence of the spectral abscissa of (1) on the delay parameters $\boldsymbol{\tau} = (\tau_1, \ldots, \tau_m)$. In general the function

$$\boldsymbol{\tau} \in (\mathbb{R}_0^+)^m \mapsto c(\boldsymbol{\tau}) \tag{11}$$

is not everywhere continuous, which carries over from the spectral properties of delay difference equations (see, e.g., [1, 8, 10]). In the light of this observation we first outline properties of the function

$$\boldsymbol{\tau} \in (\mathbb{R}_0^+)^m \mapsto c_D(\boldsymbol{\tau}) := \sup\{\Re(\lambda) : \det \Delta_D(\lambda;\, \boldsymbol{\tau}) = 0\}, \tag{12}$$

with

$$\Delta_D(\lambda;\, \boldsymbol{\tau}) := U^T A_0 V + \sum_{i=1}^m U^T A_i V e^{-\lambda \tau_i}. \tag{13}$$

Note that (13) can be interpreted as the characteristic matrix of the delay difference equation

$$U^T A_0 V z(t) + \sum_{i=1}^{m} U^T A_i V z(t-\tau_i) = 0, \tag{14}$$

associated with the neutral equation obtained by differentiating the second equation in (7).

The property that the function (12) is not continuous led in [6] to the smallest upper bound, which is '*insensitive*' to small delay changes.

Definition 1. For $\boldsymbol{\tau} \in (\mathbb{R}_0^+)^m$, let $C_D(\boldsymbol{\tau}) \in \mathbb{R}$ be defined as

$$C_D(\boldsymbol{\tau}) := \lim_{\varepsilon \to 0+} c_D^{\varepsilon}(\boldsymbol{\tau}),$$

where

$$c_D^{\varepsilon}(\boldsymbol{\tau}) := \sup\{c_D(\boldsymbol{\tau}+\delta\boldsymbol{\tau}) : \delta\boldsymbol{\tau} \in \mathbb{R}^m \text{ and } \|\delta\boldsymbol{\tau}\| \le \varepsilon\}.$$

Several properties of this upper bound on c_D, which we call the robust spectral abscissa of the delay difference equation (14), are listed below (see [9, Section 3] for an overview).

Proposition 2. *The following assertions hold:*

1. *the function*
$$\boldsymbol{\tau} \in (\mathbb{R}_0^+)^m \mapsto C_D(\boldsymbol{\tau})$$
is continuous;
2. *for every $\boldsymbol{\tau} \in (\mathbb{R}_0^+)^m$, the quantity $C_D(\boldsymbol{\tau})$ is equal to the unique zero of the strictly decreasing function*
$$\zeta \in \mathbb{R} \to f(\zeta;\boldsymbol{\tau}) - 1, \tag{15}$$
where $f : \mathbb{R} \to \mathbb{R}^+$ is defined by
$$f(\zeta;\boldsymbol{\tau}) \quad := \quad \max_{\theta \in [0,\,2\pi]^m} \rho\left(\sum_{k=1}^{m} (U^T A_0 V)^{-1}(U^T A_k V) e^{-\zeta\tau_k} e^{j\theta_k}\right); \tag{16}$$
3. *$C_D(\boldsymbol{\tau}) = c_D(\boldsymbol{\tau})$ for rationally independent* [2]
4. *for all $\boldsymbol{\tau}_1, \boldsymbol{\tau}_2 \in (\mathbb{R}_0^+)^m$, we have*
$$\operatorname{sign}(C_D(\boldsymbol{\tau}_1)) = \operatorname{sign}(C_D(\boldsymbol{\tau}_2)) := \Xi; \tag{17}$$
5. *$\Xi < 0$ (> 0) holds if and only if $\gamma_0 < 1$ (> 1) holds, where*
$$\gamma_0 := \max_{\theta \in [0,\,2\pi]^m} \rho\left(\sum_{k=1}^{m} (U^T A_0 V)^{-1}(U^T A_k V) e^{j\theta_k}\right). \tag{18}$$

We now come back to the DDAE (1), more precisely, to the properties of the spectral abscissa function (11). The following two technical lemmas make connections between the characteristic roots of (1) and the zeros of (13).

---

[2] The m components of $\boldsymbol{\tau} = (\tau_1, \dots, \tau_m)$ are rationally independent if and only if $\sum_{k=1}^{m} n_k \tau_k = 0$, $n_k \in \mathbb{Z}$ implies $n_k = 0$, $\forall k = 1, \dots, m$.

Lemma 1. *There exists a sequence* $\{\lambda_k\}_{k\geq 1}$ *of characteristic roots of (1) satisfying*

$$\lim_{k\to\infty} \Re(\lambda_k) = c_D, \quad \lim_{k\to\infty} \Im(\lambda_k) = \infty.$$

Lemma 2. *For every* $\varepsilon > 0$ *the number of characteristic roots of (1) in the half plane*

$$\{\lambda \in \mathbb{C} : \Re(\lambda) \geq C_D(\boldsymbol{\tau}) + \varepsilon\} \tag{19}$$

is finite.

The lack of continuity of the spectral abscissa function (11) leads us again to an upper bound that takes into account the effect of small delay perturbations.

Definition 2. For $\boldsymbol{\tau} \in (\mathbb{R}_0^+)^m$, let the *robust spectral abscissa* $C(\boldsymbol{\tau})$ of (1) be defined as

$$C(\boldsymbol{\tau}) := \lim_{\varepsilon\to 0+} c^{\varepsilon}(\boldsymbol{\tau}), \tag{20}$$

where

$$c^{\varepsilon}(\boldsymbol{\tau}) := \sup\{c(\boldsymbol{\tau} + \delta\boldsymbol{\tau}) : \delta\boldsymbol{\tau} \in \mathbb{R}^m \text{ and } \|\delta\boldsymbol{\tau}\| \leq \varepsilon\}.$$

The following characterization of the robust spectral abscissa (20) constitutes the main result of this section. Its proof can be found in [9].

Proposition 3. *The following assertions hold:*

1. *the function*

$$\boldsymbol{\tau} \in (\mathbb{R}_0^+)^m \mapsto C(\boldsymbol{\tau}) \tag{21}$$

is continuous;

2. *for every* $\boldsymbol{\tau} \in (\mathbb{R}_0^+)^m$, *we have*

$$C(\boldsymbol{\tau}) = \max(C_D(\boldsymbol{\tau}), c(\boldsymbol{\tau})). \tag{22}$$

In line with the sensitivity of the spectral abscissa with respect to infinitesimal delay perturbations, which has been resolved by considering the robust spectral abscissa (20) instead, we define the concept of strong stability[3].

Definition 3. The null solution of (1) is strongly exponentially stable if there exists a number $\hat{\tau} > 0$ such that the null solution of

$$E\dot{x}(t) = A_0 + \sum_{k=1}^{m} A_k x(t - (\tau_k + \delta\tau_k))$$

is exponentially stable for all $\delta\boldsymbol{\tau} \in (\mathbb{R}^+)^m$ satisfying $\|\delta\boldsymbol{\tau}\| < \hat{\tau}$ and $\tau_k + \delta\tau_k \geq 0$, $k = 1,\ldots,m$.

The following result provides necessary and sufficient conditions for exponential stability.

Theorem 2. *The null solution of (1) is strongly exponentially stable if and only if* $C(\boldsymbol{\tau}) < 0$, *or, equivalently,* $c(\boldsymbol{\tau}) < 0$ *and* $\gamma_0 < 1$, *where* γ_0 *is defined by (18).*

[3] This terminology is borrowed from the theory of neutral delay differential equations [5,6].

4 Robust Stabilization by Eigenvalue Optimization

We now consider the equations

$$E\dot{x}(t) = A_0(\mathbf{p})x(t) + \sum_{i=1}^{m} A_i(\mathbf{p})x(t-\tau_i), \tag{23}$$

where the system matrices linearly depend on parameters $\mathbf{p} \in \mathbb{R}^{n_p}$. In control applications these parameter usually correspond to controller parameters. For example, in the feedback interconnection (2)-(3) they may arise from a parameterization of the matrices $(\hat{F}_i, \hat{G}_i, \hat{H}_i, \hat{L}_i)$.

To impose exponential stability of the null solution of (23) it is necessary to find values of $\mathbf{p}$ for which the spectral abscissa is strictly negative. If the achieved stability is required to be robust against small delay perturbations, this requirement must be strengthened to the negativeness of the robust spectral abscissa. This brings us to the optimization problem

$$\inf_{\mathbf{p}} C(\boldsymbol{\tau};\, \mathbf{p}). \tag{24}$$

Strongly stabilizing values of $\mathbf{p}$ exist if the objective function can be made strictly negative. By Theorem 2 the latter can be evaluated as

$$C(\boldsymbol{\tau};\, \mathbf{p}) = \max(c(\boldsymbol{\tau};\, \mathbf{p}), C_D(\boldsymbol{\tau};\, \mathbf{p})). \tag{25}$$

An alternative approach consists of solving the constrained optimization problem

$$\inf\nolimits_{\mathbf{p}} c(\boldsymbol{\tau};\, \mathbf{p}), \text{ subject to } \gamma_0(\mathbf{p}) < \gamma, \tag{26}$$

where $\gamma < 1$. If the objective function is strictly negative, then the satisfaction of the constraint implies strong stability. Problem (26) can be solved using the barrier method proposed in [13], which is on its turn inspired by interior point methods, see, e.g., [2]. The first step consists of finding a feasible point, i.e., a set of values for $\mathbf{p}$ satisfying the constraint. If the feasible set is nonempty such a point can be found by solving

$$\min_{\mathbf{p}} \gamma_0(\mathbf{p}). \tag{27}$$

Once a feasible point $\mathbf{p} = \mathbf{p}_0$ has been obtained one can solve in the next step the unconstrained optimization problem

$$\min_{\mathbf{p}} \{c(\mathbf{p}) - r\log(\gamma - \gamma_0(\mathbf{p}))\} \tag{28}$$

where $r > 0$ is a small number and γ satisfies

$$\gamma_0(\mathbf{p}) < \gamma \leq 1.$$

The second term (the barrier) assures that the feasible set cannot be left when the objective function is decreased in a quasi-continuous way (because the objective function will go to infinity when $\gamma_0 \to \gamma$). If (28) is repeatedly solved for decreasing values of r and with the previous solution as starting value, a solution of (26) is obtained.

In [9] it has been shown that the objective functions for the optimization problem (24) and for the subproblems (27) and (28) are in general not everywhere differentiable. They might even be not everywhere Lipschitz continuous, yet they are differentiable almost everywhere. These properties preclude the use of standard optimization methods, developed for smooth problems. Instead we use a combination of BFGS with weak Wolfe line search and gradient sampling, as implemented in the MATLAB code HANSO [11]. The overall algorithm only requires the evaluation of the objective function, as well as its derivatives with respect to the controller parameters, *whenever* it is differentiable. The spectral abscissa can be computed using a spectral discretization followed by Newton corrections. The quantities C_D and γ_0 can be computed using the characterizations in Theorem 2, where the (global) maximization problems in (16) and (18) are discretized, followed by local corrections. In all cases derivatives can be obtained from the sensitivity of individual eigenvalues with respect to the free parameters. For more details and expressions we refer to [9].

5 Illustration of the Software

A MATLAB implementation of the robust stabilization algorithms is available from

`http://twr.cs.kuleuven.be/research/software/delay-control/stabilization/.`

Installation instructions can be found in the corresponding README file.

As a first example we take the system with input delay from [12]:

$$\dot{x}(t) = Ax(t) + Bu(t-\tau), \quad y(t) = x(t), \tag{29}$$

where

$$A = \begin{bmatrix} -0.08 & -0.03 & 0.2 \\ 0.2 & -0.04 & -0.005 \\ -0.06 & 0.2 & -0.07 \end{bmatrix}, \quad B = \begin{bmatrix} -0.1 \\ -0.2 \\ 0.1 \end{bmatrix}, \quad \tau = 5. \tag{30}$$

We start by defining the system:

```
>> A = [-0.08 -0.03 0.2;0.2 -0.04 -0.005;-0.06 0.2 -0.07];
>> B = [-0.1;-0.2;0.1];
>> C = eye(3);
>> plant1 = tds_create({A},0,{B},5,{C},0);
```

The uncontrolled system is unstable.

```
>> max(real(eig(A)))

ans =

    0.1081
```

We design a stabilizing dynamic controller of the form

$$\begin{cases} \dot{x}_c(t) = A_c x_c(t) + B_c y(t), \\ u(t) \;\; = C_c x_c(t) + D_c y(t), \quad x_c(t) \in \mathbb{R}^{n_c}, \end{cases} \tag{31}$$

using the approach of Section 4, where we set $\mathbf{p} = \mathrm{vec}\begin{bmatrix} A_c & B_c \\ C_c & D_c \end{bmatrix}$. Since the transfer function from u to y is strictly proper, the robust spectral abscissa equals the spectral abscissa, and the optimization problems (24) and (26) reduce to the (unconstrained) minimization of the spectral abscissa. In order to compute a controller we first specify its order,

```
>> controller_order = 2;
```

and call a routine to minimize the robust spectral abscissa

```
>> [controller1,f1] = stabilization_max(plant1,controller_order);
```

The optimized robust spectral abscissa and corresponding controller are given by:

```
controller1 =

      E: {[2x2 double]}
     hE: 0
      A: {[2x2 double]}
     hA: 0
     B1: {[2x3 double]}
    hB1: 0
     C1: {[0.2098 0.9492]}
    hC1: 0
    D11: {[0.8826 1.1548 0.6538]}
   hD11: 0
```

where empty fields of the controller are omitted for space considerations.

```
f1 =

  -0.2496
```

We define the closed-loop system

```
>> clp1 = closedloop(plant1,controller1);
```

and compute its rightmost characteristic roots (where the "l1"-field refers to the application of Newton corrections):

```
>> options = tdsrootsoptions;
>> eigenvalues1 = compute_roots_DDAE(clp1,options);
>> eigenvalues1.l1.'

ans =

   -0.2496 + 0.0251i   -0.2496 - 0.0251i
```

We can compute all eigenvalues with real part larger than -0.9 by the following code, which leads to 37 returned eigenvalues.

```
>> options.minimal_real_part = -0.9;
>> eigenvalues1 = compute_roots_DDAE(clp1,options);
>> size(eigenvalues1.l1)

ans =

    37     1
```

We plot the closed-loop characteristic roots.

```
>> p1 = eigenvalues1.l1; plot(real(p1),imag(p1),'+');
```

We now repeat the computations for a static controller:

```
>> controller_order = 0;
>> [controller2,f2] = stabilization_max(plant1,controller_order);
```

and add the optimized spectrum to our plot:

```
>> clp2 = closedloop(plant1,controller2);
>> eigenvalues2 = compute_roots_DDAE(clp2,options);
>> p2 = eigenvalues2.l1; hold on; plot(real(p2),imag(p2),'s');
```

The result is displayed in Figure 1. Note that the extra degrees of freedom in the dynamic controller lead to a further reduction of the spectral abscissa.

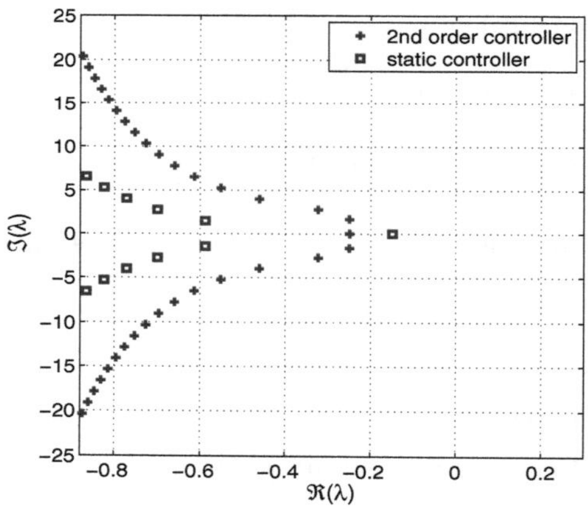

Fig. 1 Characteristic roots of the first example (29) and (31), corresponding to a minimum of the spectral abscissa, for a static controller (boxes) and a second order controller (pluses).

For the second example we assume that the measured output of system (29) is instead given by

$$\tilde{y}(t) = x(t) + \begin{bmatrix} 3 & 4 & 1 \end{bmatrix}^T u(t-2.5) + \begin{bmatrix} 2/5 & -2/5 & -2/5 \end{bmatrix}^T u(t-5). \qquad (32)$$

The difference with the previous example is that there are two feedthrough terms which are both delayed. We define the plant object

```
>> plant2 = setfield(plant1,'D11',{[3;4;1],[2/5;-2/5;-2/5]});
>> plant2.hD11 = [2.5 5];
```

Once again we design a static controller, $u(t) = D_c\tilde{y}(t)$. In this case there is a high-frequency path in the control loop. Solving the optimization problem (24) leads to

$$C = -0.0309, \quad D_c = [0.0409\ 0.0612\ 0.3837], \tag{33}$$

as can be seen from

```
>> [controller1,f1] = stabilization_max(plant2,controller_order);
>> f1

f1 =

   -0.0309
>> controller1.D11{1}

ans =

    0.0409    0.0612    0.3837
```

We compute the rightmost characteristic roots of the closed-loop system.

```
>> clp1 = closedloop(plant2,controller1);
>> eigenvalues1 = compute_roots_DDAE(clp1,options);
Warning: case C_D>=c.
Spectral discretization with 16 points (lowered if maximum size of
the eigenvalue problem is exceeded)
N= 15
>> eigenvalues1.l1.'

ans =

  -0.3740 + 7.6893i  -0.3740 - 7.6893i  -0.3788 + 5.1779i  -0.3788 - 5.1779i
  -0.3499 + 4.8863i  -0.3499 - 4.8863i  -0.3934 + 2.6712i  -0.3934 - 2.6712i
  -0.3336 + 2.3789i  -0.3336 - 2.3789i  -0.0309             -0.0309 + 0.0001i
  -0.0309 - 0.0001i  -0.3819 + 0.3603i  -0.3819 - 0.3603i
```

We conclude that the optimum is characterized by three rightmost characteristic roots. This might sound counter-intuitive because the number of degrees of freedom in the controller is also three. The explanation is related to the issued warning: we are in a situation where $C_D \geq c$. In fact the optimum of (24) is characterized by an equality between C_D and the spectral abscissa c, the latter corresponding to a rightmost root with multiplicity three. To illustrate this, we have recomputed the characteristic roots where we set N, the number of discretization points in the spectral method, to a high number in such a way that the high-frequency roots are captured. In the left pane of Figure 2 we show the rightmost characteristic roots corresponding to the minimum of the robust spectral abscissa (33). The dotted line corresponds to $\Re(\lambda) = c_D$, the dashed line to $\Re(\lambda) = C_D$. In order to illustrate that we indeed have $c = C_D$ we depict in the right pane of Figure 2 the rightmost characteristic roots after perturbing the delay value 2.5 in (32) to 2.51.

With our software we can also solve the constrained optimization problem (26). With the default parameters $r = 10^{-3}$ and $\gamma = 1 - 10^{-3}$ in the relaxation (28) we get the following result:

```
>> [controller2,f2] = stabilization_barrier(plant2,controller_order);
>> controller2.D11{1}

ans =

    0.0249    0.1076    0.3173
>> clp2 = closedloop(plant2,controller2);
>> eigenvalues2=compute_roots_DDAE(clp2,options);
Warning: case C_D>=c.
Spectral discretization with 16 points (lowered if maximum size of
the eigenvalue problem is exceeded)
N= 15
>> max(real(eigenvalues2.l1))
ans =

  -0.0345
```

Compared to (33), where we had $C = c = C_D$, a further reduction of the spectral abscissa to $c = -0.0345$ has been achieved, at the price of an increased value of C_D (equal to -0.00602). This is expected because the constraint $\gamma_0 < 1$ imposes robustness of stability, yet no bound on the exponential decay rate of the solutions.

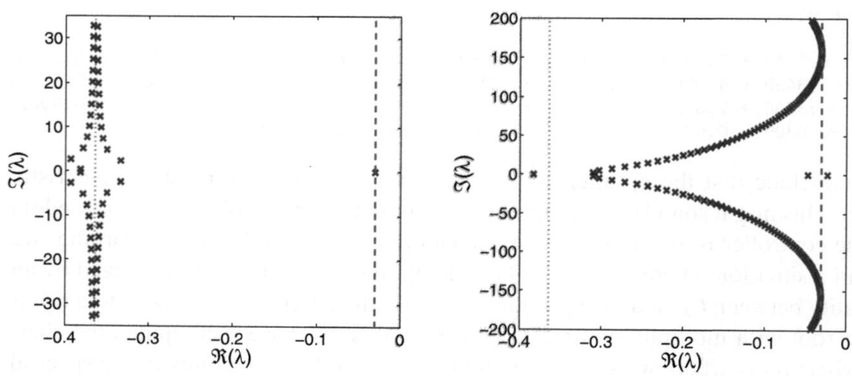

Fig. 2 (left) Characteristic roots corresponding to the minimum of the robust spectral abscissa of the second example (29) and (32), using a static controller. The rightmost characteristic roots, $\lambda \approx -0.0309$, has multiplicity three. (right) Effect on the characteristic roots of a perturbation of the delays $(2.5, 5)$ in (32) to $(2.51, 5)$.

6 Duality with the $\mathscr{H}_\infty$ Problem

In a practical control design the stabilization phase is usually only a first step in the overall design procedure. Consider now the (subsequent) fixed-order $\mathscr{H}_\infty$ synthesis problem, where the aim is to optimize the $\mathscr{H}_\infty$ norm of

$$G(\lambda) = C(\lambda E - A_0 - \sum_{i=1}^{m} A_i e^{-\lambda \tau_i})^{-1} B$$

as a function of parameters on which the system matrices depend.

It turns out the function $\boldsymbol{\tau} \mapsto \|G(j\omega;\ \boldsymbol{\tau})\|_{\mathscr{H}_\infty}$ has a very similar behavior to the spectral abscissa function (11). In particular it is not everywhere continuous. Moreover, the discontinuities are all related to the behavior of the transfer function at large frequencies (analogous to the behavior of eigenvalues with large imaginary parts in §3.2). This high frequency behavior is described by the associated *asymptotic transfer function*

$$G_a(\lambda) := -CV(U^T A_0 V + \sum_{i=1}^{m} U^T A_i V e^{-\lambda \tau_i})^{-1} U^T B,$$

which takes the role of the associated delay-difference equation (14). Finally, the sensitivity w.r.t. small delay perturbations leads to the definition of the *strong* $\mathscr{H}_\infty$ norm (analogous to strong stability), defined as:

$$|||G(j\omega;\ \boldsymbol{\tau})|||_{\mathscr{H}_\infty} := \lim_{\varepsilon \to 0+} \sup\{\|G(j\omega;\ \boldsymbol{\tau} + \delta\boldsymbol{\tau})\|_{\mathscr{H}_\infty} : \delta\boldsymbol{\tau} \in (\mathbb{R}^-)^m,\ \|\delta\boldsymbol{\tau}\|_2 < \varepsilon\}.$$

The computation of the strong $\mathscr{H}_\infty$ norm involves a tradeoff between the behavior of the transfer function at small and large frequencies, similar to the result of Theorem 2 on strong stability, and it can be optimized using the same algorithms. For the details, we refer to the article [4] and to the corresponding software available at

`http://twr.cs.kuleuven.be/research/software/delay-control/hinfopt/.`

Acknowledgements. This work has been supported by the Programme of Interuniversity Attraction Poles of the Belgian Federal Science Policy Office (IAP P6- DYSCO), by OPTEC, the Optimization in Engineering Center of the K.U.Leuven, by the project STRT1- 09/33 of the K.U.Leuven Research Council and the project G.0712.11N of the Research Foundation - Flanders (FWO).

References

1. Avellar, C.E., Hale, J.K.: On the zeros of exponential polynomials. Mathematical Analysis and Applications 73, 434–452 (1980)
2. Boyd, S., Vandenberghe, L.: Convex optimization. Cambridge University Press (2004)
3. Fridman, E., Shaked, U.: H_∞-control of linear state-delay descriptor systems: an LMI approach. Linear Algebra and its Applications 351-352, 271–302 (2002)
4. Gumussoy, S., Michiels, W.: Fixed-order H-infinity control for interconnected systems using delay differential algebraic equations. SIAM Journal on Control and Optimiza-

tion 49(5), 2212–2238 (2011)

5. Hale, J.K., Verduyn Lunel, S.M.: Strong stabilization of neutral functional differential equations. IMA Journal of Mathematical Control and Information 19, 5–23 (2002)
6. Michiels, W., Vyhlidal, T.: An eigenvalue based approach to the robust stabilization of linear time-delay systems of neutral type. Automatica 41(6), 991–998 (2005)
7. Michiels, W., Gumussoy, S.: Eigenvalue based algorithms and software for the design of fixed-order stabilizing controllers for interconnected systems with time-delays. In: 10th IFAC Workshop on Time Delay Systems, June 22-24. IFAC-PapersOnLine, pp. 144–149. Northeastern University, USA (2012), doi:10.3182/20120622-3-US-4021.00015
8. Michiels, W., Engelborghs, K., Roose, D., Dochain, D.: Sensitivity to infinitesimal delays in neutral equations. SIAM Journal on Control and Optimization 40(4), 1134–1158 (2002)
9. Michiels, W.: Spectrum based stability analysis and stabilization of systems described by delay differential algebraic equations. IET Control Theory and Applications 5(16), 1829–1842 (2011)
10. Michiels, W., Vyhlídal, T., Zítek, P., Nijmeijer, H., Henrion, D.: Strong stability of neutral equations with an arbitrary delay dependency structure. SIAM Journal on Control and Optimization 48(2), 763–786 (2009)
11. Overton, M.: HANSO: a hybrid algorithm for nonsmooth optimization (2009), `http://cs.nyu.edu/overton/software/hanso/`
12. Vanbiervliet, J., Vandereycken, B., Michiels, W., Vandewalle, S.: A nonsmooth optimization approach for the stabilization of time-delay systems. ESAIM Control, Optimisation and Calculus of Variations 14(3), 478–493 (2008)
13. Vyhlidal, T., Michiels, W., McGahan, P.: Synthesis of strongly stable state-derivative controllers for a time delay system using constrained non-smooth optimization. IMA Journal of Mathematical Control and Information 27(4), 437–455 (2010)

Computer Aided Control System Design for Time Delay Systems Using MATLAB®⋆

Suat Gumussoy and Pascal Gahinet

Abstract. Computer Aided Control System Design (CACSD) allows to analyze complex interconnected systems and design controllers achieving challenging control requirements. We extend CACSD to systems with time delays and illustrate the functionality of Control System Toolbox in MATLAB for such systems. We easily define systems in time and frequency domain system representations and build the overall complex system by interconnecting subsystems. We analyze the overall system in time and frequency domains and design PID controllers satisfying design requirements. Various visualization tools are used for analysis and design verification. Our goal is to introduce these functionalities to researchers and engineers and to discuss the open directions in computer algorithms for control system design.

1 Introduction

Time delays are frequently seen in many control applications such as process control, communication networks, automotive and aerospace, [5,9,22]. Depending on the delay length, they may limit or degrade the performance of control systems unless they are considered in the design, [1,11]. Although considerable research effort is devoted to extend classical and modern control techniques to accommodate delays, most available software packages for delay differential equations (DDE) [4,6,8,15] are restrictive and not developed for control design purposes.

We present the currently implemented framework and available functionality in MATLAB for computer-aided manipulation of linear time-invariant (LTI) models with delays. We illustrate this functionality for each important step in every practical control design:

Suat Gumussoy · Pascal Gahinet
MathWorks, 3 Apple Hill Drive, Natick, MA, 01760, USA
e-mail: {suat.gumussoy,pascal.gahinet}@mathworks.com

⋆ This work is based on and the extension of the Authors' conference paper [14].

T. Vyhlídal, J.-F. Lafay, and R. Sipahi (eds.), *Delay Systems*,
Advances in Delays and Dynamics 1,
DOI: 10.1007/978-3-319-01695-5_19,

- system representations in time and frequency domains,
- interconnections of complex systems,
- analysis tools and design techniques for time delay systems.

By introducing available functionality in Control System Toolbox, our goal is to facilitate the design of control systems with delays for researchers and engineers. Moreover, we discuss possible enhancements in CASCD for time delay systems, to illustrate the gap between the desired analysis / design techniques and the current control software implementation.

At the heart of this framework is a linear fractional transformation (LFT) based representation of time delay systems [23]. This representation handles delays in feedback loops and is general enough for most control applications. In addition most classical software tools for analyzing delay-free LTI systems are extended to this class of LTI systems with delays. Given the widespread use of linear techniques in control system design, this framework and the accompanying software tools should facilitate CACSD in the presence of delays, as well as stimulate more research into efficient numerical algorithms for assessing the properties and performance of such systems.

2 Motivation Examples

A standard PI control example is given in [16] where the plant is a chemical tank and a single-input-single-output system with an input-output delay (i.e., dead-time system),

$$P(s) = e^{-93.9s}\frac{5.6}{40.2s+1}. \tag{1}$$

In the classical feedback configuration in Figure 1, the standard PI controller is chosen as

$$C_{PI}(s) = K(1+\frac{1}{T_i s}) \tag{2}$$

where $K = 0.1$ and $T_i = 100$. The closed-loop transfer function from y_{sp} to y is

$$T_{PI}(s) = \frac{4020s^2+100s}{4020s^2+100s+(56s+0.56)e^{-93.9s}}.$$

This transfer function has an *internal* delay which can not be represented by input or output delays. Therefore, the representation for time delay systems has to capture this type of systems and to be *closed* under block diagram of operations. This plant and the controller in MATLAB are defined as

```
P = tf(5.6,[40.2 1],'OutputDelay',93.9);     % plant
Cpi = 0.1 * (1 + tf(1,[100 0]));             % PI controller
```

and the closed-loop system T_{PI} is obtained by the `feedback` command:

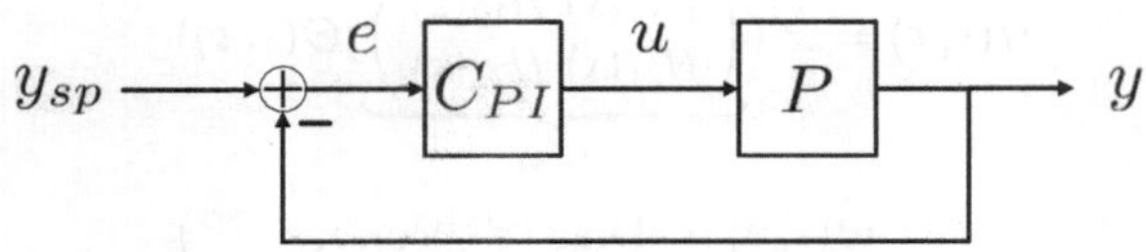

Fig. 1 Feedback loop with PI controller

```
Tpi = feedback(P*Cpi,1);          % Closed-loop transfer, ysp -> y
```

Note that these commands are natural extensions of delay-free case and are used for systems with delays without new syntax for the user.

The MIMO time delay systems may have different transport delays for each input-output channel, i.e.,

$$H(s) = \begin{pmatrix} e^{-0.1s}\frac{2}{s} & e^{-0.3s}\frac{s+1}{s+10} \\ 10 & e^{-0.2s}\frac{s-1}{s+5} \end{pmatrix}.$$

We define such systems in MATLAB by the following commands:

```
s = tf('s');
H = [2/s (s+1)/(s+10); 10 (s-1)/(s+5)];     % delay-free system
H.ioDelay = [0.1 0.3; 0 0.2];               % transport delays
```

We see on our motivation examples that the representation of time delay systems has certain challenges. Next section, we present the LFT-based representation of time delay systems to address these challenges and discuss its advantages.

3 System Representation

We represent time delay systems by the linear-fractional transformation (LFT). Recall that the LFT is defined for matrices by

$$\mathcal{L}(\begin{pmatrix} M_{11} & M_{12} \\ M_{21} & M_{22} \end{pmatrix}, \Theta) := M_{11} + M_{12}\Theta(I - M_{22}\Theta)^{-1}M_{21} .$$

The LFT has been extensively used in robust control theory for representing models with uncertainty, see [23] for details.

Consider the class *generalized LTI* (GLTI) of continuous-time LTI systems whose transfer function is of the form

$$H(s,\tau) = \mathcal{L}(\underbrace{\begin{pmatrix} H_{11}(s) & H_{12}(s) \\ H_{21}(s) & H_{22}(s) \end{pmatrix}}_{H(s)}, \Theta(s,\tau))$$

$$\Theta(s,\tau) := \mathrm{Diag}\left(e^{-\tau_1 s}, \ldots, e^{-\tau_N s}\right) \tag{3}$$

where $H(s)$ is a rational (delay free) MIMO transfer function, and $\tau = (\tau_1, \ldots, \tau_N)$ is a vector of nonnegative time delays. Systems in this class are modeled as the LFT interconnection of a delay-free LTI model and a bank of pure delays (see Figure 2). As such, they are clearly linear time-invariant. Also, pure delays are in this class since $e^{-\tau s} = \mathcal{L}(\begin{pmatrix} 0 & 1 \\ 1 & 0 \end{pmatrix}, e^{-\tau s})$.

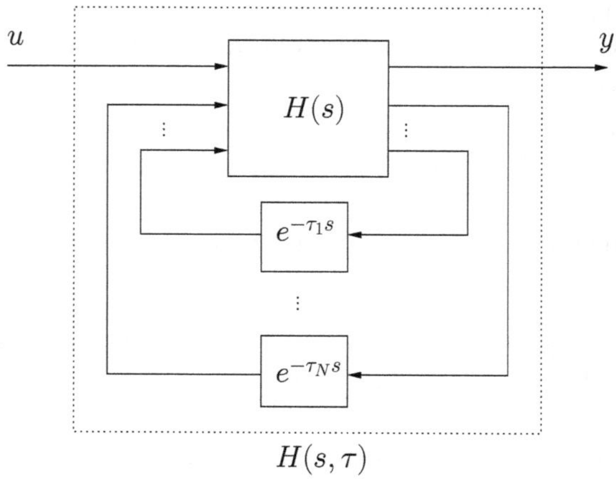

Fig. 2 LFT-based modeling of LTI systems with delays

This GLTI class has two key properties, [10]:

- Any block diagram interconnection of GLTI systems is a GLTI system. In other words, the class of GLTI systems is closed under series, parallel, and feedback connections as well as branching/summing junctions.
- The linearization of any nonlinear block diagram with time delays is a GLTI system.

These two properties show that the GLTI class is general enough to model any (linearized) system with a finite number of delays, including delays in the feedback path. For further motivation of this representation and equivalent case of discrete time systems, see [10].

The GLTI class is represented in state-space equations as follows. Let

$$\begin{pmatrix} H_{11}(s) & H_{12}(s) \\ H_{21}(s) & H_{22}(s) \end{pmatrix} = \begin{pmatrix} D_{11} & D_{12} \\ D_{21} & D_{22} \end{pmatrix} + \begin{pmatrix} C_1 \\ C_2 \end{pmatrix} (sI - A)^{-1} \begin{pmatrix} B_1 & B_2 \end{pmatrix}$$

be a minimal realization of $H(s)$ in (3). State-space equations for $H(s,\tau) = \mathscr{L}(H(s),\Theta(s,\tau))$ are readily obtained as

$$\begin{bmatrix} \dot{x}(t) \\ y(t) \\ z(t) \end{bmatrix} = \begin{bmatrix} A & B_1 & B_2 \\ C_1 & D_{11} & D_{12} \\ C_2 & D_{21} & D_{22} \end{bmatrix} \begin{bmatrix} x(t) \\ u(t) \\ w(t) \end{bmatrix} \qquad (4)$$
$$w(t) = (\Delta_\tau z)(t)$$

where $u(t), y(t)$ are the input and output vectors; $w(t)$, $z(t)$ are internal signals commensurate with the vector τ of time delays; $\Delta_\tau z$ is the vector-valued signal defined by $(\Delta_\tau z)(t) := (z_1^T(t-\tau_1), \ldots, z_N^T(t-\tau_N))^T$.

Note that standard delay-free state-space models are just a special case of (4) corresponding to $N = 0$, a handy fact when it comes to integrating GLTI models with existing software for manipulating delay-free state-space models.

Delay LTI systems of the form

$$\dot{x}(t) = A_0 x(t) + B_0 u(t) + \sum_{j=1}^{M} (A_j x(t-\theta_j) + B_j u(t-\theta_j))$$
$$y(t) = C_0 x(t) + D_0 u(t) + \sum_{j=1}^{M} (C_j x(t-\theta_j) + D_j u(t-\theta_j))$$

are often considered in the literature with various restrictions on the number and locations of the delays $\theta_1, \ldots, \theta_M$. It turns out that any model of this form belongs to the class GLTI as shown in [10]. It is possible to define a large class of time delay systems in MATLAB, both in time and frequency domains. For further details on representation of time delay systems, see [24].

4 Interconnections

Control systems, in general, are built up by interconnecting other subsystems. The most typical configuration is a feedback loop with a plant and a controller as shown in Section 2; whereas more complex configurations may have distributed systems with multiple plants, controllers and transport / internal delays.

A standard way to to obtain the closed-loop model of interconnections of systems in MATLAB is to use the `connect` command. This function requires all systems to have input and output names and summation blocks. It automatically builds the resulting closed-loop system with the given inputs and outputs. Consider the Smith Predictor control structure given in Figure 3 for the same dead-time system $P(s)$ in (1). The Smith Predictor uses an internal model to predict the delay-free response $y_p(t)$ of the plant, and seeks to correct discrepancies between this prediction and the setpoint $y_{sp}(t)$, rather than between the delayed output measurement $y(t)$ and $y_{sp}(t)$. To prevent drifting, an additional compensator $F(s)$ is used to eliminate steady-state drifts and disturbance-induced offsets.

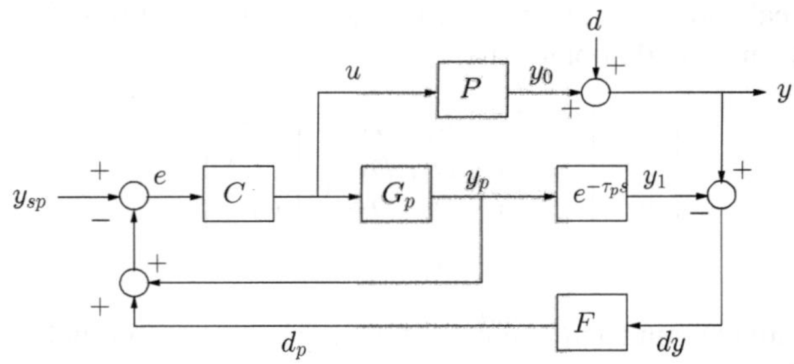

Fig. 3 Feedback loop with Smith Predictor

We first assume that the prediction model $P_p(s) = e^{-\tau_p s} G_p(s)$ matches the plant model $P(s)$ in (1), and use the following compensator settings:

$$C(s) = 0.5(1 + \frac{1}{40s}), \quad F(s) = \frac{1}{20s+1}.$$

By defining summation blocks and input and outputs names of systems, we obtain the closed-loop model T_{SP} from the input signal y_{sp} to the output signal y:

```
s = tf('s');

% LTI blocks
P = exp(-93.9*s) * 5.6/(40.2*s+1);
P.InputName = 'u'; P.OutputName = 'y';

Gp = 5.6/(40.2*s+1);
Gp.InputName = 'u'; Gp.OutputName = 'yp';

Dp = exp(-93.9*s);
Dp.InputName = 'yp'; Dp.OutputName = 'y1';

C = 0.5 * (1 + 1/(40*s));
C.InputName = 'e';  C.OutputName = 'u';

F = 1/(20*s+1);
F.InputName = 'dy'; F.OutputName = 'dp';

% Sum blocks
Sum1 = ss([1,-1,-1],'InputName',...
       {'ysp','yp','dp'},'OutputName','e');
Sum2 = ss([1,-1],...
       'InputN',{'y','y1'},'OutputN','dy');

% Build interconnection model
Tsp = connect(P,Gp,Dp,C,F,Sum1,Sum2,'ysp','y');
```

We can also construct various types of connections such as in parallel and series (`parallel` and `series`); group systems by appending their inputs and outputs (`append`); form the linear fractional transformation (`lft`). Standard system operations are also valid for time delay systems such as addition, subtraction, multiplication, division.

After we represent our subsystems and connect with each other, we easily construct the closed-loop model with time delays. Our next goal is to analyze the characteristics of the resulting closed-loop models with visualizations and compute their system properties.

5 Time / Frequency Domain Analyses and Visualizations

We analyze a plant or a closed-loop model with various interconnections and systems to understand its characteristics and properties. By simulating its time-domain response to certain inputs such as a step or tracking signals, we observe its time-domain characteristics such as rise and settling times, overshoot. On the other hand, frequency domain analysis gives us information on, for example, gain and phase margins, bandwidth and resonant peak.

In Section 3, we obtained the closed-loop system T_{PI} of the dead-time system (1) and PI controller and in Section 4 we constructed the closed-loop system T_{SP} of the same plant and the Smith Predictor. We simulate the responses of T_{PI} and T_{SP} to the tracking signal, `ref` by the following commands:

```
% time and reference signal
time = 0:.1:2000;
ref = (time>=0 & time<1000)*4 + (time>=1000 & time<=2000)*8;

% compare responses
lsim(Tsp,Tpi,ref,time);
```

The resulting responses are shown in Figure 4 (on the left). Simulation results show that PI controller has a slower response time with oscillations and the Smith Predictor has better tracking performance.

In practice, there is always a mismatch between the predicted and real plant models. We easily investigate robustness of our design to modeling uncertainties. For example, consider two perturbed plant models

$$P_1(s) = e^{-90s}\frac{5}{38s+1}, \quad P_2(s) = e^{-100s}\frac{6}{42s+1}.$$

To assess the Smith predictor robustness when the true plant model is $P_1(s)$ or $P_2(s)$ rather than the prediction model $P(s)$, simply bundle P, P_1, P_2 into an LTI array, rebuild the closed-loop model(s), and replot the responses for the tracking signal:

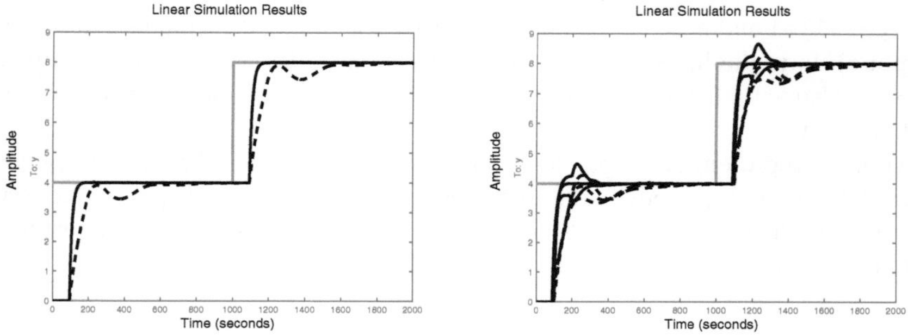

Fig. 4 (left) Responses of the Smith Predictor (–) and PI (- -) to the reference signal (gray colored). (right) Robustness of the Smith Predictor (–) to Model Mismatch.

```
P1 = exp(-90*s) * 5/(38*s+1);            % perturbed plants
P2 = exp(-100*s) * 6/(42*s+1);
Plants = stack(1,P,P1,P2);               % bundle true and perturbed plants

T = connect(Plants,Gp,Dp,C,F,Sum1,Sum2,'ysp','y'); % construct closed-loop

lsim(T,Tpi,ref,time);                    % simulate closed-loop responses
```

The resulting responses in Figure 4 (on the right) show a slight performance degradation, but the Smith predictor still retains an edge over the pure PI design.

We obtain the closed-loop frequency responses for the nominal and perturbed plants by `bode(T)` and their visualizations as shown in Figure 5. Note that the phase behavior of systems with internal delays is quite different than systems with I / O delays.

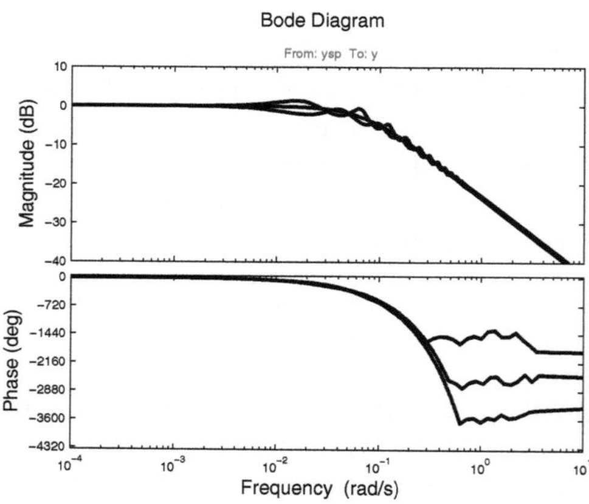

Fig. 5 Closed-Loop Response from y_{sp} to y

We numerically compute the bandwidth of the responses by `bandwidth(T)` which returns 0.0695, 0.0565, 0.0767. The gain and phase margins of the responses are calculated by `[gm, pm] = margin(T)` and their values are

$$\mathtt{gm} = [1.0835;\ \ 1.1569;\ \ 1.0304], \quad \mathtt{pm} = [180;\ \ 180;\ \ 7.1433].$$

Other well-known frequency-domain based tools are also available for the GLTI class such as `bandwidth`, `dcgain`, `nyquist`, `allmargin`.

6 Controller Design

We analyzed the closed-loop characteristics of the given PI controller and Smith Predictor. Now we design a PID controller for finite dimensional and time delay plants using `pidtune` function in The Control System Toolbox and compare its performance with other controllers. This function aims to find a PID controller stabilizing the closed-loop system and to satisfy certain performance and robustness objectives. These objectives are tracking reference changes and suppressing disturbances as rapidly as possible; designing enough phase and margins for modeling errors or variations in system dynamics.

The algorithm for tuning PID controllers helps us meet these objectives by automatically tuning the PID gains to balance the response time as a performance objective and the stability margins as robustness objectives. By default, the algorithm chooses a crossover frequency (loop bandwidth) based upon the plant dynamics, and designs for a target phase margin of 60°.

We can approximate the dead-time system $P(s)$ by a finite dimensional transfer function $P_a(s)$ using the function `pade` based on Padé approximation. The function `pidtune` designs a PID controller $C_a(s)$ for the approximate finite dimensional plant and we obtain the closed-loop system for this controller by the following commands:

```
Pa = pade(P,8);          % approximate 8th order plant
Ca = pidtune(Pa,'pid');  % design PID for Pa
Ta = feedback(P*Ca,1);   % closed-loop for Ca
```

`Pidtune` also designs a PID controller for time delay systems *without any approximation*,

```
[Cpid,info] = pidtune(P,'pid');    % design PID for P
Tpid = feedback(P*Cpid,1);         % closed-loop for Cpid
>> info

info =

                Stable: 1
     CrossoverFrequency: 0.0067
            PhaseMargin: 60.0000
```

As shown in returned `info` structure, the designed controller `Cpid` stabilizes the closed-loop and achieves 0.0067 rad/sec crossover frequency and 60° phase margin. The closed-loop step response `Tpid` of the controller `Cpid` is given Figure 6 (on the right with dashed line). Through step plot figure, we compute its transient response characteristics. The step response for this controller has 5.45% overshoot, 136 and 459 seconds rise and settling times.

We compare the closed-loop responses of the Smith Predictor, the designed PID controllers for the approximate plant $P_a(s)$ and the original plant $P(s)$ by

```
lsim(Tsp,Ta,Tpid,ref,time);
```

The responses in Figure 6 (on the left) show that the designed PID controller for the original plant offers a good compromise between the simplicity of the controller and good tracking performance compared to the Smith predictor.

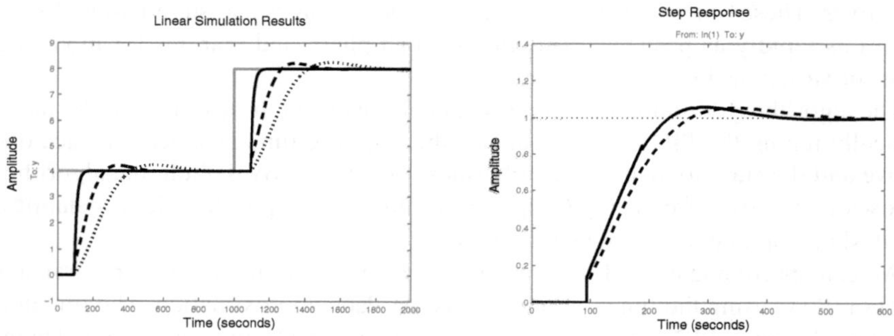

Fig. 6 (left) Responses of the Smith Predictor (–), PID (- -) for $P(s)$ and PID (:) for $P_a(s)$ to the reference signal (gray colored). (right) Closed-loop step responses for the controllers `Cpid` (–) and `Cpidf` (- -).

We can fine tune the PID controller depending on design requirements. If faster response is required, we can increase the crossover frequency slightly and obtain the controller `Cpidf` by

```
Cpidf = pidtune(P,'pid',0.0074);
Tpidf = feedback(P*Cpidf,1);
step(Tpid,Tpidf);
```

The closed-loop step responses for the controllers `Cpid` and `Cpidf` are shown Figure 6 (on the right). The new controller `Cpidf` has a faster response where its rise and settling times are 109 and 394 seconds, almost 20% and 17% faster than that of `Cpid` and its overshoot is slightly increased from 5.45% to 5.89%.

7 Possible Enhancements in CACSD

We briefly summarized the available functionality in Control System Toolbox and we discuss possible enhancements for CACSD regarding time delay systems in this section. As illustrated before, most of the functions in Control System Toolbox are extended for time delay systems. We focus on three important numerical computations for time delay systems and discuss on-going research directions on these computations.

Stability of a Time Delay System

There are various numerical methods to determine the stability of LTI systems with constant delays [5,11,21]. One idea is to compute the characteristic roots when time delay is zero and to detect characteristic roots crossing the imaginary axis from zero delay to desired time delay and determine the stability of the time delay system. This approach is applicable to only systems with commensurate time delays and quasi-polynomial form is required. Another approach is to approximate the right-most characteristic roots in the complex plane using spectral methods further explained in the next section. The computational cost in this method depends on the number of discretization points for the time delay interval, i.e., from zero to maximum delay in the system. There are some heuristic methods to choose this number and they may result in poor choices at certain cases. Lyapunov theory is another tool to determine the stability of time delay systems. The results are conservative and in general the conservatism can be reduced in the expense of the computational cost of solving larger linear matrix inequalities. Most methods in the literature can not handle time delay systems with high orders.

System Poles and Zeros

The poles and zeros of time delay systems are computed by solving a nonlinear eigenvalue problem, essentially same problem to compute the characteristic roots of time delay systems. Therefore, approximating spectrum approach for characteristic roots is also used to compute system poles and zeros.

The computations are based on either discretization of the solution operator of a delay differential equation or the infinitesimal generator of the solution operator semigroup. The solution operator approach by linear-multi-step time integration for retarded type delay differential equations is given in [6, 7]. The infinitesimal generator approach discretizes the derivative in abstract delay differential equation by Runge-Kutta or linear multi-step methods and approximates into a matrix [2, 4] for retarded type delay differential equations with multiple discrete and distributed delays. Extensions to neutral type delay differential equations and mixed-type functional differential equations are done in [3]. Numerically stable implementation of spectral methods with some heuristics is given in [28].

The computation of system poles and zeros is closely connected with the nonlinear eigenvalue problem and an eigenvalue algorithm for this is presented in [18]. A numerical method to compute all characteristic roots of a retarded or neutral quasi-polynomial on a large region in the complex-plane is proposed in [26]. The characteristic roots are calculated by finding the intersection of real and imaginary part of the characteristic equation on certain regions in complex-plane. This approach is further improved and accelerated by removing the regions outside of asymptotic chain roots in [27]. These methods consider the transfer function representation of delay differential equations which can be written as a ratio of quasi-polynomials. As noted in [27], when delay differential equations have state-space representations, transforming these systems into transfer function representation is not numerically desired, therefore in this case discretization approaches may be preferred.

$\mathscr{H}_\infty$ and $\mathscr{H}_2$ *norms*

The computation of $\mathscr{H}_\infty$ and $\mathscr{H}_2$ norms of time delay systems are quite new research topics and there are few research papers on these topics. Similar to the computation of system poles and zeros, $\mathscr{H}_\infty$ norm computation is reduced solving a nonlinear eigenvalue problem where the recent developments are applicable, [20]. The computation of $\mathscr{H}_2$ norm requires solving the delay Lyapunov equation, [19].

Note that all three computation methods are mainly used for analysis of time delay systems. Another challenging task is to design controllers and to extend classical control methods to time delay systems such as LQG, $\mathscr{H}_2$ control, $\mathscr{H}_\infty$ control, root-locus technique, model reduction methods.

There are continuing research efforts to solve these problems such as [12, 13, 17, 25]. The remaining main task is to determine numerically stable algorithms to solve control design problems for high dimensional plant with the least user interactions.

8 Concluding Remarks

We have shown that the GLTI class is suitable for computer-aided manipulation of time delay systems. We discussed various representations and interconnections of time delay systems on MATLAB. We presented the MATLAB functionality to analyze and design control systems with delays, regardless of the control structure and number of delays. Most Control System Toolbox functions have been extended to work on GLTI models, all this without additional complexity or new syntax for the user. We hope that these new tools will facilitate the design of control systems with delays and bring new insights into their behavior.

References

1. Atay, F.M. (ed.): Complex time delay Systems: Theory and Applications. Understanding Complex Systems. Springer (2010)
2. Breda, D., Maset, S., Vermiglio, R.: Pseudospectral differencing methods for characteristic roots of delay differential equations. SIAM Journal on Scientific Computing 27(2), 482–495 (2005)

3. Breda, D., Maset, S., Vermiglio, R.: Pseudospectral approximation of eigenvalues of derivative operators with non-local boundary conditions. Applied Numerical Mathematics 56(3-4), 318–331 (2006)
4. Breda, D., Maset, S., Vermiglio, R.: TRACE-DDE: a Tool for Robust Analysis and Characteristic Equations for Delay Differential Equations. In: Loiseau, J.J., Michiels, W., Niculescu, S.-I., Sipahi, R. (eds.) Topics in Time Delay Systems. LNCIS, vol. 388, pp. 145–155. Springer, Heidelberg (2009)
5. Dugard, L., Verriest, E. (eds.): Stability and control of time delay systems. LNCIS, vol. 228. Springer, Heidelberg (1998)
6. Engelborghs, K., Luzyanina, T., Samaey, G.: DDE-BIFTOOL V. 2.00: a Matlab Package for Bifurcation Analysis of Delay Differential Equations. Technical Report TW330, Department of Computer Science, K. U. Leuven, Leuven, Belgium (2001)
7. Engelborghs, K., Roose, D.: On stability of LMS methods and characteristic roots of delay differential equations. SIAM Journal on Numerical Analysis 40(2), 629–650 (2002)
8. Enright, W.H., Hayashi, H.: A delay differential equation solver based on a continuous Runge-Kutta method with defect control. Numerical Algorithms 16(3-4), 349–364 (1997)
9. Erneux, T.: Applied delay differential equations. Surveys and tutorials in the applied mathematical sciences. Springer (2009)
10. Gahinet, P., Shampine, L.F.: Software for modeling and analysis of linear systems with delays. In: Proceedings of the American Control Conference (2004)
11. Gu, K., Kharitonov, V., Chen, J.: Stability of time delay systems. Birkhuser, Boston (2003)
12. Gumussoy, S., Michiels, W.: Fixed-Order H-infinity Control for Interconnected Systems using Delay Differential Algebraic Equations. SIAM Journal on Control and Optimization 49(2), 2212–2238 (2011)
13. Gumussoy, S., Michiels, W.: Root Locus for SISO Dead-Time Systems: A Continuation Based Approach. Automatica 48(3), 480–489 (2012)
14. Gumussoy, S., Eryilmaz, B., Gahinet, P.: Working with Time-Delay Systems in MATLAB. In: 10th IFAC Workshop on Time Delay Systems, June 22-24. IFAC-PapersOnLine, pp. 108–113. Northeastern University, USA (2012), doi:10.3182/20120622-3-US-4021.00041
15. Hairer, E., Wanner, G.: RETARD: Software for delay differential equations (1995), `http://unige.ch/~hairer/software.html`
16. Ingimundarson, A., Hagglund, T.: Robust tuning procedures of dead-time compensating controllers. Control Engineering Practice 9, 1195–1208 (2001)
17. Jarlebring, E., Damm, T., Michiels, W.: Model reduction of time delay systems using position balancing and delay Lyapunov equations. Technical Report TW602, Department of Computer Science, K. U. Leuven, Leuven, Belgium (2011)
18. Jarlebring, E., Michiels, W., Meerbergen, K.: A linear eigenvalue algorithm for the nonlinear eigenvalue problem. Technical Report TW580, Department of Computer Science, K. U. Leuven, Leuven, Belgium (2011)
19. Jarlebring, E., Vanbiervliet, J., Michiels, W.: Characterizing and computing the L2 norm of time delay systems by solving the delay Lyapunov equation. IEEE Transactions on Automatic Control 56, 814–825 (2011)
20. Michiels, W., Gumussoy, S.: Characterization and computation of H-infinity norms of time delay systems. SIAM Journal on Matrix Analysis and Applications 31(4), 2093–2115 (2010)
21. Michiels, W., Niculescu, S.I.: Stability and stabilization of time delay systems. An eigenvalue based approach. In: Advances in Design and Control. vol. 12. SIAM, Philadelphia (2007)

22. Niculescu, S.I.: Delay effects on stability: A robust control approach. LNCIS, vol. 269. Springer, London (2001)
23. Skogestad, S., Postlethwaite, I.: Multivariable Feedback Control. John Wiley (1996)
24. Control System Toolbox: MathWorks Inc., Natick (2011)
25. Vanbiervliet, J., Michiels, W., Jarlebring, E.: Using spectral discretization for the optimal H2 design of time delay systems. International Journal of Control 84(2), 228–241 (2011)
26. Vyhlídal, T., Zítek, P.: Quasipolynomial mapping based rootfinder for analysis of time delay systems. In: Proceedings of the 4th IFAC Workshop on Time Delay Systems, Rocquencourt, France, pp. 227–232 (2003)
27. Vyhlídal, T., Zítek, P.: Mapping based algorithm for large-scale computation of quasi-polynomial zeros. IEEE Transactions on Automatic Control 54(1), 171–177 (2009)
28. Wu, Z., Michiels, W.: Reliably computing all characteristic roots of delay differential equations in a given right half plane using a spectral method. Technical Report TW596, Department of Computer Science, K. U. Leuven, Leuven, Belgium (2011)

Analysis and Control of Time Delay Systems Using the LambertWDDE Toolbox

Sun Yi, Shiming Duan, Patrick W. Nelson, and A. Galip Ulsoy

Abstract. This chapter provides an overview of the Lambert W function approach. The approach has been developed for analysis and control of linear time-invariant time delay systems with a single known delay. A solution in the time-domain is given in terms of an infinite series, with the important characteristic that truncating the series provides a dominant solution in terms of the rightmost eigenvalues. A solution via the Lambert W function approach is first presented for systems of order one, then extended to higher order systems using the matrix Lambert W function. Free and forced solutions are used to investigate key properties of time-delay systems, such as stability, controllability and observability. Through eigenvalue assignment, feedback controllers and state-observers are designed. All of these can be achieved using the Lambert W function-based framework. The use of the MATLAB-based open source software in the LambertWDDE Toolbox is also introduced using numerical examples.

Sun Yi
Department of Mechanical Engineering, North Carolina AT State University, Greensboro, NC 27411 USA
e-mail: `syi@ncat.edu`

Shiming Duan
Magna Powertrain, Troy, MI 48083 USA
e-mail: `shiming.duan@magnapowertrain.com`

Patrick W. Nelson
Department of Mathematics and Computer Science,
Lawrence Technological University, Southfield, MI 48075 and Department of Mathematics, University of Michigan, Ann Arbor, MI 48109 USA
e-mail: `pwn@umich.edu`

A. Galip Ulsoy
Department of Mechanical Engineering, University of Michigan,
Ann Arbor, MI 48109-2125 USA
e-mail: `ulsoy@umich.edu`

T. Vyhlídal, J.-F. Lafay, and R. Sipahi (eds.), *Delay Systems*,
Advances in Delays and Dynamics 1,
DOI: 10.1007/978-3-319-01695-5_20, © Springer International Publishing Switzerland 2014

1 Introduction

Time delay systems (TDS) arise in numerous natural and engineered systems, such as processes with transport delays, traffic flow problems, biological systems, teleoperation and many others. The literature on TDS is quite extensive, and includes several excellent books and review papers, e.g. [2, 11, 12, 14, 17, 19, 20]. This chapter focuses on a specific and recently developed approach, based on the classical Lambert W function [5], for analysis and control of linear time invariant TDS with a single known delay [27].

1.1 Motivation and Background

Consider a typical n^{th}-order system of linear time invariant (LTI) ordinary differential equations (ODEs), without delay, in standard state equation form:

$$\begin{aligned} \dot{\mathbf{x}}(t) &= \mathbf{A}\mathbf{x}(t) + \mathbf{B}\mathbf{u}(t) \\ \mathbf{y}(t) &= \mathbf{C}\mathbf{x}(t) + \mathbf{D}\mathbf{u}(t) \end{aligned} \tag{1}$$

the closed-form free and forced solutions can be obtained using the concepts of a state transition matrix and a convolution integral [4]. As the system of ODEs in Eq. (1) has a finite spectrum, the stability can be determined by examining the locations of the finite number of eigenvalues in the s-plane.

The solutions are also used to derive the controllability and observability Gramians. Then, controllability and observability can be determined by the rank of the controllability and observability matrices, respectively. If the system is controllable, a closed-loop controller can be designed by a variety of methods, including state feedback control and eigenvalue assignment. Similarly, if a system is observable, then a state estimator, or observer, can be designed, e.g., using eigenvalue assignment.

These systematic steps for analysis and control are standard for LTI systems of ODEs as in (1), because the closed-form solutions to Eq. (1) are obtained analytically. However, unlike ODEs, these steps are often difficult to achieve for LTI TDS due to lack of analytical time-domain solutions to delay differential equations. Main difficulty is due to their infinite spectrum arising from the delays. Recently methods, based on the Lambert W function, have been proposed, developed and demonstrated for the analysis and control of LTI TDS, which enable the analysis and control design steps outlined above to be applied in a manner analogous to LTI systems of ODEs [1, 28]. The key characteristic of the method is that the solution is given in terms of an infinite eigenvalue expansion, based on the branches of the Lambert W function, such that truncating the series always yields a finite dimensional representation in terms of the rightmost (i.e., dominant) eigenvalues.

There are numerous natural and engineered systems where time delays are significant (e.g., biological systems, economic models, supply chains, traffic flow, teleoperation, networked control systems, automotive control systems) [19]. Thus, benefits from the extension of the system analysis and control tools, which are standard for ODEs, to systems described by DDEs can be substantial.

1.2 *Purpose and Scope*

This chapter is intended to provide a succinct overview of the Lambert W function approach to the analysis and control of LTI TDS with a single known delay and introduction, via simple numerical examples, to the use of the open source software LambertWDDE Toolbox, which is available for downloading from the web [7]. Additional examples can be found at the same website [7] and numerous applications of the method can be found in the references cited (e.g., [8,21,24–27,32]). Note that a preliminary version of this chapter has been presented in [33].

2 Theory, Examples and Numerical Simulation

2.1 *Lambert W Function*

By definition [5, 10, 15], every function $W(s)$ that satisfies:

$$W(s)e^{W(s)} = s \tag{2}$$

is called a Lambert W function. The Lambert W function, with complex argument s, is a complex valued function with infinite branches, $k = 0, \pm 1, \pm 2, .., \pm \infty$, where s is either a scalar (i.e., scalar Lambert W function) or a matrix (i.e., matrix Lambert W function). The scalar Lambert W function is available as an embedded function in many computational software systems, e.g., see the function *lambertw* in MATLAB. The matrix Lambert W function [27] can be obtained using a similarity transformation and can be readily evaluated using the LambertWDDE Toolbox [7]. These functions are useful in combinatorics (e.g., the enumeration of trees) as well as relativity and quantum mechanics. They can be used to solve various equations involving exponentials (e.g. the maxima of the Planck, Bose-Einstein, and Fermi-Dirac distributions) as well as in the solution of delay differential equations as discussed here.

2.2 *Scalar Case*

Consider the first-order TDS [1]:

$$\dot{x}(t) = ax(t) + a_d x(t-h) + bu(t) \tag{3}$$

with constant known parameters a, b and a_d, and where h is the constant known delay. The initial condition $x(t = 0) = x_0$ and preshape function $x(t) = g(t)$ for

$-h \leq t < 0$, must be specified. The Lambert W function is applied to solve the transcendental characteristic equation of Eq. (3), which can be written as:

$$(s-a)e^{sh} = a_d \tag{4}$$

Multiplying both sides of Eq. (4) by he^{-ah} yields:

$$h(s-a)e^{h(s-a)} = a_d he^{-ah} \tag{5}$$

Based on the definition of the Lambert W function in Eq. (5) it is clear that

$$W(a_d he^{-ah})e^{W(a_d he^{-ah})} = a_d he^{-ah} \tag{6}$$

Comparing Eqs. (5) and (6)

$$h(s-a) = W(a_d he^{-ah}) \tag{7}$$

Thus, the solution of the characteristic equation in Eq. (4) can be written in terms of the Lambert W function as:

$$s = \frac{1}{h}W(a_d he^{-ah}) + a \tag{8}$$

The infinite spectrum of the scalar DDE in (3) is, thus, obtained using the infinite branches of the Lambert W function, and is given explicitly in terms of parameters a, a_d and h of the system. The roots of the characteristic equation (4), for $k = 0, \pm 1, \pm 2, .., \pm\infty$, are:

$$s_k = \frac{1}{h}W_k(a_d he^{-ah}) + a \tag{9}$$

Furthermore, for Eq. (3) stability is determined by the rightmost eigenvalue in the s-plane, which has been shown in [18] to be obtained using only the principal (i.e., $k = 0$) branch of the Lambert W function. Consequently, to ensure stability, it is not necessary to check the other eigenvalues in the infinite spectrum.

2.3 *Example 1 - Spectrum and Series Expansion in the Scalar Case*

For $a = -1$, $a_d = 0.5$, and $h = 1$ the characteristic roots are obtained using Eq. (9) and the function *lambertw* in MATLAB, and are plotted in Fig. 1. It can also be shown [28] that the total (i.e., free plus forced) solution to Eq. (3), can be represented in terms of an infinite series based on the eigenvalues in Eq. (9) as:

$$x(t) = \sum_{k=-\infty}^{+\infty} e^{s_k t} C_k^I + \int_0^t \sum_{k=-\infty}^{+\infty} e^{s_k(t-\eta)} C_k^N bu(\eta)d\eta \tag{10}$$

where

$$C_k^I = \frac{x_0 + a_d e^{-s_k h} \int_0^h e^{-s_k t} g(t-h) \mathrm{d}t}{1 + a_d h e^{-s_k h}} \tag{11}$$

and

$$C_k^N = \frac{1}{1 + a_d h e^{-s_k h}} \tag{12}$$

Note that the coefficients C_k^I are determined from the preshape function $g(t)$ and the initial state x_0, and the coefficients C_k^N are determined only in terms of the system parameters a, a_d and h. Thus, the total solution in Eq. (10) can be viewed as the sum of the free and forced solutions. Conditions for convergence of such a series solution form are discussed in [2]. A very practically important and useful aspect of this particular series representation of the solution for $x(t)$ is that truncating the series, e.g., $k = 0, \pm 1, \pm 2, .., \pm n$, yields an approximation of the solution in terms of the $(2n+1)$ rightmost, or most dominant, eigenvalues. Consequently, a simple finite dimensional approximation of the system accurately represents its dynamics.

2.4 *Example 2 - Scalar Case Approximation Response*

For $a = -1$, $a_d = 0.5$, and $h = 1$ one can obtain the values of s_k using Eq. (9) and the function *lambertw* as in Ex. 1, and the values of C_k^I and C_k^N using Eqs. (11)-(12), where $x_0 = 1$ and $g(t) = 1$ for $-h \le t < 0$. These are given in Table 1. Fig. 2 shows the total response to $u(t) = sin(t)$ and a comparison between the Lambert W function-based method (using the 7 terms in Table 1) and a numerical solution (using the function $dde23$ in MATLAB). The two plots are essentially indistinguishable.

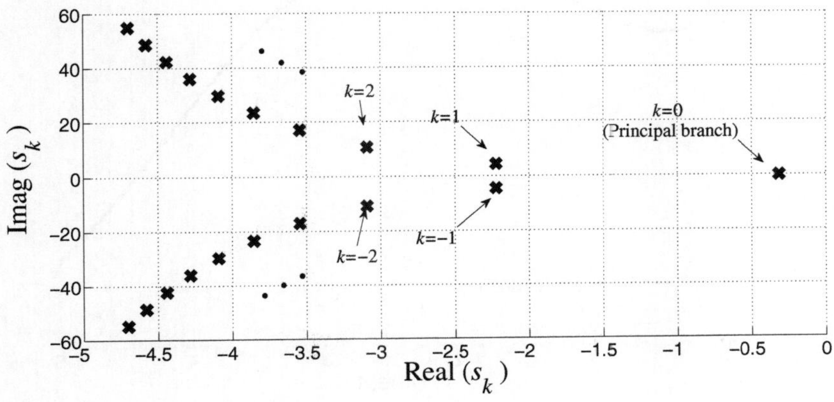

Fig. 1 Eigenvalue of Eq. (3) when $a = -1$, $a_d = 0.5$, and $h = 1$. The rightmost eigenvalue is for $k = 0$, the next pair are for $k = \pm 1$, the next for $k = \pm 2$, to $k = \pm 9$.

Table 1 The eigenvalues and coefficients in the solution for Ex. 2

k	s_k	C_k^I	C_k^N
0	-0.3149	0.9422	0.5934
± 1	$-2.2211 \pm 4.4442i$	$0.0197 \pm 0.0111i$	$-0.0112 \pm 0.2245i$
± 2	$-3.0915 \pm 10.8044i$	$0.0038 \pm 0.0015i$	$-0.0093 \pm 0.0916i$
± 3	$-3.545 \pm 17.1313i$	$0.0016 \pm 0.0005i$	$-0.0052 \pm 0.0579i$

2.5 General Case

The approach presented in the previous section has been generalized in [27] to LTI TDS of the form:

$$\begin{aligned} \dot{\mathbf{x}}(t) &= \mathbf{A}\mathbf{x}(t) + \mathbf{A_d}\mathbf{x}(t-h) + \mathbf{B}\mathbf{u}(t) \\ \mathbf{y}(t) &= \mathbf{C}\mathbf{x}(t) + \mathbf{D}\mathbf{u}(t) \end{aligned} \tag{13}$$

where $\mathbf{x}(t)$ is the state vector, $\mathbf{u}(t)$ is the input vector, $\mathbf{y}(t)$ is the output vector, $\mathbf{A}$, $\mathbf{A_d}$, $\mathbf{B}$, $\mathbf{C}$ and $\mathbf{D}$ are coefficient matrices, and h is the constant known scalar delay. The initial condition $\mathbf{x}(t=0) = \mathbf{x}_0$ and preshape function $\mathbf{x}(t) = \mathbf{g}(t)$ for $-h \le t < 0$, must also be specified. The total solution for the states is now given as:

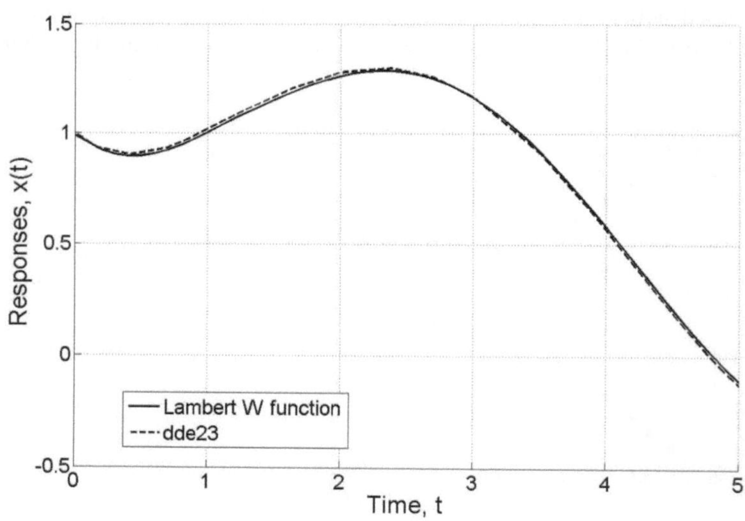

Fig. 2 Total response to $u(t) = sin(t)$, with $x_0 = 1$ and $g(t) = 1$ for $-h \le t < 0$, and comparison between the 7-term (see Table 1) Lambert W function-based method and the numerical method (function *dde23* in MATLAB). Parameters are $a = -1$, $a_d = 0.5$, and $h = 1$.

$$\mathbf{x}(t) = \sum_{k=-\infty}^{\infty} e^{\mathbf{S}_k t}\mathbf{C}_k^I + \int_0^t \sum_{k=-\infty}^{\infty} e^{\mathbf{S}_k(t-\eta)}\mathbf{C}_k^N \mathbf{B}\mathbf{u}(\eta)d\eta \tag{14}$$

where

$$\mathbf{S}_k = \frac{1}{h}\mathbf{W}_k(\mathbf{A}_d h\mathbf{Q}_k) + \mathbf{A} \tag{15}$$

and $\mathbf{Q}_k$ is obtained from numerical solution (e.g., using *fsolve* in MATLAB) of:

$$\mathbf{W}_k(\mathbf{A}_d h\mathbf{Q}_k)e^{\mathbf{W}_k(\mathbf{A}_d h\mathbf{Q}_k)+\mathbf{A}h} = \mathbf{A}_d h \tag{16}$$

This generalization is dependent on the introduction of a matrix Lambert W function, $\mathbf{W}_k$, as described in [27]. The quantities $\mathbf{Q}_k$, $\mathbf{W}_k$, $\mathbf{S}_k$, $\mathbf{C}_k^I$ and $\mathbf{C}_k^N$ in Eqs. (14)-(16) can all be computed using the software in the LambertWDDE Toolbox in terms of given h, $\mathbf{A}$, $\mathbf{A}_d$, $\mathbf{g}(t)$, $\mathbf{x}_0$, $\mathbf{B}$ and $\mathbf{u}(t)$. The main functions of the LambertWDDE Toolbox [7] are summarized in Table 2.

Table 2 Main functions of the LambertWDDE Toolbox [7]

Name	Description
lambertw_matrix	Calculate matrix Lambert W function
find_Sk	Find $\mathbf{S}_k$ and $\mathbf{Q}_k$ for a given branch
find_CI	Calculate $\mathbf{C}_I$ under specific initial conditions for a given branch
find_CN	Calculate $\mathbf{C}_N$ for a given branch
pwcont_test	Controllability test for DDEs
pwobs_test	Observability test for DDEs
cont_gramian_dde	Controllability Gramian for DDEs
obser_gramian_dde	Observability Gramian for DDEs
place_dde	Rightmost eigenvalue assignment for DDEs
stabilityradius_dde	Calculate stability radius for DDEs
examples	Lists examples for using this toolbox; each cell is a short example and can be evaluated separately (Ctr+Enter)

2.6 Example 3 - General Case Approximation

To obtain $\mathbf{S}_k$ for a particular branch, k, one needs to solve Eq. (16) for $\mathbf{Q}_k$ first, then substitute the result into Eq. (15) to obtain $\mathbf{S}_k$. The steps are carried out in the function *find_Sk*. Note that, the matrix Lambert W function $\mathbf{W}_k(\cdot)$ in Eq. (15) and (16) is calculated using the function *lambertw_matrix*, which is based on a Jordan canonical form transformation. To better understand the whole process, an example is provided here. Given

$$\mathbf{A} = \begin{bmatrix} -1 & -3 \\ 2 & -5 \end{bmatrix}; \mathbf{A_d} = \begin{bmatrix} 1.66 & -0.697 \\ 0.93 & -0.33 \end{bmatrix}; h = 1$$

For the principal branch, $k = 0$, one obtains:

$$\mathbf{S}_0 = \begin{bmatrix} 0.3055 & -1.4150 \\ 2.1317 & -3.3015 \end{bmatrix}$$

The eigenvalues of which are -1.0119 and -1.9841. Next, using *find_CI* and *find_CN*, one can obtain the coefficients for the series solution in Eq. (14). For example, if $u(t) = 0$, $\mathbf{g}(t) = 0$, and we have an abrupt change at $t = 0$ to $\mathbf{x}_0 = \begin{bmatrix} 1 & 1 \end{bmatrix}^T$, we can obtain (using *find_CI*) the coefficients for the free response for $k = 0$ as:

$$\mathbf{C}_0^I = \begin{bmatrix} 0.2635 \\ 0.4290 \end{bmatrix}$$

Thus, the single branch approximation, for $k = 0$, for the free response is:

$$\mathbf{x}(t) = \begin{Bmatrix} x_1(t) \\ x_2(t) \end{Bmatrix} = e^{\begin{bmatrix} 0.3055 & -1.4150 \\ 2.1317 & -3.3015 \end{bmatrix} t} \begin{Bmatrix} 0.2635 \\ 0.4290 \end{Bmatrix}$$

Note that in MATLAB the matrix exponential is evaluated using the function *expm*, not the scalar exponential function *exp*. To improve the approximation, this process can be repeated for additional branches, k, then an approximate series solution can be obtained using Eq. (14) with a finite number of k (see Fig. 3). For example, including the branches $k = \pm 1$ gives the additional complex conjugate $\mathbf{S}_k$ matrices:

$$\mathbf{S}_{-1,+1} = \begin{bmatrix} -0.399 \pm 4.980i & -1.6253 \pm 0.1459i \\ 2.4174 \pm 0.1308i & -5.1048 \pm 4.5592i \end{bmatrix}$$

with complex conjugate coefficients for the free response for $k = \pm 1$ as:

$$\mathbf{C}^I_{-1,+1} = \begin{bmatrix} 0.0909 \pm 0.1457i \\ 0.0435 \pm 0.1938i \end{bmatrix}$$

2.7 Observability and Controllability

In [25], the criteria for point-wise controllability and observability have been derived as follows.

Point-wise Controllability: The system of DDEs in Eq. (13) is point-wise controllable if, for any given initial conditions $\mathbf{g}(t)$ and $\mathbf{x}_0$, there exists a time t_1, $0 < t_1 < \infty$, and an admissible (i.e., measurable and bounded on a finite time interval) control segment $\mathbf{u}(t)$ for such that $\mathbf{x}(t_1; \mathbf{g}, \mathbf{x}_0, \mathbf{u}(t)) = 0$ [22]. For the scalar DDE in Eq. (3) it is point-wise controllable if and only if, for all s not at the poles of the system, $s - a - a_d e^{-sh} \neq 0$; similarly for Eq. (13) one must have linearly independent rows of $(s\mathbf{I} - \mathbf{A} - \mathbf{A}_d e^{-sh})^{-1}\mathbf{B}$. Furthermore, the controllability Gramian $\mathbf{C}(0, t_1) = \int_0^{t_1} \sum_{k=-\infty}^{\infty} e^{\mathbf{S}_k(t_1 - \eta)} \mathbf{C}_k^N \mathbf{B}\mathbf{B}^T \{e^{\mathbf{S}_k(t_1-\eta)} \mathbf{C}_k^N\}^T d\eta$ for Eq. (13) must be full rank [27].

Point-wise Observability: The system of DDEs in Eq. (13) is point-wise observable in $[0,t_1]$ if the initial point $\mathbf{x}_0$ can be uniquely determined from the knowledge of $\mathbf{u}(t)$, $\mathbf{g}(t)$, and $\mathbf{y}(t)$ [6]. For the scalar DDE in Eq. (3), it is point-wise observable if and only if, for all s not at the poles of the system; similarly, for Eq. (13) one must have linearly independent columns of $\mathbf{C}(s\mathbf{I}-\mathbf{A}-\mathbf{A}_d e^{-sh})^{-1}$. Furthermore, the observability Gramian $\mathbf{O}(0,t_1) = \int_0^{t_1} \sum_{k=-\infty}^{\infty} \{e^{\mathbf{S}_k(t_1-\eta)}\mathbf{C}_k^N\}^T \mathbf{C}^T \mathbf{C} e^{\mathbf{S}_k(t_1-\eta)}\mathbf{C}_k^N d\eta$ for Eq. (13) must be full rank [27].

2.8 Example 4 - Piecewise Observability and Controllability

Consider Eq. (13), with $\mathbf{A}$, $\mathbf{A}_d$ and h as given in Ex. 3, and

$$\mathbf{B} = \begin{Bmatrix} 1 \\ 0 \end{Bmatrix} \text{ and } \mathbf{C} = \begin{bmatrix} 0 & 1 \end{bmatrix}$$

The function *pwcontr_test* can be used to establish that the system is piecewise controllable. It examines the rank of the matrix $(s\mathbf{I}-\mathbf{A}-\mathbf{A}_d e^{-sh})^{-1}\mathbf{B}$ to determine if the system is piecewise controllable, and will display the conclusion on the screen. The Piecewise observability can be established using the function *pwobs_test* in a similar way. Furthermore, the controllability and observability Gramians over a specific time interval can be approximately computed, for $k=n$ branches, using the functions *contr_gramian_dde* and *obs_gramian_dde* respectively.

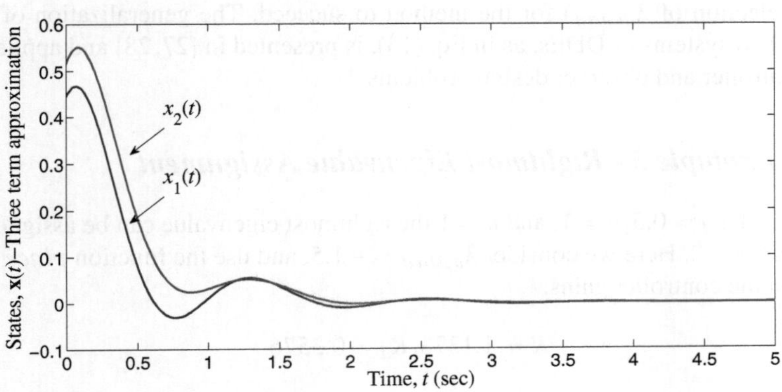

Fig. 3 Approximate (3-term) free response for the system in Ex. 3

2.9 Placement of Dominant Poles

In [28], a method for eigenvalue assignment via Lambert W function was proposed. Consider the scalar LTI TDS in Eq. (3) with the generalized feedback structure

$$u(t) = -Kx(t) - K_d x(t-h)$$

The closed-loop system becomes

$$\dot{x}(t) = (a - bK)x(t) + (a_d - bK_d)x(t-h)$$

One can use the Lambert W function approach to assign the rightmost eigenvalues of the system. The procedure for selecting the gains K and K_d can be described as:

1. Select desired rightmost eigenvalue $\lambda_{desired}$
2. Set initial gains $K = K_0$ and $K_d = K_{d0}$
3. while $\lambda(S_{0\_new} - \lambda_{desired}) > tolerance$
4. Select $K = K_{new}$ and $K_d = K_{d\_new}$
5. Let $a_{new} = (a - bK_{new}), \quad a_{d\_new} = (a_d - bK_{d\_new})$, calculate $S_{0\_new} = \frac{1}{h}W_0(a_{d\_new}he^{-a_{new}h}) + a_{new}$.
6. End

Due to the range limitation of the branches of the Lambert W function, the rightmost poles cannot be assigned to any arbitrary location in the s-plane. For scalar time-delay systems, as in Eq. (3), this can be easily seen by examining the principal branch [27].

$$\mathrm{Re}\{\frac{1}{h}W_0(a_d he^{-ah}) + a\} \geq \mathrm{Re}\{-\frac{1}{h} + a\} \geq -\frac{1}{h} + a$$

since $\mathrm{Re}\{W_0(H)\} \geq -1$. Thus the rightmost pole cannot be less than $-\frac{1}{h} + a$. For the matrix case, similar constraints apply but the relationship becomes more complicated. This feasibility constraint has to be considered in the design process (e.g., in the selection of $\lambda_{desired}$) for the method to succeed. The generalization of this approach to systems of DDEs, as in Eq. (13), is presented in [27, 28] and applied to both controller and observer design problems.

2.10 Example 5 - Rightmost Eigenvalue Assignment

For $a = -1, a_d = 0.5, b = 1$, and $h = 1$ the rightmost eigenvalue can be assigned to any value > -2. Here we consider $\lambda_{desired} = -1.5$, and use the function *place_dde* to obtain the controller gains:

$$K = 1.1378, K_d = 0.3576$$

Thus, the closed-loop LTI TDS becomes:

$$\dot{x}(t) = -2.1378x(t) + 0.1424x(t-1)$$

and the rightmost eigenvalue can be found, using $k = 0$ in the function *lambertw*, as in Ex. 1, to now be located at -1.4998 as desired.

2.11 Robust Control and Time Domain Specifications

The assignment of rightmost eigenvalues for LTI TDS can also be used for observer design, and extended to robust design in the presence of structured model uncertainties. Since the response of the LTI TDS is dominated by the rightmost eigenvalues, approximate specification of time domain characteristics (e.g., settling time, overshoot) can also be achieved [29]. The Toolbox function *stabilityradius_dde* can be used to calculate the stability radius for DDEs as described in [13, 27, 29].

2.12 Decay Function for TDS

Accurate estimation of the decay function for time delay systems has been a longstanding problem, which has recently been addressed using the Lambert W function based approach [8]. The goal is to find a tight upper bound for the decay rate, which is referred to as α-stability, as well as an upper bound for the factor K, such that the norm of the states is bounded:

$$\|\mathbf{x}(t)\| \leq Ke^{\alpha t}\Phi(h,t_0) \tag{17}$$

where $\Phi(h,t_0) = \sup_{t_0-h\leq t\leq t_0}\{\|\mathbf{x}(t)\|\}$ and $\|\cdot\|$ denotes the 2-norm. Based on the solution form in Eq. (14) in terms of the Lambert W function, an optimal estimate of α can be obtained. The estimate of the factor K is also less conservative especially for the matrix case. A less conservative estimate of the decay function leads to a more accurate description of the transient response, and more efficient control strategies based on the decay model [8].

2.13 Example 6 - Factor and Decay Rate

Consider the system in Eq. (13) with the same coefficients as in Ex. 3. From Eq. (15), with $k = 0$, the rightmost pole is found to be:

$$\begin{aligned}\alpha &= \max\{\mathrm{Re}(eig(\mathbf{S}_0))\}\\ &= \max\left\{\mathrm{Re}(eig(\tfrac{1}{h}\mathbf{W}_0(-\mathbf{A}_d h\mathbf{Q}_0)-\mathbf{A}))\right\} = -1.012\end{aligned}$$

Thus the exact decay rate is obtained and, using the solution in Eq. (14) and the approach in [8], one can also obtain a bound on K. As shown in Table 3, the results are less conservative when compared to other methods [8].

Table 3 Comparison of results for Ex. 6

	Factor, K	Decay Rate, α
Matrix Measure Approach [12]	8.019	3.053
Lyapunov Approach [16]	9.33	−0.907
Lambert-W Approach [8]	3.8	−1.012

3 Concluding Remarks

A summary of the recently developed Lambert W function approach for analysis and control of LTI TDS with a constant delay was provided in this chapter. For more details, readers are referred to the publications cited here (e.g., [27]) and the web [7]. Several numerical examples are given to illustrate the use of the *lambertw* function in MATLAB, as well as other useful functions available in the open source LambertWDDE Toolbox software package for LTI DDEs [7].

The proposed approach can be used, just as for systems of LTI ODEs as in Eq. (1), for a variety of important analysis and control tasks for LTI DDEs, such as free and forced solutions, stability, observability and controllability, controller and observer design via assignment of dominant eigenvalues, robust stability, determination of the decay function, etc. The open source software in the LambertWDDE Toolbox, as well as the accompanying documentation and examples [7], we hope will make the Lambert W function based approach more accessible and useful for those interested in applications that are well modeled as LTI TDS with a single constant delay. Numerous applications of the method (e.g., machine tool chatter, engine control, HIV dynamics, decay function estimation, DC motor control, PID control and robust control) can also be found in the references [8, 21, 24–28, 32]. Besides the LambertWDDE Toolbox, other useful software for time-delay systems based on a variety of algorithms is also available for downloading from the web [3, 9, 23].

References

1. Asl, F.M., Ulsoy, A.G.: Analysis of a system of linear delay differential equations. ASME J. Dynamic Systems, Measurement and Control 125, 215–223 (2003)
2. Bellman, R., Cooke, K.L.: Differential-Difference Equations. Academic Press, New York (1963)
3. Breda, D.,Maset, S., Vermiglio, R.: TRACE-DDE: a Tool for Robust Analysis and Characteristic Equations for Delay Differential Equations. In: Loiseau, J.J., Michiels, W., Niculescu, S.-I., Sipahi, R. (eds.) Topics in Time Delay Systems. LNCIS, vol. 388, pp. 145-155. Springer, Heidelberg (2009)
4. Chen, C.T.: Linear System Theory and Design. Oxford University Press (1998)
5. Corless, R.M., Gonnet, G.H., Hare, D.E.G., Jeffrey, D.J., Knuth, D.E.: On the Lambert W function. In: Advances in Computational Mathematics, vol. 5, pp. 329–359 (1996)
6. Delfour, M.C., Mitter, S.K.: Controllability, observability and optimal feedback control of affine hereditary differential systems. SIAM J. Control 10, 298–328 (1972)
7. Duan, S.: Supplement for time delay systems: analysis and control using the Lambert W function (2010), `http://www.umich.edu/~ulsoy/TDS_Supplement.htm`
8. Duan, S., Ni, J., Ulsoy, A.G.: Decay function estimation for linear time delay systems via the Lambert W function. Journal of Vibration and Control 18(10), 1462–1473 (2012)
9. Engelborghs, K., Luzyanina, T., Samaey, G.: DDE-BIFTOOL v. 2.00: a MATLAB package for bifurcation analysis of delay differential equations. Technical Report TW-330, 61 (2001)

10. Euler, L.: Deformulis exponentialibus replicatis. Leonardo Euleri Opera Omnia 1(15), 268–297 (1777)
11. Gu, K., Niculescu, S.I.: Stability analysis of time delay systems: a Lyapunov approach. In: Loría, A., Lamnabhi-Lagarrigue, F., Panteley, E. (eds.) Advanced Topics in Control Systems Theory. LNCIS, vol. 328, pp. 139–170. Springer, Heidelberg (2006)
12. Hale, J.K., Lunel, S.M.V.: Introduction to Functional Differential Equations. Springer (1963)
13. Hu, G., Davison, E.J.: Real stability radii of linear time-invariant time-delay systems. Systems and Control Letters 50(3), 209–219 (2003)
14. Kolmanovskii, V.B., Myshkis, A.: Introduction to the Theory and Applications of Functional Differential Equations. Kluwer Academy, Dordrecht (1999)
15. Lambert, J.H.: Observationes variea in mathesin Puram. Acta Helveticae, Physico-mathematico-anatomico-botanico-medica 3, 128–168 (1758)
16. Mondi, S., Kharitonov, V.L.: Exponential estimates for retarded time-delay systems: an LMI approach. IEEE Trans. Autom. Control 50(2) (2005)
17. Richard, J.P.: Time delay systems: an overview of some recent advances and open problems. Automatica 39(10), 335–357 (2003)
18. Shinozaki, H., Mori, T.: Robust stability analysis of linear time-delay systems by Lambert W function: Some extreme point results. Automatica 42(1), 1791–1799 (2006)
19. Sipahi, R., Niculescu, S.I., Abdallah, C.T., Michiels, W., Gu, K.: Stability and stabilization of systems with time delay: limitations and opportunities. IEEE Control Systems Magazine 31(1), 38–65 (2011)
20. Stepan, G.: Retarded Dynamical Systems: Stability and Characteristic Functions, Wiley (1989)
21. Ulsoy, A.G.: Improving stability margins via time delay control. In: ASME IDETC, Portland, OR, Paper No. DETC2013-12076 (August 2013)
22. Weiss, L.: On the controllability of delay-differential systems. SIAM J. Control 5, 575–587 (1967)
23. Vyhlidal, T., Zitek, P.: Mapping based algorithm for large-scale computation of quasi-polynomial zeros. IEEE Transactions on Automatic Control 54(1), 171–177 (2009)
24. Yi, S., Nelson, P.W., Ulsoy, A.G.: Delay differential equations via the matrix Lambert W function and bifurcation analysis: application to machine tool chatter. Mathematical Biosciences and Engineering 4(2), 355–368 (2007)
25. Yi, S., Nelson, P.W., Ulsoy, A.G.: Controllability and observability of systems of linear delay differential equations via the matrix Lambert W function. IEEE Trans. Automatic Control 53(3), 854–860 (2008)
26. Yi, S., Nelson, P.W., Ulsoy, A.G.: Eigenvalues and sensitivity analysis for a model of HIV-1 pathogenesis with an intracellular delay. In: Proc. ASME Dynamic Systems and Control Conf., Ann Arbor (October 2008)
27. Yi, S., Nelson, P.W., Ulsoy, A.G.: Time-Delay Systems: Analysis and Control Using the Lambert W Function. World Scientific (2010)
28. Yi, S., Nelson, P.W., Ulsoy, A.G.: Eigenvalue Assignment via the Lambert W Function for Control of Time-delay Systems. J. Vibration and Control 16(7-8), 961–982 (2010)
29. Yi, S., Nelson, P.W., Ulsoy, A.G.: Robust control and time-domain specifications for systems of delay differential equations via eigenvalue assignment. ASME J. Dynamic Systems, Measurement and Control 132(3) (May 2010)
30. Yi, S., Nelson, P.W., Ulsoy, A.G.: Design of observer-based feedback control for time-delay systems with application to automotive powertrain control. J. Franklin Institute 347(1), 358–376 (2010)

31. Yi, S., Nelson, P.W., Ulsoy, A.G.: Dc motor control using the Lambert W function approach. In: IFAC Workshop on Time Delay Systems, Boston (June 2012)
32. Yi, S., Nelson, P.W., Ulsoy, A.G.: PI control of first order time-delay systems via eigenvalue assignment. IEEE Trans. Control Systems Technology (in press, 2013)
33. Yi, S., Duan, S., Nelson, P.W., Ulsoy, A.G.: The Lambert W function approach to time delay systems and the LambertWDDE Toolbox. In: IFAC Workshop on Time Delay Systems, Boston (June 2012)

H_∞-Stability Analysis of (Fractional) Delay Systems of Retarded and Neutral Type with the Matlab Toolbox YALTA

David Avanessoff, André R. Fioravanti, Catherine Bonnet, and Le Ha Vy Nguyen

Abstract. YALTA is a Matlab toolbox dedicated to the H_∞-stability analysis of classical and fractional systems with commensurate delays given by their transfer function, whose binary can be downloaded at *http://team.inria.fr/disco/software/*. Delay systems of both retarded and neutral type are considered. The asymptotic position of high modulus poles is given. For a fixed known delay, poles of small modulus of standard delay systems are approximated through a Padé-2 scheme. For a delay varying from zero to a prescribed positive value, stability windows as well as root loci are given. We describe how we have circumvented the numerical issues of algorithms developed in [6, 8] and several examples are given.

1 Introduction

In this paper, we describe YALTA, a Matlab toolbox dedicated to the H_∞-stability analysis of classical and fractional systems which has been presented at the *11th IFAC Workshop on Time-Delay systems, Grenoble, 2012*.

The aim of YALTA is to give a localization of unstable poles of standard and fractional delay systems of the retarded or neutral types and a characterization of the H_∞-stability of the system.

Standard delay systems have been widely studied in the last decades (initial work can be found in [25] and [2] and more recent advances can be found in [9, 20, 21] and the survey paper [26]).

David Avanessoff · Catherine Bonnet · Le Ha Vy Nguyen
DISCO Inria Team, LSS-SUPELEC, 3 rue Joliot Curie 91192 Gif-sur-Yvette cedex, France
e-mail: `{David.Avanessoff,Catherine.Bonnet,Le-Ha-Vy.Nguyen}@inria.fr`

André R. Fioravanti
DSCE – FEEC – Unicamp, Av. Albert Einstein, 400, Cidade Universitária Zeferino Vaz, Distrito Barão Geraldo, 13.083-852, Campinas, SP, Brasil
e-mail: `fioravan@dsce.fee.unicamp.br`

T. Vyhlídal, J.-F. Lafay, and R. Sipahi (eds.), *Delay Systems*,
Advances in Delays and Dynamics 1,
DOI: 10.1007/978-3-319-01695-5_21,

Fractional order systems are also obtaining large attention in the literature of the last years, mainly because they offer an excellent fit to the data in many practical situations as, for example, in biophysics, electromagnetism, thermodynamics, rheology (see e.g. [11]). Fractional delay systems of retarded and neutral type have been introduced and partially analyzed in [3].

Asymptotic stability tests for time-delay systems have been proposed in a few papers, such as [23, 29] and others. Some computer packages able to provide insights into this question have been developed. The *Quasi-Polynomial Mapping Based Rootfinder* (QPmR, [28]) is a Matlab function for computation and analysis of spectrum of characteristic quasi-polynomials of both retarded and neutral time-delay systems, see also its update presented in the fifth chapter of this part of the book. Another Matlab package available is the *Tool for Robust Analysis and Characteristic Equations of Delay Differential Equations* (TRACE-DDE, [17]), which performs numerical stability analysis of linear autonomous systems of DDEs with several discrete and/or distributed delays. It allows for the numerical computation of the characteristic roots and then it performs a two-parameters robust stability analysis producing the so-called stability chart, i.e. the set of asymptotically stable/unstable regions in the parameters plane. Finally, the Matlab package for bifurcation analysis of delay differential equations (DDE-Biftool, [5]) can also be used for this purpose.

Most of the existing procedures for the stability study of delay systems tend to spot the crossings of poles through the imaginary axis. This fact comes from two important properties of time-delay systems, also valid for the class of fractional systems. The first one is the *root continuity argument*, which means that for any positive value of the delay, the position of the poles varies continuously with respect to delay. This means that any root crossing from the left to the right half-plane will need to pass through the imaginary axis. The second property is the invariance of the tendency of roots crossing [22]. This implies that a *manageable* number of root clusters can provide sufficient information to characterize the whole stability of the system.

The case of fractional systems is much more involved, and normally cannot be solved by the methods involving the Routh-Hurwitz table, as the one presented in [22]. To the best of the authors' knowledge, only the method of [29] can be successfully expanded to cope with fractional systems with multiple delays, see [3], but each extra commensurate term of the delay after the first one needs to be reduced, and this process potentially doubles, at each step, the degree of the polynomial we need to solve which can be a challenging and unreliable numerical problem.

Lately, there has been development of new methods dealing with the stability of fractional order system with delays. In [13], a numerical procedure based on Cauchy's integral theorem was proposed to test the stability of such systems, and in [12], a technique based on the Lambert W function was used for the same purpose. But the complete characterization of all stability windows is difficult when using those methods, and no information about the position of the unstable poles is given.

The paper is organized as follows. Section 2 contains the main functionalities of the proposed algorithm. Section 3 brings some practical implementation aspects, and in Section 4, some examples are fully presented. Finally Section 5 concludes the work.

2 Functionalities of YALTA

YALTA considers the class of delay systems with transfer function of the type:

$$G(s) = \frac{t(s) + \sum_{\kappa=1}^{N'} t_\kappa(s) e^{-\kappa s h}}{q_0(s) + \sum_{k=1}^{N} q_k(s) e^{-k s h}} = \frac{n(s)}{d(s)} \tag{1}$$

where $h > 0$, and t, q_0, q_k for all $k \in \mathbb{N}_N$, and t_κ for all $\kappa \in \mathbb{N}_{N'}$, are real polynomials in the variable s^α for $0 < \alpha \leq 1$ which satisfy $\deg p \geq \deg t$, $\deg q_0 \geq \deg t_\kappa$, $\deg q_0 \geq \deg q_k$ and such that $\deg q_0 = \deg q_k$ for at least one $k \in \mathbb{N}_N$.

The case $0 < \alpha < 1$ corresponds to fractional systems with delays while the case $\alpha = 1$ is obviously the case of standard delay systems.

When $0 < \alpha < 1$ we define an analytic branch of s^μ on the cut plane $\mathbb{C}\backslash\mathbb{R}_-$ by setting $(re^{j\theta})^\mu = r^\mu e^{j\mu\theta}$ and choosing θ with $-\pi < \theta < \pi$. The zero chains of d have the following possible form [2, 7, 24]:

1. If $\deg q_0 = \deg q_N$, then there are only chains of neutral type;
2. If $\deg q_0 = \deg q_k > \deg q_N$, for some $k \in \mathbb{N}_N$, then there are chains of both neutral and retarded types (we recall that a neutral type chain is asymptotic to a vertical axis in the complex plane and the second a retarded chain contains poles with arbitrarily large negative real part).

Let $z = e^{-sh}$, the coefficient of the highest degree term of $q_0(s) + \sum_{k=1}^{N} q_k(s) e^{-ksh}$ can then be written as a multiple of the following polynomial in z

$$\tilde{c}_d(z) = 1 + \sum_{i=1}^{N} \alpha_i z^i. \tag{2}$$

From now on, we formulate the following hypotheses

Hypothesis (H1): The roots of $\tilde{c}_d$ are of multiplicity one.

Hypothesis (H2): The polynomials $q_0(s)$ and $q_k(s)$ satisfy

$$q_0(0) + \sum_{k=1}^{N} q_k(0) \neq 0.$$

Hypothesis (H3): When dealing with neutral systems, in order to avoid the possibility of an infinite number of zero cancellations between the numerator and denominator of G, we suppose that the numerator of G satisfies either

a) $\deg t(s) > \deg t_k(s)$; or
b) $\deg t_k(s) = \deg t(s)$ for at least one k and the polynomial $\tilde{c}_n$ defined as in (2) relatively to the quasi-polynomial $n(s)$ has no root of modulus less than or equal to one, and no common root of modulus strictly greater than one with $\tilde{c}_d$.

YALTA proposes an analysis of high modulus poles, stability windows as well as a root locus when the delay varies from zero to a prescribed value τ_{max}. For those who are interested in the location of unstable small modulus poles of a classical delay system given a fixed delay, YALTA proposes an approximation of those poles through a Padé-2 scheme. This procedure allows to also determine a finite order model approximation of the initial system. Finally, YALTA gives information on the H_∞-stability of a system with fixed delay.

2.1 Asymptotic Axes and Poles of High Modulus

As retarded systems only have a finite number of unstable poles in any right half-plane, only the asymptotic location of poles of neutral delay systems is of interest. This has been described in [4] for classical delay systems and in [7] for fractional delay systems.

YALTA gives the position of the asymptotic axes in the complex plane and also in most of situations (see [4, 7] for restrictions) the asymptotic position of poles relative to the axis.

2.2 Stability Windows and Root Locus

In YALTA is implemented the numerical procedure proposed in [8] for the calculation of the so-called crossing table and root-locus. From the crossing table, stability windows are displayed. Moreover, the list of all positions of poles calculated for a set of delay values (between 0 and τ_{max}) in the root-locus procedure is available.

As explained in the following section, the suggested method in [8] for implementing root-locus has now been improved.

2.3 Approximation of Poles of Small Modulus

YALTA uses the finite-dimensional approximation (Padé-2) method proposed in [16] for strictly proper (classical) dead-time systems (i.e. systems with $\alpha = 1$).

Let $\delta = \deg q_0$. We consider $\frac{d}{(s+1)^{\delta+1}}$ which is a sum of strictly proper dead-time systems and apply the Padé-2 finite-dimensional approximation to each term of the sum.

Let

$$R_k(s) = \frac{q_k}{(s+1)^\delta} \tag{3}$$

have relative degree m_k and

$$S^{(n)}R_k = \left(\frac{1 - \frac{ksh}{2n} + \frac{1}{3}\left(\frac{ksh}{2n}\right)^2}{1 + \frac{ksh}{2n} + \frac{1}{3}\left(\frac{ksh}{2n}\right)^2} \right)^n R_k. \tag{4}$$

By [16] we have that

$$\|R_k e^{-ksh} - S^{(n)}R_k\|_\infty = \max\left(\mathscr{O}(n^{\frac{-4m_k}{5}}), \mathscr{O}(n^{-4})\right) \tag{5}$$

and a bound is given on the approximation of unstable poles of $R_k e^{-ksh}$ by those of $S^{(n)}R_k$.

2.4 H_∞-stability Analysis

This function is of use for standard neutral systems which have the imaginary axis as asymptotic axis.

For such a delay system with fixed delay, the function determines if the chain of poles clustering the imaginary axis is left or right the axis as well as the poles of small modulus (calling the Padé-2 function). The H_∞-stability depends then on the degrees of the polynomials of the system [4].

3 Practical Aspects

Most of the effort in the practical aspects concerned the management of the numerical constraints. The development of the main algorithmic results presented above was relatively straightforward. Due to the Matlab application, polynomials are coded as coefficient vectors and thus quasi-polynomials are matrices whose rows are the polynomials q_i. To the matrix is joined a vector of delay multipliers to allow a compact description of a system having by example only q_0, q_4 and q_7 as non zero polynomials. In this section, we will only develop the continuation algorithm and the Padé approximation. The continuation algorithm first used a differential approach but we had some issues with a $\frac{ds}{d\tau}$ that could reach infinite values. We thus have opted for a geometrical approach developed just below.

3.1 Continuation Algorithm

The continuation method chosen is inspired by [27]. This method is based on a predictor that uses the preceding points to compute an estimate of the following point and a corrector that will modify the estimate to put the point on the curve. We will not further develop the general details of this method.

The predictor is linear, that is the predicted point is on the line joining the two preceding points. However, the distance to the last point is dynamically evaluated depending on the results of the corrector. This will be explained in the corrector explanation part. In case of a sign change of the imaginary part between the last point and the estimated point – with some securities to avoid a simple oscillation around the real axis –, the predictor proposes a point on the real axis with a rotation in the direct sense of an angle of $\pi/2$. This allows the predictor to consider the bifurcations at real values of the zeros.

For the corrector, different methods are used. The main one computes the local minimum of the function associated to the quasi-polynomial around the predicted point at a predetermined constant distance. If the result is higher than a precision parameter, then the corrector sends back a failure and the predictor is called back with an inferior distance to the last point. The adaptation of the predictor is stored in a multiplier variable for the following predictions. If, this time, the result of the comparator is valid, the multiplier variable is increased such that the algorithm has an acceleration on almost linear parts of the curve. The second method exploits the angle change of a complex analytical function at a zero to assure the existence of a zero. Then, with the above local minimum method, we find the value of such zero. This method is heavier and takes a lot of time, but at some points on a curve or for some big systems, the evaluation of the value of the quasi-polynomial at a point leads to big numerical errors. Thus, the check of the existence of a zero validates the local minimum and avoids having too many rejections by the corrector. This second method is triggered by the value of the multiplier variable exposed just above. If the multiplier is inferior to some arbitrary value, then the second method is used up to a point in the curve where there is less rejections from the corrector and an increase of the multiplier variable.

3.2 *Padé Approximation*

Given a quasi-polynomial in matrix form as explained in the beginning of the section, we compute its Padé-2 approximation. Then, we give an estimation of the relative error in H_∞-norm, that is, the H_∞-norm of the difference between the quasi-polynomial and the approximation divided by the H_∞-norm of the quasi-polynomial. It was an arbitrary choice to use this numerical estimation instead of the theoretical error exposed above. The user may choose to have displayed the roots with positive real parts of the Padé approximation with an estimation of the absolute error between those roots and the roots of the quasi-polynomial. To compute the roots of the quasi-polynomial, we use the same idea as in the corrector explained above for the continuation algorithm.

4 Examples of YALTA Application

In this section we illustrate the use of YALTA. At the same time, they present limitations of the toolbox regarding numerical precision and computation time. The first three examples address stability analysis, and the last one shows the Padé approximation feature.

Computation time for all the examples are given by the use of an Intel Core i5 processor with the following specifications: processor speed: 2.3 GHz, total number of cores: 2, L2 cache (per core): 256 KB, L3 cache: 3 MB, memory: 8 GB.

4.1 Example 1 - Bifurcation Analysis of a Small Degree System

The first example is of small degree and small delay. This same system has been introduced and studied in [15] and [14]. Its main interest is in the bifurcation.

Below is the display of the Matlab command window for this example. Each such command window displayed in this article is in verbatim format to quickly distinguish it.

```
>> q = [6,-66,180;0,-2,12;0,6,30;0,0,-2]
q =
     6   -66   180
     0    -2    12
     0     6    30
     0     0    -2
>> DelayVector = [1,2,3]
DelayVector =
     1     2     3
>> Ex1 = delayFrequencyAnalysisMin(q,DelayVector,1,1)
Ex1 =
      AsympStability: 'There is(are) 2 unstable pole(s) in
right half plane'
                Type: 'Retarded'
        RootsNoDelay: [5.1667 + 3.1579i 5.1667 - 3.1579i]
          RootsChain: 'Roots chains only computed for neutral
systems'
       CrossingTable: 'No crossing'
      ImaginaryRoots: []
    StabilityWindows: [2x2 double]
     NbUnstablePoles: [2x2 double]
       UnstablePoles: [5.0027 + 0.0000i 5.9999 + 0.0000i]
               Error: [1.0000e-11 1.0000e-12]
           RootLoci: {2x1 cell}
```

Figure 1 shows the figure of the complete curve from a zero delay up to the final delay of 1 second. In Figure 2, the small zoom allows the user to get a better detail at the bifurcation point. Figure 3, with a high zoom, shows the limitations of the toolbox at bifurcation points where one can see the effect of the arbitrary rotation in the predictor and its consequences when the points are connected. Of course, one needs an important zoom to see these imperfections.

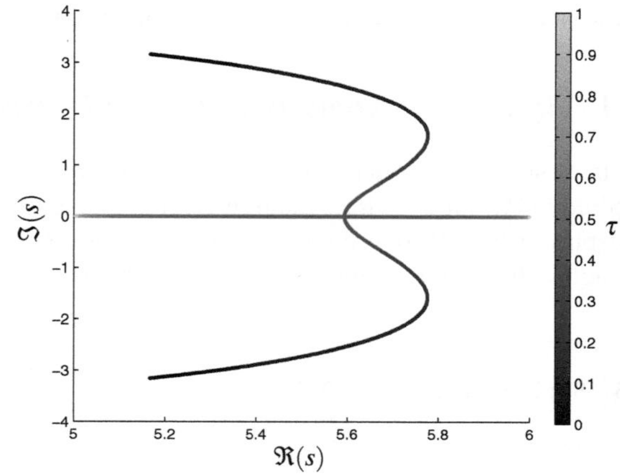

Fig. 1 First example, whole graph

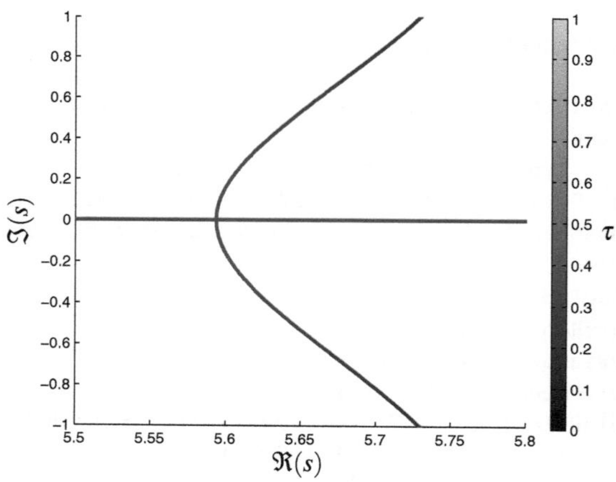

Fig. 2 First example, small zoom

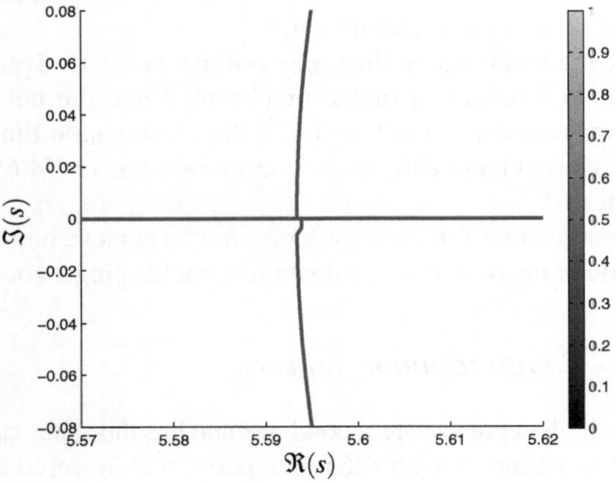

Fig. 3 First example, high zoom

4.2 Example 2 - Stability of a Fractional System

This is a fractional system with $\alpha = 1/3$.

```
>> q = ceil((rand(5,8)-0.5)*100)/100
q =
  -0.4900 -0.0600 -0.2500 -0.2700 -0.0300  0.0800  0.1400  0.4900
   0.2300 -0.4500  0.3500 -0.2700  0.1400 -0.0600  0.1300 -0.3700
  -0.1400 -0.4500  0.3600  0.0400  0.4200  0.3900 -0.1700 -0.2600
   0.2900 -0.4000  0.4700  0.2700 -0.3300 -0.1000  0.3100 -0.4700
  -0.0600  0.1000 -0.0100 -0.1500  0.2200 -0.3200  0.5000  0.1100
>> DelayVector = [1,3,4,7]; a = 1/3; h = 2;
>> Ex2 = delayFrequencyAnalysisMin(q,DelayVector,a,h,1)
Ex2 =
        AsympStability: 'There is(are) 4 unstable pole(s) in
right half plane'
                  Type: 'Neutral'
          RootsNoDelay: [0.9362 0.1330]
            RootsChain: [-0.0390 -0.0503 -0.1441 -0.3168]
         CrossingTable: [3x4 double]
        ImaginaryRoots: [2x3 double]
      StabilityWindows: [2x2 double]
       NbUnstablePoles: [2x3 double]
         UnstablePoles: [0.0680 - 0.0000i 0.7288 + 0.0000i
 0.0324 + 0.7368i 0.0324 - 0.7368i]
                 Error: [1.0000e-13 1.0000e-10 1.0000e-10

 1.0000e-10]
             RootLoci: {6x1 cell}
```

The computation time for this example is 27.678s most of which, that is 23.961s (86.4%), is spent for the root locus calculation.

In order to reduce the computation time, one can increase the advancement step of the delay leading to a reduction of the precision. When the delay step goes from 10^{-4}, which was previously used, to 10^{-3}, the computation time is lowered to 14.550s of which the root locus calculation consumes 10.857s (74.6%) while the errors are still of order 10^{-10}.

The example illustrates that for some systems we can achieve a quick result by lowering the precision of the root locus computation, yet keeping a good error level.

4.3 Example 3 - Computational Aspects

This example is of as high degree as the second one and has the same τ_{max}. However the computation is significantly longer (265.686s) even with lowered advancement of delay (10^{-3}). This is due to the fact that this system has a lot more destabilizing crossing roots (see Figure 4), thus a lot more of root loci. A technical solution to accelerate the calculation is to use parallel computation. Indeed, with the processor mentioned earlier, the computation time by two parallel cores is reduced to 90.780s.

```
>> q = ceil(rand(5,8)*1e4)/1e4
q =
  0.6232  0.6022  0.6588  0.1519  0.9665  0.4255  0.6136  0.2915
  0.0264  0.3620  0.6754  0.3807  0.8063  0.4044  0.9881  0.1884
  0.3188  0.1350  0.7446  0.8211  0.2222  0.4003  0.2200  0.0229
  0.5330  0.9139  0.8422  0.1714  0.9998  0.1120  0.3541  0.4495
  0.3268  0.6406  0.5167  0.3300  0.0638  0.4244  0.2663  0.2437
>> DelayVector = [1,3,5,8];
>> Ex3 = delayFrequencyAnalysisMin(q,DelayVector,1,5)
Ex3 =
      AsympStability: 'There is(are) 10 unstable pole(s) in
right half plane'
                Type: 'Neutral'
        RootsNoDelay: [-0.5242 0.6043 + 0.7137i
-0.8953 - 0.8817i -0.1726 + 0.9349i
-0.1726 - 0.9349i 0.6043 - 0.7137i -0.8953 + 0.8817i]
          RootsChain: [-0.0077 -0.0160 -0.0486 -0.0890]
       CrossingTable: [15x4 double]
      ImaginaryRoots: [16x3 double]
    StabilityWindows: [2x2 double]
     NbUnstablePoles: [2x10 double]
       UnstablePoles: [1x10 double]
               Error: [1.0000e-09 1.0000e-09 1.0000e-08
1.0000e-09 1.0000e-09 1.0000e-09 1.0000e-08 1.0000e-08
1.0000e-11 1.0000e-11]
           RootLoci: {26x1 cell}
```

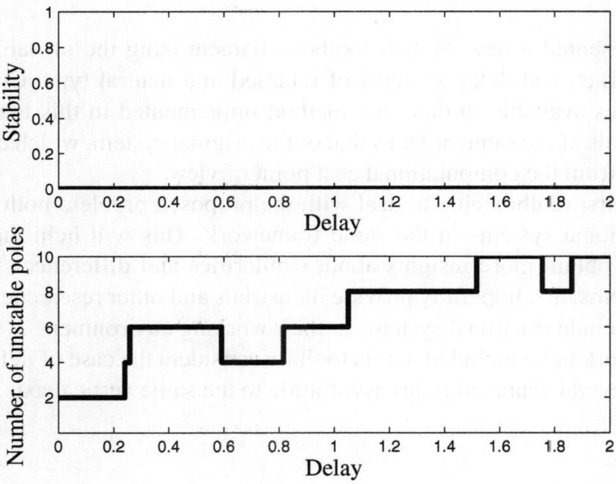

Fig. 4 Third example, stability window

4.4 *Example 4 - Padé-2 Approximation*

Here follows a Padé-2 approximation of the system introduced in [15] and [14] of order 4. Computation time is 5.94 seconds.

```
>> q=[6,-66,180;0,-2,12;0,6,30;0,0,-2]
q =
     6   -66   180
     0    -2    12
     0     6    30
     0     0    -2
>> DelayVector=[1,2,3]
DelayVector =
     1     2     3
>> Ex4 = computePade(q,3,1,DelayVector,4,'ORDER',1)
Ex4 =
       Numerator: [4x11 double]
     Denominator: [4x12 double]
       ErrorNorm: 8.7100e-04
       PadeOrder: 4
           Roots: [5.9996 5.0032]
      RootsError: [2.9745e-04 4.9066e-04]
```

At this order, there is a relative precision in norm of 8.7×10^{-4} and an absolute error of the roots given in the last line of the example. One can see that a relatively correct estimation of the quasi-polynomial can be obtained quickly.

5 Conclusion

This paper presented a new Matlab toolbox characterizing the unstable poles of standard and fractional delay systems of retarded and neutral type. Compared to others toolboxes available to date, the method implemented in this toolbox deals with polynomials of the same order as that of the original system, which can provide a huge benefit from the computational cost point of view.

This is the first toolbox able to deal with the proposed problem both from classical and fractional systems in the same framework. This will help the scientific community to obtain more insights about similarities and differences from those classes of systems, and hopefully provide industrials and other researchers the tools they need to include fractional systems in their working environment.

On going work to be included in this toolbox considers the case of delay systems which admit several chains of poles asymptotic to the same vertical axis.

References

1. Avanessoff, D., Fioravanti, A., Bonnet, C.: YALTA: a Matlab toolbox for the H_∞-stability analysis of classical and fractional systems with commensurate delays. In: 11th IFAC Workshop on Time-Delay Systems, Grenoble (2013)
2. Bellman, R., Cooke, K.: Differential-Difference Equations Academic Press (1963)
3. Bonnet, C., Partington, J.R.: Analysis of fractional delay systems of retarded and neutral type. Automatica (38), 1133–1138 (2002)
4. Bonnet, C., Fioravanti, A., Partington, J.R.: Stability of neutral systems with multiple delays and poles asymptotic to the imaginary axis. SIAM J. Control Opt. 49(2), 498–516 (2011)
5. Engelborghs, K., Luzyanina, T., Samaey, G.: DDE-BIFTOOL v. 2.00: a Matlab package for bifurcation analysis of delay differential equations, Technical Report TW-330, Department of Computer Science, K.U.Leuven, Leuven, Belgium (2001)
6. Fioravanti, A.R., Bonnet, C., Özbay, H., Niculescu, S.-I.: A numerical method to find stability windows and unstable poles for linear neutral time-delay systems. In: 9th IFAC Workshop on Time Delay Systems, Prague, Czech Republic (2010)
7. Fioravanti, A.R., Bonnet, C., Özbay, H.: Stability of fractional neutral systems with multiple delays and poles asymptotic to the imaginary axis. In: 49th IEEE Conference on Decision and Control, Atlanta, USA (2010)
8. Fioravanti, A.R., Bonnet, C., Ozbay, H., Niculescu, S.I.: A numerical method for stability windows and unstable root-locus calculation for linear fractional time-delay systems. Automatica 48(11), 2824–2830 (2012)
9. Gu, K., Kharitonov, V.L., Chen, J.: Stability of time-delay systems. Birkhauser (2003)
10. Gumussoy, S., Michiels, W.: Root-locus for SISO dead-time systems: A continuation based approach. Automatica 43(3), 480–489 (2012)
11. Hilfer, R.: Applications of Fractional Calculus in Physics. World Scientific (2000)
12. Hwang, C., Cheng, Y.C.: A note on the use of the Lambert W function in the stability analysis of time-delay systems. Automatica 41(11), 1979–1985 (2005)
13. Hwang, C., Cheng, Y.C.: A numerical algorithm for stability testing of fractional delay systems. Automatica 42(5), 825–831 (2006)
14. Lee, E.B., Gu, G., Khargonekar, P.P., Misra, P.: Finite-dimensional approximation of infinite dimensional systems. Conference on Decision and Control (1990)

15. Lee, E.B., Gu, G., Khargonekar, P.P., Misra, P.: Approximation of infinite dimensional systems. IEEE Trans. Automatic Control 34, 610–618 (1989)
16. Mäkilä, M., Partington, J.R.: Shift operator induced approximations of delay systems. SIAM J. Control Optimiz. 37(6), 1897–1912 (1999)
17. Maset, S., Vermiglio, R.: Pseudospectral differencing methods for characteristic roots of delay differential equations. SIAM J. Sci. Comput. 27(2), 482–495 (2005)
18. Matignon, D.: Stability properties for generalized fractional differential systems. ESAIM: Proceedings 5, 145–158 (1998)
19. Nguyen, L.H.V., Fioravanti, A.R., Bonnet, C.: Stability of neutral systems with commensurate delays and many chains of poles asymptotic to same points on the imaginary axis. In: 10th IFAC Workshop on Time Delay Systems, TDS 12, Boston (2012)
20. Niculescu, S.I.: Delay effects on stability: A robust control approach. LNCIS, vol. 269. Springer, London (2001)
21. Niculescu, S.-I., Gu, K.: Advances in Time-Delay Systems. Lecture Notes in Computational Science and Engineering, Vol. 38, Springer (2004)
22. Olgac, N., Sipahi, R.: An exact method for the stability analysis of time delayed LTI systems. IEEE Transactions on Automatic Control 47(5), 793–797 (2002)
23. Olgac, N., Sipahi, R.: A practical method for analyzing the stability of neutral type LTI-time delayed systems. Automatica (40), 847–853 (2004)
24. Partington, J.R.: Linear Operators and Linear Systems: An Analytical Approach to Control Theory. Cambridge University Press (2004)
25. Pontryagin, L.S.: On the zeros of some elementary transcendental functions. Amer. Math. Soc. Transl. 2(1), 95–110 (1955)
26. Richard, J.P.: Time-Delay Systems: An Overview of Some Recent Advances and Open Problems. Automatica 39, 1667–1694 (2003)
27. Seydel, R.: Practical Bifurcations and Stability Analysis. Springer (2010)
28. Vyhlídal, T., Zítek, P.: Quasipolynomial mapping based rootfinder for analysis of Time delay systems. In: Proc. of IFAC Workshop on Time-Delay Systems, TDS 2003, Rocquencourt (2003)
29. Walton, K., Marshall J.E.: Direct method for TDS stability analysis. IEE Proceedings, Part D (134), 101–107 (1987)

QPmR - Quasi-Polynomial Root-Finder: Algorithm Update and Examples

Tomáš Vyhlídal and Pavel Zítek

Abstract. An updated QPmR algorithm implementation for computation and analysis of the spectrum of quasi-polynomials is presented. The objective is to compute all the zeros of a quasi-polynomial located in a given region of the complex plane. The root-finding task is based on mapping the quasi-polynomial in the complex plane. Consequently, utilizing spectrum distribution diagram of the quasi-polynomial, the asymptotic exponentials of the retarded chains are determined. If the quasi-polynomial is of neutral type, the spectrum of associated exponential polynomial is assessed, supplemented by determining the safe upper bound of its spectrum. Next to the outline of the computational tools involved in QPmR, its Matlab implementation is presented. Finally, the algorithm is demonstrated by three examples.

1 Introduction

Due to infinite dimensionality of time delay systems, the spectrum analysis and computation have been challenging issues for decades. The spectrum properties of time delay systems of both the retarded and neutral types have been mapped by Bellman and Cooke, [1]. As for the computation of time delay system spectrum, several approaches have been proposed. A group of methods derived for approximating the stability determining right-most roots of delay systems have been based on discretization of the system solution operator, using e.g. Linear Multi-Step methods [9, 20] or Runge-Kutta methods [2]. The well known software tool DDE-BIFTOOL [10], see also [20], is the key result of this direction. Another group of methods for spectrum computation is based on a discretization of the system infinitesimal generator, using e.g. Runge-Kutta methods [3] or pseudospectral techniques [4], [25]. This strategy

Tomáš Vyhlídal and Pavel Zítek
Department of Instrumentation and Control Engineering, Faculty of Mechanical Engineering, Czech Technical University in Prague, Technická 4, 166 07 Praha 6, Czech Republic
e-mail: {tomas.vyhlidal,pavel.zitek}@fs.cvut.cz

T. Vyhlídal, J.-F. Lafay, and R. Sipahi (eds.), *Delay Systems*,
Advances in Delays and Dynamics 1,
DOI: 10.1007/978-3-319-01695-5_22,

is used in the software package TRACE-DDE [5]. Recently, based on the spectral method, the algorithm for reliable computing of all the roots of delay systems located in a given right half plane has been proposed [25] and implemented in a form of Matlab tool. Let us also mention that for simple structure time delay systems, a Lamber W function can be utilized for determining the root positions, see [26], [24] and also [15]. Alternatively, the zeros of a characteristic function of time delay system can be determined by numerical algorithms developed for computing zeros of general analytic functions, [13, 16]. The traditional approaches are the iterative schemes, e.g. Newton's method, or bisection based algorithms, see e.g. [7], [27]. The problem of computing all zeros of an analytic function can also be solved using the quadrature methods presented in [8, 16].

The results presented in this chapter are extensions of earlier works [21, 22] by the authors. In [21], we designed the Quasi-Polynomial mapping based Root-finder (QPmR) with the objective to compute all the zeros of a quasi-polynomial located in a predefined region of the complex plane. The original implementation of this algorithm (QPmR v.1) in Matlab was based on symbolic computations with the use of Symbolic Math Toolbox. The involvement of symbolic computations was one of the reasons of relatively large computational time. In the consecutive work [22], the algorithm was optimized via reducing the regions to be scanned for the root positions. With the help of spectrum distribution analysis [1] and the argument principle rule, large sub-regions are determined that are free of roots. These subregions are then omitted from the process of mapping the root positions. The SW implementation of this algorithm extension is an aQPmR (advanced QPmR) Matlab function. Even though a considerable reduction of computational time has been achieved, the complexity of the algorithm has increased considerably. Combining the user-friendly application of QPmR v.1 and some of the enhanced features of aQPmR, the QPmR v.2 was recently proposed and implemented in Matlab. The updated QPmR algorithm, which does not involve Symbolic Math Toolbox operations, is presented as the main contribution of this chapter.

1.1 Problem Formulation

The primary objective is to compute all the zeros of the quasi-polynomial

$$h(s) = \sum_{j=0}^{N} p_j(s) \mathrm{e}^{-s\alpha_j} \tag{1}$$

located in the complex plane region $\mathbb{D} \in \mathbb{C}$, with the boundaries $\beta_{min} < \Re(\mathbb{D}) < \beta_{max}$ and $\omega_{min} < \Im(\mathbb{D}) < \omega_{max}$, where $\alpha_0 > \alpha_1 > ... > \alpha_{N-1} > \alpha_N = 0$ are the delays and $p_j(s) = \sum_{k=0}^{m_j} p_{j,k} s^k$ are the polynomials with real coefficients of degree m_j, which is at most n for $p_j(s)$, $j = 0, 1, ..N$, where n is degree of the polynomial $p_N(s)$. This task will be solved in Section 2, where the mapping based algorithm for determining the root positions will be described. Notice that the function (1) can be both a numerator and a denominator of a meromorphic transfer function of a time delay system. Thus, using the root-finding algorithm described in this chapter, both the poles and zeros of a meromorphic transfer function can be computed.

As an additional objective, the asymptotic features of the spectrum of quasi-polynomial (1) are to be determined. The definition of asymptotic exponentials of the root chains together with the involved computational tools are given in Section 3. If the quasi-polynomial is of neutral type, i.e. at least one of the polynomials $p_j(s), j=0,1,...,N-1$ is of degree n, the spectrum of associated exponential polynomial and its safe upper bound are to be determined, as also presented in Section 3. In Section 4, the use of QPmR v.2 in Matlab is outlined, followed by three demonstration examples. The chapter is concluded by a brief summary section.

2 Algorithm for Spectrum Computation

The core algorithms of QPmR v.2 remain the same as in QPmR v.1 described in [21]. Considering $s=\beta+j\omega, \beta\in\mathbb{R}, \omega\in\mathbb{R}$, the characteristic quasi-polynomial $h(s)$ can be split into two real functions $R(\beta,\omega)=\Re(h(\beta+j\omega))$ and $I(\beta,\omega)=\Im(h(\beta+j\omega))$. Consequently the zero determining equation $h(s)=0$ can be split into

$$R(\beta,\omega)=0 \tag{2}$$

$$I(\beta,\omega)=0. \tag{3}$$

Notice that the root positions can easily be located as intersection points on the zero level curves of the surfaces $R(\beta,\omega)$ and $I(\beta,\omega)$ given by (2) and (3). Subsequently, the accuracy of the zeros is increased by Newton's method. The overall algorithm is given as follows:

Algorithm 1. *Mapping the quasi-polynomial spectrum*

1. The region of interest $\mathbb{D}$ is covered by a regular mesh grid

$$\Gamma=\begin{bmatrix}\beta_0+\mathrm{j}\omega_0 & \cdots & \beta_{k_{\max}}+\mathrm{j}\omega_0\\ \vdots & \beta_k+\mathrm{j}\omega_l & \vdots\\ \beta_0+\mathrm{j}\omega_{l_{\max}} & \cdots & \beta_{k_{\max}}+\mathrm{j}\omega_{l_{\max}}\end{bmatrix}, \quad \begin{aligned}\beta_k&=\beta_{\min}+k\Delta_g,\ k=0,\,1,\,....,\,k_{\max}\\ \omega_l&=\omega_{\min}+l\Delta_g,\ l=0,\,1,\,....,\,l_{\max}\end{aligned} \tag{4}$$

   with a grid step Δ_g.
2. The functions $h(s)$ is evaluated at each grid point of (4) and split into real and imaginary parts providing the matrices $\bar{R}(\beta_k,\omega_l)$ and $\bar{I}(\beta_k,\omega_l)$.
3. The zero level curves $R(\beta,\,\omega)=0$ and $I(\beta,\,\omega)=0$ are mapped applying the *contour* plotting algorithm to the matrices $\bar{R}(\beta_k,\omega_l)$ and $\bar{I}(\beta_k,\omega_l)$.
4. The intersection points of $R(\beta,\,\omega)=0$ and $I(\beta,\,\omega)=0$ are determined as the zeros of the function $I_R(\beta,\omega)=Im(h(s))$ evaluated over the points on the curve

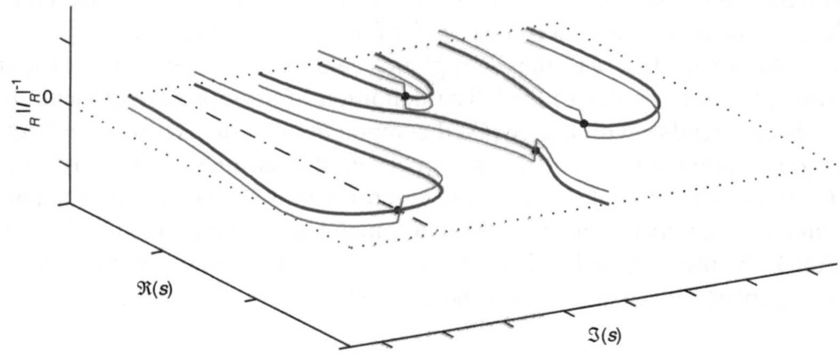

Fig. 1 Zero-level contours given by (2) - thick line, and the curves $I_R(\beta, \omega)$ normalized with respect its absolute value - thin line, • - root positions

$R(\beta, \omega) = 0$, as shown in Fig. 1. Let us remark that using this approach, all the zeros are assessed within the same accuracy determined by Δ_g.

5. Newton's iteration method is applied to increase the accuracy of each zero.

□

More details on the particular steps of Algorithm 1 can be found in [21, 23].

2.1 *Mapping the Zero Level Curves*

For mapping the zero level curves $R(\beta, \omega) = 0$ and $I(\beta, \omega) = 0$, Matlab function *contour* is used in QPmR implementation. As outlined in [6], the algorithm of Matlab *contour* function is relatively simple. It is based on covering the region of interest by a uniform grid of points. Then, for every cell of the grid, the function is evaluated. Consequently, the sign based test is involved that decides whether the contour passes the particular cell. Finally, the passing points are determined by interpolation and joined together with a straight lines providing the contour approximation segments. Obvious computational disadvantages of the *contour* algorithm are the required regularity of the grid and the use of linear approximation of the level segments over the given cell. Consequently, in order to achieve smooth zero-level contours $R(\beta, \omega) = 0$ and $I(\beta, \omega) = 0$, the grid needs to be sufficiently dense.

2.1.1 The Grid Density Adaptation

As shown in [22], the grid step $\Delta_g = \frac{\pi}{10\alpha_0}$ guarantees a sufficiently dense grid in the high frequency regions of the complex plane. However, the density of the roots close to the origin can be completely different. There can be regions with the roots very close to each other or with even multiple roots. If the selected grid size is too large, some roots can be omitted. On the other hand, if the grid size is too small, the mapping by *contour* is too time consuming. Taking these aspects into consideration, a grid adaptation rule has been implemented in the QPmR v.2. The core of the adaptation is based on independent checking the number of roots located in the region $\mathbb{D}$ by the application of the argument principle rule [1, 19, 27]

$$N_\varphi = \frac{1}{2\pi}\Delta \arg_{\varphi} h(s) = \frac{1}{2\pi \mathrm{j}}\int_\varphi \frac{h'(s)}{h(s)}ds \tag{5}$$

where φ is the counterclockwise oriented curve tightly enclosing the region boundary, and N_φ is the number of roots in the region enclosed by φ. In the QPmR code, a numerical evaluation of the integral in (5) is performed on a considerably denser grid compared to the grid used for the mapping the zero level curves. If the number of roots detected by (5) is different from the number of roots determined by Algorithm 1, it is likely that the mapping has not been performed well. Let us remark that the initial grid size is selected with respect to the size of the region as well as with respect to the spectrum distribution features so that the computation time of the first try is small. If the argument increment based test fails, the grid size is reduced and all the computations according to Algorithm 1 are performed again, including the argument increment based test. If the result of the test is still negative, the region $\mathbb{D}$ is split into four rectangular subregions and the QPmR algorithm is recursively applied to each of the subregions. If the result of the subsequent test performed for each of the subregions is still negative, the particular subregion is split into four sub-subregions and the QPmR is applied in the second recursion level. If even this does not help, the algorithm stops and the user is recommended to manually reduce the region of interest $\mathbb{D}$. Analogously, the grid size reduction procedure described above is performed whenever the Newton's iteration does not converge for any of the roots.

3 Spectrum Analysis

The spectrum asymptotic distribution features of time delay systems have already been studied already by Bellman and Cooke (1963), [1]. Determining the principal terms of the quasi-polynomial $h(s)$, the distribution properties of the roots with unlimitedly high magnitudes has been determined. As in [1, 22], let us define

$$g(s) = \sum_{j=1}^{N} p_{j,m_j} s^{m_j}(1+\varepsilon_j(s))\mathrm{e}^{s\vartheta_j} \tag{6}$$

which has the same distribution of zeros as (1), where $\vartheta_j = \alpha_0 - \alpha_j$, $0 = \vartheta_0 < \vartheta_1 < ... < \vartheta_{N-1} < \vartheta_N$, $p_{j,m_j} \neq 0\,(j = 0, 1, ..., N)$ and the functions $\varepsilon_j(s)$ have the property $\lim_{|s|\to\infty} |\varepsilon_j(s)| = 0$. As it has been shown in [1], with the points $P_j = (\vartheta_j, m_j)$, we can define the *Distribution diagram* as follows:

Proposition 1. *Quasi-polynomial spectrum distribution diagram* [1]. Given the quasi-polynomial (1) transformed to the form (6), the spectrum distribution diagram is constructed as concave polygonal line L over the points $P_j = (\vartheta_j, m_j)$, satisfying the following features:

- joins P_0 with P_N
- has vertices only at points of the set P_j
- is concave and is such that no points P_j lie above it

Let the successive segments of L be denoted by $L_1, L_2, \dots L_M$, numbered from left to right, and let μ_r denote the slopes of $L_r, r = 1..M$. For each segment of the spectrum distribution diagram L_r with $\mu_r > 0$, a retarded chain with infinitely many roots exists. The segments with $\mu_r = 0$ correspond to the neutral part of the spectrum, which is located in a vertical strip of the complex plane. □

Based on the *Distribution diagram* and utilizing further results by Bellman and Cooke [1], the asymptotic exponentials of the retarded root chains can be derived as follows:

Proposition 2. *Asymptotic exponentials of the root chains* [22]. For large magnitudes of $s = \beta + j\omega, \beta \in \mathbb{R}, \omega \in \mathbb{R}^+$, the asymptotic exponentials of the root chains of (6) can be determined as

$$\omega = \exp\left(\frac{c_r - \beta}{\mu_r}\right) \tag{7}$$

where $c_r = \mu_r \ln |w_{r_k}|$, and w_{r_k} is a zero of the polynomial

$$f_r(w) = \sum_{j=0}^{N_r} \bar{p}_j w^{\tilde{m}_j} \tag{8}$$

where $\tilde{m}_j = \bar{m}_j - \bar{m}_0$, $\bar{m}_j$ and $\bar{p}_j$ correspond to those points of the distribution diagram $P_j = (\vartheta_j, m_j)$ defined according to Proposition 1 that lie on the particular segment L_r. $N_r + 1$ is the number of points on the segment L_r. □

To sum up, first, within the spectrum analysis functionality of QPmR v.2, the spectrum distribution diagram is constructed as described in Proposition 1. Consequently, the asymptotic exponential functions (7) of the root chains are determined as described in Proposition 2.

3.1 Spectral Features of Neutral Quasi-Polynomial

If the quasi-polynomial is of neutral type, we can define an *associated exponential polynomial* as follows

$$D(s) = \sum_{j=0}^{N_M} \bar{p}_j \mathrm{e}^{-s\bar{\alpha}_j}, \tag{9}$$

where the coefficients $\bar{p}_j$ and delays $\bar{\alpha}_0 > \bar{\alpha}_1 > \dots > \bar{\alpha}_{N_M} = 0$ correspond to the points P_j on the segment L_M. The function $D(s)$ can be derived very easily as the sum of the terms in (1) corresponding to the n powers of s. Let us remark that in literature on neutral systems, the *associated exponential polynomial* is referred to as the characteristic function of the *associated difference equation* defined for the neutral system [12, 17, 18].

It can be easily proven, see e.g. [1, 17], that the infinite spectrum of (9) lies within a certain vertical strip of the complex plane. Due to this property, the function (9) can have infinitely unstable zeros. Next, the fundamental feature of a neutral quasi-polynomial of form (1) is that a part of its spectrum tends to match the spectrum of (9). Thus, the stability of (9) is a necessary condition for the stability of (1). Consequently, the neutral quasi-polynomial can have infinitely many unstable zeros, which can never happen if the quasi-polynomial is retarded. Besides, it has been shown [17] that the upper bound of the spectrum of (9) given by

$$c_D(\boldsymbol{\alpha}) = \sup\{\Re(s) : D(s) = 0\}. \tag{10}$$

can be extremely sensitive to small variations in the delays $\boldsymbol{\alpha} = [\bar{\alpha}_j], j = 1..N_M - 1$. This has led to the introduction of the concept of *strong stability* in [12]. The neutral system is strongly exponentially stable if it remains exponentially stable subject to small variations in the delays. In order to test the strong stability of the neutral system, the 'safe' upper bound $\bar{C}_D(\boldsymbol{\alpha})$ was defined in [17] as follows:

Definition 1. *Safe upper bound of the spectrum of exponential polynomial* [17]. Let $\bar{C}_D(\alpha) \in \mathbb{R}$ be defined as

$$\bar{C}_D(\boldsymbol{\alpha}) = \lim_{\varepsilon \to 0+} c_\varepsilon(\boldsymbol{\alpha}),$$

where

$$c_\varepsilon(\boldsymbol{\alpha}) = \sup\{c_D(\boldsymbol{\alpha} + \delta\boldsymbol{\alpha}) : \delta\boldsymbol{\alpha} \in \mathbb{R}^m \text{ and } \|\delta\boldsymbol{\alpha}\| \leq \varepsilon\}.$$

□

Clearly, we have $\bar{C}_D(\boldsymbol{\alpha}) \geq c_D(\boldsymbol{\alpha})$. Based on the results presented in [17, 18], the safe upper bound for the spectrum of (9) can be determined as follows

Proposition 3. *Determining* $\bar{C}_D$ [17]. The safe upper bound $\bar{C}_D$ of the spectrum of the associated exponential polynomial (9) is determined as a single zero of the strictly decreasing function

$$c \in \mathbb{R} \to \sum_{j=0}^{N_{M-1}} \left| \frac{\bar{p}_j}{\bar{p}_{N_M}} \right| \mathrm{e}^{-c\bar{\alpha}_j} - 1. \tag{11}$$

Consequently, the neutral system with the characteristic functions (1) and (9) is strongly stable if and only if $\bar{C}_D < 0$. □

Let us remark that by the test given in Proposition 3, the delays in (9) are considered as mutually independent. Particularly, if the function (1) is derived as a characteristic function of an interconnected system, there can be dependencies in the delays in (9). In this case, the result by the Proposition 3 is too conservative. Instead, a test proposed in [18] can be applied. Let us remark that a SW tool for strong stability test has recently been implemented in [14].

To sum up, in the QPmR v.2, first, the associated exponential polynomial (9) is derived from (1). Consequently, the spectrum of (9) is computed by Algorithm 1. Consequently, the safe upper bound $\bar{C}_D$ is determined as described in Proposition 3.

4 Working with QPmR v.2 in Matlab

Below, we describe how to use the QPmR[1] v.2 in Matlab. The full syntax of the QPmR Matlab function is as follows

```
[R Y]=QPmR(Region,P,D,e,ds,gr)
```

where the function inputs are

- `Region` - the region of interest $\mathbb{D}$ defined as: $[\beta_{min}, \beta_{max}, \omega_{min}, \omega_{max}]$
- `P` - a matrix of size $(N+1) \times n$ with coefficients of polynomials in (1)
- `D` - a vector of delays in (1)
- `e` - the computation accuracy (the default value, indicated by `e=-1`, is $10^{-6}\Delta_g$
- `ds` - the grid step Δ_g. If `ds=-1`, the grid step is adjusted automatically as described in Subsection 2.1.1.
- `gr` - parameter controlling graphical representation of the results. Only if `gr=1`, the results are visualized in plots.

The function outputs are:

- `R` - the vector of computed zeros (NaN indicates the algorithm failure)
- `Y` - structured variable with summary of the QPmR results. Particularly, `Y.flag` indicates the result correctness (`flag`: 1 - correct, 0 - failure indicated by the argument based test, -1 - Newton's iterations failure). Next, information on the asymptotic features of the spectrum is provided. If the quasi-polynomial is neutral, the zeros of the associated exponential polynomial and the value of the safe upper bound of its spectrum are given. Additionally, information on the final grid size and the accuracy estimates are outlined.

The quasi-polynomial can also be defined in the function handle form `Fun`. Then, the function syntax is

```
[R Y]=QPmR(Region,Fun,e,ds,gr)
```

In this QPmR mode, only the roots are determined. No additional analysis of the spectrum is performed. In this mode however, the QPmR can be used to compute roots of general analytical functions, e.g. fractional polynomials or quasi-polynomials. Let us also remark that not all the input and output arguments must be assigned. The shortest forms of the commands are `R=QPmR(Region,P,D)` or `R=QPmR(Region,Fun)`. In the following subsection we provide three demonstration examples.

[1] `http://www.cak.fs.cvut.cz/algorithms/qpmr`

4.1 Examples

The following numerical examples have been run in Matlab R2012d on PC Intel(R) Core(M) i7-3632QM CPU @2.20GHz 2.20GHz, RAM 8.00GB, 64-bit Operating System.

4.1.1 Example 1

As in the Introductory example of [21], the objective is to compute the roots of the following simple quasi-polynomial

$$h(s) = s + \exp(s), \tag{12}$$

located in the complex plane region $\mathbb{D} = [-10,2] \times [0,30]$. Let us remark that in [21], the quasi-polynomial maps have been obtained analytically. Besides, due to simplicity of the quasi-polynomial, the spectrum can be determined using the Matlab Lambert W function [26] as follows

$$s_k = W(k,-1), k = -\infty.. - 2,-1,0,1,2,..,\infty \tag{13}$$

Using QPmR with automatic adjustment of the grid size and with the default accuracy, the following command performs the task of spectrum analysis and computation

```
[R Y]=QPmR([-10 2 0 30],[1 0;0 1],[0 1],-1,-1,1).
```

The graphical results are in Fig. 2. As can be seen, there is only one chain of roots. For comparison, the spectrum computed by the Matlab Lambert W function command `R=lambertw(0:5,-1)` is shown in Fig. 2 too. If only the spectrum is of interest, the following syntax of QPmR with function handle form definition of the quasi-polynomial can be used

```
[R Y]=QPmR([-10 2 0 30],@(s)s+exp(-s),-1,-1,1),
```

providing the graphical results as shown in the right part of Fig. 2. Additionally, the computation times of the old and the new versions of QPmR are compared with the following results: QPmR v.1: 0.22*sec*, QPmR v.2: 0.25*sec* if both the spectrum analysis and computation are performed, and 0.10*sec* if the spectrum is computed only. Thus, for this simple example, the spectrum computation time was more than halved by the new version of QPmR. Let us remark that in this comparison, the adapted grid step $\Delta_g = 0.36$ was used for all the cases. Let us also remark that computation of the five roots by the Matlab Lamert W function took 0.035*sec*.

4.1.2 Example 2

As in the application example in [22], consider the retarded quasi-polynomial in the form

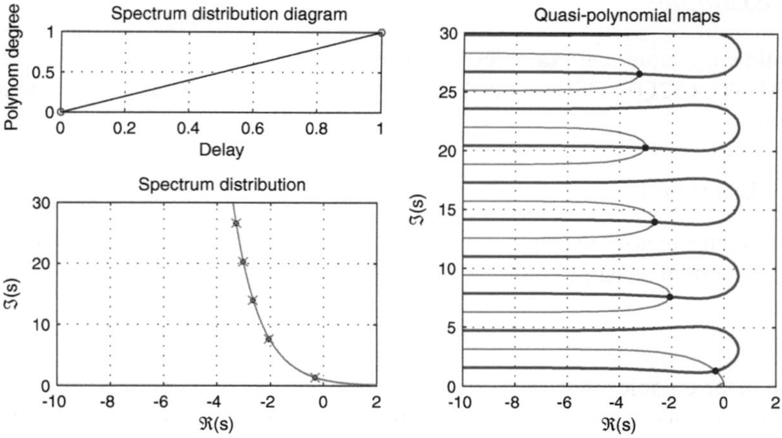

Fig. 2 Results of QPmR algorithm applied to quasi-polynomial (12). Left - upper part: spectrum distribution diagram; lower part: solid line - asymptotic exponential of the spectrum chain, • - roots determined by QPmR, × - spectrum determined using Lambert W function. Right: thick - maps of $R(\beta,\omega)=\Re(h(\beta+j\omega))=0$, thin - $I(\beta,\omega)=\Im(h(\beta+j\omega))=0$, • - roots

$$\begin{aligned}
h(s) &= 0.2s^8+1.7s^7-12.8s^5+0.01s^2-1.8s+29.1\exp(-4.61s)+ \\
&+(s^7-1.1s^5+6.7s)\exp(-8.52s)+(-8.7s^4+2.1s^2+19.3s)\exp(-10.33s)+ \\
&+(0.8s^6+0.1s^4-1.4s+7.2)\exp(-13.52s)+ \\
&+(0.15s^5+0.2s^4-0.9s^3+25.2s)\exp(-18.52s)+0.5s^3\exp(-19.9s)+ \\
&+(0.03s^3+0.04s^2-0.1s+1.5)\exp(-23.35s)+51.7\exp(-24.99).
\end{aligned} \tag{14}$$

As in [22], the grid step $\Delta_g = 0.0157$, the accuracy of the roots $1e-6$ and the region of interest $\mathbb{D} = [-4.5,3]\times[0,100]$ are considered. The Matlab code for the root computation and the spectrum analysis is as follows

```
D=[24.99;23.35;19.9;18.52;13.52;10.33;8.52;4.61;0]
P=[0 0 0 0 0 0 0 0 51.7; 0 0 0 0 0 0.03 0.04 -0.1 1.5;
   0 0 0 0 0 0.5 0 0 0; 0 0 0 0.15 0.2 -0.9 0 25.2 0;
   0 0 0.8 0 0.1 0 0 -1.4 7.2; 0 0 0 0 -8.7 0 2.1 19.3 0
   0 1 0 -1.1 0 0 0 6.7 0; 0 0 0 0 0 0 0 0 29.1
   0.2 1.7 0 -12.8 0 0 0.001 -1.8 0]
[Y A]=QPmR([-4.5 3 0 100],P,D,1e-6,0.0157,1);
```

The results are shown in Fig. 3. As can be seen, the spectrum consists of five retarded chains. Overall number of roots in $\mathbb{D}$ is 390. Also in this example, the computation time was compared with the old version of QPmR with the following result - QPmR v.1: 18.9 sec.; QPmR v.2: 15.6 sec (7 sec. if no spectrum analysis is performed). Thus, the spectrum computation by QPmR v.2 is almost three times faster compared to its first version.

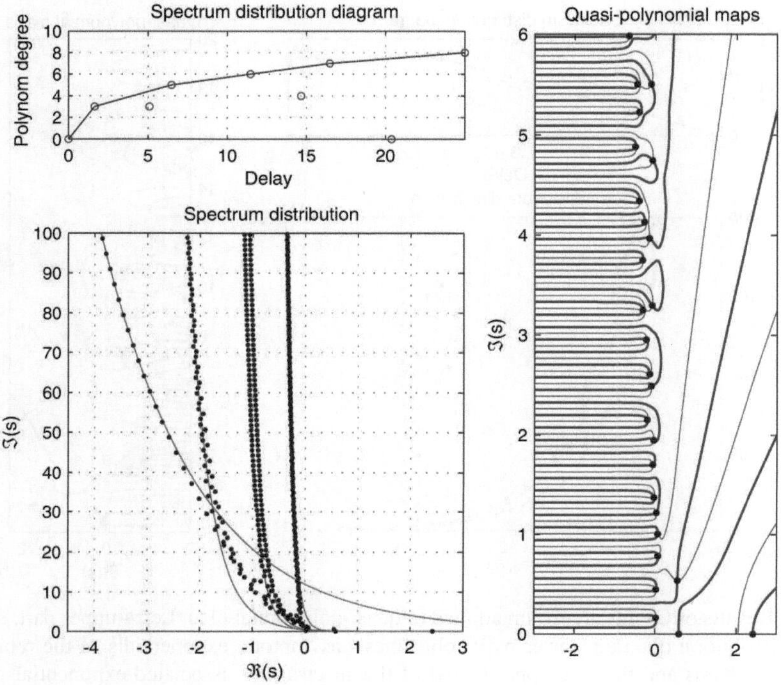

Fig. 3 Results of QPmR algorithm applied to the quasi-polynomial (14). Left - upper part: spectrum distribution diagram; lower part: solid line - asymptotic exponential of the spectrum chain, • - roots determined by QPmR. Right: thick - maps of $R(\beta,\omega) = \Re(h(\beta + j\omega)) = 0$, thin - $I(\beta,\omega) = \Im(h(\beta + j\omega)) = 0$, • - roots

4.1.3 Example 3

Consider a neutral quasi-polynomial

$$h(s) = s^4 + 0.2s^3 + 5s + 2.1 + (0.5s^4 - 2.1s)\exp(-1.5s) + \\ + (0.3s^4 + 3.2s^2)\exp(-2.2s) + (1.2s^2)\exp(-4.3s) + 3\exp(-6.3s). \quad (15)$$

The associated exponential polynomial is given by

$$D(s) = 1 + 0.5\exp(-1.5s) + 0.3\exp(-2.2s). \quad (16)$$

The task is to compute and analyze the spectrum in the region $\mathbb{D} = [-6,2] \times [0,200]$. The code to perform the task is as follows.

```
P=[1 0.2 0 5 2.1;0.5 0 0 -2.1 0;0.3 0 3.2 0 0;0 0 1.2 0 0;0 0 0 0 3]
D=[0;1.5;2.2;4.3;6.3]
[R Y]=QPmR([-6 2 0 200],P,D,-1,-1,1)
```

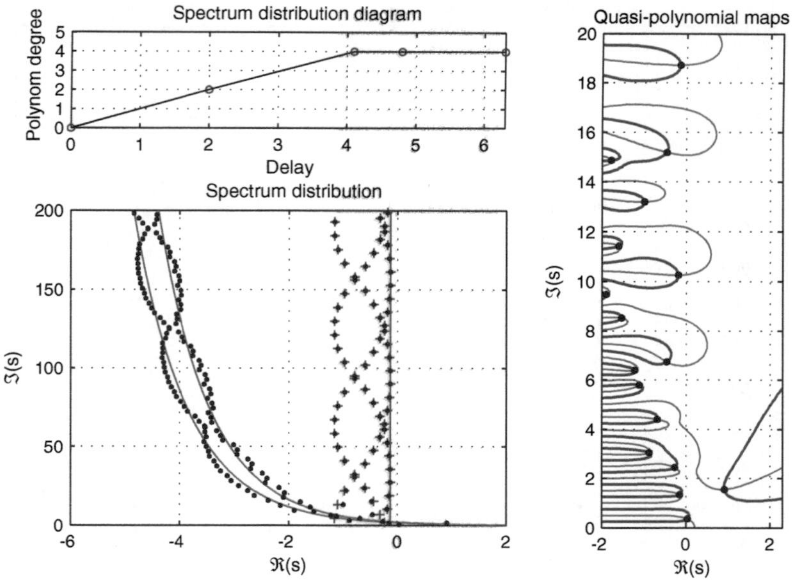

Fig. 4 Results of QPmR algorithm applied to quasi-polynomial (15). Left - upper part: spectrum distribution diagram; lower part: solid lines - asymptotic exponentials of the retarded spectrum chains and the safe upper bound of the spectrum of associated exponential polynomial (16) (vertical line), • - quasi-polynomial roots, + - spectrum of the associated exponential polynomial (16). Right - thick - maps of $R(\beta,\omega) = \Re(h(\beta + j\omega)) = 0$, thin - $I(\beta,\omega) = \Im(h(\beta + j\omega)) = 0$, • - roots

The results of the computations, which took $9.1sec.$ including the grid adaptation ($1.9sec.$ with the fixed grid size 0.1) are shown in Fig. 4. Next to the spectrum and the asymptotic functions for the retarded root chains, the spectrum of associated exponential polynomial (16) and its safe upper bound are computed and visualized in Fig. 4.

5 Conclusions

As the main result, the updated QPmR v.2 algorithm for mapping and analyzing the quasi-polynomial spectra was presented. Unlike the original QPmR described in [21], no functions of Symbolic Math toolbox are used. Besides, the main additional functionalities are: i) the recursive adaptation procedure of the the grid mesh for mapping the quasi-polynomial. ii) utilizing the results from [22], the spectrum analysis is included. As demonstrated in the application examples, QPmR v.2 is at least two times faster compared to the original QPmR version. A large part of the computation time is consumed by the Matlab function *contour* for mapping the quasi-polynomial. A possibility to further reduce the computation efficiency is to use a more advanced mapping function, e.g. *Level* with the algorithm described in [6], where an adaptive triangulation scheme is utilized. Another possibility is to follow the same procedure as in [22], where the large areas free of roots are identified and omitted from the mapping procedure. As has been shown, next to the possibility to describe the quasi-polynomial by the matrix of polynomial coefficients and the vector of delays, the Matlab function handle form can be used. In this simpler mode of QPmR, no spectrum analysis is performed. However, in this mode, the QPmR can be used for computing roots of any analytical (well-conditioned) function, e.g. fractional order polynomials or quasi-polynomials with both real and complex coefficients.

Acknowledgements. The presented research has been supported by the Ministry of Education of the Czech Republic under the project KONTAKT II - LH12066.

References

1. Bellman, R., Cooke, K.L.: Differential-difference equation. Academic Press, New York (1963)
2. Breda, D.: Solution operator approximation for characteristic roots of delay differential equations. Appl. Numer. Math. 56, 305–331 (2006)
3. Breda, D., Maset, S., Vermiglio, R.: Computing the characteristic roots for delay differential equations. IMA J. Numer. Anal. 24(1), 1–19 (2004)
4. Breda, D., Maset, S., Vermiglio, R.: Pseudospectral differencing methods for characteristic roots of delay differential equations. SIAM J. Sci. Comput. 27(2), 482–495 (2005)
5. Breda, D., Maset, S., Vermiglio, R.: TRACE-DDE: a Tool for Robust Analysis and Characteristic Equations for Delay Differential Equations. In: Loiseau, J.J., Michiels, W., Niculescu, S.I., Sipahi, R. (eds.) Topics in Time-Delay Systems: Analysis, Algorithms and Control. LNCIS, vol. 388, pp. 145–155. Springer, Heidelberg (2008), `http://users.dimi.uniud.it/~dimitri.breda/software.html`
6. Breda, D., Maset, S., Vermiglio, R.: An adaptive algorithm for efficient computation of level curves of surfaces. Numerical Algorithms 52(4), 605–628 (2009)
7. Dellnitz, M., Schutze, O., Zheng, Q.: Locating all the zeros of an analytic function in one complex variable. Journal of Computational and Applied mathematics 138, 325–333 (2002)
8. Delves, L.M., Lyness, J.N.: A numerical method for locating the zeros of analytic functions. Mathematics of Computation 21, 543–560 (1967)
9. Engelborghs, K., Roose, D.: Numerical computation of stability and detection of Hopf bifurcation of steady state solutions of delay differential equations. Advanced in Computational Mathematics 10(3-4), 271–289 (1999)

10. Engelborghs, K., Luzyanina, T., Roose, D.: Numerical bifurcation analysis of delay differential equations using DDE-BIFTOOL. ACM Transactions on Mathematical Software 28(1), 1–21 (2002), `http://twr.cs.kuleuven.be/research/software/delay/ddebiftool.shtml`
11. Engelborghs, K., Roose, D.: On stability of LMS methods and characteristic roots of delay differential equations. SIAM Journal on Numerical Analysis 40, 629–650 (2002)
12. Hale, J.K., Verduyn Lunel, S.M.: Strong stabilization of neutral functional differential equations. IMA Journal of Mathematical Control and Information 19, 5–23 (2002)
13. Householder, A.S.: The Numerical Treatment of a Single Nonlinear Equation. McGraw-Hill, New York (1970)
14. Henrion, D., Vyhlídal, T.: Positive trigonometric polynomials for strong stability of difference equations. Automatica 48(9), 2207–2212 (2012)
15. Jarlebring, E., Damm, T.: The Lambert W function and the spectrum of some multidimensional time-delay systems. Automatica 43(12), 2124–2128 (2007)
16. Kravanja, P., Van Barel, M.: Computing the zeros of analytic functions. Lecture Notes in Mathematics, vol. 1727. Springer (2000)
17. Michiels, W., Vyhlídal, T.: An Eigenvalue Based Approach for the Stabilization of Linear Time-Delay Systems of Neutral Type. Automatica 41, 991–998 (2005)
18. Michiels, W., Vyhlídal, T., Zítek, P., Nijmeijer, H., Henrion, D.: Strong Stability of Neutral Equations with an Arbitrary Delay Dependency Structure. SIAM J. Control Optim. 48(2), 763–786 (2009)
19. Suh, Y.S.: Stability of Time Delay Systems using Numerical Computation of Argument Principles. In: Proceedings of the 40th IEEE Conference on Decision and Control, Orlando, Florida USA, pp. 4738–4743 (2001)
20. Verheyden, K., Luzyanina, T., Roose, D.: Efficient computation of characteristic roots of delay differential equations using LMS methods. Journal of Computational and Applied Mathematics 214(1), 209–226 (2008)
21. Vyhlídal, T., Zítek, P.: Quasipolynomial mapping based rootfinder for analysis of Time delay systems. In: Proc. of IFAC Workshop on Time-Delay Systems, TDS 2003, Rocquencourt (2003)
22. Vyhlídal, T., Zítek, P.: Mapping based algorithm for large-scale computation of quasi-polynomial zeros. IEEE Transactions on Automatic Control 54(1), 171–177 (2009)
23. Vyhlídal, T.: Analysis and Synthesis of Time Delay System Spectrum, PhD Thesis, Faculty of Mechanical Engineering, Czech Technical University in Prague (2003)
24. Wang, Z.H., Hu, H.Y.: Calculation of the rightmost characteristic root of retarded time-delay systems via Lambert W function. Journal of Sound and Vibration 318, 757–767 (2008)
25. Wu, Z., Michiels, W.: Reliably computing all characteristic roots of delay differential equations in a given right half plane using a spectral method. Journal of Computational and Applied Mathematics 236, 2499–2514 (2012), `http://twr.cs.kuleuven.be/research/software/delay-control/roots/`
26. Yi, S., Nelson, P.W., Ulsoy, A.G.: Time-Delay Systems: Analysis and Control Using the Lambert W Function. Imperial College Press (2010)
27. Zeng, L., Hu, G.D.: Computation of Unstable Characteristic Roots of Neutral Delay Systems. Acta Automatica Sinica 31(1), 81–87 (2013)

Part V
Applications

Analysis of a New Model of Cell Population Dynamics in Acute Myeloid Leukemia

José Louis Avila, Catherine Bonnet, Jean Clairambault, Hitay Özbay, Silviu-Iulian Niculescu, Faten Merhi, Annabelle Ballesta, Ruoping Tang, and Jean-Pierre Marie

Abstract. A new mathematical model of the cell dynamics in Acute Myeloid Leukemia (AML) is considered which takes into account the four different phases of the proliferating compartment. The dynamics of the cell populations are governed by transport partial differential equations structured in age and by using the method of characteristics, we obtain that the dynamical system of equation can be reduced

José Louis Avila · Catherine Bonnet
INRIA Saclay-Île-de-France, Equipe DISCO, LSS - SUPELEC, 3 rue Joliot Curie, 91192 Gif-sur-Yvette, Cedex, France
e-mail: `jose.avila@lss.supelec.fr`, `Catherine.Bonnet@inria.fr`

Jean Clairambault
INRIA Paris-Rocquencourt, Domaine de Voluceau, B.P. 105, 78153 Le Chesnay, Cedex, France, and INSERM team U 776 "Biological Rhythms and Cancers", Hôpital Paul-Brousse, 14 Av. Paul-Vaillant-Couturier, 94807 Villejuif, Cedex, France
e-mail: `Jean.Clairambault@inria.fr`

Hitay Özbay
Department of Electrical and Electronics Engineering, Bilkent University, Ankara, 06800, Turkey
e-mail: `hitay@bilkent.edu.tr`

Silviu-Iulian Niculescu
L2S (UMR CNRS 8506), CNRS-Supélec, 3 rue Joliot Curie, 91192, Gif-sur-Yvette, France
e-mail: `Silviu.Niculescu@lss.supelec.fr`

Faten Merhi · Ruoping Tang · Jean-Pierre Marie
Tumorothèque d'Hématologie, Hôpital Saint-Antoine, AP-HP, 184 rue du Faubourg Saint Antoine, 75571 Paris cedex 12, France
INSERM U872, Université Pierre et Marie Curie, Centre de Recherche des Cordeliers, 15 Rue de l'Ecole de Médecine, 75270 Paris Cedex 06, France
e-mail: `merhi.faten@gmail.com`,
`{ruoping.tang,jean-pierre.marie}@sat.aphp.fr`

Annabelle Ballesta
INRIA Paris-Rocquencourt, Domaine de Voluceau, B.P. 105, 78153 Le Chesnay, Cedex, France
e-mail: `Annabelle.Ballesta@inria.fr`

T. Vyhlídal, J.-F. Lafay, and R. Sipahi (eds.), *Delay Systems*,
Advances in Delays and Dynamics 1,
DOI: 10.1007/978-3-319-01695-5_23, © Springer International Publishing Switzerland 2014

to two coupled nonlinear equations with four internal sub-systems involving distributed delays. Equilibrium and local stability analysis of this model are performed and several simulations illustrate the results.

1 Introduction

Acute Myeloid Leukemia (AML) is a cancer of blood cells (myeloid lineage) for which clinical progress has been quite slow in the last forty years, [13]. In this paper, we propose a new model of cell dynamics in AML in order to understand its dynamical behavior and ultimately improve its treatment in collaboration with doctors at Saint-Antoine's hospital in Paris (hematology section laboratory). The process of formation and maturation of blood cells is called hematopoiesis. Blood cells mature in the bone marrow from hematopoiteic stem cells (HSCs) until fully differentiated cells are released in blood circulation (blood cells are transported throughout the body by the plasma). The blood cell population has different components: erythrocytes, or red blood cells; leukocytes, or white blood cells; and platelets. HSCs can proliferate, self renew and differentiate into multiple lineages. The process of cell division, called proliferation or cell cycle, consists of four phases: G_1, S, G_2 and M. At the end of the M phase cell division occurs and two different daughter cells are produced:either with the same biological properties as the parent (self-renewal) or progenitors.The production of progenitors at cell division is called differentiation. One of the first mathematical models on hematopoiesis was proposed by [7].This model consists of a system of differential equations, describing haematopoietic stem cell dynamics, considering a rest (or quiescent) phase and a proliferative phase during the cell division cycle. Most recent studies of various dynamical models of hematopoiesis have been proposed and studied in the literature, see e.g. [2], [5], [6], [8] and references therein.

AML combines at least two molecular events: a blockade of the maturation and differentiation leading to the accumulation of immature myeloid cells, and an advantage of proliferation leading to the flooding of bone marrow by immature and proliferating immature cells. Recently, [2] has proposed a model for AML, that consists of a system of delay-differential equations inspired by the model of [7] with discrete maturity structure. The model takes into account the differentiation blockade that is frequently observed in AML. For the equilibrium and stability analysis (linear and nonlinear system) of this model see [4], [11], [10] and their references.

In this chapter, we consider a mathematical model originally introduced in [1]. This model modified and enriched the model of [2] in the following sense: the self-renewal phenomenon is written in two parts where fast and slow dynamics are separated (this gives us two static nonlinearities in the system); and the dynamical behavior of the proliferating cells is separated into four phases (namely the phases G_1, S, G_2 and M).

In this chapter we first recall in Sections 2 and 3 the model proposed in [1]. The equilibrium and stability analysis of this new system are performed in Section 4. A detailed academic example is presented in Section 5. The chapter ends with concluding remarks in Section 6.

2 Mathematical Model of AML

Taking into account the aforementioned properties of AML, the mathematical model may be obtained as follows. Let us consider two cell sub populations of immature cells, *proliferating* (divided in G_1, S, G_2 and M phases) and *quiescent* (in phase G_0) cells. Between the exit of the M phase and the beginning of the G_1 phase, a new phase called $\tilde{G}_0$, modelling the fast-renewal effect, is introduced here. We denote by $p_i(t,a)$, $l_i(t,a)$, $n_i(t,a)$, $m_i(t,a)$, $r_i(t,a)$ and $\tilde{r}_i(t,a)$ the cell populations of the G_1, S, G_2, M, G_0 and $\tilde{G}_0$ phases, respectively, of the i-th generation of immature cells, with age $a \geq 0$ at time $t \geq 0$.

A schematic representation of the i-th compartmental model considered is shown in Figure 1, where the subscript i is dropped for notational convenience.

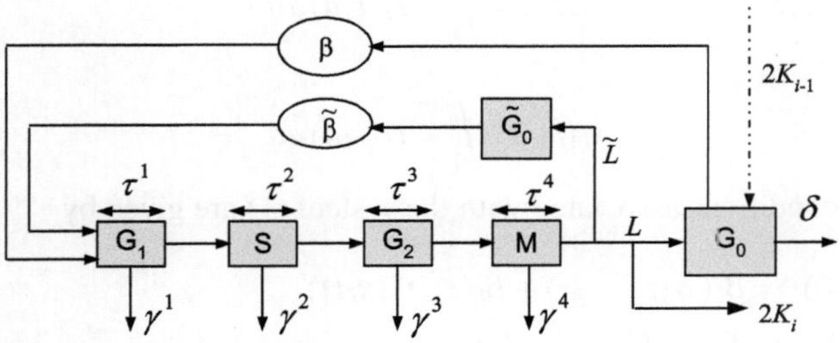

Fig. 1 The i-th compartmental model of the AML cell dynamics

The dynamical behavior of the cell populations of each phase is represented by the following system of transport equations dependent on age a:

$$\begin{cases} \dfrac{\partial p_i}{\partial t} + \dfrac{\partial p_i}{\partial a} = -\left(\gamma_i^1 + g_i^p(a)\right) p_i, & 0 < a < \tau_i^1, \ t > 0, \\[2ex] \dfrac{\partial l_i}{\partial t} + \dfrac{\partial l_i}{\partial a} = -\left(\gamma_i^2 + g_i^l(a)\right) s_i, & 0 < a < \tau_i^2, \ t > 0, \\[2ex] \dfrac{\partial n_i}{\partial t} + \dfrac{\partial n_i}{\partial a} = -\left(\gamma_i^3 + g_i^n(a)\right) n_i, & 0 < a < \tau_i^3, \ t > 0, \\[2ex] \dfrac{\partial m_i}{\partial t} + \dfrac{\partial m_i}{\partial a} = -\left(\gamma_i^4 + g_i^m(a)\right) m_i, & 0 < a < \tau_i^4, \ t > 0, \\[2ex] \dfrac{\partial r_i}{\partial t} + \dfrac{\partial r_i}{\partial a} = -\left(\delta_i + \beta_i\left(\int_0^{+\infty} r_i(t,a)\,da\right)\right) r_i, & a > 0, \ t > 0, \\[2ex] \dfrac{\partial \tilde{r}_i}{\partial t} + \dfrac{\partial \tilde{r}_i}{\partial a} = -\tilde{\beta}_i\left(\int_0^{+\infty} \tilde{r}_i(t,a)\,da\right) \tilde{r}_i & a > 0, \ t > 0, \end{cases} \tag{1}$$

where the death rate in the resting phase is $\delta_i > 0$, the re-introduction function from the resting subpopulation into the proliferative subpopulation is β_i, and the positive constant death rates in the G_1, S, G_2 and M phases are γ_i^1, γ_i^2, γ_i^3 and γ_i^4, respectively; the amount of time spent in the G_1, S, G_2 and M phases are τ_i^1, τ_i^2, τ_i^3 and τ_i^4, respectively; and, the division rates of the phases G_1, S, G_2 and M are functions that depend upon a, denoted by g_i^p, g_i^l, g_i^n and g_i^m, respectively.

Here, only the death rate is included and the birth rate is not involved in the equation; because, when individuals are born at $a = 0$, they are introduced into the population through the boundary (renewal) condition. The re-introduction rates β_i and $\tilde{\beta}_i$ are strictly positive monotonically decreasing functions depending upon the total population of resting and fast-self renewing cells denoted by $x_i(t)$ and $\tilde{x}_i(t)$, respectively, where

$$x_i(t) := \int_0^{+\infty} r_i(t,a)\,da.$$

and

$$\tilde{x}_i(t) := \int_0^{+\infty} \tilde{r}_i(t,a)\,da.$$

Boundary conditions associated with the system (1) are given by

$$\begin{cases} p_i(t,a=0) &= \beta_i(x_i(t))\,x_i(t) + \tilde{\beta}_i(\tilde{x}_i(t))\,\tilde{x}_i(t) \\ l_i(t,a=0) &= \int_0^{\tau_i^1} g_i^p(a)\,p_i(t,a)\,da, \\ n_i(t,a=0) &= \int_0^{\tau_i^2} g_i^l(a)\,l_i(t,a)\,da, \\ m_i(t,a=0) &= \int_0^{\tau_i^3} g_i^n(a)\,n_i(t,a)\,da, \\ r_i(t,a=0) &= L_i \int_0^{\tau_i^4} g_i^m(a)\,m_i(t,a)\,da + 2K_{i-1}\int_0^{\tau_{i-1}^4} g_{i-1}^m(a)\,m_{i-1}(t,a)\,da, \\ \tilde{r}_i(t,a=0) &= \tilde{L}_i \int_0^{\tau_i^4} g_i^m(a)\,m_i(t,a)\,da \end{cases}$$

where $L_i := 2\sigma_i(1-K_i)$, $\tilde{L}_i := 2(1-\sigma_i)(1-K_i)$. The positive constants K_i and σ_i represent the probability of a cell of leaving, at the stage i, the M phase and entering the phase $\tilde{G}_0$ (same stage) and G_0 (stage $i+1$), respectively. The initial age-distribution of the populations of (1) are nonnegative age-dependent functions and they are assumed to be known: $p_i(t=0,a) = p_i^0(a)$, $l_i(t=0,a) = l_i^0(a)$, $n_i(t=0,a) = n_i^0(a)$, $m_i(t=0,a) = m_i^0(a)$, $r_i(t=0,a) = r_i^0(a)$ and $\tilde{r}_i(t=0,a) = \tilde{r}_i^0(a)$. Additionally, the following assumptions are fulfilled:

1. The division rates g_i^p, g_i^l, g_i^n and g_i^m are continuous functions such that

$$\int_0^{\tau_i^j} g_i^k(a)\,da = +\infty, \qquad \text{for } (j,k) = (1,p),\ (2,l),\ (3,n),\ (4,m).$$

2. $\lim_{a\to+\infty} r_i(t,a) = 0$ and $\lim_{a\to+\infty} \tilde{r}_i(t,a) = 0$.
3. The re-introduction terms β_i and $\tilde{\beta}_i$ are differentiable and uniformly decreasing functions with $\beta_i(0) > 0$, $\tilde{\beta}_i(0) > 0$; $\beta_i(x) \to 0$ and $\tilde{\beta}_i(x) \to 0$ as $x \to \infty$.

Typically β_i and $\tilde{\beta}_i$ are Hill functions of the form

$$\beta_i(x) = \frac{\beta_i(0)}{1 + b_i x^{N_i}} \quad \text{and} \quad \tilde{\beta}_i(\tilde{x}) = \frac{\tilde{\beta}_i(0)}{1 + \tilde{b}_i \tilde{x}^{\tilde{N}_i}} \tag{2}$$

where N_i and $\tilde{N}_i$ are integers greater or equal to 2; $b_i > 0$, $\tilde{b}_i > 0$ and $\tilde{b}_i \ll 1$. Some examples of the β_i function are illustrated in Figure 2.

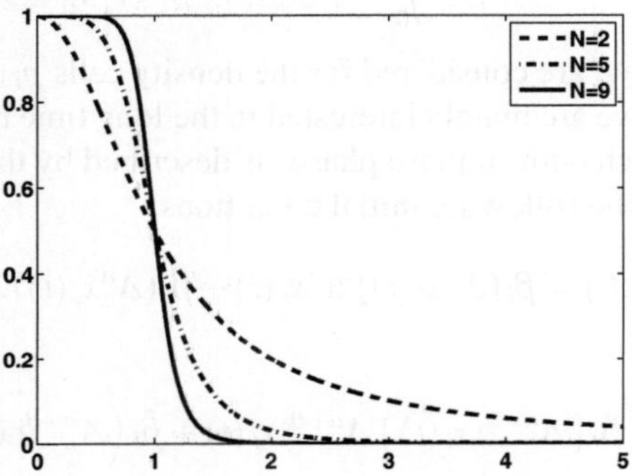

Fig. 2 Examples of three different β_i functions with $\beta_i(0) = 1$, $b_i = 1$, and $N_i = 2, 5, 9$

Now, we shall show how a system of ordinary differential distributed delayed equations is obtained from the PDE's (1).

3 Model Transformation

Using the method of characteristics (see e.g [12]), one can easily obtain an explicit formulation for $p_i(t,a)$, $l_i(t,a)$, $n_i(t,a)$ and $m_i(t,a)$ given by

$$p_i(t,a) = \begin{cases} p_i^0(a-t)\, e^{-\int_{a-t}^{a}\left(\gamma_i^1 + g_i^p(w)\right)dw}, & \text{if } t \le a, \\ p_i(t-a,0)\, e^{-\int_0^a\left(\gamma_i^1 + g_i^p(w)\right)dw} & \text{if } t > a, \end{cases}$$

$$l_i(t,a) = \begin{cases} l_i^0(a-t)\, e^{-\int_{a-t}^{a}\left(\gamma_i^2 + g_i^l(w)\right)dw}, & \text{if } t \le a, \\ l_i(t-a,0)\, e^{-\int_0^a\left(\gamma_i^2 + g_i^l(w)\right)dw} & \text{if } t > a, \end{cases}$$

$$n_i(t,a) = \begin{cases} n_i^0(a-t)\, e^{-\int_{a-t}^{a}\left(\gamma_i^3 + g_i^n(w)\right)dw}, & \text{if } t \le a, \\ n_i(t-a,0)\, e^{-\int_0^a\left(\gamma_1^3 + g_i^n(w)\right)dw} & \text{if } t > a, \end{cases}$$

$$m_i(t,a) = \begin{cases} m_i^0(a-t)\, e^{-\int_{a-t}^{a}\left(\gamma_i^4 + g_i^m(w)\right)dw}, & \text{if } t \le a, \\ m_i(t-a,0)\, e^{-\int_0^a\left(\gamma_i^4 + g_i^m(w)\right)dw} & \text{if } t > a, \end{cases}$$

where

$$p_i(t-a,0) = \beta_i(x_i(t-a))x_i(t-a) + \tilde{\beta}_i(\tilde{x}_i(t-a))\tilde{x}_i(t-a),$$

$$l_i(t-a,0) = \int_0^{\tau_i^1} g_i^p(\theta_1)\,p_i(t-a,\theta_1)\,d\theta_1,$$

$$n_i(t-a,a) = \int_0^{\tau_i^2} g_i^l(\theta_2)\,l_i(t-a,\theta_2)\,d\theta_2,$$

$$m_i(t-a,a) = \int_0^{\tau_i^3} g_i^n(\theta_3)\,n_i(t-a,\theta_3)\,d\theta_3.$$

Only the solutions $t \geq a$ are considered for the density cells $p_i(t,a)$, $l_i(t,a)$, $n_i(t,a)$ and $m_i(t,a)$ because we are mainly interested in the long time behavior of the populations; namely, the behavior of these phases is described by the second term of the above solutions with the following initial conditions

$$p_i(t-a,0) = \beta_i(\Delta^a x_i(t))\,\Delta^a x_i(t) + \tilde{\beta}_i(\Delta^a\tilde{x}_i(t))\,\Delta^a\tilde{x}_i(t)$$

$$l_i(t-a,0) = \int_0^{\tau_i^1}\left(\beta_i\left(\Delta^{a+\theta_1}x_i(t)\right)\Delta^{a+\theta_1}x_i(t) + \tilde{\beta}_i\left(\Delta^{a+\theta_1}\tilde{x}_i(t)\right)\Delta^{a+\theta_1}\tilde{x}_i(t)\right)$$
$$\cdot\, f_i^p(\theta_1)\,e^{-\gamma_i^1\theta_1}d\theta_1$$

$$n_i(t-a,0) = \int_0^{\tau_i^2}\left(\int_0^{\tau_i^1}\left(\beta_i\left(\Delta^{a+\theta_1+\theta_2}x_i(t)\right)\cdot\Delta^{a+\theta_1+\theta_2}x_i + \tilde{\beta}_i\left(\Delta^{a+\theta_1+\theta_2}\tilde{x}_i(t)\right)\right.\right.$$
$$\left.\left.\cdot\Delta^{a+\theta_1+\theta_2}\tilde{x}_i(t)\right)f_i^p(\theta_1)\,e^{-\gamma_i^1\theta_1}d\theta_1\right)\cdot f_i^l(\theta_2)\,e^{-\gamma_i^2\theta_2}d\theta_2$$

$$m_i(t-a,0) = \int_0^{\tau_i^3}\left[\int_0^{\tau_i^2}\left(\int_0^{\tau_i^1}\left(\beta_i\left(\Delta^{a+\theta_1+\theta_2+\theta_3}x_i(t)\right)\cdot\Delta^{a+\theta_1+\theta_2+\theta_3}x_i(t)\right.\right.\right.$$
$$\left. + \tilde{\beta}_i\left(\Delta^{a+\theta_1+\theta_2+\theta_3}\tilde{x}_i(t)\right)\cdot\Delta^{a+\theta_1+\theta_2+\theta_3}\tilde{x}_i(t)\right)$$
$$\left.\left.\cdot f_i^p(\theta_1)\,e^{-\gamma_i^1\theta_1}d\theta_1\right)f_i^l(\theta_2)\,e^{-\gamma_i^2\theta_2}d\theta_2\right]\cdot f_i^n(\theta_3)\,e^{-\gamma_i^3\theta_3}d\theta_3$$

where

$$f_i^k(t) = g_i^k(t)\,e^{-\int_0^t g_i^k(w)dw} \quad \text{if } 0 < t < \tau_j$$
$$= 0 \quad \text{otherwise,}$$

for $(j,k) = (1,p),\ (2,l),\ (3,n),\ (4,m)$, and the shift operator Δ is defined by

$$\Delta^a x_i(t) := x_i(t-a)\ .$$

The functions f_i^k, are density functions, i.e. $\int_0^{\tau_i^1} f_i^k(t)\,dt = 1$, for $k = p,l,n,m$.

Finally, integrating the last two equations in (1) with respect to the age variable a, between $a = 0$ and $a = +\infty$ one obtains:

$$\dot{x}_i(t) = -(\delta_i + \beta_i(x_i(t)))x_i(t) + L_i \int_0^{\tau_i^4} g_i^m(a)\, m_i(t,a)\, da \quad (3)$$
$$+ 2K_{i-1} \int_0^{\tau_{i-1}^4} g_{i-1}^m(a)\, m_{i-1}(t,a)\, da$$

and

$$\dot{\tilde{x}}_i(t) = -\tilde{\beta}_i(\tilde{x}_i(t))\tilde{x}_i(t) + \tilde{L}_i \int_0^{\tau_4} g_i^m(\theta_1)\, m_i(t,\theta_1)\, d\theta_1 \quad (4)$$

where

$$m_i(t,a) = m_i(t-a,0)\, e^{-\int_0^a (\gamma_i^4 + g_m^i(w))dw}.$$

It is important to point out that equations (3) and (4) depend explicitly on each other, because of the term $m_i(t-a,0)$, which contains the expressions $\Delta x_i^{a+\theta_1+\theta_2+\theta_3}(t)$ and $\Delta \tilde{x}_i^{a+\theta_1+\theta_2+\theta_3}(t)$; that is, the equations form a coupled system.

Defining $h_i^1(t) := f_i^p(t)e^{-\gamma_i^1 t}$, $h_i^2(t) := f_i^l(t)e^{-\gamma_i^2 t}$, $h_i^3(t) := f_i^n(t)e^{-\gamma_i^3 t}$ and $h_i^4(t) := f_i^m(t)e^{-\gamma_i^4 t}$, we rewrite the equations (3) and (4) as

$$\dot{x}_i(t) = -(\delta_i + \beta_i(x_i(t)))x_i(t) + L_i\left(h_i^4 * h_i^3 * h_i^2 * h_i^1 * \omega_i\right)(t)$$
$$+ 2K_{i-1}\left(h_{i-1}^4 * h_{i-1}^3 * h_{i-1}^2 * h_{i-1}^1 * \omega_{i-1}\right)(t)$$
$$\dot{\tilde{x}}_i(t) = -\tilde{\beta}_i(\tilde{x}_i(t))\tilde{x}_i(t) + \tilde{L}_i\left(h_i^4 * h_i^3 * h_i^2 * h_i^1 * \omega_i\right)(t)$$

where $\omega_i(t) := \beta_i(x_i(t))x_i(t) + \tilde{\beta}_i(\tilde{x}_i(t))\tilde{x}_i(t)$ and $*$ is the convolution operator.

4 Analysis of the i-th Compartmental Model

4.1 Equilibrium Points

Let us denote by x_i^e and $\tilde{x}_i^e$, the equilibrium points of (3) and (4), respectively; namely, the trajectories that satisfy $\frac{dx_i^e}{dt} = 0$ and $\frac{d\tilde{x}_i^e}{dt} = 0$. The i-th equilibrium point is the solution of

$$-\bar{u}_{i-1} = -(1 - L_i H_i(0))\,\beta_i(x_i^e)x_i^e - \delta_i x_i^e + L_i H_i(0)\,\tilde{\beta}_i(\tilde{x}_i^e)\tilde{x}_i^e \quad (5)$$

$$0 = \tilde{L}_i H_i(0)\,\beta_i(x_i^e)x_i^e - \left(1 - \tilde{L}_i H_i(0)\right)\tilde{\beta}_i(\tilde{x}_i^e)\tilde{x}_i^e \quad (6)$$

where

$$\bar{u}_{i-1} = \begin{cases} 0 & \text{if } i = 1 \\ 2K_{i-1}H_{i-1}(0)\left(\beta_{i-1}(x_{i-1}^e)x_{i-1}^e + \tilde{\beta}_{i-1}(\tilde{x}_{i-1}^e)\tilde{x}_{i-1}^e\right) & \text{if } i > 1 \end{cases}$$

and

$$H_i(s) = H_i^1(s) \cdot H_i^2(s) \cdot H_i^3(s) \cdot H_i^4(s)$$

with $H_i^j(s) = \int_0^{\tau_i^j} h_i^j(t)\, e^{-st} dt$, for $j = 1, \ldots, 4$.

We can readily note that the points $x_i^e = 0$ and $\tilde{x}_i^e = 0$ satisfy (5) and (6). We will refer to this equilibrium point as the trivial equilibrium point. From (5) and (6), a non-trivial equilibrium point satisfy

$$\beta_i(x_i^e) = \begin{cases} \frac{\delta_1}{\alpha_1} & \text{if } i = 1 \\ \frac{\delta_i}{\alpha_i} - \left(\frac{\bar{u}_{i-1}}{\alpha_i}\right) \frac{1}{x_i^e} & \text{if } i > 1 \end{cases} \tag{7}$$

$$\tilde{\beta}_i(\tilde{x}_i^e) = \begin{cases} \left(\frac{\delta_1 x_1^e}{\tilde{\alpha}_1}\right) \frac{1}{\tilde{x}_1^e} & \text{if } i = 1 \\ \left(\frac{\delta_i x_i^e - \bar{u}_{i-1}}{\tilde{\alpha}_i}\right) \frac{1}{\tilde{x}_i^e} & \text{if } i > 1 \end{cases} \tag{8}$$

where

$$\alpha_i := \frac{2(1-K_i)H_i(0) - 1}{1 - 2(1-\sigma_i)(1-K_i)H_i(0)} \quad \text{and} \quad \tilde{\alpha}_i := \frac{2(1-K_i)H_i(0) - 1}{2(1-\sigma_i)(1-K_i)H_i(0)}$$

recall $L_i := 2\sigma_i(1-K_i)$ and $\tilde{L}_i := 2(1-\sigma_i)(1-K_i)$.

The result stated below deals with existence and uniqueness of positive equilibrium points x_i^e.

Theorem 1. *If* $1 < 2(1-K_i)H_i(0) < \frac{1}{1-\sigma_i}$ *for all i, and* $\beta_1(0) > \frac{\delta_1}{\alpha_1}$, *then we have a unique positive equilibrium point* x_i^e.

Proof. First, note that α_i is non-negative for every i by hypothesis. For $i = 1$, $\beta_1(x_1^e) = \frac{\delta_1}{\alpha_1}$; the existence and uniqueness is guaranteed if $\beta_1(0) > \frac{\delta_1}{\alpha_1}$. For $i \geq 2$, let $\psi_i : \mathbb{R}_+ \setminus \{0\} \to \mathbb{R}$ be given by $\psi_i(\bar{x}) = \frac{\delta_i}{\alpha_i} - \frac{b_i}{\alpha_i}\frac{1}{\bar{x}}$. The non-negativeness of α_i implies the following: ψ_i is strictly increasing ($\frac{d}{d\bar{x}}\psi_i(\bar{x}) > 0$), $\lim_{\bar{x}\to+\infty} \psi_i(\bar{x}) = \frac{\delta_i}{\alpha_i}$ is positive, $\lim_{\bar{x}\to 0-} \psi_i(\bar{x}) = -\infty$, and the difference $\beta_i - \psi_i$ is strictly decreasing with the properties $\lim_{\bar{x}\to 0^-} (\beta_i(\bar{x}) - \psi_i(\bar{x})) = \infty$ and $\lim_{\bar{x}\to+\infty} (\beta_i(\bar{x}) - \psi_i(\bar{x})) = -\frac{\delta_i}{\alpha_i}$. Therefore, there is a positive real number x^* such that $\beta_i(x^*) - \psi_i(x^*) = 0$. In other words, (5) has a unique positive solution for every i. □

We now discuss the existence and uniqueness conditions for $\tilde{x}_i^e$. Since

$$\tilde{\beta}_i(\tilde{x}_i^e) = \frac{1}{\tilde{x}_i^e}\left(\frac{2(1-\sigma_i)(1-K_i)H_i(0)}{(1-2(1-\sigma_i)(1-K_i)H_i(0))}\beta_i(x_i^e)x_i^e\right),$$

it is easy to see that for a suitable function $\tilde{\beta}_i$, there will be an intersection point between the functions $\tilde{x}_i^e \mapsto \tilde{\beta}_i(\tilde{x}_i^e)$ and $\tilde{x}_i^e \mapsto c/\tilde{x}_i^e$ where c is the constant

$$c = \frac{2(1-\sigma_i)(1-K_i)H_i(0)}{(1-2(1-\sigma_i)(1-K_i)H_i(0))}\beta_i(x_i^e)x_i^e.$$

If we assume that $\tilde{\beta}_i$ is a Hill function of the form (2) for every i, then $\tilde{x}_i^e$ is the solution of

$$\tilde{b}_i\left(\tilde{x}_i^e\right)^{\tilde{N}_i} - \frac{\tilde{\beta}_i(0)}{c}\,\tilde{x}_i^e + 1 = 0\,. \tag{9}$$

By applying Descartes rule of signs, we obtain that the maximum number of positive solutions of (9) is two. In fact, the parameters of the system may be such that we have, one, or two, or no equilibrium as the positive solution of (9) for $\tilde{x}_i^e$. Figure 3 illustrates this point by showing possible intersections of $\tilde{\beta}_i\left(\tilde{x}_i^e\right)$ and $c/\tilde{x}_i^e$ for different numerical values. Thus, a positive equilibrium, when it exists, may not be unique for the sub-compartments of $\tilde{x}_i$. This poses an extra challenge for the stability analysis.

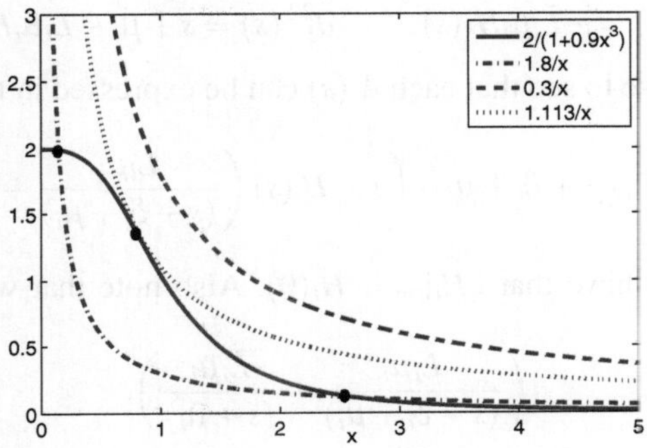

Fig. 3 Examples of intersections of $\tilde{\beta}_i\left(\tilde{x}_i^e\right)$ and $c/\tilde{x}_i^e$

4.2 Model Linearization and Stability

Let us define a perturbed trajectory around the equilibrium points of (3) and (4) by $X_i(t) := x_i(t) - x_i^e(t)$ and $\tilde{X}_i(t) := \tilde{x}_i(t) - \tilde{x}_i^e(t)$, for every i. The linearization of (3) and (4) around their equilibrium points is

$$\begin{aligned}\tfrac{d}{dt}X_i(t) = &-(\delta_i+\mu_i)X_i(t)\\ &+L_i\mu_i\left[h_i^4*\left(h_i^3*\left(h_i^2*\left(h_i^1*X_i\right)\right)\right)\right](t)\\ &+L_i\tilde{\mu}_i\left[h_i^4*\left(h_i^3*\left(h_i^2*\left(h_i^1*\tilde{X}_i\right)\right)\right)\right](t)\\ &+2K_{i-1}\mu_{i-1}\left[h_{i-1}^4*\left(h_{i-1}^3*\left(h_{i-1}^2*\left(h_{i-1}^1*X_{i-1}\right)\right)\right)\right](t)\\ &+2K_{i-1}\tilde{\mu}_{i-1}\left[h_i^4*\left(h_i^3*\left(h_i^2*\left(h_i^1*\tilde{X}_{i-1}\right)\right)\right)\right](t)\end{aligned} \tag{10}$$

and

$$\begin{aligned}\frac{d}{dt}\tilde{X}_i(t) = &-\tilde{\mu}_i\tilde{X}_i(t) + \tilde{L}_i\mu_i\left[h_i^4*\left(h_i^3*\left(h_i^2*\left(h_i^1*X_i\right)\right)\right)\right](t)\\ &+\tilde{L}_i\tilde{\mu}_i\left[h_i^4*\left(h_i^3*\left(h_i^2*\left(h_i^1*\tilde{X}_i\right)\right)\right)\right](t)\end{aligned} \tag{11}$$

where

$$\mu_i = \frac{d}{dx}(\beta_i(x)x)\Big|_{x=x_i} \quad \text{and} \quad \tilde{\mu}_i = \frac{d}{dx}\left(\tilde{\beta}_i(x)x\right)\Big|_{x=\tilde{x}_i}$$

Taking the Laplace transform of (10) and (11), we can see that the characteristic equation of the system represented by (10) and (11) is given by

$$\prod_{i=1}^{n} A_i(s) = 0 \tag{12}$$

where $A_i(s) = d_i^{11}(s)\,d_i^{22}(s) - d_i^{12}(s)\,d_i^{21}(s)$ with

$$d_i^{11}(s) = s + \delta_i + \mu_i - L_i\mu_i H_i(s), \qquad d_i^{12}(s) = -L_i\tilde{\mu}_i H_i(s),$$

$$d_i^{21}(s) = -\tilde{L}_i\mu_i H_i(s), \qquad d_i^{22}(s) = s + \tilde{\mu}_i - \tilde{L}_i\tilde{\mu}_i H_i(s)$$

It is a simple exercise to see that each $A_i(s)$ can be expressed in the form

$$A_i(s) = (s+\tilde{\mu}_i)(s+\delta_i+\mu_i)\cdot\left(1 - H_i(s)\left(\frac{L_i\mu_i}{(s+\delta_i+\mu_i)} + \frac{\tilde{L}_i\tilde{\mu}_i}{(s+\tilde{\mu}_i)}\right)\right)$$

Since $h_i(a) > 0$ we have that $\|H_i\|_\infty = H_i(0)$. Also note that when $\mu_i > -\delta_i$ and $\tilde{\mu}_i > 0$ the function

$$\left(\frac{L_i\mu_i}{(s+\delta_i+\mu_i)} + \frac{\tilde{L}_i\tilde{\mu}_i}{(s+\tilde{\mu}_i)}\right)$$

is a low pass filter whose H_∞-norm is attained at $s = 0$. Thus, by the Nyquist stability criterion, all the roots of $A_i(s) = 0$ are in the open left half plane if and only if

$$H_i(0)\left(\frac{L_i\mu_i}{\delta_i+\mu_i} + \tilde{L}_i\right) < 1.$$

Using the definition of L_i and $\tilde{L}_i$, after some algebraic manipulations, this condition can be re-written as

$$H_i(0) < \frac{\delta_i+\mu_i}{2(1-K_i)(\mu_i+(1-\sigma_i)\delta_i)}. \tag{13}$$

An interesting observation is that (13) is independent of $\tilde{\mu}_i$.

5 Numerical Example and Simulation Results

In what follows we take d to be the unit of time. Hence the time delays τ_i^1, τ_i^2, τ_i^3 and τ_i^4 are in d and the rates δ_i, γ_i, $\beta_i(0)$ and $\tilde{\beta}_i(0)$ are in d^{-1}. The remaining parameters are normalized to have no unit. The parameters given below are chosen to illustrate various possibilities for the equilibrium and stability properties.

We consider the following general form for the division rates h_i^j:

$$h_i^j(a) = \frac{m_i^j}{e^{m_i^j \tau_i^j} - 1} e^{(m_i^j - \gamma_i^j)a} \quad \text{and} \quad \gamma_i^j << m_i^j, \quad j \in \{1,2,3,4\}.$$

Let us study a system with two compartments with $\delta_1 = 1.3$, $\delta_2 = 0.9$, $K_1 = 0.05$, $K_2 = 0.1$, $\sigma_1 = 0.9$, $\sigma_2 = 0.8$, and the other parameters as indicated in Table 1. The resulting equilibrium points and computed values of μ_i, α_i and other computed values are shown in the third part of Table 1.

Table 1 Simulation parameters and resulting equilibrium points

i	$\beta_i(0)$	$\tilde{\beta}_i(0)$	b_i	$\tilde{b}_i$	N_i	$\tilde{N}_i$	m_i^1	m_i^2	m_i^3	m_i^4
1	2	1	1	0.1	2	2	3	1	2	4
2	1	1	1	0.3	4	4	3	1	2	2

i	τ_i^1	τ_i^2	τ_i^3	τ_i^4	γ_i^1	γ_i^2	γ_i^3	γ_i^4
1	0.3	0.1	0.2	0.4	0.03	0.010	0.020	0.04
2	0.3	0.1	0.2	1.0	0.05	0.015	0.085	0.05

i	$H_i(0)$	x_i^e	$\tilde{x}_i^e$	α_i	$\tilde{\alpha}_i$	μ_i	$\tilde{\mu}_i$
1	0.982	0.7992	0.2249	1.0652	4.6420	0.2691	0.9849
2	0.950	0.8100	0.2950	1.0794	2.0763	−0.1424	0.9887

Time domain simulation, performed in Matlab Simulink, shows that with the initial conditions

$$x_1(\tau) = 0.6 \quad \text{for all} \quad -0.4 \leq \tau \leq 0; \qquad x_2(\tau) = 0.1 \quad \text{for all} \quad -1.0 \leq \tau \leq 0;$$

and $\tilde{x}_1(\tau) = \tilde{x}_2(\tau) = 0$, for all $\tau \leq 0$, the states converge to the equilibrium points

$$\left[\ x_1^e \ \ x_2^e\ \right] = \left[\ 0.7992 \ \ 0.8100\ \right] \quad \text{and} \quad \left[\ \tilde{x}_1^e \ \ \tilde{x}_2^e\ \right] = \left[\ 0.2250 \ \ 0.2950\ \right] \tag{14}$$

see Figs. 4 and 5.

Indeed it can be verified that with the parameters in Table 1 the local stability condition stated in (13) is satisfied for both $i = 1$ and $i = 2$ (note that $\tilde{\mu}_i > 0$ and $\delta_i + \mu_i > 0$ for both compartments):

$$H_1(0) = 0.982 < \frac{\delta_1 + \mu_1}{2(1-K_1)(\mu_1 + (1-\sigma_1)\delta_1)} = 4.1388$$

$$H_2(0) = 0.950 < \frac{\delta_2 + \mu_2}{2(1-K_2)(\mu_2 + (1-\sigma_2)\delta_2)} = 22.3926$$

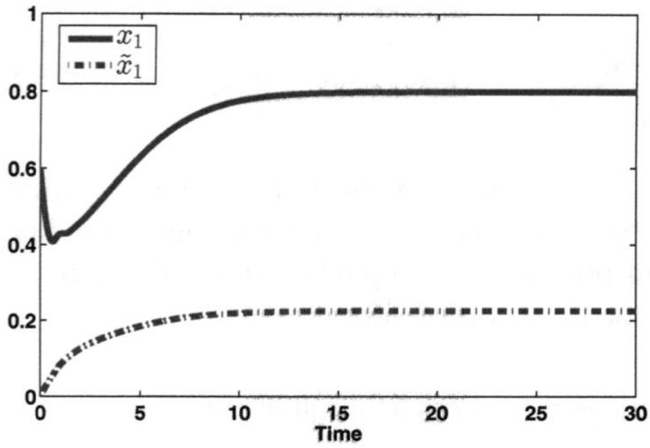

Fig. 4 Trajectories of the states x_1 and $\tilde{x}_1$

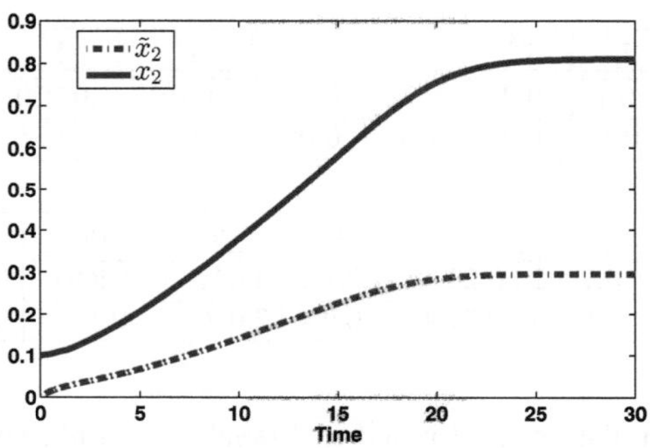

Fig. 5 Trajectories of the states x_2 and $\tilde{x}_2$

Note that with the above system parameters, we have multiple solutions of (9). For $\tilde{x}_1^e$ the other solution is 44.4544 and for $\tilde{x}_2^e$ the other solution is 2.1374. Convergence to any one of these depend on the local stability conditions at these points as well as the initial conditions. Further simulations showed that (14) is the only stable equilibrium point. For example, with the initial conditions

$$x_1(0) = 1.4 \quad x_2(0) = 0.9 \quad \tilde{x}_1(0) = 1.5 \quad \tilde{x}_2(0) = 3.0$$

and $x_1(\tau) = x_2(\tau) = \tilde{x}_1(\tau) = \tilde{x}_2(\tau) = 0$, for all $\tau < 0$, we observe that $\tilde{x}_2(t)$ diverges from its previous equilibrium 0.2950, that causes $x_2(t)$ to move away from its equilibrium 0.8100; see Figures 6 and 7.

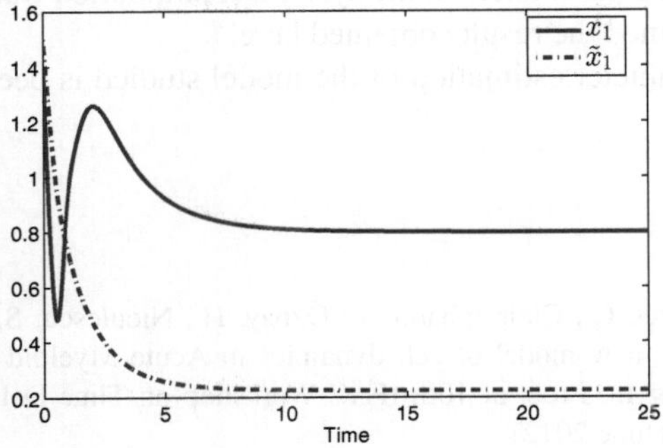

Fig. 6 Trajectories of the states x_1 and $\tilde{x}_1$

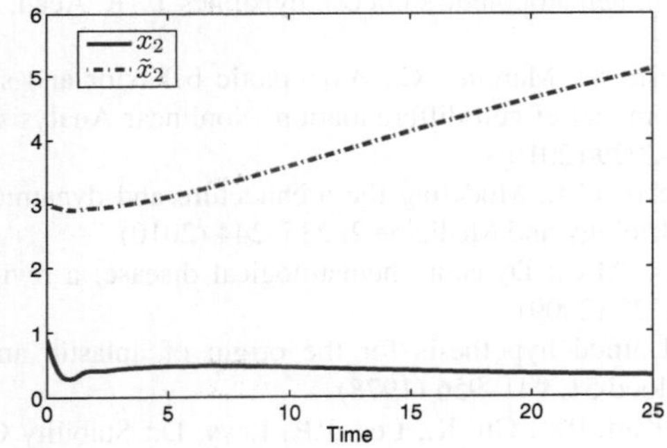

Fig. 7 Trajectories of the states x_2 and $\tilde{x}_2$

6 Conclusions

In this chapter we have re-visited the new model of cell population dynamics introduced in [1] for AML. The new model was derived by using the PDEs representing the cell dynamics for the phases G_1, S, G_2, M, G_0 and $\tilde{G}_0$ and by using a model reduction. This gave us a pair of coupled nonlinear systems involving distributed delays for the dynamical cell behavior in AML.

Here, we derived conditions for having positive equilibrium points and a linear system is obtained around these positive solutions. It is observed that having possibly non-unique equilibrium makes the problem of finding the region of attraction of locally stable equilibrium more challenging than the previous models considered in [2] and [11].

The characteristic equation is derived for the local stability analysis; then by using classical methods of control theory we obtained a new local stability condition which depends on the parameters of the system. A simulation study of two compartment system confirmed the results obtained here.

Currently, a parameter estimation of the model studied is been performed using biological data.

References

1. Avila, J.L., Bonnet, C., Clairambault, J., Özbay, H., Niculescu, S.-I., Merhi, F., Tang, R., Marie, J.P.: A new model of cell dynamics in Acute Myeloid Leukemia involving distributed delays. In: Proc. of 10th IFAC Workshop on Time Delay Systems, Boston, USA, pp. 55–60 (June 2012)
2. Adimy, M., Crauste, F., El Abdllaoui, A.: Discrete maturity-structured model of cell differentiation with applications to acute myelogenous leukemia. J. Biological Systems 16(3), 395–424 (2008)
3. Adimy, M., Crauste, F., El Abdllaoui, A.: Boundedness and Lyapunov function for a nonlinear system of hematopoietic stem cell dynamics. C. R. Acad. Sci. Paris, Ser. I 348, 373–377 (2010)
4. Adimy, M., Crauste, F., Marquet, C.: Asymptotic behavior and stability switch for a mature-immature model of cell differentiation. Nonlinear Analysis: Real World Applications 11, 2913–2929 (2010)
5. Dingli, D., Pacheco, J.M.: Modeling the architecture and dynamics of hematopoiesis. WIREs Systems Biology and Medicine 2, 235–244 (2010)
6. Foley, C., Mackey, M.C.: Dynamic hematological disease: a review. J. Mathematical Biology 58, 285–322 (2009)
7. Mackey, M.C.: Unified hypothesis for the origin of aplastic anaemia and periodic hematopoiesis. Blood 51, 941–956 (1978)
8. Niculescu, S.-I., Kim, P.S., Gu, K., Lee, P.P., Levy, D.: Stability Crossing Boundaries of Delay Systems Modeling Immune Dynamics in Leukemia. Discrete and Continuous Dynamical Systems Series B 13, 129–156 (2010)
9. Özbay, H., Bonnet, C., Clairambault, J.: Stability Analysis of Systems with Distributed Delays and Application to Hematopoietic Cell Maturation Dynamics. In: Proc. of the 47th IEEE Conference on Decision and Control, Cancun, Mexico, pp. 2050–2055 (December 2008)
10. Özbay, H., Benjelloun, H., Bonnet, C., Clairambault, J.: Stability Conditions for a System Modeling Cell Dynamics in Leukemia. In: IFAC Workshop on Time Delay Systems, TDS 2010, Prague, Czech Republic (June 2010)
11. Özbay, H., Bonnet, C., Benjelloun, H., Clairambault, J.: Stability Analysis of Cell Dynamics in Leukemia. Mathematical Modelling of Natural Phenomena 7(1), 203–234 (2012)
12. Perthame, B.: Transport equations in biology. In: Frontiers in Mathematics. Birkhäuser Verlag (2007)
13. Rowe, J.: Why is clinical progress in acute myelogenous leukemia so slow? Best Practice & Research Clinical Haematology 21, 1–3 (2008)

The Influence of Time Delay on Crane Operator Performance

Joshua Vaughan and William Singhose

Abstract. Cranes are used extensively in many industries throughout the world. These cranes operate in a wide array of environments, including some that are hazardous to humans. The vast majority of cranes are directly controlled by human operators. However, in some cases, it is necessary to remove the human operator from hazardous operating conditions, creating a crane that must be remotely operated. This, however, introduces additional challenges for the operator. The operator must now control the oscillatory payload while suffering from decreased perception of the environment and the potential time delays caused by remote operation. A number of studies of crane operator performance with varying time delays are presented here. The compiled results show that the type of crane control and duration of the communication delay directly influence task completion time and difficulty. Input shaping control is shown to improve completion times over a large range of operating conditions and communication time delays.

1 Introduction

Cranes are essential to a large number of industries throughout the world. Bridge cranes, like the one shown in Fig. 1, and gantry cranes serve as the primary heavy lifters at ports and factories. Tower cranes and boom cranes, like the the one shown in Fig. 2, often aid in the construction (and demolition) of buildings. Cranes typically have large delays between operator commands and payload response, stemming from a combination of the large inertial properties of the crane structure and the low natural frequency of payload oscillation. In addition, any movement of the

Joshua Vaughan
University of Louisiana at Lafayette, Lafayette, LA USA
e-mail: joshua.vaughan@louisiana.edu

William Singhose
Georgia Institute of Technology, Atlanta, GA USA
e-mail: Singhose@gatech.edu

T. Vyhlídal, J.-F. Lafay, and R. Sipahi (eds.), *Delay Systems*,
Advances in Delays and Dynamics 1,
DOI: 10.1007/978-3-319-01695-5_24,

Fig. 1 Bridge Crane at Logan Aluminum

overhead support point necessarily causes some oscillation of the payload. To accurately and safely position the payload, operators must account for both the response delays and the payload oscillation.

Fig. 2 Boom Crane Working on the Olympic Pool

Cranes are also used in places that are dangerous to human operators, such as nuclear facilities, areas of high temperature, and disaster sites. In these areas, it is desirable to remotely control the crane, distancing operators from the hazards. Remote operation, however, increases the difficulty of accurately and safely moving the crane payload. Communication delays increase the system lag, while the remote location limits the ability of the operator to survey the work area and plan a safe payload path. In addition, remote operation makes it more difficult for the operator to manually monitor and control payload oscillation. To allow the safe and efficient operation of remote cranes, both the remote operation and payload oscillation challenges must be addressed.

Remote operation, or teleoperation, has been an active area of research for some time. For a thorough review of teleoperation, the reader is pointed to three surveys by Sheridan and the references contained within [23–25]. Other surveys of the field are presented by Niemeyer [16] and Hokayem [6]. The recent research is a large advancement from the first studies of Ferrell, who showed that human operators would adopt a "move-and-wait" strategy in the presence of sufficiently large time delays and that teleoperation task performance is a function of the delay time [3]. The results from one study suggested that not only is task performance a function of the delay time, but the variance between repeated operations is as well (task completion time and variance increased as delay time increased) [1].

Teleoperation via the Internet has been a topic of much recent research [4, 13, 15, 17, 20–22, 29, 36, 37]. The Internet provides a cheap, readily available medium for teleoperation, but its packet-based nature presents additional challenges. For example, the delay time can change greatly and rapidly. Despite this fact, numerous systems have been successfully controlled via the Internet, including several cranes [7, 11, 12].

The work presented here utilizes a different approach to teleoperation than considered in most previous papers. The vast majority of teleoperation research has been conducted on systems where the communication path between the user and the remote system is part of the system's computerized feedback controller. A large portion of this research has addressed the stability issues of the communication time delay in the feedback loop. However, there are many systems for which bilateral operation does not suit the system, such as teleoperated cranes. The approach presented here seeks to improve teleoperated systems by reducing their dynamic complexity, namely reducing the vibration of the remote system. This approach avoids the stability concerns of the force-feedback methods prevalent in the literature.

The method used to reduce the dynamic complexity of the remotely operated system is input shaping. Input shaping limits system vibration by intelligently shaping the reference command [26, 27, 30]. The original, unshaped reference command is convolved with a series of impulses, called an input shaper. The resulting shaped command moves the system with little residual oscillation. The process is shown in Fig. 3. Input shaping has been successfully implemented on many vibratory systems including bridge [31, 32], tower [2, 12], and boom [18, 19] cranes, coordinate measurement machines [8, 28], robotic arms [14, 26], and de-mining robots [5].

In addition, studies have been conducted that indicate input shaping improves the performance of crane operators, both local [9, 10] and remote [33, 35]. The work presented here investigates the factors that influence operator performance on teleoperated cranes with input shaping.

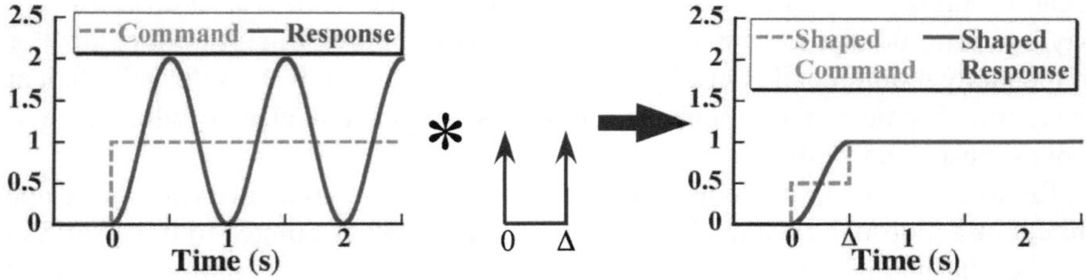

Fig. 3 The Input-Shaping Process

The next section will present a study of the influence of time delay on bridge crane operator performance, with and without input shaping. Then, in Sect. 3 several studies of human operators moving a tower crane payload through an obstacle course under local and remote operation are presented and their data compiled and analyzed. Finally, conclusions summarize the key results.

2 The Influence of Communication Delay on Bridge Crane Operators

This section presents a study of the influence of pure time delays on crane operator performance. The operators in this test controlled the crane directly, *i.e.* they were local to the crane. However, their commands were artificially delayed to analyze the influence of time delay on crane operator performance, while eliminating other effects of remote operation. As such, this study attempts to isolate the effects of the communication delay present in teleoperated systems, with and without input shaping.

2.1 Experimental Protocol

Four operators completed a manipulation task using the 10-ton industrial bridge crane shown in Fig. 4. The crane has a workspace that is 6 meters high, 5 meters wide and 42 meters long. It is controlled using a Siemens programmable logic controller (PLC), which receives operator commands from the control pendent. Commands from the PLC are sent to Schneider Electric motor drives, which ensure accurate execution of the commands. To measure the payload response, the crane is also equipped with a Siemens vision system.

The manipulation task that the users were asked to complete is shown schematically in Fig. 5. The task required the operator to move the crane payload from the 0.5m square start-zone around an obstacle to the 0.75m diameter circular end-zone 2.0m away. This motion is represented in the figure by "Nominal Path". The obstacle course was completed three times without input shaping, with time delays of 0s, 2s, and 4s. Each operator also completed the course ten times with input shaping enabled, using both ZV [26, 30] and EI-shaped [27] commands. Time delays of

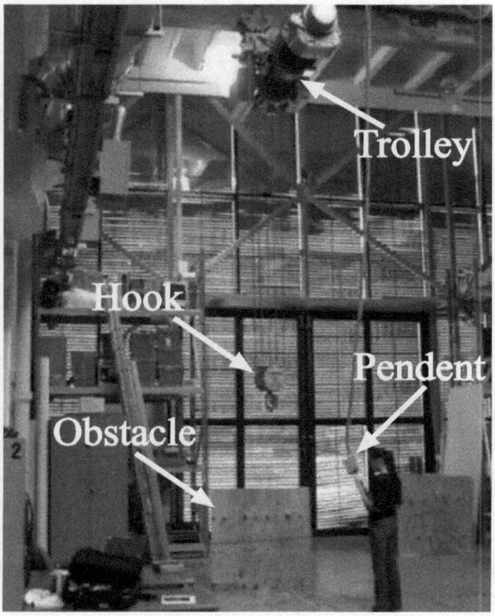

Fig. 4 10-ton Industrial Bridge Crane at Georgia Tech

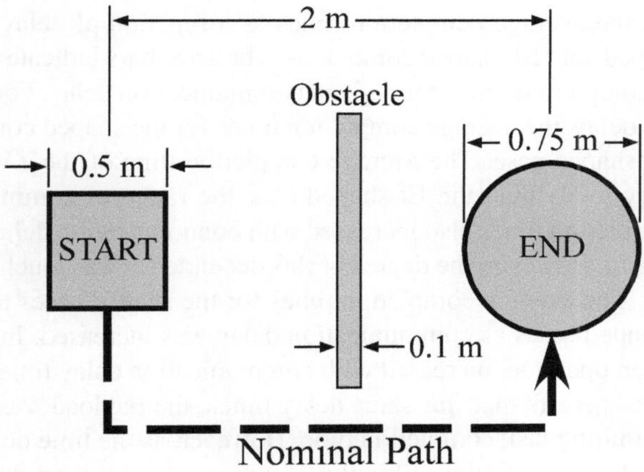

Fig. 5 The Manipulation Task for Bridge Crane Operator Study

0 – 4s, at 1s intervals, were used with both input shapers. For all commands, the time delays were enforced as pure time delays of the operator commands. The resulting control system is summarized in Fig. 6, where G_p represents the crane, *IS* represents the input shaper, and G_h is the human operator. Notice that the human operates as a feedback controller, modifying the commands based on the error

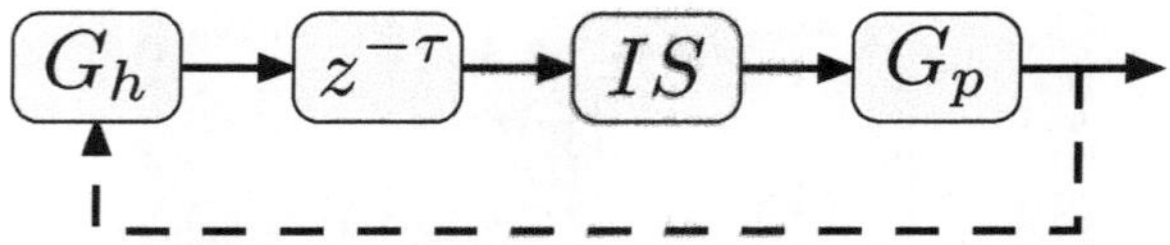

Fig. 6 Block Diagram of the Enforced Time Delay

between the desired crane location and its current position. The control inputs from the human operator are delayed by τ before being passed to the input shaper for modification and issue to the system.

For each test, the crane suspension cable length was set to 5.88m, creating a payload oscillation frequency of approximately 0.205Hz (and a period 4.88s). For each trial, the completion time, number of pendent button pushes, and obstacle collisions were recorded. The completion time was measured from the time of the first operator command to the time when the payload reached, and remained within, the circular end-zone. Button pushes are used as an indicator of operator effort and stress. An operator that pressed the buttons a larger number of times has made a larger number of control decisions, indicating a higher cognitive load than an operator who completed the task with fewer button pushes.

2.2 Experimental Results

Figure 7 shows the average completion times as a function of delay time for unshaped, ZV-shaped, and EI-shaped commands. The error bars indicate the minimum and maximum completion times for a given communication delay. For all values of communication delay, the average completion times for the shaped commands were less than the unshaped case. The average completion time of the ZV-shaped case was also slightly lower than the EI-shaped over the range of communication delays tested. Completion times also increased with communication delay for all three methods of control. However, the degree of this dependence was much lower for the unshaped cases. The average completion times for the shaped cases tended toward that of the unshaped cases as communication delay was increased. In addition, the variation between operators increased with communication delay time.

These results suggests that, for short delay times, the payload oscillation is the primary factor limiting task completion time. However, as the time delay increases, its influence on task completion difficulty increases. As the time delay increases further, it eventually becomes the dominant limiting factor.

These results also suggest that input shaping provides some benefit for a large range of communication time delays. Figure 8 overlays linear, least-squared curve-fits for each control method. The linear curve-fits provide good agreement with the data for all three control methods, with $R^2 \geq 0.95$ for all. Using these curve fits to predict completion times for longer communication delay times, it is not until communication delay time reaches nearly 8s that completion times with input shaping

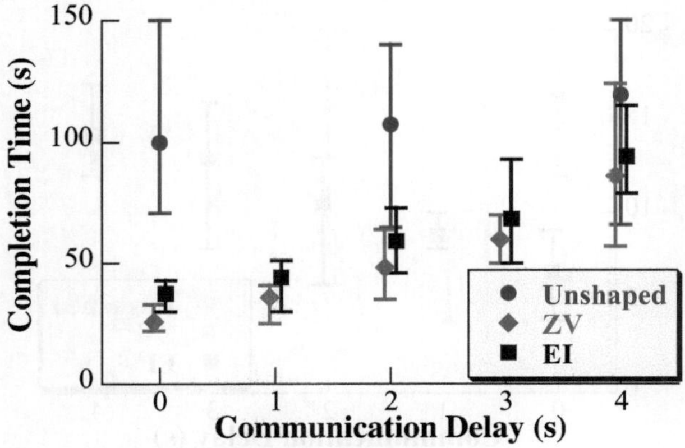

Fig. 7 Average Completion Time as a Function of Communication Delay

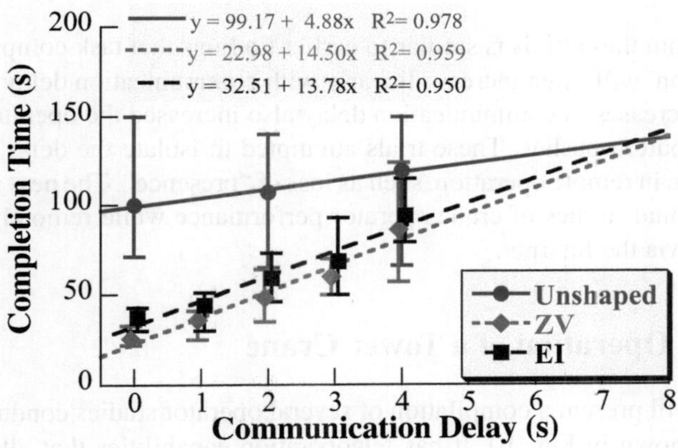

Fig. 8 Extrapolated Completion Time for Higher Communication Delays

enabled equal those of the unshaped case. At delay times this high, the time delay is the dominant factor determining task completion time.

The average number of button pushes needed to complete the maneuvers is shown in Fig. 9. The error bars indicate the minimum and maximum number of button pushes for a given communication delay. With short delay times, the number of button pushes needed with input shaping was dramatically less than without shaping. As delay time increased, however, the number of button pushes needed with input shaping approached that of the unshaped case. Over the range of parameters tested, there was a fairly linear relationship between the number of button pushes and communication delay time for the two shapers tested.

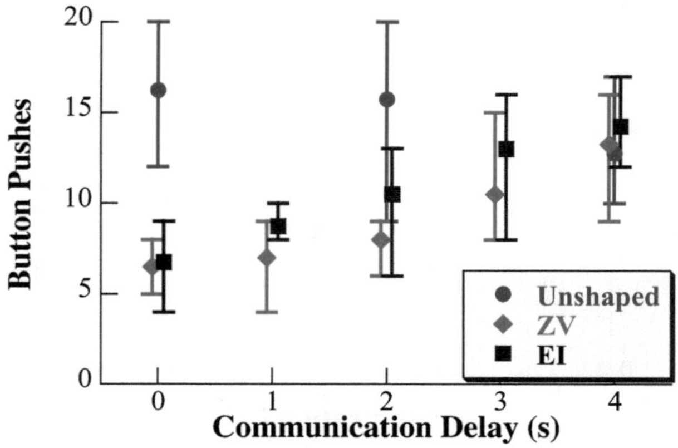

Fig. 9 Average Number of Button Pushes as a Function of Communication Delay

The data from these trials is similar to earlier findings that task completion time, and its variation, will often increase linearly with communication delay time [1, 3]. In addition, increases in communication delay also increased the operator effort, as measured by button pushes. These trials attempted to isolate the delay from other issues common in remote operation, such as loss of "presence". The next section will present additional studies of crane operator performance while remotely operating a tower crane via the Internet.

3 Remote Operation of a Tower Crane

This section will present a compilation of several operator studies conducted on the tower crane shown in Fig. 10. It has teleoperation capabilities that allow it to be operated in real-time from anywhere in the world via the Internet [12]. The crane is approximately 2m tall with a 1m jib arm. It is actuated by Siemens synchronous, AC servomotors. The jib is capable of 340° rotation about the tower. The trolley moves radially along the jib, and a hoisting motor controls the suspension cable length. A Siemens digital camera is mounted to the trolley and records the swing deflection of the payload [12]. Table 1 summarizes the performance characteristics of the tower crane.

This crane has been used in a number of crane operator studies [33–35]. Each of these studies contained completion of tasks both local and remote to the tower crane, often over a variety of controllers and crane payload configurations. The obstacle courses used in the studies are shown in Fig. 11.

For each of these studies, the operators were asked to navigate from a start location to some final position in the workspace. In [33] and [35], this move was a simple point-to-point move through the crane workspace. For the studies in [34], the operators moved through the workspace, retrieved a payload, then moved this payload to

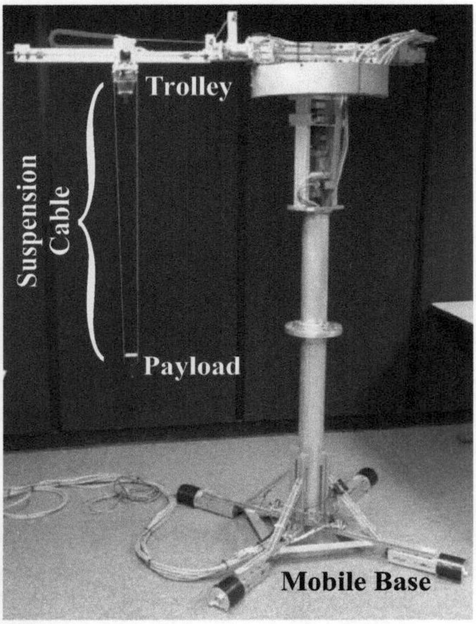

Fig. 10 A Remotely Operable Tower Crane

Table 1 Mobile Tower Crane Performance Limits

Parameter	Min	Max	Units
Cable Length	0.45	1.70	m
Slew Velocity	-0.35	0.35	rad/s
Slew Acceleration	-0.70	0.70	rad/s^2
Radial Velocity	-0.14	0.14	m/s
Radial Acceleration	-1.20	1.20	m/s^2

the final target location. As a result, the trials in this study are longer than the other two. For all three studies, the task completion time was measured, beginning with first operator command and ending when the payload reached and remained within the end-zone. Button pushes and obstacle collisions were also recorded during each trial. The payload configuration differed between the three studies.

These three studies represent 168 trials from 42 different operators over varying operating conditions. Despite the differences between them, each of the studies contained trials that were both local and remote from the tower crane. In addition, each study also examined the performance with and without input shaping enabled. As such, they provide a rich data set to examine the influence of remote operation on crane operator performance.

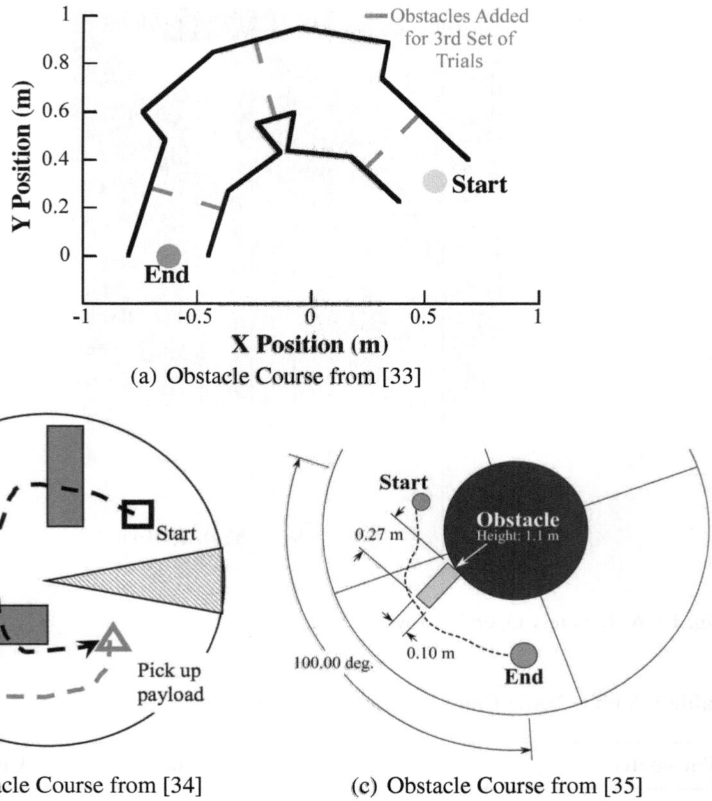

Fig. 11 Obstacle Courses for Crane Operator Studies

To enable comparison between the three operator studies, the task completion times were normalized by the average unshaped completion time when the crane was locally controlled for that study. Figure 12 shows the normalized completion times for all 46 operators. This figure shows few operators completed any trial faster without input shaping than with it (Using the symbols on the plot, there are very few triangles above squares for a given locale.).

The task completion times are summarized in Fig. 13, which shows the normalized average completion times. This plot highlights the local and remote unshaped cases and the local and remote input-shaped cases for each study. Each operator study examined at least two sub-tasks. These are displayed separately in the plot.

As seen in Fig. 13, for every operator study presented in [33–35], input shaping reduced the average completion time from the unshaped control case. This occurred both when the operators were local to the crane and when they were remote from it. The percentage improvement in average completion time provided by input shaping when the crane was controlled locally ranged from 13% to 94%. The smaller

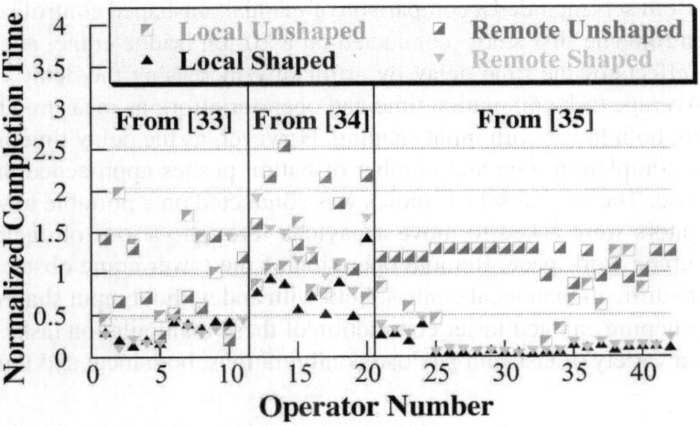

Fig. 12 Normalized Completion Times for All 46 Operators

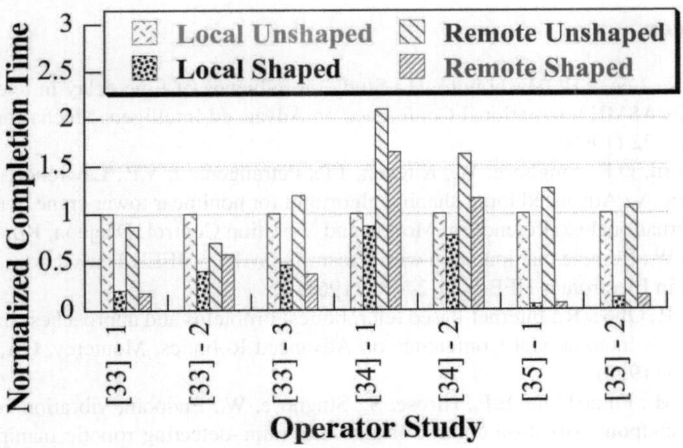

Fig. 13 Normalized Average Completion Times

improvements came in the longer-duration tasks examined in [34]. Across all the local-operation studies, input shaping reduced completion time by an average of 58%. The percentage reductions in completion time for the remote trials were remarkably similar, ranging from 17% to nearly 95%. Input shaping reduced completion time by an average of 57% across all the remote trials.

4 Conclusions

This work investigated the influence of time delay on human-operated cranes. Two sets of studies of crane operator performance in the presence of time delays were presented. Both sets included a comparison of manual, unshaped control, and input-shaping control. The first study, conducted on a 10-ton bridge crane, attempted to isolate the effects of the time delay by artificially increasing the delay of system response. Average task completion time and operator effort, as measured by button pushes, were both lower with input shaping. However, as the delay time increased, the average completion time and number of button pushes approached that of the unshaped case. The second set of studies was conducted on a portable tower crane, where operators were asked to move a payload through various obstacle courses within the crane workspace. Remote operation of the tower crane obstacle course proved more difficult than local control, both with and without input shaping. However, input shaping enabled faster completion of these manipulation tasks, on average, across a variety of task and payload configurations, both local and remote.

Acknowledgements. The authors would like to thank Siemens Energy and Automation and Boeing Research and Technology for their support of this work.

References

1. Ando, N., Lee, J.H., Hashimoto, H.: Study on influence of time delay in teleoperation. In: IEEE/ASME International Conference on Advanced Intelligent Mechatronics, AIM, pp. 317–322 (1999)
2. Blackburn, D.F., Singhose, W., Kitchen, J.P., Petrangenaru, V.P., Lawrence, J., Kamoi, T., Taura, A.: Advanced input shaping algorithm for nonlinear tower crane dynamics. In: 8th International Conference on Motion and Vibration Control, Daejeon, Korea (2006)
3. Ferrell, W.: Remote manipulation with transmission delay. IEEE Transactions on Human Factors in Electronics HFE-6(1), 24–32 (1965)
4. Fiorini, P., Oboe, R.: Internet-based telerobotics: Problems and approaches. In: Proceedings of the International Conference on Advanced Robotics, Monterey, CA, USA, pp. 765–770 (1997)
5. Freese, M., Fukushima, E.F., Hirose, S., Singhose, W.: Endpoint vibration control of a mobile endpoint vibration control of a mobile mine-detecting robotic manipulator. In: Proceedings of 2007 American Control Conference, New York, NY, United states, pp. 7–12 (2007)
6. Hokayem, P.F., Spong, M.W.: Bilateral teleoperation: An historical survey. Automatica 42(12), 2035–2057 (2006)
7. Huey, J., Fortier, J., Wolff, S., Singhose, W., Haraldsson, H.B., Sasaki, S.K., Watari, E.: Remote manipulation of cranes via the internet. In: Proceedings of International Conference on Motion and Vibration Control, Daejeon, Korea (2006)
8. Jones, S., Ulsoy, A.G.: An approach to control input shaping with application to coordinate measuring machines. J. of Dynamics, Measurement, and Control 121, 242–247 (1999)
9. Khalid, A., Huey, J., Singhose, W., Lawrence, J., Frakes, D.: Human operator performance testing using an input-shaped bridge crane. Journal of Dynamic Systems, Measurement and Control 128(4), 835–841 (2006)
10. Kim, D., Singhose, W.: Performance studies of human operators driving double-pendulum bridge cranes. Control Engineering Practice 18(6), 567–576 (2010)

11. Kim, J.: A TCP/IP-based remote control system for yard cranes in a port container terminal. Robotica 24, 613–620 (2006)
12. Lawrence, J., Singhose, W., Weiss, R., Erb, A., Glauser, U.: An internet-driven tower crane for dynamics and controls education. In: 7th IFAC Symposium on Advances in Control Education, Madrid, Spain (2006)
13. Lim, J., Ko, J., Lee, J.: Internet-based teleoperation of a mobile robot with force-reflection. In: Proceedings of 2003 IEEE Conference on Control Applications, Istanbul, Turkey, vol. 1, pp. 680–685 (2003)
14. Magee, D.P., Book, W.J.: Filtering micro-manipulator wrist commands to prevent flexible base motion. In: Proceedings of the American Controls Conference, Seattle, WA, vol. 2, pp. 474–479 (1995)
15. Munir, S., Book, W.J.: Control techniques and programming issues for time delayed internet based teleoperation. Journal of Dynamic Systems, Measurement and Control 125(2), 205–214 (2003)
16. Niemeyer, G., Slotine, J.J.E.: Telemanipulation with time delays. International Journal of Robotics Research 23(9), 873–890 (2004)
17. Oboe, R., Fiorini, P.: Design and control environment for internet-based telerobotics. International Journal of Robotics Research 17(4), 433–449 (1998)
18. Parker, G., Groom, K., Hurtado, J., Feddema, J., Robinett, R., Leban, F.: Experimental verification of a command shaping boom crane control system. In: American Control Conference, San Diego, CA, USA, vol. 1, pp. 86–90 (1999)
19. Parker, G.G., Groom, K., Hurtado, J., Robinett, R.D., Leban, F.: Command shaping boom crane control system with nonlinear inputs. In: Proceedings of IEEE Conference on Control Applications, Kohala Coast, HI, USA, vol. 2, pp. 1774–1778 (1999)
20. Rosch, O., Schilling, K., Roth, H.: Haptic interfaces for the remote control of mobile robots. Control Engineering Practice 10(11), 1309–1313 (2002)
21. Schilling, K., Roth, H.: Control interfaces for teleoperated mobile robots. In: Proceedings of 7th IEEE International Conference on Emerging Technologies and Factory Automation, Barcelona, Spain, vol. 2, pp. 1399–1403 (1999)
22. Schilling, K., Roth, H., Spilca, C.: A tele-experiment on rover motor control via internet. J. Robot. Syst. 22(3), 123–130 (2005)
23. Sheridan, T.: Telerobotics. Automatica 25(4), 487–507 (1989)
24. Sheridan, T.: Teleoperation, telerobotics and telepresence: a progress report. Control Engineering Practice 3(2), 205–214 (1995)
25. Sheridan, T.B.: Space teleoperation through time delay: Review and prognosis. IEEE Transactions on Robotics and Automation 9(5), 592–606 (1993)
26. Singer, N.C., Seering, W.P.: Preshaping command inputs to reduce system vibration. Journal of Dynamic Systems, Measurement, and Control 112, 76–82 (1990)
27. Singhose, W., Seering, W., Singer, N.: Residual vibration reduction using vector diagrams to generate shaped inputs. ASME J. of Mechanical Design 116, 654–659 (1994)
28. Singhose, W., Singer, N., Seering, W.: Improving repeatability of coordinate measuring machines with shaped command signals. Precision Engineering 18, 138–146 (1996)
29. Slawinski, E., Postigo, J.F., Mut, V.: Bilateral teleoperation through the internet. Robotics and Autonomous Systems 55(3), 205–215 (2007)
30. Smith, O.J.M.: Posicast control of damped oscillatory systems. Proceedings of the IRE 45, 1249–1255 (1957)
31. Sorensen, K., Singhose, W., Dickerson, S.: A controller enabling precise positioning and sway reduction in bridge and gantry cranes. Control Engineering Practice 15(7), 825–837 (2007)

32. Starr, G.P.: Swing-free transport of suspended objects with a path-controlled robot manipulator. Journal of Dynamic Systems, Measurement and Control 107, 97–100 (1985)
33. Vaughan, J., Kim, D., Singhose, W.: Control of tower cranes with double-pendulum payload dynamics. IEEE Transactions on Control Systems Technology 18(6), 1345–1358 (2010)
34. Vaughan, J., Peng, K.C.C., Singhose, W., Seering, W.: Influence of remote-operation time delay on crane operator performance. In: Proc. of 10th IFAC Workshop on Time Delay Systems. IFAC Papers Online, Boston, USA, vol. 10, Part I, pp. 85–90 (2012)
35. Vaughan, J., Smith, A., Kang, S.J., Singhose, W.: Predictive graphical user interface elements to improve crane operator performance. IEEE Transactions on Systems, Man and Cybernetics, Part A: Systems and Humans 41(2), 323–330 (2011)
36. Wang, M., Liu, J.N.: Interactive control for internet-based mobile robot teleoperation. Robotics and Autonomous Systems 52(2-3), 160–179 (2005)
37. Xue, X., Yang, S.X., Meng, M.Q.H.: Remote sensing and teleoperation of a mobile robot via the internet. In: Proceedings of 2005 International Conference on Information Acquisition, Hong Kong, China, pp. 537–542 (2005)

Decomposing the Dynamics of Delayed Hodgkin-Huxley Neurons

Gábor Orosz

Abstract. The effects of time delays on the nonlinear dynamics of neural networks are investigated. A decomposition method is utilized to derive modal equations that allow one to analyze the dynamics around synchronous states. The D-subdivision method is used to study the stability of equilibria while the stability of periodic orbits is investigated using Floquet theory. These methods are applied to a system of delay coupled Hodgkin-Huxley neurons to map out stable and unstable synchronous states. It is shown that for sufficiently strong coupling there exist delay ranges where stable equilibria coexist with stable oscillations which allow neural systems to respond to different environmental stimuli with different spatiotemporal patterns.

1 Introduction

Since Hodgkin and Huxley has constructed the first biophysical model of a neuron more than six decades ago [6] the field of neuroscience has gone through a enormous development. This led to detailed understanding of the dynamical phenomena underlying signal generation and propagation on neural membranes [4]. When modeling these processes, nonlinear ordinary differential equations (ODEs) are used to describe the voltage changes and ion transport at a given location of the membrane, while to describe the activity on the surface of the entire neuron partial differential equations (PDEs) are required. Additional ODEs can be used to describe the chemical processes at the synapses where signals are transmitted between neurons. Indeed, such detailed models are not feasible when modeling the behavior of populations of neurons. In this case, simplifications are often made so that neurons are considered to be "point-wise" and the couplings are considered to be instantaneous, that is, the infinite-dimensional dynamics of signal propagation is neglected. In this paper, we

Gábor Orosz
Department of Mechanical Engineering, University of Michigan,
Ann Arbor, MI 48109 USA
e-mail: orosz@umich.edu

T. Vyhlídal, J.-F. Lafay, and R. Sipahi (eds.), *Delay Systems*,
Advances in Delays and Dynamics 1,
DOI: 10.1007/978-3-319-01695-5_25,

consider an extended modeling framework for neural networks where neurons are still point-wise but signal propagation is modeled by inserting time delays into the coupling functions. This leads to delay differential equations (DDEs) which retain the essential infinite dimensional dynamics of signal transmission while the models remain scalable for large numbers of neurons and connections.

In order to understand the behavior of the resulting large interconnected delayed systems, we decompose the dynamics and derive modal equations in the vicinity of synchronous states. In particular, we focus on synchronous equilibria and periodic oscillations. The decomposition method used here can also be extended to more general cluster states [11] while other methods may be used when studying traveling wave solutions [2, 8]. Similar decomposition methods have been used to investigate the synchronized states in neural networks and laser networks [3,5] and to study the dynamics of communication protocols [1,9]. We remark that in the former case stability is usually determined numerically by calculating Lyapunov exponents while the latter case focuses on linear systems. In this paper, we apply rigorous mathematical techniques from dynamical systems theory to analyze the nonlinear dynamics of large interconnected systems.

As the result of the modal decomposition we obtain linear delay differential equations of small size. One of these modal equations describes the stability of synchronous states within the infinite-dimensional synchronization manifold which is called *tangential stability*. The other modal equations correspond to braking the synchrony, that is, they describe the so-called *transversal stability*; see [11, 12]. When considering the modal equations around equilibria, they have time-independent coefficients and consequently the D-subdivision method and Stépán's formulae [15] can be used to derive analytical stability charts. Each modal equation produces a set of stability curves and crossing these curves lead to different oscillatory solutions. For periodic oscillations the modal equations have time-periodic coefficients, that is, one needs to use Floquet theory [7] to evaluate stability. Tangential stability can be evaluated by restricting the dynamics to the synchronization manifold. For transversal stability, augmented systems are created so that each system consists of the nonlinear synchronous equation and a linear transversal modal equation.

In this paper, we apply these techniques to systems of delay coupled Hodgkin-Huxley neurons considering different connectivity structures. We derive the bifurcation structure arising within the synchronization manifold and show that this structure is independent of the connectivity and the number of oscillators in the system when the coupling strength is scaled appropriately. On the other hand, transversal bifurcations of equilibria and periodic orbits are influenced by the coupling structure. We demonstrate that for appropriate values of time delay and coupling strength, stable synchronous equilibrium coexist with stable oscillations. In this case, applying different external perturbations the system approaches different spatiotemporal patterns that can be exploited when encoding information.

2 Decomposition of Delayed Networks around Synchronous States

In this paper, we consider a system consisting of N identical oscillators coupled by identical couplings:

$$\dot{x}_i(t) = f\big(x_i(t)\big) + \frac{1}{N}\sum_{j=1}^{N} a_{ij}\, g\big(x_i(t), x_j(t-\tau)\big)\,, \tag{1}$$

for $i = 1,\ldots,N$, where the internal state of node i is described by the vector $x_i \in \mathbb{R}^n$ and the internal dynamics consist of a set of nonlinear ODEs $\dot{x}_i = f(x_i)$. The couplings are described by the function $g(x_i, x_j)$ that depends on the states of the interacting nodes. The time delay τ is the time needed for the signal transmission processes to take places. The coupling structure of the system is captured by a directed graph, whose elements are represented by the coefficients of the N-dimensional adjacency matrix

$$a_{ij} = \begin{cases} 1 & \text{if node } j \text{ is connected to node } i, \\ 0 & \text{otherwise,} \end{cases} \tag{2}$$

for $i, j = 1,\ldots,N$. Referring to the graph representation of the network, the oscillators are often called nodes and the connections between them called edges. Here, we use the abbreviated notation $A_N = [a_{ij}]$ and assume that A_N is diagonalizable, that is, if an eigenvalue has algebraic multiplicity m then it also has geometric multiplicity m, resulting in m linearly independent eigenvectors. The methods presented below may still be used when this condition does not hold but the algebraic calculations become more involved. We remark that equation (1) requires an infinite dimensional state space and the initial conditions are functions on the time interval $[-\tau, 0]$.

In this paper, we focus on the synchronous state

$$x_i(t) = x_s(t)\,, \quad i = 1,\ldots,N\,. \tag{3}$$

Substituting (3) into (1) results in the delay differential equation

$$\dot{x}_s(t) = f\big(x_s(t)\big) + \frac{M}{N}\, g\big(x_s(t), x_s(t-\tau)\big)\,, \tag{4}$$

where the row sum

$$M = \sum_{j=1}^{N} a_{ij}\,, \tag{5}$$

must be the same for every i to ensure the existence of synchronous solutions. We remark that equation (3) still requires an infinite dimensional state space, that is, the *synchronization manifold* defined by (3) is infinite dimensional. Equation (4) may produce a variety of different behaviors, e.g., equilibria, periodic orbits, and even chaotic motion. Here we focus on the first two cases. Synchronized equilibria are defined by

$$x_s(t) \equiv x_s^*, \tag{6}$$

and substituting this into (4) results in

$$0 = f(x_s^*) + \frac{M}{N} g(x_s^*, x_s^*), \tag{7}$$

that is, the delay does not influence the location of equilibria (but may influence their stability). On the other hand, synchronous periodic oscillations satisfy

$$x_s^p(t) = x_s^p(t + T_p), \tag{8}$$

where T_p represents the period. These can be determined by solving the boundary value problem comprised of (4) and (8) and the shape and stability of these orbits are influenced by the delay.

We define the perturbations $y_i = x_i - x_s$ for $i = 1, \ldots, N$, so the linearization of (1) about the synchronous solution (3) can be written as

$$\dot{y}_i(t) = L y_i(t) + R \sum_{j=1}^{N} a_{ij} y_j(t - \tau), \tag{9}$$

for $i = 1, \ldots, N$. When linearizing about the synchronous equilibrium (6), the $n \times n$ matrices L, R are time-independent, that is,

$$L^* = Df(x_s^*) + \frac{M}{N} D^{(1)} g(x_s^*, x_s^*), \qquad R^* = \frac{1}{N} D^{(2)} g(x_s^*, x_s^*), \tag{10}$$

where and $D^{(1)}$ and $D^{(2)}$ represent derivatives with respect to the first and second variables, respectively. In this case, (9) gives a linear time-invariant system allowing the use of analytical techniques like the D-subdivision method and Stépán's formulae [15] to determine the stability of the equilibrium. However, when linearizing about synchronous oscillations (8), the matrixes L, R in (9) become time-periodic with period T_p, that is,

$$\begin{aligned} L(t) &= Df\big(x_s^p(t)\big) + \frac{M}{N} D^{(1)} g\big(x_s^p(t), x_s^p(t - \tau)\big) = L(t + T_p), \\ R(t) &= \frac{1}{N} D^{(2)} g\big(x_s^p(t), x_s^p(t - \tau)\big) = R(t + T_p). \end{aligned} \tag{11}$$

Thus, one must use Floquet theory to evaluate the stability of oscillations. The corresponding monodromy operators usually cannot be written in closed form and consequently, numerical techniques like full discretization [14] or semi-discretization [7] are needed.

Using $\mathbf{y} = \mathrm{col}\,[y_1 \;\; y_2 \;\; \ldots \;\; y_N] \in \mathbb{R}^{nN}$ the linear system (9) can be rewritten as

$$\dot{\mathbf{y}}(t) = (I_N \otimes L)\,\mathbf{y}(t) + (A_N \otimes R)\,\mathbf{y}(t - \tau), \tag{12}$$

where I_N is the N-dimensional unit matrix and A_N is the adjacency matrix. In order to decompose system (12) we construct the coordinate transformation

$$\mathbf{y} = (T_N \otimes I)\mathbf{z}, \tag{13}$$

where $\mathbf{z} = \mathrm{col}[z_1 \; z_2 \; \dots \; z_N] \in \mathbb{R}^{nN}$, I is the $n \times n$ unit matrix, while $T_N = [e_1 \; e_2 \; \dots \; e_N]$ where e_i is the i-th eigenvector of the adjacency matrix A_N. This transformation yields the linear modal equations

$$\dot{z}_i(t) = L z_i(t) + \Lambda_i R z_i(t-\tau), \tag{14}$$

for $i = 1, \dots, N$, where Λ_i is the i-th eigenvalue of the adjacency matrix A_N. Note the due to the constant row sum (5), we have $\Lambda_1 = M$ and $e_1 = \mathrm{col}[1 \; \dots \; 1]$. The corresponding modal equation is indeed the linearization of (4) and it describes the *tangential stability*: stability against perturbations that keep the synchronous configuration. The other modal equations for $i = 2, \dots N$ describe *transversal stability*: stability against perturbations that split the synchronous configuration; see [11, 12]. We remark that for $\Lambda_{i,i+1} = \Sigma_i \pm \mathrm{i}\Omega_i \in \mathbb{C}$, defining $\xi_i = \mathrm{Re}\, z_i, \eta_i = -\mathrm{Im}\, z_i$ and taking the real and imaginary parts of (14) leads to the $2n$-dimensional real system

$$\begin{bmatrix} \dot{\xi}_i(t) \\ \dot{\eta}_i(t) \end{bmatrix} = \begin{bmatrix} L & O \\ O & L \end{bmatrix} \begin{bmatrix} \xi_i(t) \\ \eta_i(t) \end{bmatrix} + \begin{bmatrix} \Sigma_i R & \Omega_i R \\ -\Omega_i R & \Sigma_i R \end{bmatrix} \begin{bmatrix} \xi_i(t-\tau) \\ \eta_i(t-\tau) \end{bmatrix}, \tag{15}$$

where O represents the n-dimensional matrix with zero elements.

In this paper, we consider $N = 5$. For all-to-all coupling (without self coupling) the adjacency matrix, its eigenvalues, and the transformation matrix (given by the eigenvectors) can be written as

$$A_5 = \begin{bmatrix} 0&1&1&1&1 \\ 1&0&1&1&1 \\ 1&1&0&1&1 \\ 1&1&1&0&1 \\ 1&1&1&1&0 \end{bmatrix} \Rightarrow \begin{cases} \Lambda_1 = 4, \\ \Lambda_2 = -1, \\ \Lambda_3 = -1, \\ \Lambda_4 = -1, \\ \Lambda_5 = -1, \end{cases} T_N = \begin{bmatrix} 1&1&1&1&1 \\ 1&-1&0&0&0 \\ 1&0&-1&0&0 \\ 1&0&0&-1&0 \\ 1&0&0&0&-1 \end{bmatrix}. \tag{16}$$

Here the row sum is $M = N - 1 = 4$ (cf. (5)) and the transversal eigenvalue has multiplicity M. We will also consider the adjacency matrix

$$A_5 = \begin{bmatrix} 0&1&0&1&1 \\ 0&0&1&1&1 \\ 1&1&0&0&1 \\ 1&0&1&0&1 \\ 1&1&0&1&0 \end{bmatrix} \Rightarrow \begin{cases} \Lambda_1 = 3, \\ \Lambda_2 = 0, \\ \Lambda_3 = -1, \\ \Lambda_4 = -1+\mathrm{i}, \\ \Lambda_5 = -1-\mathrm{i}, \end{cases} T_N = \begin{bmatrix} 1&1&1&1&1 \\ 1&-2&1&\frac{-2+6\mathrm{i}}{5}&\frac{-2-6\mathrm{i}}{5} \\ 1&-2&1&\frac{-1-7\mathrm{i}}{5}&\frac{-1+7\mathrm{i}}{5} \\ 1&1&1&\frac{-8-\mathrm{i}}{5}&\frac{-8+\mathrm{i}}{5} \\ 1&1&-3&1&1 \end{bmatrix}, \tag{17}$$

where $M = 3$ and all eigenvalues are distinct.

2.1 Stability of Synchronous Equilibria and Periodic Orbits

As mentioned above, when linearizing about synchronous equilibria (6), the matrices L, R are time independent, cf. (10). Thus, in order to determine stability, the trial solutions $z_i(t) = Z_i \mathrm{e}^{\lambda t}$, $\lambda \in \mathbb{C}$, $Z_i \in \mathbb{C}^n$ are substituted into (14), which result in the characteristic equations

$$\det\left(\lambda I - L - \Lambda_i R \mathrm{e}^{-\lambda \tau}\right) = 0, \tag{18}$$

for $i = 1, \ldots, N$. When all the infinitely many characteristic roots λ are located in the left-half complex plane for $i = 1$ and $i = 2, \ldots N$, the equilibrium is tangentially and transversally stable, respectively. Substituting $\lambda = \mathrm{i}\omega$, $\omega \geq 0$ into the above equation one may obtain the tangential and transversal stability boundaries that divide the parameter space into stable and unstable domains. For each domain, stability can be evaluated by applying Stépán's formulae [15]. When crossing a tangential stability boundary the synchronized configuration is kept by the arising oscillations while crossing a transversal boundary gives rise to asynchronous oscillatory solutions.

When linearizing about the synchronized oscillations (8), the matrices L, R are time-periodic with period T_p, cf. (11). Instead of exponential trial solutions one must use Floquet theory to determine stability [7]. This requires the reformulation (14) using the state variables $z_{i,t}(\theta) = z_i(t+\theta)$, $\theta \in [-\tau, 0]$ that are contained by the infinite-dimensional space of continuous functions. These states can be obtained from the initial functions as

$$z_{i,t} = \mathscr{U}_i(t)\, z_{i,0}\,, \tag{19}$$

using the solutions operators $\mathscr{U}_i(t)$ for $i = 1, \ldots, N$. The eigenvalues of the monodromy operators $\mathscr{U}_i(T_\mathrm{p})$, called Floquet multipliers, determine the tangential and transversal stability of oscillations. If all these multipliers are smaller than 1 in magnitude, then the periodic solution is stable. As the monodromy operators cannot be written into closed form, one needs to use numerical techniques to determine the stability boundaries. First, we compute the periodic orbit which is the solution of the boundary value problem (4,8) using numerical collocation. Then using arc-length continuation we find the orbit when parameters are varied; see [14] for details. For $i = 1$, (14) is the linearization of (4) and collocations provide a discretization of the monodromy operator $\mathscr{U}_1(T_\mathrm{p})$ in (19) which allows the computation of the tangential Floquet multipliers. However, for $i = 2, \ldots N$, to obtain the transversal Floquet multipliers (i.e., the eigenvalues of $\mathscr{U}_i(T_\mathrm{p})$ for $i = 2, \ldots N$), the matrices in (14) or (15) have to be evaluated at the periodic solution. Thus, we create augmented systems consisting of (4,8) and a chosen equation of (14) or (15). The corresponding $2n$ or $3n$ dimensional equations possess a periodic orbit: the first n variables are equal to x_s^p while $z_i \equiv 0$ or $\xi_i = \eta_i \equiv 0$.

3 Synchrony of Delay Coupled Hodgkin-Huxley Neurons

Since the original work of Hodgkin and Huxley [6] a large number different models have been proposed to describe voltage activity and ion transport at the neural membrane (e.g., FitzHugh-Nagumo model, Morris-Lecar model); see [4]. As a matter of fact, these all originate form the Hodgkin-Huxley model which is still considered to be an etalon in neuro-dynamics. Here we consider Hodgkin-Huxley neurons coupled via direct electronic coupling called gap junctions; see [13]. (For the same model with synaptic coupling see [8].)

The time evolution of the system is given by the delay differential equations

$$\begin{aligned}
C\dot{V}_i(t) &= I - g_{\mathrm{Na}} m_i^3(t) h_i(t) \big(V_i(t) - V_{\mathrm{Na}}\big) - g_{\mathrm{K}} n_i^4(t) \big(V_i(t) - V_{\mathrm{K}}\big) \\
&\quad - g_{\mathrm{L}} \big(V_i(t) - V_{\mathrm{L}}\big) + \frac{\kappa}{N} \sum_{j=1}^{N} a_{ij} \big(V_j(t-\tau) - V_i(t)\big), \\
\dot{m}_i(t) &= \alpha_m\big(V_i(t)\big)\big(1 - m_i(t)\big) - \beta_m\big(V_i(t)\big) m_i(t), \\
\dot{h}_i(t) &= \alpha_h\big(V_i(t)\big)\big(1 - h_i(t)\big) - \beta_h\big(V_i(t)\big) h_i(t), \\
\dot{n}_i(t) &= \alpha_n\big(V_i(t)\big)\big(1 - n_i(t)\big) - \beta_n\big(V_i(t)\big) n_i(t),
\end{aligned} \tag{20}$$

for $i = 1,\ldots,N$, where the time t is measured in ms, the voltage of the i-th neuron at the soma V_i is measured in mV, and the dimensionless gating variables $m_i, h_i, n_i \in [0,1]$ characterize the "openness" of the sodium and potassium ion channels embedded in the cell membrane. The conductances g_{Na}, g_{K}, g_{L} and the reference voltages V_{Na}, V_{K}, V_{L} for the sodium channels, potassium channels and the so-called leakage current are given together with the membrane capacitance C and the driving current I in the appendix of [13]. The equations for m_i, h_i, n_i are based on measurements and the nonlinear functions $\alpha_m(V)$, $\alpha_h(V)$, $\alpha_n(V)$, $\beta_m(V)$, $\beta_h(V)$, $\beta_n(V)$ are also given in the appendix of [13]. The coupling term $\frac{\kappa}{N}\big(V_j(t-\tau) - V_i(t)\big)$ represents a direct electronic connection between the axon of the j-th neuron and the dendrites of the i-th neuron. Here $V_i(t)$ is the postsynaptic potential, $V_j(t-\tau)$ is the presynaptic potential, κ is the conductance of the gap junction, and τ represents the signal propagation time along the axon of the j-th neuron (dendritic delays are omitted here). That is, the presynaptic potential is equal to what the potential of the soma of the j-th neuron was τ time before.

For $\kappa = 0$ the neurons are uncoupled. In this case, there exist a unique stable oscillatory state where neurons spike periodically (with period $T_{\mathrm{p}} \approx 11.57$ ms), see the green curves in Fig. 1. The equation that describes the dynamics on the infinite dimensional synchronization manifold can be obtained by substituting

$$[V_i \;\; m_i \;\; h_i \;\; n_i] = [V_{\mathrm{s}} \;\; m_{\mathrm{s}} \;\; h_{\mathrm{s}} \;\; n_{\mathrm{s}}], \qquad i = 1,\ldots,N, \tag{21}$$

into (20); cf. (3,4). In this case, the coupling term becomes $\kappa\frac{M}{N}\big(V_{\mathrm{s}}(t-\tau) - V_{\mathrm{s}}(t)\big)$. Notice that this term disappears for $\tau = 0$, that is, the synchronized motion is the same as the uncoupled one. However, this does not hold for $\tau > 0$.

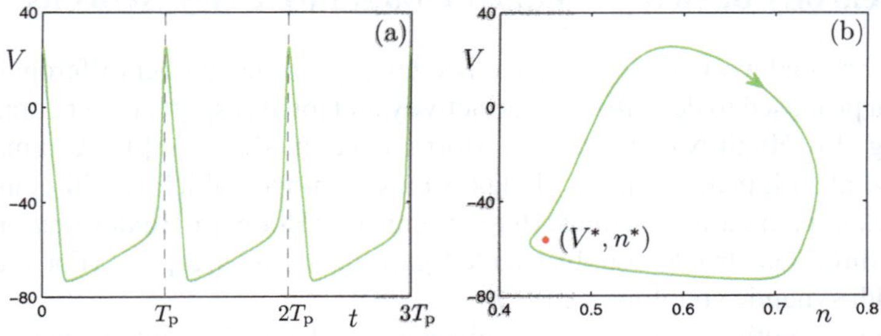

Fig. 1 Stable periodic solution of model (20) without coupling $\kappa = 0$. Panel (a) shows the periodic variation of the membrane voltage V as a function of time t (with period $T_p \approx 11.57$ ms) while panel (b) depicts the periodic orbit in state space. The red dot represents the unstable equilibrium.

When linearizing (20) about (21) one obtains the matrixes

$$L = \begin{bmatrix} -p - \frac{M}{N}\frac{\kappa}{C} & -a_1 & -a_2 & -a_3 \\ -b_1 & -c_1 & 0 & 0 \\ -b_2 & 0 & -c_2 & 0 \\ -b_3 & 0 & 0 & -c_3 \end{bmatrix}, \quad R = \begin{bmatrix} \frac{1}{N}\frac{\kappa}{C} & 0 & 0 & 0 \\ 0 & 0 & 0 & 0 \\ 0 & 0 & 0 & 0 \\ 0 & 0 & 0 & 0 \end{bmatrix}, \tag{22}$$

where

$$\begin{aligned} &p = (g_{\mathrm{Na}} m_{\mathrm{s}}^3 h_{\mathrm{s}} + g_{\mathrm{K}} n_{\mathrm{s}}^4 + g_{\mathrm{L}})/C, \\ &a_1 = g_{\mathrm{Na}} 3 m_{\mathrm{s}}^2 h_{\mathrm{s}} (V_{\mathrm{s}} - V_{\mathrm{Na}})/C, \qquad b_1 = -\alpha_m'(V_{\mathrm{s}})(1 - m_{\mathrm{s}}) + \beta_m'(V_{\mathrm{s}}) m_{\mathrm{s}}, \\ &a_2 = g_{\mathrm{Na}} m_{\mathrm{s}}^3 (V_{\mathrm{s}} - V_{\mathrm{Na}})/C, \qquad b_2 = -\alpha_h'(V_{\mathrm{s}})(1 - h_{\mathrm{s}}) + \beta_h'(V_{\mathrm{s}}) h_{\mathrm{s}}, \\ &a_2 = g_{\mathrm{K}} 4 n_{\mathrm{s}}^3 (V_{\mathrm{s}} - V_{\mathrm{K}})/C, \qquad b_3 = -\alpha_n'(V_{\mathrm{s}})(1 - n_{\mathrm{s}}) + \beta_n'(V_{\mathrm{s}}) n_{\mathrm{s}}, \\ &c_1 = \alpha_m(V_{\mathrm{s}}) + \beta_m(V_{\mathrm{s}}), \quad c_2 = \alpha_h(V_{\mathrm{s}}) + \beta_h(V_{\mathrm{s}}), \quad c_3 = \alpha_n(V_{\mathrm{s}}) + \beta_n(V_{\mathrm{s}}), \end{aligned} \tag{23}$$

that appear in the linear equation (9,12) as well as in the modal equations (14,15).

3.1 Stability of Synchronous Equilibria

Let us consider synchronized equilibria, that is, $[V_{\mathrm{s}}(t) \;\; m_{\mathrm{s}}(t) \;\; h_{\mathrm{s}}(t) \;\; n_{\mathrm{s}}(t)] \equiv [V_{\mathrm{s}}^* \;\; m_{\mathrm{s}}^* \;\; h_{\mathrm{s}}^* \;\; n_{\mathrm{s}}^*]$. At this state the coupling term disappears and consequently the synchronized equilibrium is the same as the equilibrium of an uncoupled neuron. For parameters defined in [13] we have a unique equilibrium as shown by the red dot in Fig. 1(b). Moreover, the matrices (22) become constant (cf. (10)) and the characteristic equation (18) leads to

$$\lambda^4 + d_1\lambda^3 + d_2\lambda^2 + d_3\lambda + d_4 + \tfrac{1}{N}\tfrac{\kappa}{C}\big(M - \Lambda_i \mathrm{e}^{\lambda\tau}\big)\big(\lambda^3 + \tilde{c}_1\lambda^2 + \tilde{c}_2\lambda + \tilde{c}_3\big) = 0, \tag{24}$$

where

$$
\begin{aligned}
&\tilde{c}_1 = c_1 + c_2 + c_3\,, \qquad \tilde{c}_2 = c_1c_2 + c_1c_3 + c_2c_3\,, \qquad \tilde{c}_3 = c_1c_2c_3\,,\\
&d_1 = p + \tilde{c}_1\,,\\
&d_2 = p\tilde{c}_1 + \tilde{c}_2 - \big(a_1b_1 + a_2b_2 + a_3b_3\big)\,,\\
&d_3 = p\tilde{c}_2 + \tilde{c}_3 - \big(a_1b_1(c_2+c_3) + a_2b_2(c_1+c_3) + a_3b_3(c_1+c_2)\big)\,,\\
&d_4 = p\tilde{c}_3 - \big(a_1b_1c_2c_3 + a_2b_2c_1c_3 + a_3b_3c_1c_2\big)\,,
\end{aligned}
\tag{25}
$$

that are evaluated at $V_s(t) \equiv V_s^*$. Substituting $\lambda = \mathrm{i}\omega$ into (24), separating the real and imaginary parts, and using some algebraic manipulations one may obtain the stability boundaries in the (τ,κ)-plane parameterized by the angular frequency ω. In particular, considering $\Lambda_1 = M$ results in the tangential boundaries

$$
\begin{aligned}
\tau &= \frac{2}{\omega}\left\{ \arctan\left[-\frac{\alpha(\omega)}{\beta(\omega)} \right] + \ell\pi \right\}, \qquad \ell = 0,1,2,\ldots\\
\kappa &= -\frac{CN}{2M}\,\frac{\alpha^2(\omega)+\beta^2(\omega)}{\alpha(\omega)\gamma(\omega)}\,,
\end{aligned}
\tag{26}
$$

where

$$
\begin{aligned}
\alpha(\omega) &= (d_1 - \tilde{c}_1)\omega^6 + (d_2\tilde{c}_1 + \tilde{c}_3 - d_3 - d_1\tilde{c}_2)\omega^4 + (d_3\tilde{c}_2 - d_2\tilde{c}_3 - d_4\tilde{c}_1) + d_4\tilde{c}_3\,,\\
\beta(\omega) &= -\omega^7 + (d_2 + \tilde{c}_2 - d_1\tilde{c}_1)\omega^5 + (d_1\tilde{c}_3 + d_3\tilde{c}_1 - d_4 - d_2\tilde{c}_2) + (d_4\tilde{c}_2 - d_3\tilde{c}_3)\omega\,,\\
\gamma(\omega) &= \omega^6 + (\tilde{c}_1^2 - 2\tilde{c}_2)\omega^4 + (\tilde{c}_2^2 - 2\tilde{c}_1\tilde{c}_3)\omega^2 + \tilde{c}_3^2\,.
\end{aligned}
\tag{27}
$$

Similarly for $\Lambda_i \in \mathbb{R}$ one may obtain the transversal boundaries

$$
\begin{aligned}
\tau &= \frac{2}{\omega}\left\{ \arctan\left[\frac{1}{M+\Lambda_i}\,\frac{\alpha(\omega)}{\beta(\omega)}\Big(-\Lambda_i \pm \sqrt{\Delta}\Big) \right] + \ell\pi \right\}, \qquad \ell = 0,1,2,\ldots\\
\kappa &= \frac{CN}{M^2-\Lambda_i^2}\,\frac{\alpha(\omega)}{\gamma(\omega)}\Big(-M \mp \sqrt{\Delta}\Big)\,,
\end{aligned}
\tag{28}
$$

where

$$
\Delta = \Lambda_i^2 - \big(M^2 - \Lambda_i^2\big)\frac{\beta^2(\omega)}{\alpha^2(\omega)}\,. \tag{29}
$$

For $\Lambda_i = \Sigma_i + \mathrm{i}\Omega_i \in \mathbb{C}$ the first equation in (28) changes to

$$
\tau = \frac{2}{\omega}\left\{ \arctan\left[\frac{1}{M+\Sigma_i+\Omega_i\frac{\alpha(\omega)}{\beta(\omega)}}\,\frac{\alpha(\omega)}{\beta(\omega)}\Big(-\Sigma_i + \Omega_i\frac{\beta(\omega)}{\alpha(\omega)} \pm \sqrt{\Delta}\Big) \right] + \ell\pi \right\}, \tag{30}
$$

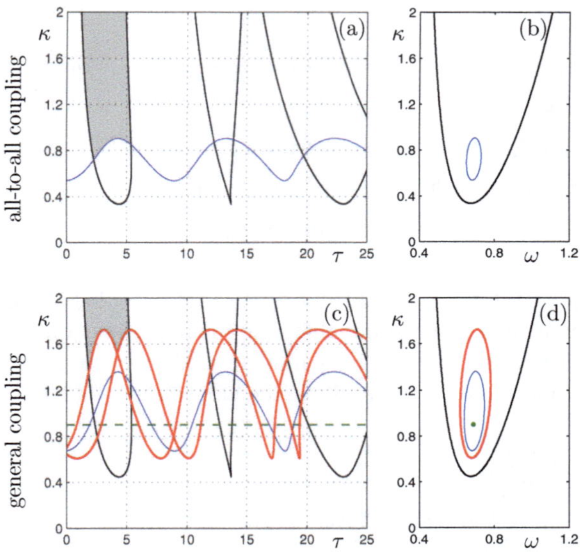

Fig. 2 Stability charts corresponding to the formulae (26,28,30,31) are shown in panels (a,c) and the corresponding angular frequencies are shown in panels (b,d). The top and bottom rows correspond the adjacency matrices (16) and (17), respectively. The stable domain is shaded, black "lobe shaped" curves represent tangential boundaries for $\Lambda_1 = M$, and colored "wavy" curves represent transversal boundaries for $\Lambda_2 = -1$ — thin blue, $\Lambda_3 = 0$ — dashed green, and $\Lambda_{4,5} = -1 \pm \mathrm{i}$ — thick red.

and $\Lambda_i^2 = \Sigma_i^2 + \Omega_i^2$. Finally, we remark that for $\Lambda_i = 0$ the boundary is given by

$$\kappa = -\frac{CN}{M}\frac{\alpha(\omega^*)}{\gamma(\omega^*)}, \tag{31}$$

where ω^* is the solution of $\beta(\omega^*) = 0$, that is, this boundary is delay independent.

The corresponding curves are plotted for $N = 5$ in the (τ, κ)-plane in Fig. 2(a) and (c) for the coupling matrixes (16) and (17), respectively. The tangential stability boundaries are shown as black curves and these form lobes. One may observe that in (26) τ is independent of the number of oscillators N and the row sum M while κ is proportional to N/M. Corresponding to this the lobes in Fig. 2(c) are the "stretched" versions of the lobes in Fig. 2(a). The transversal boundaries are shown as colored curves and for each transversal eigenvalue Λ_i the boundary appears as a "wavy" curve. For all-to-all coupling there is only one transversal curve corresponding to the multiplicity of the modal eigenvalues in (16). For general coupling (17) there are four distinct curves: the horizontal dashed green line corresponds to the zero modal eigenvalue, the thin blue curve corresponds to the real modal eigenvalue while the thick red curves correspond to the complex conjugate pair of modal eigenvalues. Notice that the larger the magnitude of the transverse modal eigenvalue is, the larger the "amplitude" of the "wavy" curve is.

The stability of the synchronized equilibrium changes via Andronov-Hopf bifurcation when crossing either a tangential or a transversal stability curve. That is a pair of complex conjugate eigenvalues crosses the imaginary axis leading to oscillations. The corresponding frequencies are shown in Fig. 2(b) and (d). One may observe that along the tangential lobes the frequency changes with κ in the interval $\omega \in [0,\infty)$. On the other hand, frequencies along tangential boundaries are contained in a closed interval that increases with the magnitude of the transversal eigenvalue. Applying the analytical stability criteria [15], it can be shown that the system is tangentially stable within in the leftmost lobe while transversal stability of each mode can be guaranteed when choosing parameters above the corresponding "wavy" curve. Thus the equilibrium is linearly stable in the shaded domain that appears to be larger for the all-to-all coupled network.

3.2 Stability of Synchronous Periodic Orbits

As shown above, the modal equations (14,15) with matrices (22) allow one to determine the stability of equilibria in a systematic way. However, neural systems encode information using rhythmic patterns which correspond to periodic oscillations. To derive synchronous oscillations, one must solve (20,21) when considering $[V_s(t)\ m_s(t)\ h_s(t)\ n_s(t)] = [V_s(t+T_p)\ m_s(t+T_p)\ h_s(t+T_p)\ n_s(t+T_p)]$, where T_p is the period of oscillations. We use the numerical continuation package DDE-BIFTOOL [14] to follow branches of periodic solutions when varying parameters. To evaluate the stability of these solutions we use the nonlinear equation describing the dynamics on the synchronization manifold (i.e., (20) with restriction (21)) and augment this with a modal equation from (14) or (15).

The left and right columns of Fig. 3 show the bifurcation diagrams for the adjacency matrices (16) and (17), respectively. Each panel depicts the peak-to-peak voltage amplitude $|V_s|$ as a function of the time delay τ while the value of the coupling strength κ is indicated on each panel. The horizontal axis represents the synchronized equilibrium. Solid green and dashed red curves represent stable and unstable states, respectively. Bifurcations are marked as stars (Hopf and Neimark-Sacker), crosses (fold and pitchfork), and diamonds (period doubling). The color of the marker indicates which mode becomes unstable: black symbols indicate tangential stability losses while green, blue and red symbols corresponds to zero, real and complex conjugate modal eigenvalues, cf. Fig. 2. For simplicity we only mark the bifurcations where the stability of a mode changes.

Notice that the structure of the bifurcation diagrams and the tangential stability losses are the same for all-to-all and general coupling when rescaling the κ values by $4/3$; cf. (4). For weak coupling the equilibrium is tangentially unstable while the periodic orbit is tangentially stable for all values of τ; see panels (a) and (f). For stronger coupling the equilibrium may loose its tangential stability via Hopf bifurcations (black stars) corresponding to the lobes in Fig. 2 and the arising oscillations stay within the synchronization manifold; see panels (c–e) and (h–j). Oscillatory solutions may undergo tangential fold bifurcations (black crosses) leading to

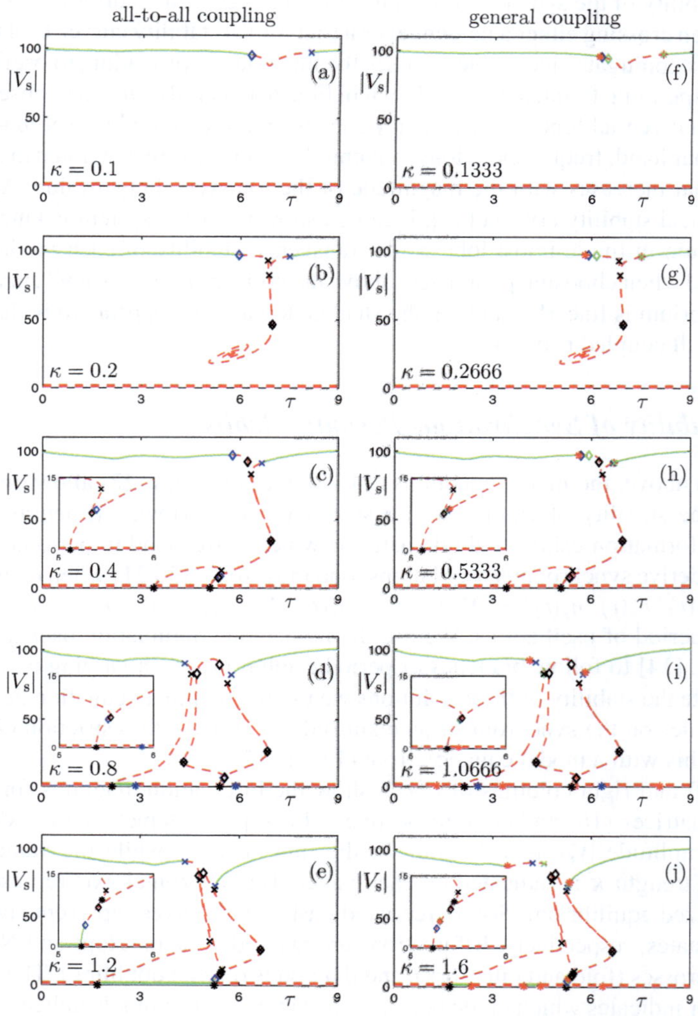

Fig. 3 Bifurcation diagrams showing the peak-to-peak voltage amplitude $|V_s|$ of synchronized oscillations as a function of the delay τ for different values of the coupling strength κ. The left and right columns correspond to the adjacency matrices (16) and (17), respectively. The horizontal axis represents the equilibrium. Stable and unstable states are depicted by solid green and dashed red curves. Stars represent Hopf bifurcations of equilibria or Neimark-Sacker bifurcations of periodic orbits, crosses represent fold or pitchfork bifurcations, and diamonds denote period doubling bifurcations. The color of symbols distinguishes between the modes; see the caption of Fig. 2.

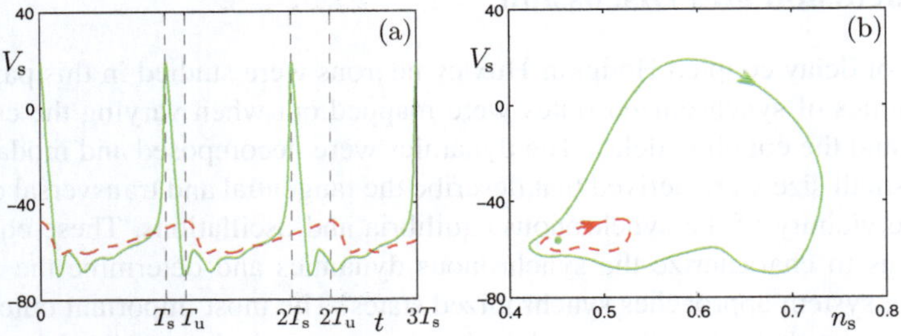

Fig. 4 Stable (solid green) and unstable (dashed red) periodic orbits corresponding to $\tau = 3$ ms and $\kappa = 1.2$ mS/cm^2; cf. Fig. 3(e). The corresponding periods are $T_s \approx 15.39$ ms and $T_u \approx 17.78$ ms.

coexisting T_p-periodic solutions. For a range of κ a cascade of fold bifurcations is observed that culminate in single point where $T_p \to \infty$; see the spiral in panels (b) and (g). Tangential period doubling (flip) bifurcations (black diamonds) give rise to branches of $2T_p$-periodic oscillations that are typically tangentially unstable. (These are not depicted in the figures). Also, co-dimension two fold-flip bifurcations can be observed when the coupling is sufficiently strong; see panels (b–e) and (g–h). For strong coupling tangential Neimark-Sacker bifurcations (black stars) result in quasi-periodic oscillations; see panels (e) and (j). (Such oscillations cannot be traced with the current state-of-the-art techniques).

The equilibrium may also loose its transversal stability via Hopf bifurcations (colored stars) corresponding to the "wavy" curves in Fig. 2. The oscillations arising trough these bifurcations brake the synchrony. For synchronous oscillations, transversal stability losses may occur via pitchfork bifurcations (colored crosses), via period doubling bifurcations (colored diamonds), and via Neimark-Sacker bifurcations (colored stars). Corresponding to the multiplicity of modal eigenvalues each transversal boundary found for (16) splits into three for (17), so that stability is typically lost to the mode with complex conjugate modal eigenvalues (that are the largest in magnitude) via Neimark-Sacker bifurcation (red stars).

For strong coupling one may observe domains where stable synchronized equilibrium coexist with stable and unstable synchronized oscillations. These domains arise via subcritical Hopf bifurcations and, depending on the initial conditions, the system may approach the equilibrium or the periodic solution. Such orbits are depicted in Fig. 4. When comparing this figure with Fig. 1 one may notice the changes in the shape of the stable (solid green) periodic orbit. Moreover, the period of stable oscillations ($T_s \approx 15.39$ ms) and the period of unstable oscillations ($T_u \approx 17.78$ ms) exceed the period of the uncoupled oscillations ($T_p \approx 11.57$ ms). We remark that there exist additional stable periodic solutions corresponding to different cluster states that can be approached by the system for certain initial conditions but these are not investigated in this paper; see [11] for more details.

4 Conclusion and Discussion

Systems of delay coupled Hodgkin-Huxley neurons were studied in this paper and the dynamics of synchronized states were mapped out when varying the coupling strength and the coupling delay. The dynamics were decomposed and modal equations of small size were derived that describe the tangential and transversal dynamics in the vicinity of the synchronous equilibria and oscillations. These equations allowed us to characterize the synchronous dynamics and determine the regions where the system approaches synchronized states. The most important outcome of the analysis is that when the coupling is strong enough there exist delay ranges where stable synchronized equilibria coexist with stable oscillations. We remark that for simplified neural models such multi-stability may not occur [10] which emphasizes the importance of models that are based on biophysical measurements.

While in this paper we only mapped out synchronous oscillations, detailed investigations show that many different cluster oscillations may also exist in these domains [11]. That is, depending on the initial conditions, the neural system may approach the synchronous equilibrium (which is a homogenous rest state) or different oscillatory states corresponding to different spatiotemporal patterns. As external stimuli can "reset the initial condition", the multi-stable dynamics discovered allow the neural system to respond to different external stimuli with different spatiotemporal patterns which is crucial for encoding environmental information. Note that such domains only exist for sufficiently large time delays which emphasizes that delays cannot be neglected when modeling neural networks. In fact, our results suggest that nature may tune the delays in large interconnected biological systems so that the information encoding capabilities of organisms are maximized.

References

1. Cepeda-Gomez, R., Olgac, N.: An exact method for the stability analysis of linear consensus protocols with time delay. IEEE Transactions on Automatic Control 56(7), 1734–1740 (2011)
2. Coombes, S.: Neuronal networks with gap junctions: a study of piecewise linear planar neuron models. SIAM Journal on Applied Dynamical Systems 7(3), 1101–1129 (2008)
3. D'Huys, O., Fischer, I., Danckaert, J., Vicente, R.: Role of delay for the symmetry in the dynamics of networks. Physical Review E 83(4), 046223 (2011)
4. Ermentrout, G.B., Terman, D.H.: Mathematical Foundations of Neuroscience. In: Interdisciplinary Applied Mathematics., vol. 35, Springer (2010)
5. Flunkert, V., Yanchuk, S., Dahms, T., Schöll, E.: Synchronizing distant nodes: a universal classification of networks. Physical Review Letters 105(25), 254101 (2010)
6. Hodgkin, A.L., Huxley, A.F.: A quantitative description of membrane current and its application to conduction and excitation in nerve. Journal of Physiology 117(4), 500–544 (1952)
7. Insperger, T., Stépán, G.: Semi-discretization for Time-delay Systems: Stability and Engineering Applications. Applied Mathematical Sciences, vol. 178. Springer (2011)
8. Kantner, M., Yanchuk, S.: Bifurcation analysis of delay-induced patterns in a ring of Hodgkin-Huxley neurons. Philosophical Transactions of the Royal Society A (to appear, 2013)

9. Olfati-Saber, R., Murray, R.M.: Consensus problems in networks of agents with switching topology and time-delays. IEEE Transactions on Automatic Control 49(9), 1520–1533 (2004)
10. Orosz, G.: Decomposing the dynamics of delayed networks: equilibria and rhythmic patterns in neural systems. In: Sipahi, R. (ed.) 10th IFAC Workshop on Time Delay Systems, pp. 173–178. IFAC (2012)
11. Orosz, G.: Decomposition of nonlinear delayed networks around cluster states with applications to neuro-dynamics. SIAM Journal on Applied Dynamcal Systems (submitted, 2013)
12. Orosz, G., Moehlis, J., Ashwin, P.: Designing the dynamics of globally coupled oscillators. Progress of Theoretical Physics 122(3), 611–630 (2009)
13. Orosz, G., Moehlis, J., Murray, R.M.: Controlling biological networks by time-delayed signals. Philosophical Transactions of the Royal Society A 368(1911), 439–454 (2010)
14. Roose, D., Szalai, R.: Continuation and bifurcation analysis of delay differential equations. In: Krauskopf, B., Osinga, H.M., Galan-Vioque, J. (eds.) Numerical Continuation Methods for Dynamical Systems, Understanding Complex Systems, pp. 359–399. Springer (2007)
15. Stépán, G.: Retarded Dynamical Systems: Stability and Characteristic Functions. Pitman Research Notes in Mathematics, vol. 210. Longman (1989)

Practical Delay Modeling of Externally Recirculated Burned Gas Fraction for Spark-Ignited Engines

Delphine Bresch-Pietri, Thomas Leroy, Jonathan Chauvin, and Nicolas Petit

Abstract. In this chapter, the authors provide an overview and study of the low-pressure burned gas recirculation in spark-ignited engines for automotive powertrain. It is shown, at the light of supportive experimental results, that a linear delay system permits to capture the dominant effects of the system dynamics. The modeled transport delay is defined by implicit equations stemming from first principles and can be calculated online. This model is shown to be sufficiently accurate to replace a sensor that would be difficult and costly to implement on commercial engines.

1 Introduction

In this chapter, the authors focus on an application problem in the area of automotive powertrain control. Indeed, in the past decades, the still more stringent norms on fuel consumption and pollutant emissions for automotive engines have substantially increased the architecture of thermal engines and, consequently, the complexity of the control task. In this context, the treatment of time-delay systems constitutes an important design consideration, as delays are an often encountered phenomenon in powertrain systems, as highlighted by numerous studies [9].

Delphine Bresch-Pietri
Department of Mechanical Engineering, Massachussetts Institute of Technology,
77 Massachussetts Avenue, Cambridge MA 02139, USA
e-mail: dbp@mit.edu

Thomas Leroy · Jonathan Chauvin
IFP Energies nouvelles, Département Contrôle, Signal et Système,
1 et 4 avenue du Bois-Préau, 92852 Rueil-Malmaison, France
e-mail: {thomas.leroy,jonathan.chauvin}@ifpen.fr

Nicolas Petit
MINES Paristech, Centre Automatique et Systèmes, Unité Mathématiques et Systèmes,
60 Bd St-Michel, 75272 Paris, Cedex 06 France
e-mail: nicolas.petit@mines-paristech.fr

T. Vyhlídal, J.-F. Lafay, and R. Sipahi (eds.), *Delay Systems*,
Advances in Delays and Dynamics 1,
DOI: 10.1007/978-3-319-01695-5_26, © Springer International Publishing Switzerland 2014

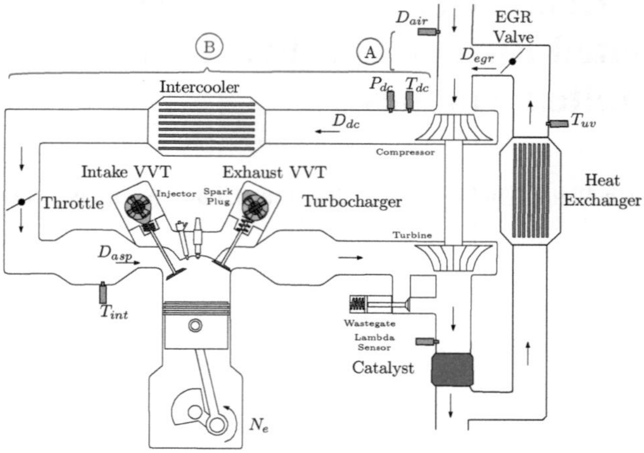

Fig. 1 Scheme of a turbocharged SI engine equipped with direct injection, VVT and a low-pressure EGR loop

In this chapter, the particular application under consideration is exhaust gas recirculation (EGR) through a low-pressure (LP) circuit for a Spark-Ignited (SI) engine. It is shown in this chapter that this technology introduces a significant delay transport which should be taken into account to accurately estimate and control the (distributed) composition of the gas inside the intake line. Before detailing this point, a few elements of context are given to motivate the use of EGR.

1.1 Why Exhaust Gas Recirculation ?

One of the main issues when dealing with downsized[1] SI engine is the prevention of the malicious knock phenomenon. This unwanted self-ignition of the gaseous mixture which appears at high load, due to high resulting thermodynamical conditions, may cause the engine to stall and eventually damage the combustion chamber.

One of the solutions considered in the automotive industry to handle this phenomenon consists in using EGR through a low-pressure circuit [8, 18]. In such a configuration, exhaust burned gas are picked up downstream of the catalyst and reintroduced upstream of the compressor. A typical implementation is represented in Fig. 1. Indeed, the addition of exhaust gas into the gaseous blend leads to an increase of the auto-ignition delay: intermixing the incoming air with recirculated exhaust gas dilutes the mixture with inert gas, increases its specific heat capacity and, consequently, lowers the peak combustion temperature. Then, the net effect of

[1] Downsizing consists in the reduction of the engine size to operate on more efficient points. To provide similar performances to much larger engines, these engines are usually equipped with a turbocharger and direct injection devices. This technology has appeared in the last decade as a major solution to reduce fuel consumption [14].

EGR is a prevention of knock which leads to potential substantial improvements of overall combustion efficiency [6].

1.2 Necessity of a Virtual Composition Sensor

Yet, EGR has some downsides. During tip-outs (defined as a transient mode during which the torque demand is suddenly decreased), the presence of burned gases in the intake manifold and later in the combustion chamber seriously impacts the combustion process and may cause the engine to stall. Further, EGR has strong interactions with simultaneously operating engine controllers such as the regulation of Fuel-to-Air Ratio (FAR) to stoichiometry (see [9]) or the spark advance. Indeed, EGR impacts the fresh air quantity which is aspirated inside the cylinder at each stroke. Therefore, to counteract the impact of intake burned gas, a solution would be to modify the feedforward action on the cascaded controllers (fuelpath controller and ignition path controller) based on a real-time estimate $\hat{x}$ of the intake burned gas rate. Nevertheless, no real-time sensor of this variable is embedded in any real-world vehicle and obtaining such an estimate is not an easy task. The approach that we advocate here is to substitute one such sensor with a model[2].

For the considered low-pressure gas recirculation circuit, the amount of reintroduced burned gases is controlled by the EGR Valve, an actuator which is located upstream of the compressor. Consequently, the relative long distance between the compressor and the inlet manifold leads to a large transport delay (up to several seconds depending on the engine specifications). Most importantly, this delay depends on the gas flow rate and therefore is time-varying to a large extent.

1.3 Comparison with Diesel EGR

In the seemingly similar context of automotive Diesel engines[3], numerous solutions for the discussed control issues have been developed in the last decades (see for example [1, 21, 23] and the references therein). Yet, none of these strategies includes a transport delay model, which as has been discussed is non-negligible for SI engines. Indeed, on top of using a low-pressure EGR circuit configuration (which substantially increases the transport lag compared to high-pressure configuration studied in [12, 20]), SI engines combustion constraints significantly increase the scale of the delay: (i) first, SI engines operations require a stoichiometric FAR, which results into a fraction of burned gas close to one in the exhaust line. Consequently, to obtain a given intake fraction of burned gas, the amount of exhaust burned gas to be reintroduced at steady state is substantially lower than the corresponding one for Diesel engines; (ii) besides, on the contrary of Diesel engines, SI engines may

[2] Other works (see [5]) investigate the potential of using a cylinder pressure sensor signal. Yet, due to stringent cost constraints, such a sensor is not currently commercially embedded.

[3] The use of EGR for Diesel engines has been widely investigated for a different purpose, to decrease the emissions of nitrogenous oxides emissions.

operate at intake pressure under atmospheric values (low loads). Then, on this operating range, the steady-state gas flow rates are considerably less important.

For these reasons, modeling this transport delay is a milestone in the design of controllers for SI engines.

In this chapter, a model of the intake burned gas rate is presented, accounting explicitly for transport time-varying delay and its dependency on the history of gas flow rates in a way which compensates for thermal exchanges and induced gas velocity changes. It is then used as a "software" sensor. This estimation is based on a practical delay calculation methodology which is experimentally validated on a test bench. The purpose of this chapter, based on the previous contribution [4], is to present this model along with its practical validation, to enhance the role of the variable transport delay in the modeling and to illustrate how this estimate can be used to coordinate the controllers. Experimental FAR control tests stress the relevance of the estimate.

This chapter is organized as follows. In Section 2, we present a model of the intake burned gas rate dynamics, under the form of a linear time-varying system with a time-dependent delay output. The practical usage of this model is discussed. Implementation and experimental results are provided in Section 3. We conclude by briefly sketching potential directions of work for combustion control improvements.

2 Modeling

Consider the airpath of a turbocharged SI engine equipped with intake throttle, wastegate, dual independent VVT actuators and a low-pressure external gas recirculation (EGR) loop as depicted in Fig. 1. Such a setup is usually considered for downsized engines (see [10]). Acronyms and notations used below are listed in Table 1.

Formally, the in-cylinder burned gas fraction x_{cyl} is defined as the ratio between the in-cylinder burned gas mass originated from the EGR loop m_{bg} and the total mass of gas in the cylinder volume $m_{asp} = m_{air} + m_{bg}$ in which m_{air} is the aspirated mass of fresh air, i.e.

$$x_{cyl} = \frac{m_{bg}}{m_{air} + m_{bg}}$$

From now on, this variable is considered equal to x the intake burned gas fraction [4].

2.1 *Dilution Dynamics and Transport Delay*

Defining x_{lp} as the burned gas rate upstream of the compressor, the EGR dynamics can be expressed in terms of the mass flow rates of air D_{air} and recirculated burned gas D_{egr} as

---

[4] Actually, this relation depends mainly on the VVT control strategy. We neglect this influence here for sake of clarity.

$$\dot{x}_{lp} = \alpha \left[-(D_{egr}(t) + D_{air}(t))x_{lp}(t) + D_{egr}(t) \right] \quad (1)$$

$$x(t) = x_{lp}(t - \tau(t)) \quad (2)$$

where $\tau(t)$, the delay between this ratio and the intake composition, can be implicitly defined according to the integral equation (Plug-Flow assumption for the gas composition along the intake line, see [16])

$$\int_{t-\tau(t)}^{t} v_{gas}(s)ds = L_P \quad (3)$$

where L_P is the pipe length from the compressor down to the intake manifold and v_{gas} stands for the gas velocity.

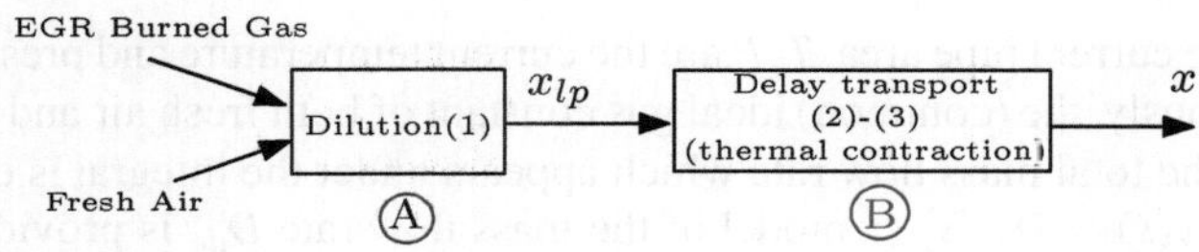

Fig. 2 Scheme of the intake burned gas fraction dynamics

Equation (1) is a balance equation on the volume downstream of the EGR valve, using the fact that the EGR circuit is totally filled with burned gas[5]. Depending on engine setups, the thermodynamics constant α appearing in (1) is either measured or known.

The integral delay model (3) is representative of a wide class of systems involving transport phenomena [2,17,22]. This delay can be understood as a propagation time for a variable velocity v_{gas}. In particular, one can observe that at steady-state this delay is inversely proportional to the gas speed, which is a more intuitive modeling one can think of. Alternatively, PDE models can be used to represent more accurately the induced transport dynamics. However, the induced computational burden discard them from real-time implementation.

In a nutshell, following the proposed model, which is pictured in Fig. 2, the intake burned gas fraction is the result of a first order dynamics coupled with a transport delay.

For sake of clarity, the approach used to model the mass flow rate quantities (D_{egr},...) used through (1)-(3) is not detailed here and given in Appendix. Using the approach presented in Appendix, one can now assume that they are known quantities. To provide an implementable open-loop estimate of x based on the model (1)-(3), a practical calculation methodology of the delay τ, using only real-time measurements, remains to be developed. This point is now addressed.

---

[5] For SI engines, the FAR is regulated to its stoichiometric value (see [7]), which results into an exhaust burned gas fraction close to unity.

2.2 Transport Delay Description

Equation (3) implicitly determines the delay according to the gas speed along the intake line, which, on top of being a distributed parameter, is not measured in practice. Further, the thermodynamical transformations the gas is submitted to in the intake line modify this distributed velocity. This complexity can be handled by a relatively fair delay description.

Indeed, using the ideal gas law (as is classically done for engine gas flows, e.g. in [7]), one can relate this speed to current thermodynamical conditions and mass flow rates, which are measured/modeled. Namely,

$$\forall t \geq 0, \quad v_{gas}(t) = \frac{1}{S(t)} \frac{rT(t)}{P(t)} [D_{air}(t) + D_{egr}(t)]$$

where S is the current pipe area, T, P are the current temperature and pressure values, r is, as previously, the (common) ideal gas constant of both fresh air and burned gas. In practice, the total mass flow rate which appears under the integral is estimated as $D_{air}(s) + D_{egr}(s) = D_{dc}(s)$ (a model of the mass flow rate D_{dc} is provided in (7) in Appendix).

A thermal contraction of the gas occurs inside the intake cooler, resulting in spatial changes of the gas velocity v_{gas}. To model this, we split the intake line into three main sections with three respective and cumulative transport delays τ_1, τ_2 and τ_3 such that $\tau = \tau_1 + \tau_2 + \tau_3$. This decomposition, pictured in Fig. 3, is as follows:

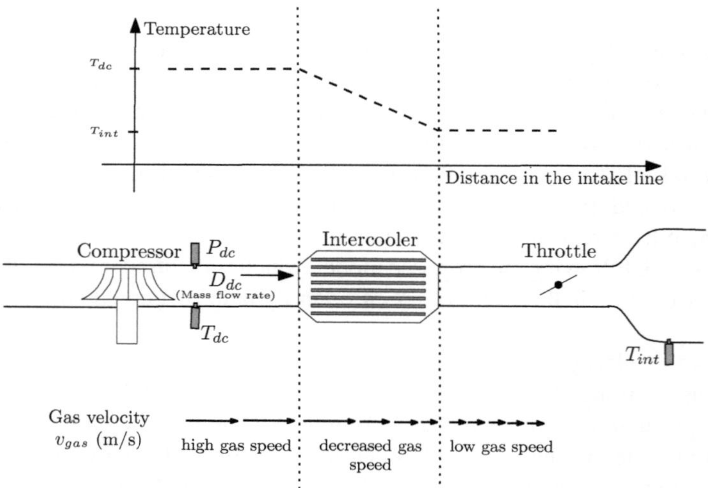

Fig. 3 The intake line is split into three parts to account for the spatial differences of the gas velocity. The temperature decreases, which results into an increase of velocity which is analytically determined by the ideal gas low fed with measurements from temperature and pressure sensors located along the line.

- *downstream of the compressor to the intercooler :* in this part, the current pressure and the temperature are measured and one can write

$$\int_{t-\tau_1(t)}^{t} \frac{rT_{dc}}{P_{dc}} D_{dc}(s)ds = V_1 \tag{4}$$

  with V_1 the corresponding volume.
- *inside the intercooler:* considering boundary conditions, the pressure inside the intercooler can reasonably be assumed as constant and equal to the input one P_{dc}. Further, we assume that the spatial profile of the inside temperature is affine with respect to the spatial variable, with measured boundary conditions T_{dc} and T_{int}[6]. Under this assumption, equation (3) can be reformulated on this section as

$$\int_{t-\tau_2(t)-\tau_1(t)}^{t-\tau_1(t)} \frac{r}{P_{dc}} D_{dc}(s)ds = S_2 \int_0^{L_2} \frac{dx}{T(x)} = \frac{V_2}{T_{int}-T_{dc}} \ln\left(\frac{T_{int}}{T_{dc}}\right) \tag{5}$$

  where L_2, S_2 and V_2 are the corresponding length, area and volume.
- *downstream of the intercooler to the intake manifold:* in this section, the temperature can be approximated by the intake manifold temperature, which yields

$$\int_{t-\tau_3(t)-\tau_2(t)-\tau_1(t)}^{t-\tau_2(t)-\tau_1(t)} \frac{rT_{int}}{P_{dc}} D_{dc}(s)ds = V_3 \tag{6}$$

  with V_3 the corresponding volume.

Knowing intermediate volumes V_1, V_2 and V_3, one can calculate the delay in a very straightforward manner, solving, one after the other, (4), (5) and finally (6). The transport delay is then simply deduced as $\tau(t) = \tau_1(t) + \tau_2(t) + \tau_3(t)$.

The involved numerical solving is based on the observation that the integrand is strictly positive and that the integral is then an increasing function of the delay τ_i ($i \in \{1,2,3\}$) appearing in its lower bound. Then, by simply sampling and evaluating the integral at increasing values of τ_i starting from 0, one can obtain a numerical evaluation of the corresponding delay. All these calculations are on-line compliant[7].

2.3 Estimation Strategy with Practical Identification Procedure

An estimation strategy of the model above is summarized on Fig. 4. Real-time measurements of temperatures and pressures serve to determine the value of the delay. These information are commonly available using (cheap) embedded sensors. Values for physical volumes (V_{lp}, V_1, V_2 and V_3) can be used to calibrate the model.

It is worth noticing that splitting the intake line as has been proposed in the previous section has been motivated mainly by the engine embedded instrumentation and in particular by the availability of temperature (and pressure) sensors. It can

[6] i.e. $T(x) = \frac{T_{int}-T_{dc}}{L_2}x + T_{dc}$.

[7] This approach is directly inspired of [17] for modeling plug flows in networks of pipes problem.

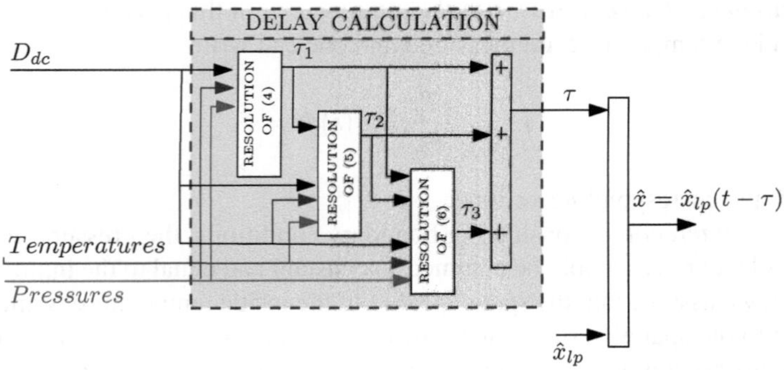

Fig. 4 Scheme of the proposed delay calculation strategy for intake burned gas fraction estimate x. The numerical solving of implicit integral equations (4)-(6) can be obtained by sampling and calculating the integrals at increasing values of τ_i starting from 0, which are real-time compliant calculations.

be easily adapted to any considered engine. In particular, if no temperature and no pressure sensors are available downstream of the compressor, they can be efficiently approximated by the intake ones at the expense of slight updates of the volumes values in the fit. Indeed, the two pressures are sufficiently close and these equations are of moderate temperature sensitivity. In such a case, the delay can be directly determined by one equation of type (4).

3 Experimental Results

The proposed model is now used as a "software" sensor. The obtained estimate is embedded into a real-time control target and employed at test-bench. The experiments aim at validating the model presented in Section 2 and in particular the delay modeling.

3.1 Experimental Setup and Indirect Validation Methodology from FAR Measurements

The engine under consideration is a $1.8L$ four cylinder SI engine with direct injection (see [13] for details). The airpath consists of a turbocharger controller with a wastegate, an intake throttle, an intercooler and a LP-EGR loop. This engine setup is consistent with the scheme reported in Fig. 1.

To validate the proposed estimation strategy, as *no real-time information of the intake burned gas fraction is available* for this engine, we focus on the open-loop response of the normalized FAR. This quantity is formally defined in terms of the fuel and fresh air mass aspirated inside the cylinder at each stroke as $\phi = \frac{1}{FAR_{st}} \frac{m_{inj}}{m_{air}}$. It has to be regulated to the unity to maximize the efficiency of after-treatment devices.

Usually, this control is realized with the injection path[8], using the measurements of a dedicated sensor located downstream of the turbine (Lambda Sensor).

Here, the FAR is simply controlled by a feedforward strategy on the mass of fuel injected in the cylinder, namely

$$m_{inj} = FAR_{st} m_{air}$$

The additional feedback term that is usually used is purposely omitted.

When no burned gas is recirculated, the in-cylinder air mass is accurately estimated with the model presented in Appendix (see [15]), i.e. $m_{air} = m_{asp}$. When burned gas are reintroduced, one can formally write $m_{air} = m_{asp} - m_{bg} = m_{asp}(1-x)$ and, consequently, estimate the in-cylinder air mass as $m_{asp}(1-\hat{x})$ where $\hat{x}$ is the intake burned gas fraction estimate provided by the proposed model.

With this setup, it is possible *to qualitatively relate the FAR variations to the intake burned gas fraction*. Indeed, if the estimation is accurate, the normalized FAR remains close to unity and, in turn, one then obtains an indirect validation of the intake burned gas rate estimation. Any offset reveals a steady-state estimation error while any temporary undershoot (or overshoot) reveals a mis-estimation of the delay.

3.2 *First Validation : Variation of the Amount of Reintroduced EGR (Constant Delay)*

The first scenario under consideration here is a variation of the amount of reintroduced burned gas for a given operating point : constant engine speed $N_e = 2000$ rpm for a requested torque of 12.5 bar. This scenario is of particular interest for validation as the intake mixture composition is the only varying variable.

Fig.5(c) pictures the intake burned gas fraction estimates corresponding to the EGR valve variations pictured in (a). The corresponding delay is constant and simply not reported.

With Burned Gas Feedforward Correction, i.e. Considering $m_{air} = (1-\hat{x})m_{asp}$
The corresponding normalized FAR evolution is pictured in black, in Fig.5(b). One can easily observe that the normalized FAR remains satisfactorily close to the unity. This behavior reveals a good fit between the real intake manifold burned gas rate and the estimate one provided in Fig.5(c).

For sake of comparisons, the FAR response with a burned gas fraction estimate computed neglecting the delay is also provided (red dotted curves). Neglecting the delay leads to a transient overestimation of the burned gas fraction and, consequently, to a significant FAR undershoot. This stresses the importance of the delay into the burned gas rate dynamics and the relevance of the proposed model.

Without Burned Gas Correction, i.e. Considering $m_{air} = m_{asp}$
In that case, as the in-cylinder mass air is overestimated, the injected mass of fuel turns to be too large. This results into a deviation of the normalized FAR up to 1.09 (blue dotted curve in Fig.5). A feedback control would reasonably eliminate

[8] The airpath being then dedicated to meet torque requests.

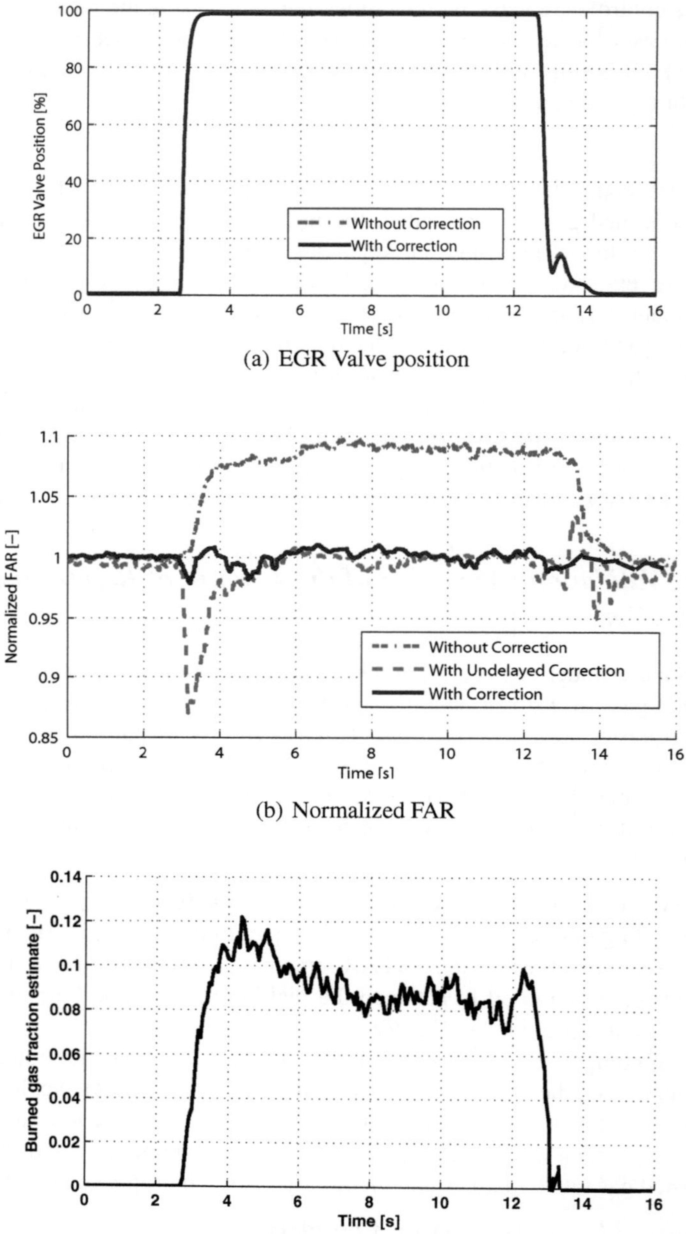

(a) EGR Valve position

(b) Normalized FAR

(c) Burned gas fraction estimate

Fig. 5 Experimental results for constant engine speed ($N_e = 2000$ rpm) and torque request ($IMEP = 8$ bar). The EGR valve position is pictured in (a). Blue dotted curve : gas composition transient without estimation. Black curve : gas composition transient with estimation and feedforward correction.

this offset, but, as the obtained FAR measurement is delayed (see [3, 11] for a FAR dynamics details), an important overshoot would still be present.

3.3 Second Validation : Torque Transients (Varying Delay)

The second scenario under consideration is a torque transient requested by the driver, a step from 6 bar to 12.5 bar. This tip-in is a typical driving situation case study, which defines an increase in the in-cylinder air mass set point and consequently on the total gas flow rate. Then, both dilution dynamics (1) and the delay are varying.

Further, this also implies a variation of the requested amount of reintroduced burned gas, as the initial operating point is low loaded and does not require any EGR. Without dedicated control structure, we simply consider here the EGR valve position either fully closed or fully opened. Its variations are pictured in Fig. 6(c).

The corresponding calculated delay is given in Fig. 6(d). As the total mass flow rate increases during the transient, the delay decreases, as expected.

Finally, as in the previous scenario, the FAR remains close to the unity. This validates the burned gas fraction estimate variations depicted in Fig. 6(b).

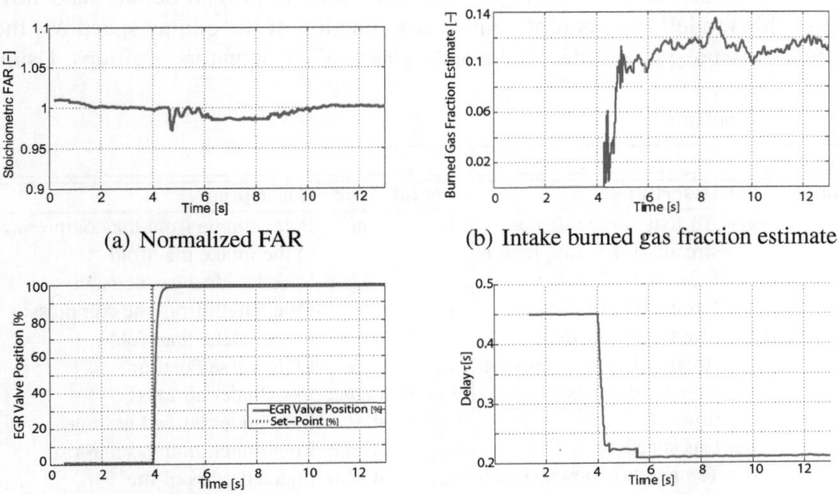

(a) Normalized FAR (b) Intake burned gas fraction estimate

(c) EGR valve position (d) Transport delay

Fig. 6 Experimental results for constant engine speed ($N_e = 2000$ rpm) and transient torque request (step from $IMEP = 6$ bar to 12.5 bar), resulting into a delay variation. The normalized FAR response pictured in (a) uses the intake burned gas fraction estimation pictured in (b), obtained with the on-line estimation of the delay (d).

4 Conclusion and Perspectives

In this chapter, it has been shown that low-pressure burned gas recirculation for SI engines can be accurately represented as a first order linear dynamics with a time-varying delay. The value of the delay is determined by an implicit integral relation in which data from commonly available sensors (temperatures, pressures) come into play through the ideal gas law. Experiments conducted at test bench validate both this model and the proposed delay calculation methodology and highlight the key rule played by the delay in the overall dynamics.

This result opens new perspectives in term of engine control applications : coordination of low-level controllers, advanced feedforward compensation (FAR, Spark Advance),... Yet, because the delay is varying, new techniques are required, especially if one wishes to take advantage of the known source of delay variability.

In particular, the input-dependency of the delay brings new challenges in term of control. As mentioned in [19], while a few number of works have investigated open-loop tracking for input-dependent delay systems, closed-loop control is still an open problem.

Appendix : Flow Rates Model

In-Cylinder and Downstream Compressor Mass Flow Rates

We use the model of in-cylinder gas mass presented in [15] to define mass flow rates. In this model, D_{asp} is represented as a function of the engine speed N_e, the manifold pressure P_{int} and the intake and exhaust VVT actuators positions. Using

Table 1 Nomenclature

Notations

Symbol	Unit	Description	Symbol	Unit	Description
D_{air}	kg/s	Air mass flow rate upstream of the compressor	V_P	m^3	Pipe volume from the compressor to the intake manifold
D_{egr}	kg/s	EGR mass flow rate through the EGR valve	r	J/kg/K	Specific ideal gas constant
			L_P	m	Pipe length from the compressor to the intake manifold
D_{dc}	kg/s	Mass flow rate downstream of the compressor	θ_{egr}	%	EGR Valve Position
D_{asp}	kg/s	In -cylinder mass flow rate	m_{air}	mg/str	In-cylinder air mass
			m_{bg}	mg/str	In-cylinder burned gas mass
v_{gas}	m/s	Gas speed	m_{asp}	mg/str	In-cylinder total gas mass
T_{dc}	K	Temperature downstream of the compressor	m_{inj}	mg/str	Injected mass of fuel
			FAR_{st}	-	Stoichiometric FAR
P_{dc}	Pa	Pressure downstream of the compressor	x	-	Intake burned gas fraction
			x_{lp}	-	Burned gas fraction upstream of the compressor
T_{atm}	K	Atmospheric temperature			
P_{atm}	Pa	Atmospheric pressure	x_{cyl}	-	In-cylinder burned gas fraction
V_{lp}	m^3	Volume between the EGR valve and the compressor			

Acronyms

EGR	Exhaust Gas Recirculation
FAR	Fuel-to-Air Ratio
IMEP	Indicated Mean Effective Pressure
SI	Spark Ignited

the ideal gas law, this flow rate is dynamically related to the flow rates through the throttle and downstream of the compressor as

$$D_{thr} = D_{asp}(N_e, P_{int}, VVT) + \frac{V_{int}}{rT_{int}}\dot{P}_{int}, \quad D_{dc} = D_{thr} + \frac{V_P}{rT_{dc}}\dot{P}_{dc} \tag{7}$$

where $r = r_{air} = r_{bg}$ is the (common) ideal gas constant. The variables used in these two last equations are either known or measured.

EGR Mass Flow Rate. Assuming that an intake mass air flow sensor is available on the engine, only the mass flow rate D_{egr} remains to be expressed. Neglecting the mis-synchronization of the flows signals, we simply write (with a projection operator forcing the flow rate to be zero when the valve is closed)

$$\hat{D}_{egr}(t) = \mathrm{Proj}_{\theta_{egr}>0}\{D_{dc}(t) - D_{air}(t)\}$$

References

1. Ammann, M., Fekete, N.P., Guzzella, L., Glattfelder, A.H.: Model-based control of the VGT and EGR in a turbocharged common-rail Diesel engine: theory and passenger car implementation. SAE Transactions 112(3), 527–538 (2003)
2. Bresch-Pietri, D.: Robust control of variable time-delay systems. Theoretical contributions and applications to engine control. PhD thesis, MINES ParisTech (2012)
3. Bresch-Pietri, D., Chauvin, J., Petit, N.: Adaptive backstepping controller for uncertain systems with unknown input time-delay. application to SI engines. In: Proc. of the Conference on Decision and Control (2010)
4. Bresch-Pietri, D., Leroy, T., Chauvin, J., Petit, N.: Practical delay modeling of externally recirculated burned gas fraction for Spark-Ignited engines. In: Proc. of the 11th IFAC Workshop on Time-Delay Systems (2013)
5. Caicedo, M.A.R., Witrant, E., Sename, O., Higelin, P., et al.: A high gain observer for enclosed mass estimation in a spark ignited engine. In: Proceedings of the 2012 American Control Conference (2012)
6. Cairns, A., Blaxill, H.: The effects of combined internal and external gas recirculation on gasoline controlled auto-ignition. In: Proc. of the Society of Automotive Engineering World Congress, number 2005-01-0133 (2005)
7. Heywood, J.B.: Internal combustion engine fundamentals. McGraw-Hill, New York (1988)
8. Hoepke, B., Jannsen, S., Kasseris, E., Cheng, W.K.: EGR effects on boosted SI engine operation and knock integral correlation. SAE International Journal of Engines 5(2), 547–559 (2012)
9. Jankovic, M., Kolmanovsky, I.: Developments in control of time-delay systems for automotive powertrain applications. In: Balachandran, B., Kalmar-Nagy, T., Gilsinn, D.E. (eds.) Delay Differential Equations, Recent Advances and New Directions, pp. 55–92. Springer Science (2009)
10. Kiencke, U., Nielsen, L.: Automotive control systems. Springer, Berlin (2000)
11. Lauber, J., Guerra, T.M., Dambrine, M.: Air-fuel ratio control in a gasoline engine. International Journal of Systems Science 42(2), 277–286 (2011)

12. Lauber, J., Guerra, T.M., Perruquetti, W.: IC engine: tracking control for an inlet manifold with EGR. SAE Transactions Journal of Passenger Cars: Electronic and Electrical Systems 20, 913–917 (2002)
13. Le Solliec, G., Le Berr, F., Colin, G., Corde, G., Chamaillard, Y.: Engine control of a downsized spark ignited engine: from simulation to vehicle. In: Proc. of ECOSM Conference (2006)
14. Lecointe, B., Monnier, G.: Downsizing a gasoline engine using turbocharging with direct injection. In: Proc. of the Society of Automotive Engineering World Congress, number 2003-03-0542 (2003)
15. Leroy, T., Chauvin, J., Le Berr, F., Duparchy, A., Alix, G.: Modeling fresh air charge and residual gas fraction on a dual independant variable valve timing SI engine. SAE International Journal of Engines 1(1), 627–635 (2009)
16. Perry, R.H., Green, D.W., Maloney, J.O.: Perry's chemical engineers' handbook, vol. 7. McGraw-Hill, New York (1984)
17. Petit, N., Creff, Y., Rouchon, P.: Motion planning for two classes of nonlinear systems with delays depending on the control. In: Proceedings of the 37th IEEE Conference on Decision and Control, pp. 1007–1011 (1998)
18. Potteau, S., Lutz, P., Leroux, S., Moroz, S., et al.: Cooled EGR for a turbo SI engine to reduce knocking and fuel consumption. SAE Technical Paper, Paper No. 2007-01-3978 (2007)
19. Richard, J.-P.: Time-delay systems; an overview of some recent advances and open problems. Automatica 39, 1667–1694 (2003)
20. Stotsky, A., Kolmanovsky, I.: Application of input estimation techniques to charge estimation and control in automotive engines. Control Engineering Practice 10(12), 1371–1383 (2002)
21. Van Nieuwstadt, M.J., Kolmanovsky, I.V., Moraal, P.E., Stefanopoulou, A., Jankovic, M.: EGR-VGT control schemes: experimental comparison for a high-speed diesel engine. IEEE Control Systems Magazine 20(3), 63–79 (2000)
22. Zenger, K., Niemi, A.J.: Modelling and control of a class of time-varying continuous flow processes. Journal of Process Control 19(9), 1511–1518 (2009)
23. Zheng, M., Reader, G.T., Hawley, J.G.: Diesel engine exhaust gas recirculation–a review on advanced and novel concepts. Energy Conversion and Management 45(6), 883–900 (2004)

Design and Control of Force Feedback Haptic Systems with Time Delay

Quoc Viet Dang, Antoine Dequidt, Laurent Vermeiren, and Michel Dambrine

Abstract. Time-delay effects on the properties of stability and transparency for a haptic device are well known and are far from to be negligible if high performances are expected. This chapter deals with the issues of design and control for a force feedback haptic device in the case of the interaction with a virtual wall. The stability condition expressed in term of Linear Matrix Inequality allows considering the effects of a constant time-delay as well as other mechanical parameters such as vibration mode. This result will be applied in the first part of the chapter to the optimal design method for an electromechanical haptic device with high performances (stability and low inertia). The second part will deal with the implementation of the virtual wall and the observer-based force feedback architecture in order to improve the stability of haptic system taking into account variations of the communication delay.

1 Introduction

Haptic devices are used to produce a kinesthetic or tactile stimulus of the interaction between a user and a virtual environment. This chapter is about a force feedback haptic system that allows the user to manipulate objects in the virtual environment while feeling a reaction force when there is contact with an obstacle, e.g. a virtual wall. Stability and transparency are the two most important performance measures for the design of a haptic device. Transparency implies low inertia, little friction, and no constraints on motion imposed by the device kinematics so that the user must have the impression of directly manipulating virtual objects without feeling the dynamics of haptic device. Ideally, the weight, inertia, friction, etc. of haptic

Quoc Viet Dang · Antoine Dequidt · Laurent Vermeiren · Michel Dambrine
LAMIH, UMR CNRS 8201, University of Valenciennes et du Hainaut-Cambresis,
Mont Houy, 59300 Valenciennes, France
e-mail: dqviet2212@gmail.com, {Antoine.Dequidt,Laurent.Vermeiren,
Michel.Dambrine}@univ-valenciennes.fr

T. Vyhlídal, J.-F. Lafay, and R. Sipahi (eds.), *Delay Systems*,
Advances in Delays and Dynamics 1,
DOI: 10.1007/978-3-319-01695-5_27,

device should not be perceived by the user. In addition, the haptic system must be able to reproduce a high contact stiffness so that the user should not have capacity to go through the virtual wall. However, a too high contact stiffness of the virtual wall can lead to a destabilization of the haptic device. Any unstable behavior occurring during a virtual contact can damage the haptic device or injure the user. The worst case of the stability study is the case of the contact between a manipulated object and a virtual wall.

In free motion, the improvement of haptic device's transparency can be obtained by using a position feedback loop in order to reduce the device's dynamics and compensators are used to cancel friction forces and gravity [1]. In order to assure the stability of the haptic device in the interaction with a virtual wall, the haptic system is described as a sampled-data dynamic system in which the haptic device is modeled by a one dimensional mechanical system consisting of a rigid mode (a mass/inertia m and a damping coefficient b), the virtual wall is characterized by a linear spring-damper system having virtual stiffness K_{ve} and damping coefficient B_{ve}. In the traditional approach, the backward finite difference method is used to estimate the velocity from the measured position. The linear inequality (1) has been proposed by Colgate and Schenkel [2] as a guideline for engineers in choosing design specifications of haptic system:

$$b > \frac{K_{ve} T_s}{2} + B_{ve} \tag{1}$$

where T_s is the sampling period. By using the analysis method on the basis of the frequency domain criteria such as: Routh-Hurwitz or Nyquist, the stability region of haptic system is defined in the virtual damping-stiffness plan. By linearizing the stability region around the origin, Gil et al. [3] proposed a generalized stability condition for the linear haptic system taking into account the effect of time-delay T_d as follows:

$$K_{ve} < \frac{b + B_{ve}}{\frac{T_s}{2} + T_d} \tag{2}$$

Moreover, the human operator's mechanical impedance can be modified over a wide dynamics range [4], the stability of haptic systems depends on how the human operator interacts with the haptic device. The stability condition

$$B_{ve} + b + b_h > \frac{(K_{ve} + k_h)\, T_s}{2} \tag{3}$$

given by Minsky et al. [5] includes the effects of the damping b_h and stiffness k_h of the human operator. However, this condition was derived from the passivity criterion of haptic system that is a very conservative design requirement [6] and the time delay was ignored. The impedance model of human operator was also included in some studies [7–9] to infer the evolution of stability condition. These results have shown that the human operator tends to make the haptic system more stable.

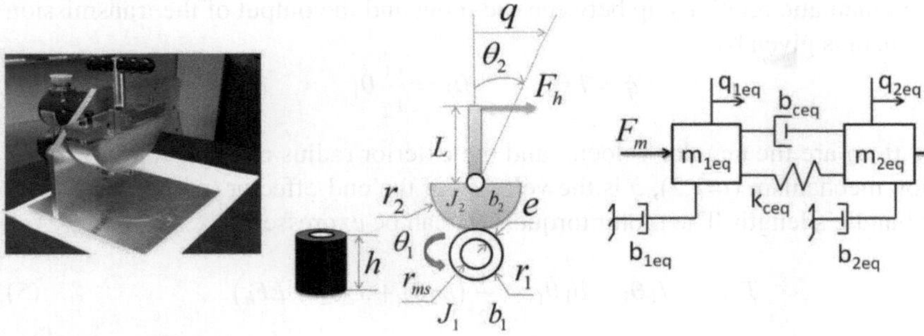

Fig. 1 Cable-driven 1DoF haptic device and mechanical model

The aim of this chapter is to address the issues of design and control for haptic devices in the case of the interaction with a virtual wall. The main contribution is firstly to present an optimal design method for an electromechanical haptic device with high performances (stability and low inertia). Using the backward finite difference method, the necessary and sufficient stability condition in term of Linear Matrix Inequality (LMI) is proposed to allow taking into account the effects of the time-delay as well as other mechanical parameters such as vibration mode in the same mechanical pre-design process. Next, an example for designing a cable-driven haptic device using the optimal design method is presented. Finally, an augmented state observer-based force feedback architecture will be considered for the implementation of the virtual wall in order to improve the haptic system stability. Time-varying communication delays will be considered. The numerical analysis show that the proposed force feedback architecture is quite efficient for expanding the stability region of haptic system.

2 Optimal Design Method for Haptic Device

2.1 *Dynamic Constraint*

Cable-driven mechanisms have attracted the attention of researchers in the field of haptics because of their advantages such as: low inertia, lost cost, large workspace, etc. Previous studies [8, 10, 11] have shown that the stability of a haptic system depends on the nature of flexibility of the haptic device's transmission mechanism. Thus, the dynamic behavior of haptic device taking into account the vibration modes must be studied in the interaction of the haptic device with the virtual wall. In this section, a cable-driven one degree-of-freedom(dof) haptic device as illustrated in Fig 1 is considered. The first transmission includes a motor and a cable spool mounted on the motor shaft, the second transmission comprises a handle connected rigidly with a semi-drum.

The kinematic relationship between the input and the output of the transmission mechanism is given by:

$$\dot{q} = L\dot{\theta}_2\,, \qquad \dot{\theta}_2 = \frac{r_1}{r_2}\dot{\theta}_1 \tag{4}$$

where $\dot{\theta}_i$, r_i are the angular velocity and the exterior radius of the ith part of transmission mechanism (i=1,2), $\dot{q}$ is the velocity of the end-effector (end of handle), L is the handle's length. The motor torque $T(t)$ can be expressed as:

$$T(t) = J_1\ddot{\theta}_1 + b_1\dot{\theta}_1 + \frac{r_1}{r_2}\left(b_2\dot{\theta}_2 + J_2\ddot{\theta}_2 + LF_h\right) \tag{5}$$

where F_h is the force applied from the user's hand, J_i ,b_i are the total inertia and the total damping coefficient of the ith part of transmission (i=1,2), F_h is the force applied by the human operator. It is noted that $J_1 = J_m + J_{sp}$, $J_{sp} = \frac{1}{2}\pi\rho h\left(r_1^4 - {r_{ms}}^4\right)$, $r_1 = r_{ms} + e$; J_m and r_{ms} are the inertia of motor and the radius of motor shaft; J_{sp}, ρ, e, and h are the inertia, mass density, thickness and height of cable spool; in respectively. From (4) and (5), we have:

$$F_h(t) = \frac{r}{L}T(t) - M_{app}\ddot{q} - B_{app}\dot{q} \tag{6}$$

where $r = \frac{r_2}{r_1}$, $M_{app} = \frac{J_1 r^2 + J_2}{L^2}$, $B_{app} = \frac{b_1 r^2 + b_2}{L^2}$; r is the transmission rapport, M_{app} and B_{app} are the apparent mass and damping perceived by the user while manipulating the handle of the haptic device.

For force feedback haptic systems, in steady-state regime with a stable contact with the virtual wall, i.e. $\dot{q} = 0$, the motor must be able to generate an enough high torque to equalize the force exerted by the user. This leads to following constraint:

$$T_m \geq \frac{L}{r}F_{h\max} \tag{7}$$

where F_{hmax} is the allowable maximal force exerted by the user, T_m is the nominal torque of motor.

2.2 Mechanical Model of the Haptic Device

The haptic device can be modeled using the following state-space representation:

$$\begin{cases} \dot{\mathbf{x}} = \mathbf{A}_c\mathbf{x} + \mathbf{B}_c F_m + \mathbf{E}_c F_h \\ y = \mathbf{C}_c\mathbf{x} \end{cases} \tag{8}$$

where $\mathbf{x} = \begin{bmatrix} q_{1eq} & \dot{q}_{1eq} & q_{2eq} & \dot{q}_{2eq} \end{bmatrix}^T$ is the state vector and $y = q_{1eq}$ is the output vector. The matrices expressions are given by:

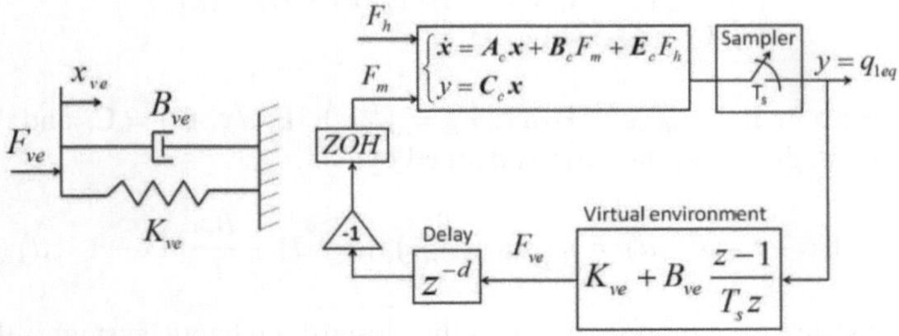

Fig. 2 Contact model in the virtual environment and overall dynamic model of a force feedback haptic system

$$\mathbf{A}_c = \begin{bmatrix} 0 & 1 & 0 & 0 \\ -\frac{k_{ceq}}{m_{1eq}} & -\frac{b_{1eq}+b_{ceq}}{m_{1eq}} & \frac{k_{ceq}}{m_{1eq}} & \frac{b_{ceq}}{m_{1eq}} \\ 0 & 0 & 0 & 1 \\ \frac{k_{ceq}}{m_{2eq}} & \frac{b_{ceq}}{m_{2eq}} & -\frac{k_{ceq}}{m_{2eq}} & -\frac{b_{2eq}+b_{ceq}}{m_{2eq}} \end{bmatrix}, \quad \begin{array}{l} \mathbf{B}_c = \left[0\ \frac{1}{m_{1eq}}\ 0\ 0\right]^T \\ \mathbf{E}_c = \left[0\ 0\ 0\ \frac{1}{m_{2eq}}\right]^T \\ \mathbf{C}_c = \left[1\ 0\ 0\ 0\right] \end{array}$$

where k_c and b_c are the stiffness and damping coefficient of transmission cable, and

$$\begin{array}{llll} m_{1eq} = J_1\left(\frac{r}{L}\right)^2 & b_{1eq} = b_1\left(\frac{r}{L}\right)^2 & q_{1eq} = \frac{L}{r}\theta_1 & F_m = \frac{r}{L}T(t) \\ m_{2eq} = \frac{J_2}{L^2} & b_{2eq} = \frac{b_2}{L^2} & b_{ceq} = b_c\left(\frac{r_1 r}{L}\right)^2 & k_{ceq} = k_c\left(\frac{r_1 r}{L}\right)^2 \end{array}$$

2.3 Necessary and Sufficient Stability Condition

The critical case for stability occurs when the manipulated object encounters a virtual wall (or constrained motion). The penetration x_{ve} of the virtual wall is given from the actuator position θ_1 by using an encoder mounted on the motor shaft, the velocity $\dot{x}_{ve}$ is approximated by using the backward finite-difference method. When the virtual wall is penetrated, the human operator should feel the reaction force F_{ve} from this wall. The feedback force F_m is calculated from the virtual contact model. The model of virtual contact includes a virtual stiffness K_{ve} and a virtual damping B_{ve}, as shown in the left side of Fig 2. The reaction force F_{ve} of virtual contact is given by:

$$F_{ve} = \begin{cases} K_{ve}x_{ve} + B_{ve}\dot{x}_{ve} & \text{if} \quad x_{ve} \geq 0 \\ 0 & \text{if} \quad x_{ve} < 0 \end{cases} \tag{9}$$

The right side of Fig 2 presents the overall dynamic model of the haptic system including the continuous-time state-space representation (8) of the haptic device, the sampler (Analog/Digital board) and the Zero-Order Holder (ZOH-Digital/Analog board) with the sampling period T_s, and the time-delay $T_d = dT_s$ is assumed here to be constant, d is a positive integer. The discrete-time state-space representation of the haptic device taking into account the effect of the ZOH can then be written:

$$\begin{cases} \mathbf{x}(k+1) = \mathbf{A}_d x(k) + \mathbf{B}_d F_m(k) + \mathbf{E}_d F_h(k) \\ y(k) = \mathbf{C}_d \mathbf{x}(k) \end{cases} \tag{10}$$

where: $\mathbf{A}_d = e^{\mathbf{A}_c T_s}$, $\mathbf{B}_d = \int_0^{T_s} e^{\mathbf{A}_c \tau} \mathbf{B}_c d\tau$, $\mathbf{E}_d = \int_0^{T_s} e^{\mathbf{A}_c \tau} \mathbf{E}_c d\tau$, $\mathbf{C}_d = \mathbf{C}_c$ and the feedback force F_m in discrete-time form is defined by:

$$F_m(k) = -F_{ve}(k-d) = -(K_{ve} + \frac{B_{ve}}{T_s}) y(k-d) + \frac{B_{ve}}{T_s} y(k-1-d) \tag{11}$$

The discrete state-space representation of the closed-loop haptic system is then:

$$\begin{cases} \mathbf{x}(k+1) = \mathbf{A}_d \mathbf{x}(k) - (K_{ve} + \frac{B_{ve}}{T_s}) \mathbf{B}_d \mathbf{C}_d \mathbf{x}(k-d) \\ \qquad + \frac{B_{ve}}{T_s} \mathbf{B}_d \mathbf{C}_d \mathbf{x}(k-1-d) + \mathbf{E}_d F_h(k) \\ y(k) = \mathbf{C}_d \mathbf{x}(k) \end{cases} \tag{12}$$

In the steady regime of stable contact with the virtual wall, the force F_h applied by the user can be considered as an external signal. By introducing the augmented state variable $\overline{\mathbf{x}}(k)^T = \left[\mathbf{x}(k)^T \ \mathbf{x}(k-1)^T \ ... \ \mathbf{x}(k-d)^T \ \mathbf{x}(k-1-d)^T \right]$, the discrete-time state-space representation of the closed-loop haptic system (12) can be rewritten:

$$\begin{cases} \overline{\mathbf{x}}(k+1) = \overline{\mathbf{A}} \overline{\mathbf{x}}(k) + \mathbf{E}_d F_h(k) \\ y(k) = \overline{\mathbf{C}} \overline{\mathbf{x}}(k) \end{cases} \tag{13}$$

where the augmented matrix $\overline{\mathbf{A}}$ of the $(d+2)n \times (d+2)n$ dimensions and $\overline{\mathbf{C}}$ of the $1 \times (d+2)n$ dimension are given by:

$$\overline{\mathbf{A}} = \begin{bmatrix} \mathbf{A}_d & \mathbf{0}_{n\times n} & \cdots & \mathbf{0}_{n\times n} & -(K_{ve} + \frac{B_{ve}}{T_s}) \mathbf{B}_d \mathbf{C}_d & \frac{B_{ve}}{T_s} \mathbf{B}_d \mathbf{C}_d \\ \mathbf{I}_{n\times n} & \mathbf{0}_{n\times n} & \cdots & \mathbf{0}_{n\times n} & \mathbf{0}_{n\times n} & \mathbf{0}_{n\times n} \\ \mathbf{0}_{n\times n} & \mathbf{I}_{n\times n} & \cdots & \mathbf{0}_{n\times n} & \mathbf{0}_{n\times n} & \mathbf{0}_{n\times n} \\ \vdots & \vdots & \ddots & \vdots & \vdots & \vdots \\ \mathbf{0}_{n\times n} & \mathbf{0}_{n\times n} & \cdots & \mathbf{I}_{n\times n} & \mathbf{0}_{n\times n} & \mathbf{0}_{n\times n} \\ \mathbf{0}_{n\times n} & \mathbf{0}_{n\times n} & \cdots & \mathbf{0}_{n\times n} & \mathbf{I}_{n\times n} & \mathbf{0}_{n\times n} \end{bmatrix}$$

and $\overline{\mathbf{C}} = \left[\mathbf{C}_d \ \mathbf{0}_{1\times(d+1)n} \right]$ with n is the size of vector $\mathbf{x}$. Based on Lyapunov's stability theorem, the discrete-time haptic system (13) is asymptotically stable if and only if there exists a positive definite matrix $\mathbf{P}$ satisfying the LMI constraint :

$$\overline{\mathbf{A}}^T \mathbf{P} \overline{\mathbf{A}} - \mathbf{P} < 0 \tag{14}$$

2.4 *Optimal Design Method for Haptic Device*

For designing a haptic device, we would like to minimize the apparent mass M_{app} perceived by the user. The optimal design of subparts (e.g. the second transmission) can be carried out independently as a local optimization process. So that, this section proposes a global optimal design method from its mechanical subparts for choosing the motor (J_m, T_m, b_m, r_{ms}) and the transmission rapport (r) of the haptic device. The objective of this method is to obtain a haptic device that has a low inertia (transparency performance) and must be able to keep stability during a contact to a virtual wall having a desired stiffness K_{ve_des} for all virtual damping $B_{ve} \in [B_{ve\min}, B_{ve\max}]$. Especially, this method allows to design a haptic device in the interaction to a virtual wall taking into account the effects of the vibration mode and the time delay. The optimal design problem of haptic device is described:

Original optimisation problem:

Finding: the motor (J_m, T_m, b_m, r_{ms}) and the transmission rapport (r).
Objective: minimize $M_{app} = \frac{J_1 r^2 + J_2}{L^2}$.
Constraints: dynamic constraint (7) and stability condition (14).

By denoting:

$$\alpha = \frac{T_m}{\sqrt{J_1}}, \qquad \beta = r\sqrt{J_1} \tag{15}$$

we have

$$\begin{aligned} M_{app} &= \frac{\beta^2 + J_2}{L^2}\; k_{ceq} = \frac{k_c}{J_1}\left(\frac{r_1\beta}{L}\right)^2 m_{1eq} = \left(\frac{\beta}{L}\right)^2 b_{1eq} = \frac{b_1}{J_1}\left(\frac{\beta}{L}\right)^2 \\ q_{1eq} &= \frac{L\sqrt{J_1}}{\beta}\theta_1\; b_{ceq} = \frac{b_c}{J_1}\left(\frac{r_1\beta}{L}\right)^2 m_{2eq} = \frac{J_2}{L^2}\; b_{2eq} = \frac{b_2}{L^2}\; F_m = \frac{\beta}{L\sqrt{J_1}}T(t) \end{aligned} \tag{16}$$

Therefore, the optimal design problem can be transformed into :

New optimisation problem:

Finding: the value of α.
Objective: minimize β.
Constraints: dynamic constraint $\beta \geq \frac{LF_{h\max}}{\alpha}$ and stability condition (14) with the physical parameters in (16).

In order to illustrate the proposed optimal design method, a list of motors in Table 1 is considered and the desired parameters are given in Table 2. A description for the optimal design method is shown in Fig 3. Each upright line corresponds to a motor. The intersection of each upright line with the limit curve gives a "feasible value" of β. In addition, for each motor, a lowest "stability value" of β at which the haptic device is stable in the case of the contact with a desired virtual wall is calculated by using the stability condition (14). As a result, corresponding to each motor, we obtain two values of β (i.e. a "feasible value" and a "stability value"). The maximum of these two values gives a solution. And so, a set of solutions S_i $(i = 1, \ldots, n_u)$ including n_u values of β from a list of n_u motors is obtained. The minimum of these values corresponds to the optimal solution. Obviously, from the optimal solution

Table 1 List of motors

Model	Torque(mNm)	Inertia(gcm^2)	Damping (Ns/m)	Shaft radius (mm)
283870	167	101	1.9536×10^{-3}	3
393023	161	53.8	2.5742×10^{-3}	3
136200	190	119	1.5909×10^{-3}	3
305015	129	33.3	1.9821×10^{-3}	2.5
370356	420	542	14.339×10^{-3}	4
148867	170	139	2.9764×10^{-3}	3
353295	485	1290	37.283×10^{-3}	6
136209	347	209	5.4286×10^{-3}	4
272765	63.8	21.9	0.5252×10^{-3}	2
218013	189	123	2.9639×10^{-3}	3

Table 2 Desired parameters for designing haptic device

Parameters	Variable	Value
Handle's length(m)	L	0.1
Total inertia(kgm^2)	J_2	1.55×10^{-3}
Total damping(Ns/m)	b_2	10^{-5}
Mass density(kg/m^3)	ρ	7800
Thickness(m)	e	0.002
Height(m)	h	0.02
Stiffness(N/m)	k_c	5060
Damping(Ns/m)	b_c	0
Virtual stiffness(N/m)	K_{ve_des}	5000
Minimal virtual damping(Ns/m)	B_{vemin}	0
Maximal virtual damping(Ns/m)	B_{vemax}	50
Maximal force applied by the user(N)	$Fhmax$	10
Communication time-delay(ms)	d	1
Sampling period(ms)	T_s	1

Table 3 Optimal solution

Model	Torque(mNm)	Inertia(gcm^2)	Damping (Ns/m)	Shaft radius (mm)
305015	129	33.3	1.9821×10^{-3}	2.5
Transmission rapport			r	11.54

of β, the designed parameters for the first transmission (i.e. motor parameters and exterior radius of cable spool) are completely determined by choosing the value of α. From the chosen motor and β_{opt}, the transmission rapport r is derived by using (15). The result of the optimal solution is given in the Table 3.

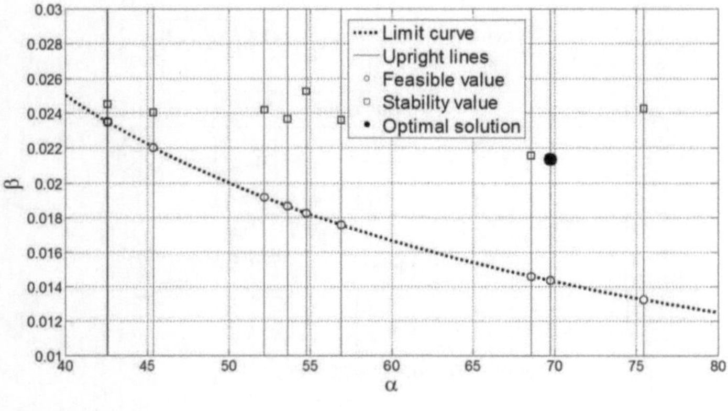

Fig. 3 An illustration for the optimal design method

3 Proposed Force Feedback Architecture

3.1 Design of the Virtual Wall and the State Observer

In the case of contact to the virtual wall, most force feedback architectures of haptic system in the literatures [3, 8, 12] took the measured position of the motor shaft θ_1 in order to calculate the penetration depth x_{ve} of the virtual wall. Its velocity $\dot{x}_{ve}$ is obtained by using the backward finite difference method. In order to improve the stability of the discrete-time haptic system under the time-varying communication delay, we consider an alternative force feedback architecture using an augmented state observer (see Fig 4) instead of the backward finite difference method in Fig 2. In the general case of interaction between a haptic device and a virtual environment, the communication delay $d(k)$ is a time-varying function that can be assumed to vary between two integers $d_{\min}$ and $d_{\max}$ with $d_{\min} < d_{\max}$:

$$0 < d_{\min} \leq d(k) \leq d_{\max} \tag{17}$$

It is recalled that the force F_h applied by the user can be considered as a constant external excitation signal in the steady regime of stable contact between the manipulated object and a virtual wall. Denote

$$\begin{aligned} \overline{\mathbf{X}}(k) &= \begin{bmatrix} \mathbf{x}(k) \\ F_h(k) \end{bmatrix} \quad u(k) = F_m(k) \\ \overline{\mathbf{A}} &= \begin{bmatrix} \mathbf{A}_d & \mathbf{E}_d \\ \mathbf{0}_{1\times n} & 1 \end{bmatrix} \quad \overline{\mathbf{B}} = \begin{bmatrix} \mathbf{B}_d \\ 0 \end{bmatrix} \quad \overline{\mathbf{C}} = \begin{bmatrix} \mathbf{C}_d & 0 \end{bmatrix} \end{aligned} \tag{18}$$

Then, the augmented, discrete-time state representation of the haptic device can be rewritten

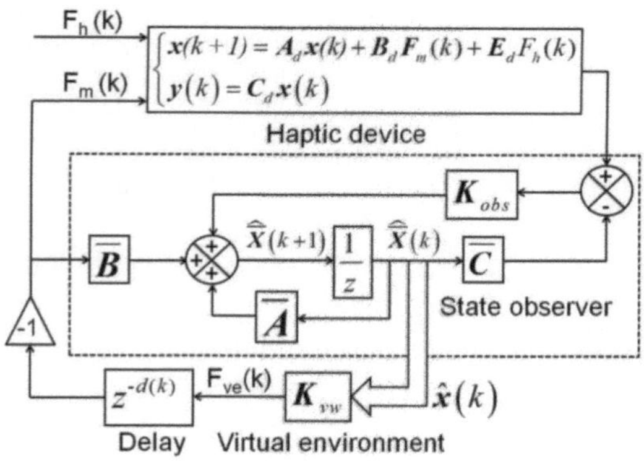

Fig. 4 Augmented state observer-based force feedback architecture for haptic system

$$\begin{cases} \overline{\mathbf{X}}(k+1) = \overline{\mathbf{A}}\,\overline{\mathbf{X}}(k) + \overline{\mathbf{B}}u(k) \\ \mathbf{Y}(k) = \overline{\mathbf{C}}\,\overline{\mathbf{X}}(k) \end{cases} \tag{19}$$

So that, the dynamic model of the discrete-time state observer is the following one:

$$\begin{cases} \hat{\overline{\mathbf{X}}}(k+1) = \overline{\mathbf{A}}\hat{\overline{\mathbf{X}}}(k) + \overline{\mathbf{B}}u(k) + \mathbf{K}_{obs}\left(\mathbf{Y}(k) - \hat{\mathbf{Y}}(k)\right) \\ \hat{\mathbf{Y}}(k) = \overline{\mathbf{C}}\hat{\overline{\mathbf{X}}}(k) \end{cases} \tag{20}$$

where K_{obs} is the real gain matrix of the state observer to be determined. When this state observer is used, the delayed feedback force can be set in the following form:

$$F_m(k) = u(k) = -\left[\mathbf{K}_{vw}\ 0\right]\hat{\overline{\mathbf{X}}}(k - d(k)) = -\mathbf{K}_{vw}\hat{\mathbf{x}}(k - d(k)) \tag{21}$$

where $\mathbf{K}_{vw} = \left[K_{ve}\ B_{ve}\ \mathbf{0}_{1\times(n-2)}\right]$, K_{ve} and B_{ve} are, respectively, the stiffness and the damping coefficient of the virtual wall. Then, the discrete-time state-space representation of the closed-loop haptic system can be written as:

$$\begin{cases} \mathbf{x}(k+1) = \mathbf{A}_d\mathbf{x}(k) - \mathbf{B}_d\mathbf{K}_{vw}\hat{\mathbf{x}}(k - d\,(k)) + \mathbf{E}_d F_h(k) \\ y(k) = \mathbf{C}_d\mathbf{x}(k) \end{cases} \tag{22}$$

As the global system is linear, the separation principle applies and so, the design of gain matrices $\mathbf{K}_{vw}$ of the virtual wall and $\mathbf{K}_{obs}$ of the state observer can be done separately. The gain matrix $\mathbf{K}_{vw}$ of the virtual wall is firstly chosen from the stability analysis of the discrete-time system (23):

$$\begin{cases} \mathbf{x}(k+1) = \mathbf{A}_d\mathbf{x}(k) - \mathbf{B}_d\mathbf{K}_{vw}\mathbf{x}(k - d\,(k)) \\ \mathbf{x}(k) = \boldsymbol{\phi}(k),\ k = -d_{max}, -d_{max}+1, \ldots, 0 \end{cases} \tag{23}$$

where $\boldsymbol{\phi}(.)$ is the initial value sequence. The stability property of the system (23) is checked in by using Theorem 1.

Theorem 1. *If there exists real symmetric positive definite matrices* $\boldsymbol{P}, \boldsymbol{Q}_1, \boldsymbol{Q}_2, \boldsymbol{R}_1, \boldsymbol{R}_2$ *and free-weighting matrices* $\boldsymbol{M}_1, \boldsymbol{M}_2, \boldsymbol{N}_1, \boldsymbol{N}_2$ *of appropriate dimensions such that the following LMI conditions are feasible:*

$$\begin{bmatrix} \boldsymbol{\Theta}_{11} & \boldsymbol{\Theta}_{12} & \boldsymbol{R}_1 + \boldsymbol{M}_1 & -\boldsymbol{N}_1 & d\boldsymbol{X}_1^i \\ * & \boldsymbol{\Theta}_{22} & \boldsymbol{M}_2 & -\boldsymbol{N}_2 & d\boldsymbol{X}_2^i \\ * & * & \boldsymbol{Q}_2 - \boldsymbol{Q}_1 - \boldsymbol{R}_1 & \boldsymbol{0}_{n\times n} & \boldsymbol{0}_{n\times n} \\ * & * & * & -\boldsymbol{Q}_2 & \boldsymbol{0}_{n\times n} \\ * & * & * & * & -d\boldsymbol{R}_2 \end{bmatrix} < \boldsymbol{0}, \qquad i = 1, 2 \tag{24}$$

where $\boldsymbol{X}^1 = \boldsymbol{M}$, $\boldsymbol{X}^2 = \boldsymbol{N}$, *and*

$$\begin{aligned} \boldsymbol{\Theta}_{11} &= \boldsymbol{A}_d{}^T \boldsymbol{P} \boldsymbol{A}_d - \boldsymbol{P} + \boldsymbol{Q}_1 + (\boldsymbol{A}_d - \boldsymbol{I}_{n\times n})^T \boldsymbol{\Psi} (\boldsymbol{A}_d - \boldsymbol{I}_{n\times n}) - \boldsymbol{R}_1 \\ \boldsymbol{\Theta}_{12} &= -\boldsymbol{A}_d{}^T \boldsymbol{P} \boldsymbol{B}_d \boldsymbol{K}_{vw} - (\boldsymbol{A}_d - \boldsymbol{I}_{n\times n})^T \boldsymbol{\Psi} \boldsymbol{B}_d \boldsymbol{K}_{vw} - \boldsymbol{M}_1 + \boldsymbol{N}_1 \\ \boldsymbol{\Theta}_{22} &= (\boldsymbol{B}_d \boldsymbol{K}_{vw})^T \boldsymbol{P} (\boldsymbol{B}_d \boldsymbol{K}_{vw}) + (\boldsymbol{B}_d \boldsymbol{K}_{vw})^T \boldsymbol{\Psi} \boldsymbol{B}_d \boldsymbol{K}_{vw} - \boldsymbol{M}_2 - \boldsymbol{M}_2{}^T + \boldsymbol{N}_2 + \boldsymbol{N}_2{}^T \\ \boldsymbol{\Psi} &= d_{\min}{}^2 \boldsymbol{R}_1 + d\boldsymbol{R}_2, d = d_{\max} - d_{\min} \end{aligned}$$

then, the discrete-time system (23) is asymptotically stable for any time-delay $d(k)$ *satisfying (17).*

A complete proof of this result can be found in [13].

From Theorem 1, it is interesting to find the critical virtual stiffness K_{ec} of virtual wall at which the haptic system begins becoming unstable for each value of virtual damping B_{ve}. As a result, the stability boundary of haptic system can be reconstructed in the virtual damping-stiffness (B_{ve}-K_{ve}) plan. The feedback gain matrix $\mathbf{K}_{vw}$ can be determined from this stability region. Let the state observer error be $\boldsymbol{\eta}(k)^T = \left[(\mathbf{x}(k) - \hat{\mathbf{x}}(k))^T \; (F_h(k) - \hat{F}_h(k))^T \right]$. Then, the error dynamics of the state observer can be written

$$\boldsymbol{\eta}(k+1) = (\overline{\mathbf{A}} - \mathbf{K}_{obs}\overline{\mathbf{C}})\, \boldsymbol{\eta}(k) \tag{25}$$

The design of the observer gain is standard. The observer gain matrix $\mathbf{K}_{obs}$ can then be obtained by using the pole placement method [14].

3.2 Numerical Simulation Results

In this section, the designed haptic device with physical parameters in the Table 4 is examined. The sampling period is $T_s = 1(ms)$. We consider first the case of a constant delay ($d = 1$) for which necessary and sufficient stability conditions exist. The stability boundaries of the haptic system using the backward finite difference method and the augmented state observer in the force feedback architecture are shown in Fig 5. This figure shows the interest of the proposed approach. In other words, it allows considering a higher stiffness virtual wall with respect to the traditional backward finite difference method. In addition, the stability region of the haptic system using the backward finite difference method validates the proposed

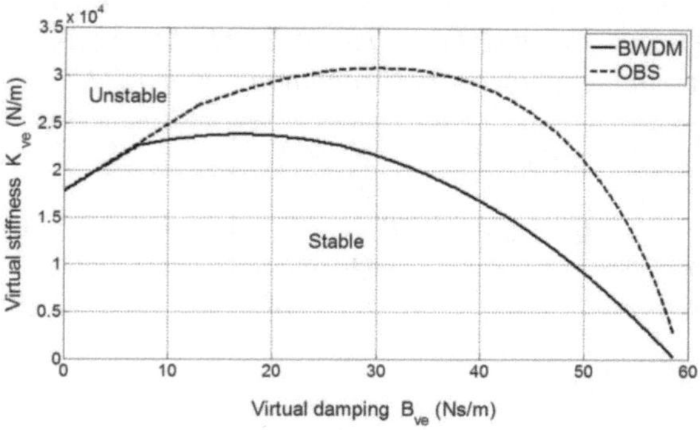

Fig. 5 Stability boundaries of the haptic system in the case of the time-constant delay d=1, **BWFD**: Backward finite difference method, **OBS**: Augmented state observer

optimal design method. Obviously, the haptic device with designed physical parameters is stable for the desired stiffness $K_{ve} = 5000(N/m)$ and all virtual damping $B_{ve} \in [0, 50]$ of the virtual wall.

Table 4 Physical parameters of the designed haptic device

Parameters	Variable	Value
First transmission's mass(kg)	m_{1eq}	0.04553
First transmission's damping(Ns/m)	b_{1eq}	26.381
Linkage stiffness(N/m)	k_{ceq}	1363.7
Linkage damping(Ns/m)	b_{ceq}	0
Second transmission's mass(kg)	m_{2eq}	0.155
Second transmission's damping(Ns/m)	b_{2eq}	0.001

The next section presents some numerical simulation results for the designed haptic system using the augmented state observer in the force feedback architecture. Based on Theorem 1, the feasibility region of the LMIs in the case of this haptic system under the time-varying communication delay satisfied: $1 = d_{\min} \leq d(k) \leq d_{\max} = 10$ is shown in Fig 6. This result was obtained by using LMI Toolbox in the Matlab software. From the stability region in Fig 6, the parameters of virtual wall corresponding to a point on the stability boundary can be chosen as follow:

$$\boldsymbol{K}_{vw} = \begin{bmatrix} 3000 \; 4.41 \; 0 \; 0 \end{bmatrix} \tag{26}$$

Next, the gain matrix $\boldsymbol{K}_{obs}$ of the augmented state observer is determined by using the pole placement method. As a rule of thumb, the gain matrix of state observer is chosen so that its transient response is faster than the one of haptic

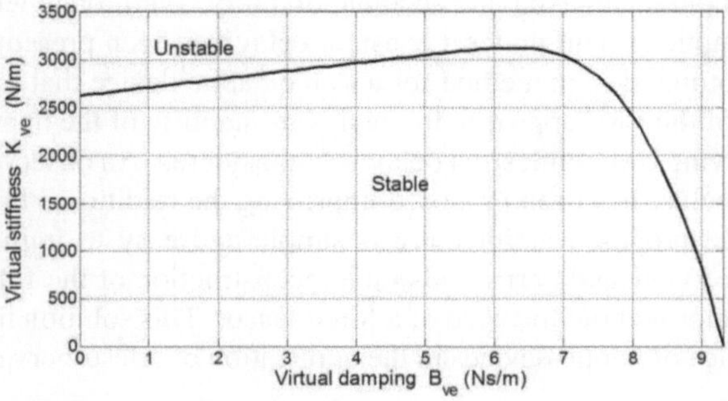

Fig. 6 Stability region of the haptic system under the time-varying communication delay: $1 = d_{\min} \leq d(k) \leq d_{\max} = 10$

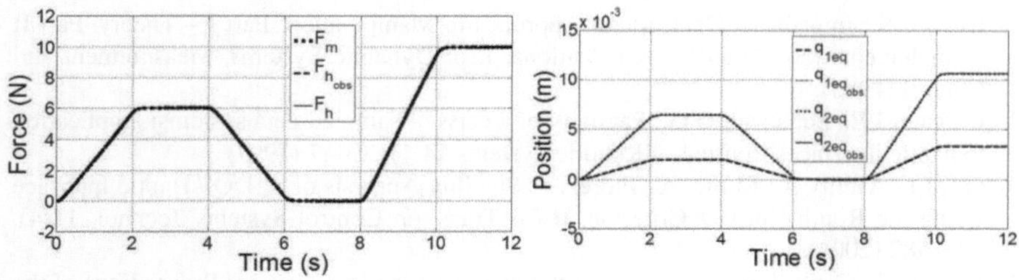

Fig. 7 Time response of the haptic system

control system. One of possible solutions for observer poles can be chosen as $\boldsymbol{P}_{obs} = [0.4\ 0.3\ 0.35\ 0.25\ 0.5]$ giving:

$$\boldsymbol{K}_{obs} = \begin{bmatrix} 2.7286\ 2217.3\ 79.636\ 27870\ 699930 \end{bmatrix}^T \tag{27}$$

The force and position responses of the haptic system are presented in Fig 7. For the haptic system, the feedback force F_m generated by the motor should equalize to the force F_h applied by the human operator in the steady regime of stable stiffness contact between the manipulated object and the virtual wall. In addition, it is obvious that the haptic system is stable for the chosen parameters of the augmented state observer and of the virtual wall. Another noteworthy interest of using an augmented state observer in the force feedback architecture is that the force F_h applied by the human operator can also be reconstructed without the need of a force sensor.

4 Conclusions

In this chapter, the model of a haptic device in interaction with a virtual wall is presented where the effects of mechanical parameters such as vibration mode and of tthe zero-order holder are taken into account. Stability condition in term of LMI for analyzing the haptic system under a constant delay has been presented. From this condition, an optimal design method for a stable haptic device that has low inertia. The advantage of the LMI approach for analyzing stability of the haptic system under the time-varying communication delay was considered. An observer-based force feedback architecture has been proposed improving the traditional finite-difference architecture. The proposed architecture is simple and easy to implement from a practical point of view and permits also the reconstruction of the force applied by the human operator without the need of a force sensor. This solution may contribute to reduce the price of haptic devices in the application of telesurgery systems in the future.

References

1. Hogan, N.: Impedance Control: An Approach to Manipulation: Part I – Theory, Part II – Implementation, Part III – Applications. J. of Dynamic Systems, Measurement and Control (1985).
2. Colgate, J.E., Schenkel, G.G.: Passivity of a class of sampled-data systems: Application to haptic interfaces. Journal of Robotic Systems 14(1), 37–47 (1997)
3. Gil, J.J., Avello, A., Rubio, A., Florez, J.: Stability Analysis of a 1DOF Haptic Interface Using the Routh-Hurwitz Criterion. IEEE Trans. on Control Systems Technol. 12(4), 583–588 (2004)
4. Hogan, N.: Controlling impedance at the man/machine interface. In: Proceedings of the IEEE Int Conf. on Robotics and Automation, Scottsdale, AZ, vol. 3, pp. 1626–1631 (May 1989)
5. Minksy, M., Ouh-young, M., Steele, O., Brooks Jr., Frederick, P., Behensky, M.: Feeling and seeing: Issues in force display. SIGGRAPH Computer Graph 24(2), 235–241 (1990)
6. Colgate, J.E., Brown, J.M.: Factors affecting the Z-width of a haptic display. In: Proc. of the IEEE Int. Conf. on Robotics and Automation, San Diego, CA, vol. 4, pp. 3205–3210 (1994)
7. Hulin, T., Preusche, C., Hirzinger, G.: Stability boundary for haptic rendering: Influence of human operator. In: Proceedings of the IEEE/RSJ International Conference on Intelligent Robots and Systems, Nice, France, pp. 3483–3488 (September 2008)
8. Diaz, I., Gil, J.J.: Influence of vibration modes and human operator on the stability of haptic rendering. IEEE Transactions on Robotics 26(1), 160–165 (2010)
9. Dang, Q.V., Dequidt, A., Vermeiren, L., Fratu, A., Dambrine, M.: Stability analysis of haptic interfaces: effects of dynamic parameters. In: Proceedings of the 8th International Conference on Remote Engineering and Virtual Instrumentation, Brasov, Romania, June 28-July 1 (2011)
10. Gil, J.J., Puerto, M.J., Diaz, I., Sanchez, E.: On the Z-Width Limitation due to the Vibration Modes of Haptic Interfaces. In: Proceedings of the IEEE/RSJ International Conference on Intelligent Robots and Systems, Taipei, Taiwan, October 18-22, pp. 5054–5059 (2010)
11. Dang, Q.V., Dequidt, A., Vermeiren, L., Dambrine, M.: Stability analysis of 1DOF haptic interface: time delay and vibration modes effects. In: Int. Conf. on Integrated Modeling and Analysis in Applied Control and Automation, Rome, Italy, pp. 234–240 (2011)

12. Diaz, I., Gil, J.J., Hulin, T.: Stability Boundary and Transparency for Haptic Rendering. In: Zadeh, M.H. (ed.) Advances in Haptics. InTech (2010)
13. Dang, Q.V., Vermeiren, L., Dequidt, A., Dambrine, M.: Control of Haptic Systems with Time-varying Delay: a Novel Approach. In: Proceedings of 11th Workshop on Time-Delay Systems, Part of 2013 IFAC Joint Conference SSSC, Grenoble, France, February 4-6, pp. 440–445 (2013)
14. Franklin, G.F., Powell, J.D., Workman, M.: Digital Control of Dynamic Systems, 3rd edn. Addison-Wesley (1998)

Engineering a Genetic Oscillator Using Delayed Feedback

Edward Lambert, Edward J. Hancock, and Antonis Papachristodoulou

Abstract. Oscillators are one of the best studied synthetic genetic circuits and a focus of the emerging field of Synthetic Biology. A number of different feedback arrangements that can produce oscillations have been proposed; the two most important constructs involve a single gene with negative feedback including delay and three genes in negative feedback forming a structure called a repressilator. Each of these has a different range of performance characteristics and different design rules. In this book chapter we discuss how oscillators of the first type can be designed to meet frequency and amplitude requirements. We also discuss how coupling heterogeneous populations of delayed oscillators can produce oscillations with robust amplitude and frequency. The analysis and design is rooted in techniques from control theory and dynamical systems.

1 Introduction

Oscillators are present in many biological systems in nature. Examples include the circadian clock which regulates the activity of a wide variety of organisms with a period of ~24 hours. This is entrained, usually by varying light levels, to the real day/night cycle. The presence of an underlying oscillator is suggested by the continuation of this rhythm even in constant darkness [1]. The core oscillators are often but

Edward Lambert
Lincoln College, University of Oxford, Turl Street, Oxford OX1 3DR, U.K.
e-mail: edward.lambert@lincoln.ox.ac.uk

Edward J. Hancock
Department of Engineering Science, University of Oxford, Parks Road, Oxford, OX1 3PJ, U.K.
e-mail: edward.hancock@eng.ox.ac.uk

Antonis Papachristodoulou
Department of Engineering Science, University of Oxford, Parks Road, Oxford, OX1 3PJ, U.K.
e-mail: antonis@eng.ox.ac.uk

T. Vyhlídal, J.-F. Lafay, and R. Sipahi (eds.), *Delay Systems*,
Advances in Delays and Dynamics 1,
DOI: 10.1007/978-3-319-01695-5_28,

not exclusively comprised of transcription-translation feedback loops [1,2]. These natural systems are robust to temperature variation and other environmental disturbances and persist through generations of cells.

Synthetic Biology is a new field that uses an Engineering approach to design biological circuits with predictable properties [3,4]. A synthetic oscillator is an extremely useful component in this new field, for similar reasons to those in natural biological systems or even electronic circuits: as a timing device, it can coordinate other processes and the order in which they occur. Their ubiquity in nature from plants to humans makes their utility apparent. Using oscillators to pace processes, an engineered organism could be programmed to cease division during the day to protect it from DNA damaging due to ultraviolet radiation or, combined with a counter [5], to remain dormant for a fixed time period.

There are a number of different feedback arrangements that can produce oscillations. One approach is to have three genes, each one repressing the transcription of the next one in feedback. This structure is called the *repressilator* and was the first oscillator to be designed [6]. Another approach is to use a single gene, with negative feedback including delay. This is the oscillator that we will consider in this paper, and we will discuss ways to choose the delay to produce oscillations of desired frequency and how the amplitude of the oscillations can be estimated. Moreover, we will consider how coupling delayed oscillators together can produce more robust oscillations.

This book chapter is organized as follows. Some background on modelling genetic networks is given in Section 2. In Section 3 we will analyze the oscillator frequency and amplitude for the case of a delay oscillator. In Section 4 we discuss the mechanisms and benefits for coupling delay oscillators, before concluding this book chapter.

2 Background

In this section we review briefly the modelling approach that will be used throughout the book chapter.

Mathematical models can be used to describe the flow of information from DNA to mRNA (transcription) and from mRNA to protein (translation) [7]. Transcription refers to the process whereby RNA polymerase binds with the DNA, unwinding it and creating a messenger RNA (mRNA). The mRNA (after diffusing out of the nucleus in eukaryotes) then binds with a ribosome which builds the protein described by the mRNA. This is called translation. Some proteins – called transcription factors – can bind the DNA at specific sites, the promoter region, regulating the rate at which an associated gene is transcribed [7]. Figure 1 shows this process.

The concentration of a particular protein within a cell at any point in time is a result of a continuous production through the transcription-translation process, dilution due to cell expansion, and protein degradation as proteins denature over time. To model this mathematically, the rate of production of mRNA is assumed to consist of a basal rate β_0, an added rate dependent on the concentration of the transcription

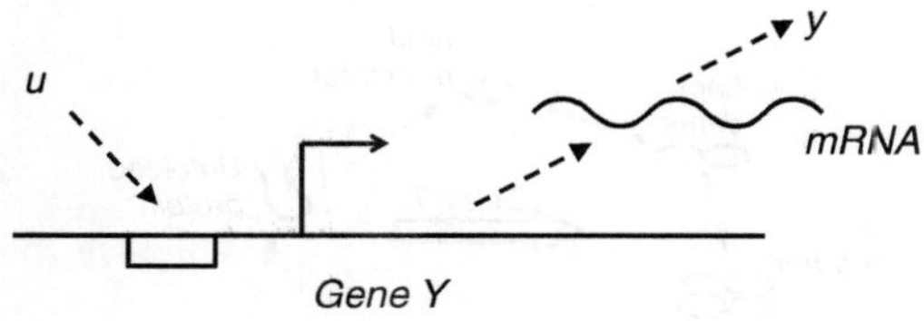

Fig. 1 Transcription factor u binds to the promoter of Gene Y causing increased transcription of mRNA and hence production of protein y

factor u and a dilution/degradation term. Moreover, the translation process can be modelled using a first order process. For simplicity, a single non-linear equation (1) is used to describe both transcription and translation:

$$\dot{y} = \beta_0 + \beta \frac{(\frac{u}{k})^n}{1 + (\frac{u}{k})^n} - \alpha y \tag{1}$$

This standard form is widely used [7] as it corresponds well to experiments and it is simple to compute and amenable to analysis: the non-linearities provide sufficient complexity to give rise to interesting behaviour. In this model, y is the concentration of the protein of interest, u is the concentration of the transcription factor, which in this model is activating - otherwise the term in the middle (called the 'Hill Function') would be replaced by $\beta \frac{1}{1+(\frac{u}{k})^n}$. Also, β_0 is the basal production rate and β, k and n describe the Hill function: the maximal activity, activation coefficient and steepness. Finally, the constant α accounts for protein dilution and degradation.

A model that we will need to consider in the sequel is that of negative auto-regulation. To achieve this, we can set $u = y$ in the repressor model:

$$\dot{y} = \beta_0 + \beta \frac{1}{1 + (\frac{y}{k})^n} - \alpha y \tag{2}$$

3 Oscillations Using Delayed Negative Feedback

A single gene with delayed self-repression will oscillate, in much the same way that a first order system with delayed negative feedback can oscillate if the delay is large enough. This was demonstrated in a recent paper by [8], using the *lacI* gene.

A synthetic circuit based on this idea requires tunability of the frequency and amplitude of the oscillations, which can be achieved by adding lengths of redundant code or junk DNA, creating a time lag after activation before any mRNA is read out. The negative autoregulator model can be modified by the introduction of τ, a time delay, to create a delay differential equation of the form

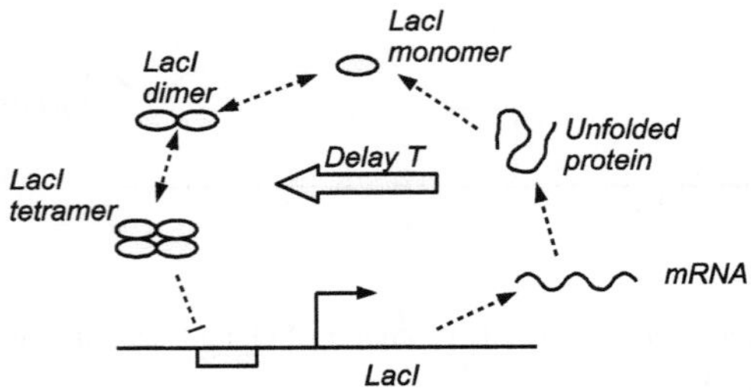

Fig. 2 Additional reaction steps for the *lacI* gene leading to delay between RNA transcription and the tetrameric form which binds to the promoter [8]

$$\dot{y}(t) = \beta_0 + \beta \frac{1}{1+\left(\frac{y(t-\tau)}{k}\right)^n} - \alpha y(t) \tag{3}$$

The effect of the delay parameter on the stability, period and amplitude is the subject of the next section. The other parameters used in simulation are assumed fixed by the type of gene chosen [9].

3.1 Period

We proceed by linearizing the delay differential equation around an equilibrium. Our aim is to use the resulting linear delay differential equation to find the delay required for instability (the delay value when a Hopf bifurcation occurs) as well as the period of oscillation. The resulting model takes the form:

$$G(s) = \frac{\gamma e^{-s\tau}}{s+\alpha} \tag{4}$$

with unity negative feedback, where γ is a lumped parameter that depends on the rest of the parameters in the model. Using this model we can calculate the point of instability as well as the frequency using the following two equations:

$$\omega = \sqrt{\gamma^2 - \alpha^2} \tag{5}$$

$$\tau_{cr} = \frac{\pi - \arctan\frac{\omega}{\alpha}}{\omega} \tag{6}$$

where τ_{cr} is the critical time delay and ω is the frequency of oscillations. In practice, and to ensure robustness of oscillations, the delay chosen should be larger than the critical value.

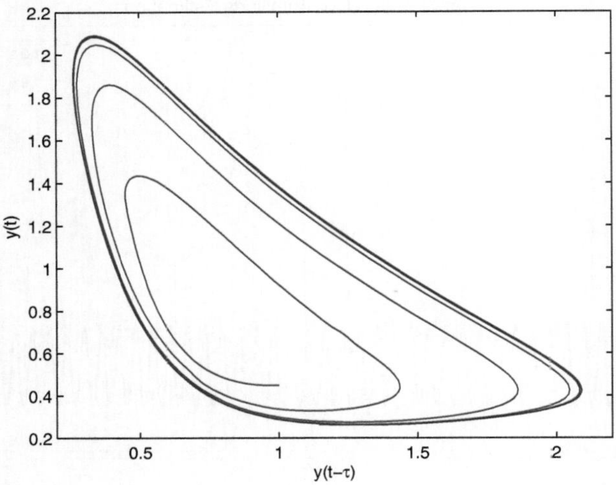

Fig. 3 The limit cycle using standard parameters and a delay equal to $2\tau_c$

3.2 *Amplitude*

One of the key questions in designing an oscillator, is to ensure robustness with respect to the amplitude and the shape of the limit cycle. As mentioned above, the limit cycle is a result of a Hopf bifurcation [10]. This limit cycle is stable, and to estimate its amplitude we consider two approaches: an iterative method and Linstedt's method.

3.2.1 Iterative Method

Assuming a periodic form for the solution, the amplitude can be approximated using a simple iterative estimation algorithm, based on the fact that at the maxima and minima of the periodic solution we must have $\dot{y} = 0$. Unlike for the case of an equilibrium, when this condition needs to hold for all time, we consider an iteration in which we assume that initially $y(t+\theta) > 0$ for all $\theta \in [-\tau, 0]$. Denote by y_L the lower bound and by y_U the upper bound of the periodic solution. To find an improved estimate for y_L we need to find the minimum of

$$y = \frac{1}{\alpha}\left(\beta_0 + \frac{\beta}{1+\left(\frac{y(t-\tau)}{K}\right)^n}\right) \tag{7}$$

over all feasible values of $y(t+\theta)$, $\theta \in [-\tau, 0]$. Similarly, an improved estimate of y_U is given by the maximum of (7). These two updates form the basis of

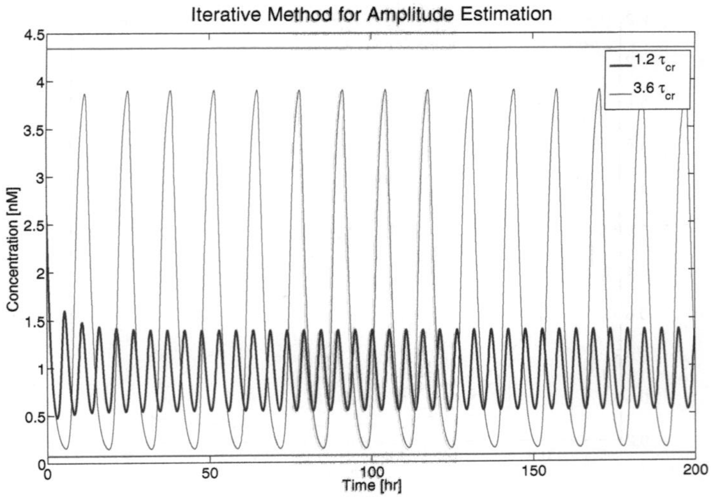

Fig. 4 Non-linear simulation results for large and small τ bounded by four iterations of the amplitude estimate

Algorithm 1. The iteration can be initialized for large enough values of y_U; alternatively, the first assignment can be carried out by hand giving $y_L = \beta_0/\alpha$.

The lines in algorithm 1 need only be evaluated once symbolically as thereafter they involve a series of substitutions. This allows the algorithm to be reused for any parameter values without recalculation. This would be advantageous if many iterations are required but in our case, our algorithm converged within three or four iterations for $n = 2$ and was much faster for higher n.

Algorithm 1. Iterative amplitude solution. y_U and y_L are the upper and lower bounds respectively. We iterate for N steps to get the desired accuracy.

$y_U \leftarrow \infty$
$y_L \leftarrow 0$
for I = 1:N **do**
  $y_U \leftarrow \frac{1}{\alpha}(\beta_0 + \beta(1 + (\frac{y_L}{K})^n)^{-1})$
  $y_L \leftarrow \frac{1}{\alpha}(\beta_0 + \beta(1 + (\frac{y_U}{K})^n)^{-1})$
end for

The first observation is that there is no delay dependence in the expressions for the bounds despite the fact that in simulation the amplitude increases at larger delays. Figure 4 shows the upper bounds calculated in this way compared to simulation results with large and small time delay. The bound is appropriate for large delays but less useful close to the critical delay.

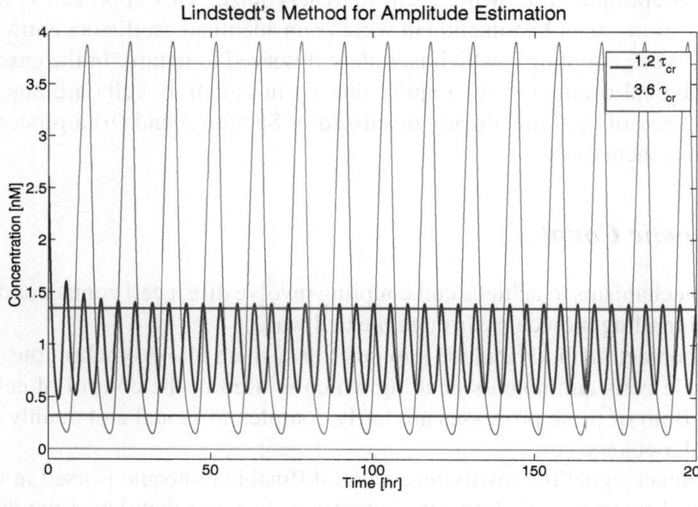

Fig. 5 The amplitude estimation using Lindstedt's method, from Equation (8), compared to the output of the simulation for two different delays

3.2.2 Linstedt's Method

An alternative method to obtain an analytical result for the amplitude involves the use of Linstedt's Method [10] [11]. This is a perturbation method where we assume a sinusoidal form for the solution, which is substituted into the governing equation. The additional information about amplitude can be extracted by setting the coefficients of unbounded terms to zero. See the Appendix for the full derivation. The approximate solution for the amplitude A given in Equation (8) is dependent on the applied delay τ.

$$A^2 = \frac{4(\tau - \tau_{cr})\gamma_1 \omega \sin(\omega \tau_{cr})}{3\gamma_1\gamma_3\tau_{cr} - (3\gamma_3 + 4\gamma_2 m_a)\cos(\omega\tau_{cr}) + 4\gamma_2 m_1 \sin(\omega\tau_{cr}) + 4\gamma_1\gamma_2\tau_{cr} m_a} \tag{8}$$

In this expression, γ_1, γ_2 and γ_2 are given by (15),(16) and (17) respectively, $m_a = 2m_3 + m_2$ with m_1, m_2, m_3 given by (24)–(26) and τ_{cr} is the critical time-delay, given by (6). Lindstedt's method is a clear improvement over the iterative method for delays closer to critical.

4 Coupled Delay Oscillators

The approaches for estimating the oscillator amplitude and period reveal that both are sensitive to uncertainties in parameters. One way of increasing the robustness of oscillations in a population of cells, to parameter variation in each cell, is to introduce coupling and allowing them to synchronize. This approach is based on synchronization – the phenomenon in which non-identical oscillators entrain in frequency by weak coupling – which is widely observed in nature. In the case of synthetic biological oscillators we require that (i) in isolation, cells are autonomous oscillators, e.g. of the type already discussed in Section 3 and (ii) a process which weakly links them exists.

4.1 Genetic Coupling

Possible mechanisms to achieve cell coupling involve direct cell contact or diffusion of small signalling molecules through the cell wall.

Direct contact methods include gap junctions which allow synchronous firing of heart muscle cells and notch signalling which is used in the control of cell differentiation. Both of these processes are fairly complex to model and mainly occur in multicellular eukaryotes.

Non-contact signalling involving a small diffusible molecule is used in a variety of biological systems including some species of bacteria that detect the density of the population of which they are a part. This process is known as quorum sensing. Each component has a gene which expresses the signalling molecule and a gene which is regulated by it. The detected levels of the gene are low until the population reaches a certain density which allows the small molecule to move from cell to cell in significant quantity. This process has been used experimentally to program an artificial population control circuit [12].

As a possible mechanism for direct contact would be difficult to implement and model, in this section we only consider the case of diffusion.

4.2 Coupled Delay Oscillators

The network of oscillators we will consider will consist of coupled delay oscillators of the type discussed in Section 3. In order to couple them together, a diffusive small molecule must be introduced which is driven by the oscillator in each cell. The simplest way this could be achieved is if the protein product of the oscillator gene or downstream of it is diffusible across the cell membrane [13].

4.2.1 Single Gene Delay Coupling

Assuming it is feasible to create a negative feedback delay oscillator which expresses a diffusible molecule, the analysis in Section 3 can be used to predict its frequency and amplitude.

To model coupled such oscillators, we extend the earlier model to include gene product diffusion and coupling. The dynamics of the system shown in Figure 6 are given by (9–10).

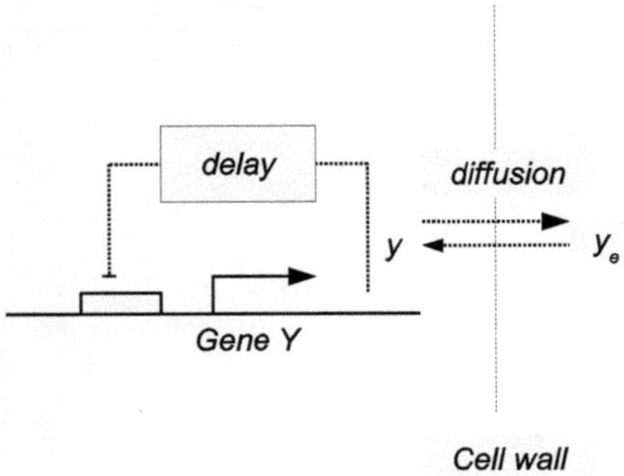

Fig. 6 Schematic of the genetic circuit to achieve coupling. The core oscillator is the delay type. The product y diffuses out of the cell providing the mechanism for coupling. y_e is the local mean field concentration outside the cell.

$$\dot{y}_i = \beta_{0i} + \frac{\beta_i}{1+\left(\frac{y_i(t-\tau_i)}{K}\right)^n} - \alpha_i y_i + Q(y_e - y_i); \tag{9}$$

$$y_e(t) = \frac{1}{N}\sum_{j=1}^{N} A_{ij} y_j(t) \tag{10}$$

Here Q is the strength of the coupling, which may be interpreted as the density of the population of cells. In sparse cells, coupling strength and the density are weakly coupled as the signalling molecules they produce will more diluted at a greater distance.

In (10), A is a connection matrix describing the effect of other nearby internal concentrations on the local signal y_e. This allows different topologies to be investigated while keeping the model simple. If the cells are well mixed then A will be full. This indicates that every cell contributes to the local mean for every other cell. If the cells were constrained to lie in a plane then this would form a lattice. Alternatively the cells might lie in a channel and so have only two neighbours, corresponding to a line graph. The connection strengths could also vary to represent cell distance.

4.2.2 Order Parameter

In order to evaluate the success of synchronization in a simulated group of cells a measure of the variability in their time series is required. One such measure is the mean field amplitude [14]. Consider N cells oscillating at the same frequency and in phase; the sum of the amplitudes is equal to the amplitude of the sum of the time series. If the oscillators are out of phase then for large N the mean field amplitude will be zero. For different frequencies the mean field amplitude will again be close

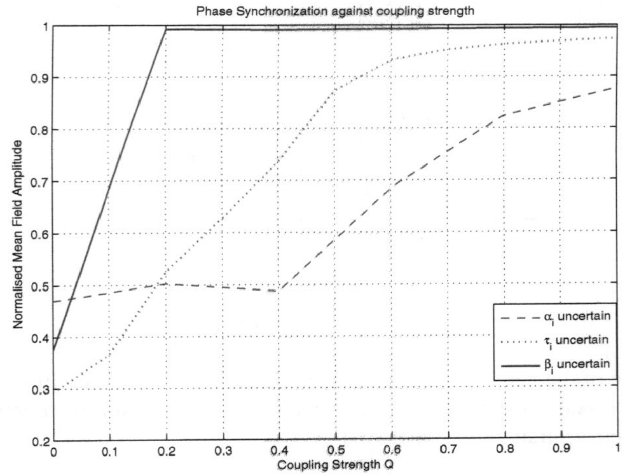

Fig. 7 Phase synchronization against coupling strength. Dotted: The value of the delay τ_i for each oscillator is drawn from a normal distribution centred at the design value ($\tau = 3$ hr) with standard deviation 0.1τ. Other parameters were identical. Solid: The value of β_i for each oscillator is drawn from a normal distribution centred at the design value ($\beta = 2$ nM/h) with standard deviation 0.3β. Other parameters were identical. Dashed: The value of α_i for each oscillator is drawn from a normal distribution centred at the design value ($\alpha = 0.7$ /hr) with standard deviation 0.3α. Other parameters were identical.

to zero. The order parameter R is therefore defined as the normalised mean field amplitude. The calculation can be performed as follows:

$$m(t) = \sum_{i=1}^{N} \frac{y_i(t)}{NP_i} \tag{11}$$

$$R = N\frac{\max_t(m(t)) - \min_t(m(t))}{\sum_{i=1}^{N} P_i} \tag{12}$$

where P_i is the amplitude of oscillator i, $m(t)$ is the mean field and R is the order parameter.

4.2.3 Connected Graph

Assuming that all cells are connected, y_e is the global mean. In this case, simulation was carried out on 25 cells with the time delay of each perturbed around the design value with standard deviation of 10%. Figure 7 shows the variation of the Order parameter with Q, the coupling strength. As we can see, synchronization occurs for values of $Q > 0.5$ and this demonstrates that uncertainties in the parameters of individual oscillators can be compensated for by coupling the oscillators together.

5 Conclusion

In this book chapter we have used two methods for oscillator amplitude calculation for delay differential equations, for the case of genetic oscillators. By coupling the oscillators, uncertainties in the oscillator parameters can be compensated for and robust oscillators can be obtained.

Appendix: Applying Lindstedt's Method for Estimating the Amplitude of the Delay Oscillator

The following analysis is based on Lindstedt's Method as applied in [10]. Consider the delay differential equation

$$\dot{y} = \beta_0 + \frac{\beta}{1+\left(\frac{y(t-\tau)}{k}\right)^n} - \alpha y \tag{13}$$

We first approximate (13) with a third order Taylor expansion about its equilibrium y^*. We set $z = y - y^*$, such that

$$\dot{z} = -\gamma_1 z(t-\tau) + \gamma_2 z^2(t-\tau) + \gamma_3 z^3(t-\tau) - \alpha z \tag{14}$$

where $\gamma_1, \gamma_2, \gamma_3$ are given by

$$\gamma_1 = \frac{\beta n y^{*(n-1)}}{k^n(1+(\frac{y^*}{k})^n)^2} \tag{15}$$

$$\gamma_2 = \left(\frac{n\beta}{k^2}\right)\frac{(2n(\frac{y^*}{k})^{(2n-2)} - (1+(\frac{y^*}{k})^n)(n-1)(\frac{y^*}{k})^{(n-2)})}{(1+(\frac{y^*}{k})^n)^3} \tag{16}$$

$$\gamma_3 = \frac{4\beta n(n^2-1)(\frac{y^*}{k})^{2n} - \beta n(\frac{y^*}{k})^n(3n(\frac{y^*}{k})^{2n} - 3n + (n^2+2)(1+(\frac{y^*}{k})^{2n}))}{y^{*3}((\frac{y^*}{k})^n+1)^4} \tag{17}$$

Scaling time using $\rho = \Omega t$ and setting ρ as the independent variable, then (14) becomes

$$\Omega\frac{dz}{d\rho} = -\gamma_1 z(\rho - \Omega\tau) + \gamma_2 z(\rho - \Omega\tau)^2 + \gamma_3 z(\rho - \Omega\tau)^3 - \alpha z(\rho) \tag{18}$$

We introduce the small parameter ε with the scaling $z = \varepsilon u$:

$$\Omega\frac{du}{d\rho} = -\gamma_1 u(\rho - \Omega\tau) + \varepsilon\gamma_2 u(\rho - \Omega\tau)^2 + \varepsilon^2\gamma_3 u(\rho - \Omega\tau)^3 - \alpha u(\rho) \tag{19}$$

The delay τ is chosen to be close to the critical value τ_{cr} such that $\tau = \tau_{cr} + \varepsilon^2\delta$, where the linear term is ignored due to later cancellation. We wish to determine

Ω, which is the natural frequency of the solution to (19). Expanding Ω as series in terms of ε with undetermined coefficient K_2 then

$$\Omega = \omega + \varepsilon^2 K_2 + O(\varepsilon^3) \tag{20}$$

We similarly expand u as a power series in ε such that

$$u(\rho) = u_0(\rho) + \varepsilon u_1(\rho) + \varepsilon^2 u_2(\rho) + O(\varepsilon^3) \tag{21}$$

Expanding the delayed solution with (20) and $\tau - \tau_{cr} = \varepsilon^2 \delta$ then

$$u(\rho - \Omega\tau) = u(\rho - \omega\tau_{cr}) - \varepsilon^2 (K_2 \tau_{cr} + \delta\omega) \frac{du(\rho - \omega\tau_{cr})}{d\rho} + O(\varepsilon^4)$$

Expanding $u(\rho - \omega\tau_{cr})$ as a power series gives

$$u_\tau = u_{\tau 0} + \varepsilon u_{\tau 1} + \varepsilon^2 u_{\tau 2} + O(\varepsilon^3)$$

where $u_\tau = u(\rho - \omega\tau_{cr})$, $u_{\tau 0} = u_0(\rho - \omega\tau_{cr})$, $u_{\tau 1} = u_1(\rho - \omega\tau_{cr})$ and $u_{\tau 2} = u_2(\rho - \omega\tau_{cr})$. Substitution of expressions for ω, $u(\rho)$ and $u(\rho - \Omega\tau)$ into (19) gives

$$\begin{aligned}
&(\omega + \varepsilon^2 K_2)\left(\frac{du_0}{d\rho} + \varepsilon \frac{du_1}{d\rho} + \varepsilon^2 \frac{du_2}{d\rho}\right) + \alpha\left(u_0 + \varepsilon u_1 + \varepsilon^2 u_2\right) \\
&\quad = -\gamma_1 \left(u_\tau - \varepsilon^2 (K_2 \tau_{cr} + \delta\omega) \frac{du_\tau}{d\rho}\right) + \varepsilon \gamma_2 u_\tau^2 + \varepsilon^2 \gamma_3 u_\tau^3 + O(\varepsilon^3)
\end{aligned}$$

Equating coefficients of ε terms results in

$$\omega \frac{du_0}{d\rho} + \alpha u_0 + \gamma_1 u_{\tau 0} = 0$$

$$\omega \frac{du_1}{d\rho} + \alpha u_1 + \gamma_1 u_{\tau 1} = \gamma_2 u_{\tau 0}^2 \tag{22}$$

$$\omega \frac{du_2}{d\rho} + \alpha u_2 + \gamma_1 u_{\tau 2} = \gamma_3 u_{\tau 0}^3 - K_2 \frac{du_0}{d\rho} + \gamma_1 \frac{du_{\tau 0}}{d\rho}(K_2 \tau_{cr} + \delta\omega) + 2\gamma_2 u_{\tau 0} u_{\tau 1} \tag{23}$$

We propose solutions

$$\begin{aligned}
u_0 &= \hat{A}\cos(\rho) \\
u_1 &= \hat{A}^2 (m_1 \sin(2\rho) + m_2 \cos(2\rho) + m_3)
\end{aligned}$$

which we substitute into (22) and compare cosine and sine coefficients to obtain

$$\begin{aligned}
2\omega m_1 + \alpha m_2 - \gamma_1 m_1 \sin(2\omega\tau_{cr}) + \gamma_1 m_2 \cos(2\omega\tau_{cr}) &= \frac{\gamma_2}{2}\cos(2\omega\tau_{cr}) \\
-2\omega m_2 + \alpha m_1 + \gamma_1 m_1 \cos(2\omega\tau_{cr}) + \gamma_1 m_2 \sin(2\omega\tau_{cr}) &= \frac{\gamma_2}{2}\sin(2\omega\tau_{cr}) \\
\alpha m_3 + \gamma_1 m_3 &= \frac{\gamma_2}{2}
\end{aligned}$$

Solving simultaneously for m_1 and m_2 gives m_1, m_2 and m_3 in terms of τ_{cr} and ω

$$m_1 = \frac{\gamma_2}{2}\frac{(\alpha\sin(2\omega\tau_{cr}) + 2\omega\cos(2\omega\tau_{cr})}{B} \tag{24}$$

$$m_2 = \frac{\gamma_2}{2}\frac{(\gamma_1 + \alpha\cos(2\omega\tau_{cr}) - 2\omega\sin(2\omega\tau_{cr})}{B} \tag{25}$$

$$m_3 = \frac{\gamma_2}{2}\frac{1}{(\alpha+\gamma_1)} \tag{26}$$

$$B = \alpha^2 + 4\omega^2 + \gamma_1^2 + 2\alpha\gamma_1\cos(2\omega\tau_{cr}) - 4\gamma_1\omega\sin(2\omega\tau_{cr})$$

We similarly substitute the proposed solutions into (23), to obtain

$$\begin{aligned} RHS = {} & K_2\hat{A}\sin(\rho) - \gamma_1\hat{A}(K_2\tau_{cr} + \delta\omega)\sin(\rho - \omega\tau_{cr}) \\ & + \gamma_2\hat{A}^3(2m_3\cos(\rho - \omega\tau_{cr}) + m_1(\sin(\rho - \omega\tau_{cr}) + \sin(3\rho - 3\omega\tau_{cr})) \\ & + m_2(\cos(\rho - \omega\tau_{cr}) + \cos(3\rho - 3\omega\tau_{cr}))) \\ & + \frac{1}{4}\gamma_3\hat{A}^3(3\cos(\rho - \omega\tau_{cr}) + \cos(3(\rho - \omega\tau_{cr}))) \end{aligned}$$

We set any resonant terms from the singular perturbation ($\cos(\rho)$ and $\sin(\rho)$ terms) to zero, such that

$$\begin{aligned} & K_2 - \gamma_1(K_2\tau_{cr} + \delta\omega)\cos(\omega\tau_{cr}) + \frac{3}{4}\gamma_3\hat{A}^2\sin(\omega\tau_{cr}) \\ & + \gamma_2\hat{A}^2(2m_3\sin(\omega\tau_{cr}) + m_1\cos(\omega\tau_{cr}) + m_2\sin(\omega\tau_{cr})) = 0 \end{aligned} \tag{27}$$

$$\begin{aligned} & \gamma_1(K_2\tau_{cr} + \delta\omega)\sin(\omega\tau_{cr}) + \gamma_2\hat{A}^2(2m_3\cos(\omega\tau_{cr}) \\ & - m_1\sin(\omega\tau_{cr}) + m_2\cos(\omega\tau_{cr})) + \frac{3}{4}\gamma_3\hat{A}^2\cos(\omega\tau_{cr}) = 0 \end{aligned} \tag{28}$$

We solve (28) and (27) as simultaneous equations in $\hat{A}^2$ and K_2 to give

$$\begin{aligned} K_2 &= -\frac{\delta\gamma_1\omega(3\gamma_3 + 4\gamma_2 m_a)}{3\gamma_1\gamma_3\tau_{cr} - (3\gamma_3 + 4\gamma_2 m_a)\cos(\omega\tau_{cr}) + 4\gamma_2 m_1\sin(\omega\tau_{cr}) + 4\gamma_1\gamma_2\tau_{cr}m_a} \\ \hat{A}^2 &= \frac{4\delta\gamma_1\omega\sin(\omega\tau_{cr})}{3\gamma_1\gamma_3\tau_{cr} - (3\gamma_3 + 4\gamma_2 m_a)\cos(\omega\tau_{cr}) + 4\gamma_2 m_1 sin(\omega\tau_{cr}) + 4\gamma_1\gamma_2\tau_{cr}m_a} \end{aligned} \tag{29}$$

where $m_a = 2m_3 + m_2$ with m_1, m_2, m_3 given by (24)–(26). We finally recover the initial amplitude A of the unscaled variable $y = \varepsilon u$, where $A = \hat{A}\varepsilon$. Using $\tau - \tau_{cr} = \varepsilon^2\delta$, we finally obtain

$$A^2 = \frac{4(\tau - \tau_{cr})\gamma_1\omega\sin(\omega\tau_{cr})}{3\gamma_1\gamma_3\tau_{cr} - (3\gamma_3 + 4\gamma_2 m_a)\cos(\omega\tau_{cr}) + 4\gamma_2 m_1\sin(\omega\tau_{cr}) + 4\gamma_1\gamma_2\tau_{cr}m_a} \tag{30}$$

Acknowledgements. This work was part of EL's undergraduate masters project work. EJH and AP were supported by EPSRC project EP/I031944/1. AP would like to thank Dr Elias August and Ms Sofia Piltz for useful discussions.

References

1. Zordan, M., Costa, R., Macino, G., Fukuhara, C., Tosini, G.: Circadian clocks: what makes them tick. Chronobiology International 17(4), 433–451 (2000)
2. Dunlap, J.C.: Genetic and molecular analysis of circadian rhythms. Annual Review of Genetics 30, 579–601 (1996)
3. Purcell, O., Savert, N., Grierson, C., di Bernardo, M.: A comparative analysis of synthetic genetic oscillators. J. R. Soc. Interface 7(52), 1503–1524 (2010)
4. Anderson, J., Strelkowa, N., Stan, G.B., Douglas, T., Savulescu, J., Barahona, M., Papachristodoulou, A.: Engineering and ethical perspectives in synthetic biology. EMBO Reports 13(7), 584–590 (2012)
5. Friedland, A.E., Lu, T.K., Wang, X., Shi, D., Church, G., Collins, J.J.: Synthetic gene networks that count. Science 324, 1199–1202 (2009)
6. Elowitz, M., Leibler, S.: A synthetic oscillatory network of transcriptional regulators. Nature 403, 335–338 (2000)
7. Alon, U.: An Introduction to Systems Biology. Chapman and Hall (2007)
8. Stricker, J., Cookson, S., Bennett, M., Mather, W., Tsimring, L., Hasty, J.: A fast, robust and tunable synthetic gene oscillator. Nature 456, 516–519 (2008)
9. Dolan, J., Anderson, J., Papachristodoulou, A.: A loop shaping approach for designing biological circuits. In: Proc. of IEEE CDC (2012)
10. Verdugo, A., Rand, R.: Hopf bifurcation in a DDE model of gene expression. Communications in Nonlinear Science and Numerical Simulation 13(2), 235–242 (2006)
11. Rand, R.: Lecture notes on nonlinear vibrations (2005), `http://www.tam.cornell.edu/randdocs/nlvibe52.pdf`
12. You, L., Cox, R.S., Weiss, R., Arnold, F.H.: Programmed population control by cell-cell communication and regulated killing. Nature 428, 868–871 (2004)
13. Danino, T., Mondragon-Palomino, O., Tsimring, L., Hasty, J.: A synchronized quorum of genetic clocks. Nature 463, 326–330 (2010)
14. Yuan, Z., Zhou, T., Zhang, J., Chen, L.: Synchronization of genetic oscillators. Chaos 18, 037126 (2008)

Index

MIX
Papier aus verantwortungsvollen Quellen
Paper from responsible sources
FSC® C105338

If you have any concerns about our products,
you can contact us on
ProductSafety@springernature.com

In case Publisher is established outside the EU,
the EU authorized representative is:
Springer Nature Customer Service Center GmbH
Europaplatz 3, 69115 Heidelberg, Germany

Printed by Libri Plureos GmbH
in Hamburg, Germany